Landmarks in Intracellular Signalling

Landmarks in Intracellular Signalling

Edited by

R.D. Burgoyne and O.H. Petersen

PORTLAND PRESS
London and Miami

Published by Portland Press Ltd, 59 Portland Place,
London W1N 3AJ, U.K.

In North America orders should be sent to Ashgate Publishing Co., Old Post Road, Brookfield, VT 05036-9704, U.S.A.

ISBN 1 85578 101 8

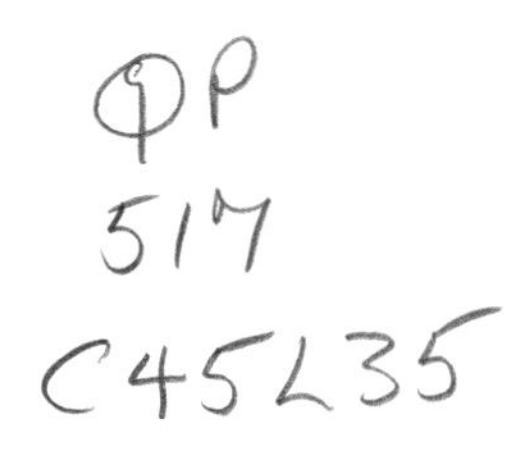

British Library Cataloguing-in-Publication Data

A catalogue record for this book is available from the British Library

Typeset by Unicus Graphics Ltd, Horsham, UK
Printed in Great Britain by Information Press Ltd, Eynsham, UK

Contents

Preface ix

Acknowledgements x

Abbreviations xi

1 Cyclic AMP

Commentary 1

Rall, T.W., Sutherland, E.W. and Berthet, J. (1957) The relationship of epinephrine and glucagon to liver phosphorylase. *J. Biol. Chem.* **224**, 463–475 5

Sutherland, E.W., Rall, T.W. and Menon, T. (1962) Adenyl cyclase I. Distribution, preparation and properties. *J. Biol. Chem.* **237**, 1220–1227 19

Walsh, D.A., Perkins, J.P. and Krebs, E.G. (1968) An adenosine 3′,5′-monophosphate-dependent protein kinase from rabbit skeletal muscle. *J. Biol. Chem.* **243**, 3763–3765 27

2 Identification and characterization of G-proteins

Commentary 31

Rodbell, M., Birnbaumer, L., Pohl, S.L. and Krans, M.J. (1971) The glucagon-sensitive adenyl cyclase system in plasma membranes of rat liver. *J. Biol. Chem.* **246**, 1877–1882 35

Northup, J.K., Sternweis, P.C., Smigel, M.D., Schleifer, L.S., Ross, E.M. and Gilman, A.G. (1980) Purification of the regulatory component of adenylate cyclase. *Proc. Natl. Acad. Sci. U.S.A.* **77**, 6516–6520 41

3 Structure of the β-adrenergic receptor

Commentary 47

Dixon, R.A.F., Kobilka, B.K., Strader, D.J., Benovic, J.L., Dohlman, H.G., Frielle, T., Bolanowski, M.A., Bennett, C.D., Rands, E., Diehl, R.E., Mumford, R.A., Slater, E.E., Sigal, I.S., Caron, M.G., Lefkowitz, R.J. and Strader, C.D. (1986) Cloning of the gene and cDNA for mammalian β-adrenergic receptor and homology with rhodopsin. *Nature (London)* **321**, 75–79 51

4 The patch-clamp technique

Commentary 57

Hamill, O.P., Marty, A., Neher, E., Sakmann, B. and Sigworth, F.J. (1981) Improved patch-clamp techniques for high-resolution current recording from cells and cell-free membrane patches. *Pflugers Arch.* **391**, 85–100 61

5 The nicotinic acetylcholine receptor channel

Commentary 77

Imoto, K., Busch, C., Sakmann, B., Mishina, M., Konno, T., Nakai, J., Bujo, H., Mori, Y., Fukuda, K. and Numa, S. (1988) Rings of negatively charged amino acids determine the acetylcholine receptor channel conductance. *Nature (London)* **335**, 645–648 81

6 The role of cyclic GMP in photoreceptors

Commentary 85

Fesenko, E.E., Kolesnikov, S.S. and Lyubarsky, A.L. (1985) Induction by cyclic GMP of cationic conductance in plasma membrane of retinal rod outer segment. *Nature (London)* **313**, 310–313 89

7 Voltage-sensitive Ca^{2+} channels

Commentary 93

Nowycky, M.C., Fox, A.P. and Tsien, R.W. (1985) Three types of neuronal calcium channel with different calcium agonist sensitivity. *Nature (London)* **316**, 440–443 95

8 Ca^{2+} uptake into intracellular stores

Commentary 99

Ebashi, S. and Lipmann, F. (1962) Adenosine triphosphate-linked concentration of calcium ions in a particulate fraction of rabbit muscle. *J. Cell Biol.* **14**, 389–400 101

9 Inositol 1,4,5-trisphosphate releases Ca^{2+} from intracellular stores

Commentary 113

Hokin, M.R. and Hokin, L.E. (1953) Enzyme secretion and the incorporation of P^{32} into phospholipids of pancreatic slices. *J. Biol. Chem.* **203**, 967–977 117

Streb, H., Irvine, R.F., Berridge, M.J. and Schulz, I. (1983) Release of Ca^{2+} from a nonmitochondrial intracellular store in pancreatic acinar cells by inositol-1,4,5-trisphosphate. *Nature (London)* **306**, 67–69 129

10 Cytosolic Ca^{2+} spiking

Commentary 131

Woods, N.M., Cuthbertson, K.S.R. and Cobbold, P.H. (1986) Repetitive transient rises in cytoplasmic free calcium in hormone-stimulated hepatocytes. *Nature (London)* **319**, 600–602 135

11 Protein kinase C

Commentary 139

Takai, Y., Kishimoto, A., Kikkawa, U., Mori, T. and Nishizuka, Y. (1979) Unsaturated diacylglycerol as a possible messenger for the activation of calcium-activated, phospholipid-dependent protein kinase system. *Biochem. Biophys. Res. Commuun.* **91**, 1218–1224 143

Parker, P.J., Coussens, L., Totty, N., Rhee, L., Young, S., Chen, E., Stabel, S., Waterfield, M.D. and Ullrich, A. (1986) The complete primary structure of protein kinase C — the major phorbol ester receptor. *Science* **233**, 853–859 151

12 Calmodulin

Commentary 159

Cheung, W.Y. (1970) Cyclic 3′,5′-nucleotide phosphodiesterase. Demonstration of an activator. *Biochem. Biophys. Res. Commun.* **38**, 533–538 163

13 The site of action of Ca^{2+} in exocytosis in neurons and neuroendocrine cells

Commentary 169

Katz, B. and Miledi, R. (1967) The timing of calcium action during neuromuscular transmission. *J. Physiol. (London)* **189**, 535–544 173

Baker, P.F. and Knight, D.E. (1978) Calcium-dependent exocytosis in bovine adrenal medullary cells with leaky plasma membranes. *Nature (London)* **276**, 620–622 183

14 Endothelium-derived relaxing factor is nitric oxide

Commentary **187**

Palmer, R.M.J., Ferrige, A.G. and Moncada, S. (1987) Nitric oxide release accounts for the biological activity of endothelium-derived relaxing factor. *Nature (London)* **327**, 524–526 **191**

15 Src: tyrosine phosphorylation and src-homology domains in signal transduction pathways

Commentary **195**

Hunter, T. and Sefton, B.M. (1980) Transforming gene product of Rous sarcoma virus phosphorylates tyrosine. *Proc. Natl. Acad. Sci. U.S.A.* **77**, 1311–1315 **199**

Sadowski, I., Stone, J.C. and Pawson, T. (1986) A noncatalytic domain conserved among cytoplasmic protein-tyrosine kinases modifies the kinase function and transforming activity of Fujinami sarcoma virus P130$^{gag\text{-}fps}$. *Mol. Cell. Biol.* **6**, 4396–4408 **205**

16 Phosphatidylinositol 3-kinase in signal transduction and intracellular membrane traffic

Commentary **219**

Whitman, M., Downes, C.P., Keeler, M., Keller, T. and Cantley, L. (1988) Type I phosphatidylinositol kinase makes a novel inositol phospholipid, phosphatidylinositol-3-phosphate. *Nature (London)* **332**, 644–646 **223**

17 The Ras/MAP kinase pathway

Commentary **227**

Ray, L.B. and Sturgill, T.W. (1987) Rapid stimulation by insulin of a serine/threonine kinase in 3T3-L1 adipocytes that phosphorylates microtubule-associated protein 2 *in vitro*. *Proc. Natl. Acad. Sci. U.S.A.* **84**, 1502–1506 **231**

18 Growth factor receptors and oncogenes

Commentary **237**

Downward, J., Yarden, Y., Mayes, E., Scrace, G., Totty, N. Stockwell, P., Ullrich, A., Schlessinger, J. and Waterfield, M.D. (1984) Close similarity of epidermal growth factor receptor and v-*erb*-B oncogene protein sequences. *Nature (London)* **307**, 521–527 **239**

19 Control of transcription by nuclear receptors

Commentary **247**

Hollenberg, S.M., Weinberger, C., Ong, E.S., Cerelli, G., Oro, A., Lebo, R., Thompson, E.B., Rosenfeld, M.G. and Evans, R.M. (1985) Primary structure and expression of a functional human glucocorticoid receptor cDNA. *Nature (London)* **318**, 635–641 **251**

20 Cyclic AMP regulation of gene transcription

Commentary **259**

Montminy, M.R. and Bilezikjian, L.M. (1987) Binding of a nuclear protein to the cyclic-AMP response element of the somatostatin gene. *Nature (London)* **328**, 175–178 **261**

Subject index **265**

Preface

The intracellular signalling pathways that control cell function have been, and still are, one of the most intensively studied aspects of biology. In recent years the detailed characterization of the multiple cell signalling pathways by many laboratories has resulted in a bewildering increase in knowledge in this field. For this reason, it has become very easy to lose sight of the key findings that have illuminated this area and that still stand as landmarks in the complex landscape of cell regulation. We have, therefore, set out to select a series of the key papers that have resulted in significant advances in one or other aspect of intracellular signalling. It has been a difficult task to choose the selection of papers illustrated, but some kind of arbitrary limit on the number of papers had to be set. The idea behind *Landmarks in Intracellular Signalling* was to provide a full reproduction of each chosen paper and a short commentary to put each paper into context, discussing how the findings related to knowledge at the time that the work was carried out, and briefly outlining subsequent progress. Our aim, however, was that the focus should remain on the original papers and that this collection would provide ready access to this original literature.

It is inevitable, since this collection covers such a wide area, that almost every reader will think of other papers that we should have included and perhaps prefer alternatives to the ones we have chosen. We have tried to mention other important work in the commentaries, but apologise now to any authors who feel slighted by the omission of their own papers. We could only have done full justice to all of the potential landmark papers in this field in a considerably larger collection. Responsibility for the choice of paper lies, of course, entirely with us, but we wish to express our gratitude to the many people, consulted by Portland Press, who commented and advised upon our early lists of papers and who helped us reach our final choice. We should also note that we deliberately decided to exclude any papers published after 1990 since we wanted to consider only those whose full impact on subsequent work is clearly apparent and non-controversial.

Finally, we thank the authors of the chosen landmark papers for their permission to reproduce them here and for their comments on early drafts of our commentaries. The suggestions and corrections made were invaluable. We hope that readers of this collection will gain as many new insights as we have in the reading or re-reading of these landmark papers, and a new appreciation of the quality of the original research that has led to our current understanding of intracellular signalling.

Robert D. Burgoyne
Ole H. Petersen
Liverpool, 1997

Acknowledgements

We gratefully acknowledge the kind permission of the various publishers and authors for allowing us to reproduce the original articles in this volume: *Biochem. Biophys. Res. Commun.* papers are reproduced by permission of Academic Press, Inc.; *J. Biol. Chem.* papers are reproduced by permission of the American Society for Biochemistry and Molecular Biology; the *J. Cell Biol.* paper is reproduced by permission of the Rockefeller University Press; the *J. Physiol. (London)* paper is reproduced by permission of the Physiological Society; the *Mol. Cell. Biol.* paper is reproduced by permission of the American Society for Microbiology; *Nature (London)* papers are reproduced by permission of Macmillan Magazines Ltd; the *Pflugers Arch.* paper is reproduced by permission of Springer-Verlag; *Proc. Natl. Acad. Sci. U.S.A.* papers are reproduced by permission of the authors; the *Science* paper is reproduced by permission of the American Association for the Advancement of Science.

Abbreviations

ACh	acetylcholine
βARK	β-adrenergic receptor kinase
$[Ca^{2+}]_i$	cytosolic free calcium ion concentration
CaBP	Ca^{2+}-binding protein
cAMP	cyclic adenosine 3′,5′-monophosphate
CBP	CREB-binding protein
CRE	cyclic AMP-response element
CREB	cyclic AMP-response element-binding protein
DAG	diacylglycerol
EDRF	endothelium-derived relaxing factor
EGF	epidermal growth factor
ER	endoplasmic reticulum
ERK	extracellular-signal regulated kinase
FFA	free fatty acid
G-protein	GTP-binding protein
HRE	hormone-response element
HVA Ca^{2+} current	high voltage-activated Ca^{2+} current
Ins(1,4,5)P_3	inositol 1,4,5-trisphosphate
LVA Ca^{2+} current	low-voltage-activated Ca^{2+} current
LysoPC	Lysophosphatidylcholine
MAP	microtubule-associated protein
MAPKK	MAP kinase kinase
NO	nitric oxide
PDGF	platelet-derived growth factor
PI 3-kinase	phosphatidylinositol 3-kinase
PKA	cyclic AMP-dependent protein kinase
PKC	protein kinase C
PLC	phospholipase C
PMA	phorbol 12-myristate 13-acetate
PMCA	plasma membrane Ca^{2+}-ATPase
PtdIns	phosphatidylinositol
PtdIns(4,5)P_2	phosphatidylinositol 4,5-bisphosphate
Ptalns (3,4,5)-P_3	phosphatidylinositol 3,4,5-trisphosphate
RGS	regulators of G-protein signalling
SERCA	sarcoplasmic/endoplasmic reticulum Ca^{2+}-ATPase
SH	src-homology domain
TGF	transforming growth factor
TPA	12-O-tetradecanoylphorbol-13 acetate

Cyclic AMP

The importance of the discovery of cyclic adenosine 3′,5′-monophosphate (cyclic AMP) is enormous. In many ways it was the starting point for the whole field of cellular transduction processes, and cyclic AMP was certainly the first example of a second messenger. Two of the papers we have chosen [1,2] are from a very substantial series published by Sutherland and his collaborators, which describes the discovery of cyclic AMP, its production, metabolism and actions. The third paper [3] outlines the purification from skeletal muscle of a protein kinase that catalyses a cyclic AMP-dependent phosphorylation of casein and protamine.

The discovery of cyclic AMP began with a study of the mechanism by which adrenaline or glucagon induced the formation of active liver phosphorylase, the enzyme that catalyses the breakdown of glycogen to glucose, from dephosphophosphorylase (Figure 1.1). Through the elegant studies of Sutherland and his co-workers, summarized in his Nobel Prize Lecture [4], it became clear that adrenaline and glucagon act on a system in the liver plasma membrane to cause the activation of an enzyme, adenylate cyclase, that catalyses the conversion of ATP to cyclic AMP. This water-soluble molecule in turn starts an amplification cascade where enzymes act upon further enzymes (Figure 1.1).

Following the discovery that adrenaline- or glucagon-evoked cyclic AMP formation initiates the intracellular cascade that ultimately leads to the release of glucose into the blood from the

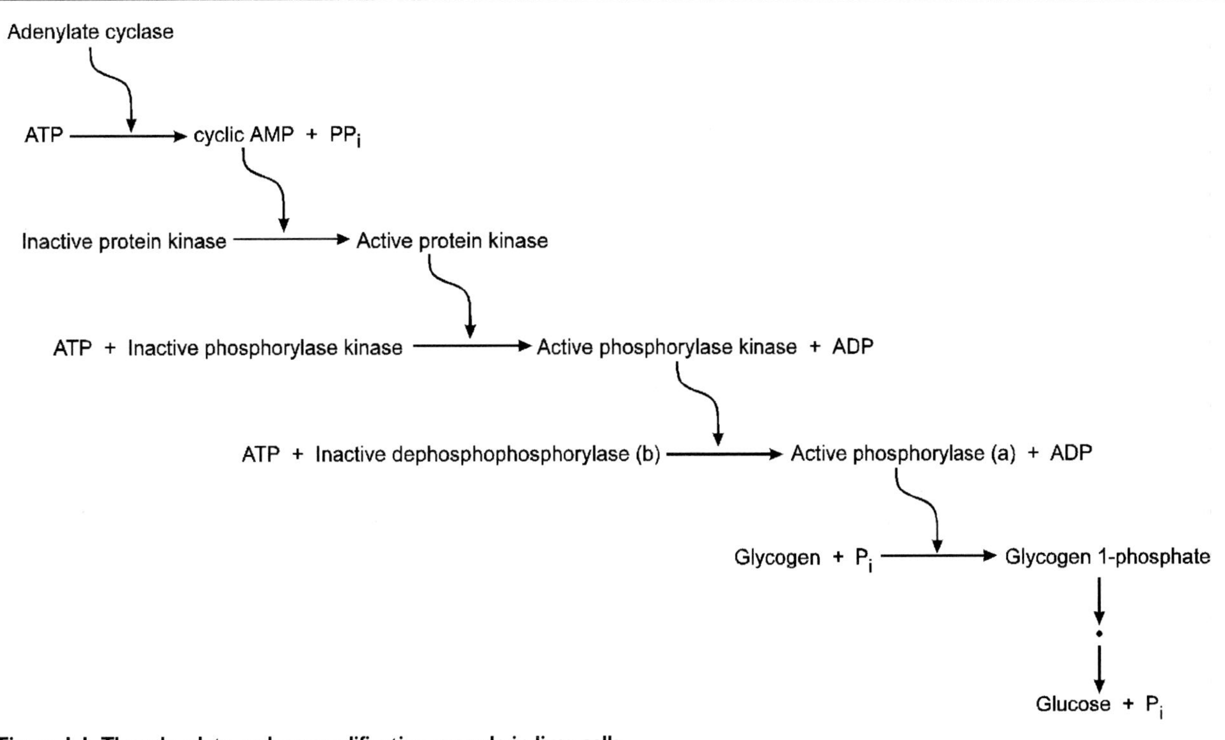

Figure 1.1 The adenylate cyclase amplification cascade in liver cells.

liver cell, it was quickly realized that cyclic AMP functions as a second messenger for very many different primary messengers (hormones, neurotransmitters). Histamine, serotonin, dopamine, vasopressin, parathyroid hormone, thyroid-stimulating hormone, secretin and vasoactive intestinal polypeptide are examples of agonists activating adenylate cyclase in their respective target cells, thereby inducing a number of different changes in cell function. The specificity of the system lies in the specificity of the hormone receptors and the fact that cyclic AMP is formed inside cells, and, because it is water soluble, cannot easily leave the cells in which it has been formed. The particular action in a particular cell type depends on the molecular target for cyclic AMP-dependent phosphorylation [5].

In order for cyclic AMP to function as a messenger it must be formed rapidly when the appropriate primary signal reaches the cell membrane, and further work on this problem led to the discovery of GTP-binding transducer proteins (see Section 2). The messenger must also be broken down quickly inside the cell, and an enzyme, cyclic 3′,5′-nucleotide phosphodiesterase, was identified that is responsible for this essential process [5]. Further studies on the control of this enzyme led to the discovery of calmodulin (see Section 12). Cyclic AMP was the first second messenger to be discovered, but several others followed many years later and are discussed elsewhere in this book (see Sections 6, 9 and 11).

As in the central nervous system there is both divergence and convergence in the endocrine system. One hormone may act on receptors in different cell types and may also act on different types of receptors to produce different second messengers. On the other hand different hormones may act on the same cell through different receptors, but nevertheless produce the same second messenger and therefore produce the same effect on cell function. Sometimes the appropriate physiological response in a target cell depends on the activation of more than one messenger system. Adrenaline and noradrenaline are of course known to act on both adrenergic α- and β-receptors. In salivary gland cells, for example, excitation of the β-receptors activates adenylate cyclase to produce cyclic AMP, which, in this system, is the major stimulator of exocytosis [6]. Excitation of α-receptors in the same cells activates phospholipase C (PLC) to produce inositol 1,4,5-trisphosphate [Ins(1,4,5)P_3], which in turn releases Ca^{2+} from internal stores opening ion channels crucial for fluid secretion [6]. In this case the final physiological response, fluid and protein secretion, is evoked by one first messenger, noradrenaline, acting through two separate second messenger pathways.

In some cases cross-talk between different messenger pathways has been discovered. In the liver cells, for example, it is now known that glucagon can activate both adenylate cyclase and the PLC pathways [7]. PLC action will lead to protein kinase C activation and there is evidence indicating that this may induce desensitization of the glucagon-stimulated adenylate cyclase. Conversely adenylate cyclase activation may lead to inhibition of Ins(1,4,5)P_3 production [7].

It is abundantly clear that, as Sutherland and colleagues predicted [2], cyclic AMP has emerged as a second messenger of enormous importance in many biological systems. However, it has also become clear that cyclic AMP is not responsible for all hormone or neurotransmitter actions and that other messenger systems, notably those induced by PLC activation, play an equally important role in accounting for non-steroidal hormone effects.

References

1. Rall, T.W., Sutherland, E.W. and Berthet, J. (1957) *J. Biol. Chem.* **224**, 463–475
2. Sutherland, E.W., Rall, T.W. and Menon, T. (1962) *J. Biol. Chem.* **237**, 1220–1227
3. Walsh, D.A., Perkins, J.P. and Krebs, E.G. (1968) *J. Biol. Chem.* **243**, 3763–3765
4. Sutherland, E.W. (1992) in *Nobel Lectures in Physiology and Medicine* (Lindstein, J., ed.), pp. 5–23, Nobel Foundation, World Scientific, Singapore
5. Robison, G.A., Butcher, R.W. and Sutherland, E.W. (1971) *Cyclic AMP*, Academic Press, London and New York
6. Schramm, M. and Selinger, Z. (1974) *Adv. Cytopharmacol.* **2**, 29–32
7. Petersen, O.H. and Bear, C. (1986) *Nature (London)* **323**, 18

Rall et al. (1957) J. Biol. Chem. 224, 463–475

THE RELATIONSHIP OF EPINEPHRINE AND GLUCAGON TO LIVER PHOSPHORYLASE

IV. EFFECT OF EPINEPHRINE AND GLUCAGON ON THE REACTIVATION OF PHOSPHORYLASE IN LIVER HOMOGENATES*

By T. W. RALL, EARL W. SUTHERLAND, and JACQUES BERTHET†

(From the Department of Pharmacology, School of Medicine, Western Reserve University, Cleveland, Ohio)

(Received for publication, July 16, 1956)

The concentration of active phosphorylase in liver represents a balance between inactivation by liver phosphorylase phosphatase (inactivating enzyme) and reactivation by dephosphophosphorylase kinase. The enzymatic inactivation of phosphorylase proceeds with the release of inorganic phosphate (2, 3), while the reactivation of dephosphophosphorylase requires magnesium ions and ATP[1] and proceeds with the transfer of phosphate to the enzyme protein (4).

It has been shown in liver slices that epinephrine and glucagon displace this balance in favor of the active phosphorylase (5, 6). This report is concerned with the demonstration of a similar effect in cell-free liver homogenates; *i.e.*, an increased formation of active phosphorylase occurred in cell-free homogenates in the presence of sympathomimetic amines and glucagon. The relative activities of the sympathomimetic amines in homogenates were found to be similar to the relative activities determined by liver slice technique or by injection into intact animals.

It has been possible to show that the response of the homogenates to the hormones occurred in two stages. In the first stage, a particulate fraction of homogenates produced a heat-stable factor in the presence of the hormones; in the second stage, this factor stimulated the formation of liver phosphorylase in supernatant fractions of homogenates in which the hormones themselves were inactive.

* This research was supported in part by grants from Eli Lilly and Company and from the Cleveland Area Heart Society. A preliminary report was presented at the meeting of the American Society of Biological Chemists, Atlantic City, April, 1956 (1).

† Fellow of the Rockefeller Foundation. Present address, Department of Physiological Chemistry, University of Louvain, Belgium.

[1] The following abbreviations are used: ATP, adenosine triphosphate; ADP, adenosine diphosphate; 5-AMP, adenosine-5-phosphate; Tris, tris(hydroxymethyl)-aminomethane; TCA, trichloroacetic acid; LP, liver phosphorylase; dephospho-LP, liver dephosphophosphorylase; phosphokinase, dephosphophosphorylase kinase.

Methods

Preparation of Liver Homogenates—Mature dogs were killed by severing the arteries in the neck under deep secobarbital anesthesia. Mature cats were similarly killed under chloroform anesthesia. The livers were perfused with 0.9 per cent NaCl and sliced as previously described (4). The slices were rinsed with 5 volumes of 0.9 per cent NaCl and were shaken in air at 37° for 15 minutes in 2 to 3 volumes of a mixture containing 0.12 M NaCl plus 0.04 M glycylglycine buffer plus 0.001 M potassium phosphate buffer at pH 7.4. At the end of the incubation, the medium was decanted and the slices were rinsed twice with 3 to 4 volumes of cold 0.33 M sucrose. The slices (in 15 to 20 gm. portions) were then homogenized in 2 volumes of 0.33 M sucrose in an all-glass homogenizer. The homogenates were routinely centrifuged at 900 × *g* for 1 minute before use.

Fractionation of Homogenates—Low speed centrifugations (up to 1200 × *g*) were conducted in a cold room at 3°, with the horizontal yoke (head No. 240) on the International centrifuge No. 2. Approximately 25 ml. portions of homogenate were placed in 45 ml. Lusteroid tubes and centrifuged for 10 minutes at the specified centrifugal force. The supernatant fluid (1200 × *g* supernatant fraction) was removed by aspiration. The precipitate was rehomogenized in an equal volume of 0.25 M sucrose, and the suspension diluted to the original volume of the homogenate. For experiments in recombination, these suspensions were centrifuged in 25 ml. portions at successively higher speeds, and the resulting precipitate fractions were suspended in the 1200 × *g* supernatant fraction. For other experiments, these suspensions were centrifuged at 1200 × *g*, and the resulting precipitate was suspended in an equal volume of 0.25 M sucrose (washed liver particles).

The 11,000 × *g* supernatant fraction was prepared by centrifugation of either the 1200 × *g* supernatant fraction or the whole homogenate for 15 minutes at 11,000 × *g* on the Spinco preparative ultracentrifuge; for some experiments, this fraction was centrifuged at either 50,000 × *g* for 1 hour or 100,000 × *g* for 45 minutes to remove the formed elements. In all cases, the supernatant fluid was removed by aspiration. The 100,000 × *g* supernatant fraction at times was dialyzed *versus* 150 volumes of distilled water for 3 hours with shaking.

Assay of LP in Homogenates and Fractions of Homogenates—Aliquots of a homogenate or fraction were added to iced culture tubes containing various additions, bringing the final volume to 0.20 to 0.25 ml. The basic phosphorylase assay reagent (2.8 ml.), containing glucose-1-phosphate, glycogen, and 5-AMP (7), was added either immediately or after 5 to 10 minutes of shaking at 30°. After addition of the assay reagent, the tubes were incubated 10 minutes at 37°, and the assay was terminated by the addition of 1.0 ml. of 15 per cent TCA. The inorganic phosphate present in an equivalent of 0.15 ml. of reaction mixture was determined by the method of Fiske and Subbarow (8), as adapted to the Klett-Summerson photometer. Units of phosphorylase activity were calculated as defined previously (7).

Materials—Dephospho-LP was prepared from dog liver as described previously (4). Amorphous glucagon samples (about 50 per cent pure) were donated by Eli Lilly and Company. *l*(−)-Epinephrine bitartrate, *d*(+)-epinephrine, *l*(−)-arterenol bitartrate (*l*(−)-norepinephrine), and *d*(+)-arterenol bitartrate (*d*(+)-norepinephrine) were kindly supplied by M. L. Tainter. Amphetamine (Benzedrine) was obtained as the sulfate salt and ATP as the crystalline disodium salt. Tris was recrystallized before use (7).

Results

Effects of Epinephrine and Glucagon in Whole Homogenates—Aliquots of homogenates were incubated at 30° with buffer, magnesium ions, and ATP in the absence and in the presence of epinephrine or glucagon. Phosphorylase activity was assayed before and after a 10 minute incubation (Fig. 1, left-hand bars). Since the homogenate was derived from preincubated slices, the initial level of active LP was low, most of the phosphorylase being present as dephospho-LP. In the absence of the hormones, only a small amount of dephospho-LP was converted to LP during the incubation of the homogenate. However, in the presence of the hormones, the formation of LP was increased nearly 4-fold. When the homogenate was supplemented with purified dephospho-LP, the effect of the hormones was magnified so that the formation of LP in the presence of the hormones was nearly 7 times that in their absence (Fig. 1). The formation of LP in homogenates in either the absence or presence of the hormones required the addition of both ATP and magnesium ions.

Increased formation of LP in the presence of epinephrine and glucagon also occurred in homogenates which had been frozen and thawed (Fig. 2). Some preparations (dog liver homogenates) have been frozen and stored at the temperature of solid CO_2 for a few weeks without appreciable change in properties, except those ascribable to the initial freezing process. The principal effect of freezing or other methods of storage of homogenates was an increased formation of phosphorylase in the absence of the hormones, with only a small diminution of the formation of phosphorylase in their presence.

The assumption that an increase in the phosphorylase activity of a homogenate corresponded to an increase in the amount of LP formed was substantiated by an experiment in which the phosphorylase activity of homogenates incubated with and without epinephrine or glucagon was assayed before and after precipitation with ammonium sulfate. The increase in phosphorylase activity after incubation with the hormones was still present after the protein was precipitated twice at 0.67 saturation with ammonium sulfate.

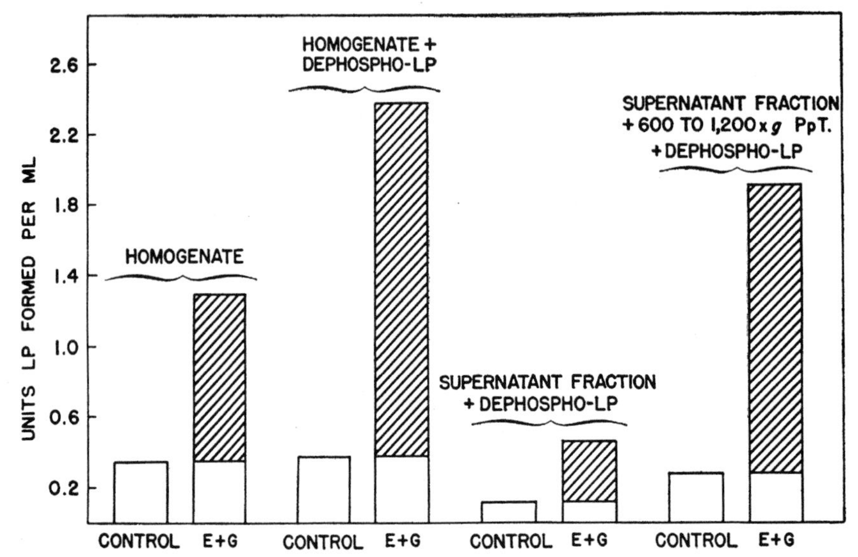

FIG. 1. The effect of epinephrine and glucagon on LP formation in a whole and fractionated cat liver homogenate. 0.14 ml. of homogenate or homogenate fraction was incubated with 2×10^{-2} M Tris buffer (pH 7.4), 2.5×10^{-3} M $MgSO_4$, and 1.7×10^{-3} M ATP in the presence and absence of 0.4 γ of *l*-epinephrine plus 1.0 γ of glucagon. The final volume was 0.20 ml. Dephospho-LP (4.2 units per ml.) was added where indicated. The supernatant fraction used in this experiment was the $1200 \times g$ supernatant fraction. LP activity was assayed before and after 10 minutes incubation at 30°. The bars represent the amount of LP formed during the incubation period; the cross-hatched portions of the bars represent the increased LP formation above that of the control.

Participation of Particulate Fractions Other Than Intact Cells in Response of Liver Homogenates—The probability that the response to epinephrine and glucagon in liver homogenates was restricted to unbroken cells remaining in the homogenates was small because, first, partially purified dephospho-LP added to liver homogenates participated in the response to the hormones (Fig. 1), and, second, the response to the hormones in homogenates survived the process of freezing (Fig. 2). Furthermore, it was possible to observe a good hormone response in preparations which contained no microscopically detectable intact cells. The preparation used in the experiment of Fig. 1 (right-hand bars) was composed of a washed particulate fraction collected at 600 to 1200 × g and the 1200 × g supernatant fraction. Microscopic examination of this preparation, with use of Wright's stain or Leishman's stain, did not reveal the presence of either intact cells or intact nuclei.

Centrifugation of homogenates at 1200 × g or more virtually abolished the hormone response in the resultant supernatant fraction (Figs. 1 and 3).

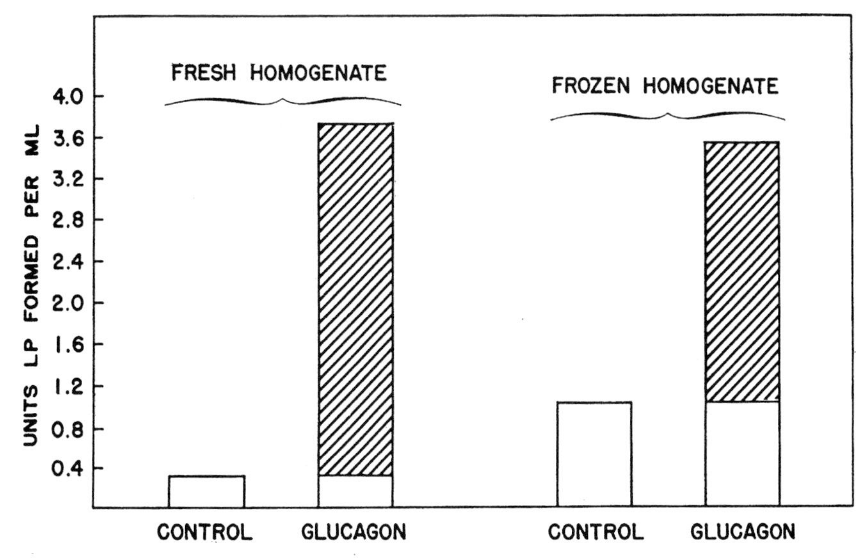

Fig. 2. The effect of glucagon on LP formation in a fresh and a frozen homogenate of dog liver. 0.15 ml. of homogenate was incubated for 10 minutes at 30° with 4 × 10^{-2} M Tris buffer (pH 7.4), 4 × 10^{-3} M $MgSO_4$, 2.8 × 10^{-3} M ATP, 4.2 units per ml. of dephospho-LP, and 0.1 mg. per ml. of casein in the presence and absence of 2.5 γ of glucagon. The final volume was 0.25 ml. The experiment depicted in the left-hand bars was performed immediately after preparation of the homogenate, and the experiment depicted in the right-hand bars with an aliquot of the homogenate which had been frozen in a dry ice-alcohol bath and stored 24 hours in solid CO_2.

The recombination of any portion of the particulate fractions sedimenting at 1200 × g or less with supernatant fractions resulted in preparations which responded to the hormones. In the example shown in Fig. 1 (right-hand portion), the addition of a small amount of a washed particulate fraction collected at 600 to 1200 × g to a 1200 × g supernatant fraction restored to a large extent the response to the hormones observed in the whole homogenate. In this experiment, particulate fractions collected at 0 to 300 × g and 300 to 600 × g were equally effective in restoring the hormone response in the 1200 × g supernatant fraction. In another experiment, a mixture of washed particles collected at 1200 × g and the 11,000 × g

supernatant fraction exhibited the same hormone response as the homogenate from which the fractions were derived.

These experiments have not established any of the cell fractions obtainable by differential centrifugation as the locus of the particle primarily responsible for the response to the hormones. Intact cells and intact nuclei appear to be excluded by microscopic examination of active preparations. Furthermore, results to date have not indicated a close association of the active particles to the mitochondria. Supernatant fractions, prepared by

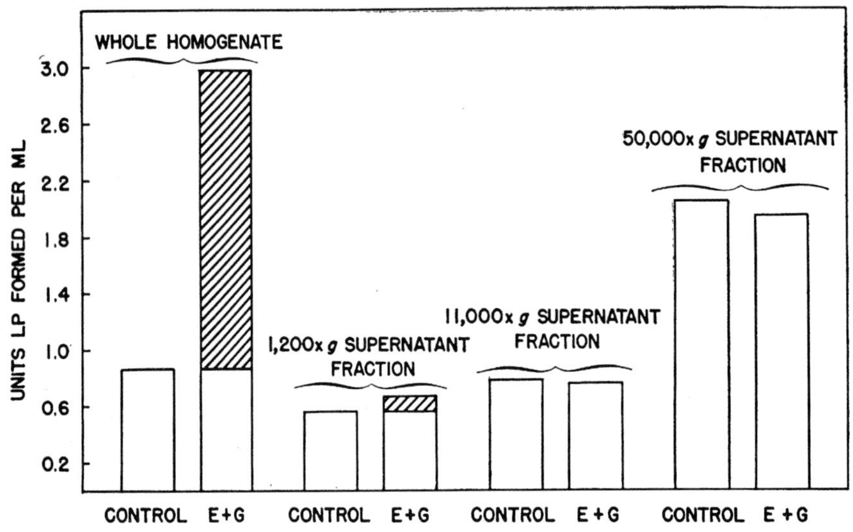

FIG. 3. The effect of epinephrine and glucagon on LP formation in fractions of a cat liver homogenate. 0.14 ml. of homogenate or homogenate fraction was incubated 10 minutes at 30° with 4×10^{-2} M Tris buffer (pH 7.4), 2.5×10^{-3} M $MgSO_4$, 1.7×10^{-3} M ATP, and 4.2 units per ml. of dephospho-LP in the presence and absence of 0.4 γ of *l*-epinephrine plus 2.0 γ of glucagon. The final volume was 0.20 ml.

centrifuging homogenates at $1200 \times g$, had little or no ability to respond to the hormones (Figs. 1 and 3); these fractions would be expected to contain the major portion of the mitochondria. However, cytochrome oxidase determinations by the method of Cooperstein and Lazarow (9) indicated that $1200 \times g$ supernatant fractions contained only about 30 to 35 per cent of the cytochrome oxidase activity of the whole homogenate. Since such a large proportion of the mitochondria appeared to be in the particulate fractions collected at $1200 \times g$, it is difficult to rule out the possibility that recombination procedures raised the ratio of mitochondria to other cell fractions above some critical level necessary for the hormone response. In any event, it is evident that preparations derived from liver homogenates required the presence of some particulate fraction in order to show a significant increase in the formation of LP in the presence of epinephrine or glucagon.

Relative Activities of Sympathomimetic Amines and Glucagon in Liver Homogenates—The magnitude of the increased formation of phosphorylase in the presence of the hormones was related to the amount of epinephrine or glucagon added to the homogenates. The half maximal response occurred at a concentration below 1×10^{-8} M with glucagon and 1×10^{-7} M with *l*-epinephrine.[2] It was of interest to estimate the relative activities of compounds related to *l*-epinephrine in the liver homogenate system,

TABLE I

Relative Activities of Sympathomimetic Amines in Vitro and in Vivo

For the liver homogenate assay, 0.15 ml. of frozen dog liver homogenate, diluted 2.5-fold with 0.33 M sucrose after thawing, was incubated 10 minutes at 30° with 5×10^{-2} M Tris buffer (pH 7.4), 5×10^{-3} M $MgSO_4$, 3.5×10^{-3} M ATP, dephospho-LP (4.8 units per ml.), and bovine serum albumin (8 γ per ml.) in a final volume of 0.25 ml. The increase in LP formation owing to the addition of *l*-epinephrine (5.4×10^{-8} M to 3.2×10^{-7} M) was compared to that owing to the addition of various amounts of the other sympathomimetic amines listed below. The values express the potency of these compounds relative to that of *l*-epinephrine calculated on a molar basis.

Sympathomimetic amine	Relative activity		
	Liver homogenate assay	Liver slice assay*	Intact animal assay
l-Epinephrine	100	100	100†
l-Norepinephrine	10	16	12†
d-Epinephrine	12	16	
d-Norepinephrine	0.4	2	0.6†
Amphetamine	0.0006	0.0	0.0

* Calculated from the data of Sutherland and Cori (5).
† Determined by McChesney *et al.* (10).

since these compounds vary in potency *in vivo*. In Table I are listed the relative activities of *l*-epinephrine, *d*-epinephrine, *l*-norepinephrine, *d*-norepinephrine, and amphetamine in stimulating the net formation of LP in liver homogenates. Included in Table I for comparison are relative activities of these compounds in stimulating glucose output of liver slices and in causing hyperglycemia in the intact animal. It can be seen that the activities of the compounds relative to that of *l*-epinephrine in liver homogenates are similar to those observed in the other systems.

[2] The adaptation of the homogenate system to the measurement of small amounts of epinephrine and glucagon involved modification of the conditions recorded in Fig. 1. The details of these modifications, as well as some applications of this assay system, will be reported in a subsequent publication.

Production of Factor Active in LP Formation by Particulate Fractions of Homogenates in Presence of Hormones—The observation that particulate fractions of liver homogenates were essential for the effect of epinephrine and glucagon on LP formation prompted experiments in which the washed particulate material was incubated with the hormones. Fig. 4 depicts the results of a typical experiment. Aliquots of a suspension of washed liver particles, collected at 1200 × *g*, were incubated with ATP and magnesium ions in the absence and presence of a mixture of epinephrine and glucagon. The entire incubation mixtures were heated in boiling water, chilled, and centrifuged. Aliquots of the resulting supernatant fluid (referred to as "boiled extract" below) were incubated with ATP, magnesium ions, and an 11,000 × *g* supernatant fraction. It can be seen from Fig. 4 that, in the presence of the boiled extract derived from particles incubated with the hormones, the formation of LP was increased and that the magnitude of this increase was related to the amount of boiled extract added to the incubation mixture. The boiled extract derived from particles incubated in the absence of the hormones, as well as a mixture of the hormones themselves, had only a small effect on LP formation. The addition of magnesium ions and ATP was found to be essential for production of the active principle in the presence of the hormones and liver particles and also for formation of LP in the 11,000 × *g* supernatant fraction, either in the absence or presence of active preparations of the boiled extract.

Properties of Active Factor—The stimulation of LP formation in the 11,000 × *g* supernatant fraction (Fig. 4) was used to estimate the amount of the unknown factor in crude or purified preparations. Before attempting purification procedures, some general information about the stability of the factor was gathered. The factor survived heating in boiling water for 3 minutes at pH 7.4 during preparation of the boiled extracts, as well as incubation for 24 hours at 25° in 0.1 N HCl. After being heated for 30 minutes in boiling water in 0.05 N HCl, factor preparations retained their original activity. It was also determined that the factor was dialyzable and was not extracted from aqueous solutions at either pH 7 or pH 1 by shaking with *n*-butanol or diethyl ether.

Attempts to chromatograph factor preparations on ion exchange resins not only resulted in extensive purification of the active principle, but also revealed more of its chemical properties. It was found that the factor was adsorbed on Dowex 2 chloride from active boiled extract preparations at neutral pH and subsequently was eluted with dilute HCl (0.02 N to 0.1 N). Under similar conditions, ATP and ADP remained adsorbed on the resin; 5-AMP was eluted earlier than the active factor. In 0.05 N HCl, the factor was adsorbed weakly to Dowex 50 (hydrogen form) and could be eluted by further washing of the resin with 0.05 N HCl. Under similar conditions, 5-AMP was not eluted from the resin, and ADP and ATP were eluted considerably before the factor. By adsorption and elution on ion exchange resins, it has been possible to purify the factor by about 500-fold over the

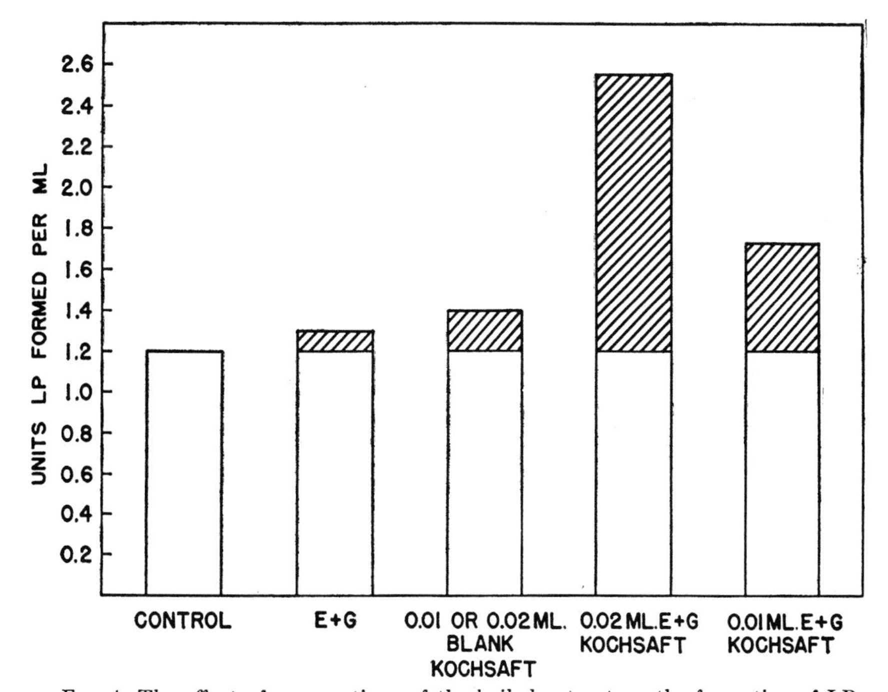

FIG. 4. The effect of preparations of the boiled extract on the formation of LP. Two 25 ml. portions of a suspension of washed particles in 0.25 M sucrose, derived from about 25 gm. of cat liver slices, were incubated with 2×10^{-2} M Tris buffer (pH 7.4), 2.5×10^{-3} M $MgSO_4$, and 1.7×10^{-3} M ATP in a final volume of 30 ml. One vessel contained 150 γ of both *l*-epinephrine and glucagon (E + G), while the other contained no added hormones (blank). After shaking for 5 minutes at 30°, the flasks were heated in boiling water for 3 minutes and then chilled to 0°. The flask contents were centrifuged at $15{,}000 \times g$ for 15 minutes in the cold. 0.02 ml. and 0.01 ml. aliquots of the supernatant fluids (boiled extracts) were incubated with 0.13 ml. of an $11{,}000 \times g$ supernatant fraction of a dog liver homogenate in 4×10^{-2} M Tris buffer (pH 7.4), 2.5×10^{-3} M $MgSO_4$, and 1.7×10^{-3} M ATP. The final volume was 0.20 ml. Control experimental vessels contained either water or 1.2 γ of both *l*-epinephrine and glucagon in place of the boiled extracts. LP activity was assayed before and after 5 minutes incubation at 30°, and the bars represent the amount of LP formed during this incubation period.

boiled extract, as judged by the lowering of optical density at 258 mμ in relation to activity in stimulating LP formation in the $11{,}000 \times g$ supernatant fraction.[3] As yet, no consistent differences in properties have been

[3] The active factor recently has been purified to apparent homogeneity. From ultraviolet spectrum, the orcinol reaction, and total phosphate determination, the

noted between factor preparations derived from the incubation of liver particles with epinephrine and those derived from incubation with glucagon.

Action of Factor in Supernatant Fractions of Liver Homogenates—Factor preparations were active in increasing LP formation in supernatant fractions of liver homogenates obtained by centrifugation at either 1200 or 11,000 × *g*. The effect of factor preparations was also apparent in supernatant fractions obtained by centrifugation for 45 minutes at 100,000 × *g*,

TABLE II

Effect of Factor Preparations on Formation of LP in 100,000 × g Supernatant Fractions

0.13 ml. portions of the fractions from a homogenate of dog liver were incubated with 3.6×10^{-2} M Tris buffer, pH 7.4, 2.3×10^{-3} M $MgSO_4$, 1.6×10^{-3} M ATP, dephospho-LP (5.5 units per ml.), and the additions listed below. The final volume was 0.22 ml. The blank and active factor preparations were corresponding fractions collected during ion exchange chromatography of boiled extracts derived from the incubation of washed liver particles in the absence and presence of epinephrine, respectively. The boiled extracts were prepared as described (Fig. 4). LP activity was determined before and after 5 minutes incubation at 30°.

Fraction of homogenate	Additions	LP formed, units per ml.	Δ, units per ml.
11,000 × *g* supernatant fraction	Water	0.94	
	0.6 γ glucagon + 0.6 γ epinephrine	1.00	+0.06
	Blank factor preparation	0.97	+0.03
	Active factor preparation	1.48	+0.54
100,000 × *g* supernatant fraction	Water	3.00	
	0.6 γ glucagon + 0.6 γ epinephrine	3.09	+0.09
	Blank factor preparation	2.93	−0.07
	Active factor preparation	3.68	+0.68
Dialyzed 100,000 × *g* supernatant fraction	Water	3.00	
	0.6 γ glucagon + 0.6 γ epinephrine	3.07	+0.07
	Blank factor preparation	3.26	+0.26
	Active factor preparation	3.85	+0.85

either before or after dialysis. In Table II are recorded the results of an experiment in which LP formation in a 100,000 × *g* supernatant fraction before and after dialysis was compared with that in the 11,000 × *g* supernatant fraction from which the 100,000 × *g* supernatant fractions were derived. It can be seen that the formation of LP in the three fractions of

active factor appeared to contain adenine, ribose, and phosphate in a ratio of 1:1:1. Neither inorganic phosphate formation nor diminution of activity resulted when the factor was incubated with various phosphatase preparations, including prostatic and intestinal phosphatase and Russell's viper venom. However, the activity of the factor was rapidly lost upon incubation with extracts from dog heart, liver, and brain.

homogenates was increased by the active factor preparation, and the net increase in LP formation was nearly the same in all three cases. Thus it appeared that the formed elements were not required in the action of the factor on LP formation, nor was any readily dialyzable component of the supernatant fraction necessary.

It is of interest to note that the removal of the formed elements of liver homogenates by centrifugation at 100,000 × g greatly increased the formation of LP in the resulting supernatant fraction. In preliminary experiments, the precipitate collected by centrifugation of homogenates between 11,000 × g and 100,000 × g (presumably consisting primarily of microsomes and glycogen) was found to inhibit strongly the formation of LP from dephospho-LP catalyzed by preparations of partially purified liver phosphokinase.

DISCUSSION

Differential centrifugation and microscopic examination of fractions obtained by centrifugation have shown that intact cells are not necessary components in the response of liver phosphorylase concentration to sympathomimetic amines and glucagon. The participation of added dephosphophosphorylase in the activation process and the ability to demonstrate hormone effects in previously frozen homogenates were considered additional evidence that intact cells were not necessarily involved. The absolute and relative activities of the sympathomimetic amines and glucagon in homogenates, in liver slices, and in intact animals indicate that the phenomena observed in homogenates may be related to the physiological activity of these agents. Although the demonstration that epinephrine and glucagon stimulate the net formation of LP in cell-free preparations surmounts the problem of dealing with intact cells, the analysis of the mechanism of action of the hormones is still complex.

It has been shown that the response to the hormones occurs in two stages, each of which may be eventually broken down into several steps. In the first stage, some portion of the particulate fraction of liver homogenates produces a heat-stable, dialyzable factor in the presence of the hormones. The identity of the particulate fraction to date has not been revealed by simple differential centrifugation experiments. The active factor produced by the particles in the presence of the hormones has been purified considerably, and it seems reasonable that identification of the active factor will yield important clues to the process involved in its production and to the mechanism by which it acts. The problem of identification of this substance is complicated by the probability that its molar concentration is extremely small in biological preparations.

In the second stage, this factor somehow influences the reactivation or inactivation reactions occurring in the soluble fractions of homogenates, resulting in an increase in the formation of LP. In liver homogenates, the reactivation process (conversion of dephospho-LP to LP) is opposed by the action of LP phosphatase, and also may be inhibited by various components of homogenates, including the microsomal fraction. To date, data have not been conclusive enough to distinguish between a stimulation of the reactivation process by the hormones via the active factor and an inhibition of the inactivation of LP. Preliminary experiments have not shown reproducible effects of factor preparations on either purified phosphokinase or LP phosphatase; it is possible that the factor may undergo metabolic alteration before participating in or affecting one of the two processes.

It has been shown that heart contains enzymes capable of catalyzing the interconversion of LP and dephospho-LP as well as the interconversion of the heart phosphorylases (11). Recent experiments have shown that factor preparations from either heart or liver increased the conversion of dephospho-LP to LP when this reaction was catalyzed by extracts of dog heart. This suggests that tissues other than liver may possess some or all of the components involved in the response of liver homogenates to epinephrine.[4]

SUMMARY

1. The formation of liver phosphorylase from dephosphophosphorylase in cell-free homogenates of dog and cat liver was increased markedly in the presence of either epinephrine or glucagon in low concentration.

2. The relative activities of sympathomimetic amines in homogenates were similar to those observed in liver slices and in the intact animal.

3. The response to the hormones in liver homogenates was separated into two phases: first, the formation of an active factor in particulate fractions in the presence of the hormones and, second, the stimulation by the factor of liver phosphorylase formation in supernatant fractions of homogenates in which the hormones themselves had no effect.

4. The active factor was heat-stable, dialyzable, and was purified considerably by chromatography on anion and cation exchange resins.

The authors wish to thank Miss Arleen M. Maxwell, Mr. James W. Davis, and Mr. Robert H. Sharpley for technical assistance in these studies.

[4] Active factor prepared from muscle particles of dog heart behaved in a manner similar to that of the factor from liver when chromatographed on ion exchange resins. In addition, it has been possible to observe production of an active factor in particulate preparations from dog skeletal muscle in the presence of epinephrine.

BIBLIOGRAPHY

1. Rall, T., Sutherland, E. W., and Berthet, J., *Federation Proc.*, **15**, 334 (1956).
2. Sutherland, E. W., and Wosilait, W. D., *Nature*, **175**, 169 (1955).
3. Wosilait, W. D., and Sutherland, E. W., *J. Biol. Chem.*, **218**, 469 (1956).
4. Rall, T. W., Sutherland, E. W., and Wosilait, W. D., *J. Biol. Chem.*, **218**, 483 (1956).
5. Sutherland, E. W., and Cori, C. F., *J. Biol. Chem.*, **188**, 531 (1951).
6. Sutherland, E. W., *Ann. New York Acad. Sc.*, **54**, 693 (1951).
7. Sutherland, E. W., and Wosilait, W. D., *J. Biol. Chem.*, **218**, 459 (1956).
8. Fiske, C. H., and Subbarow, Y., *J. Biol. Chem.*, **66**, 375 (1925).
9. Cooperstein, S. J., and Lazarow, A., *J. Biol. Chem.*, **189**, 665 (1951).
10. McChesney, E. W., McAuliff, J. P., and Blumberg, H., *Proc. Soc. Exp. Biol. and Med.*, **71**, 220 (1949).
11. Rall, T. W., Wosilait, W. D., and Sutherland, E. W., *Biochim. et biophys. acta*, **20**, 69 (1956).

Sutherland et al. (1962) J. Biol. Chem. **237**, 1220–1227

Adenyl Cyclase

I. DISTRIBUTION, PREPARATION, AND PROPERTIES*

EARL W. SUTHERLAND, T. W. RALL, AND TARA MENON†

WITH THE TECHNICAL ASSISTANCE OF JAMES W. DAVIS AND ARLEEN M. MAXWELL

From the Department of Pharmacology, School of Medicine, Western Reserve University, Cleveland 6, Ohio

(Received for publication, August 21, 1961)

Studies of the mechanism of action of sympathomimetic amines and glucagon led to the discovery of a cyclic adenine ribonucleotide in animal tissues (1–4). This cyclic nucleotide was identical with adenosine 3′,5′-phosphate (cyclic 3′,5′-AMP) found by Cook, Lipkin, and Markham (5). The chemical properties of this compound have been reported recently by Lipkin, Cook, and Markham (6). In this series of papers some observations are reported regarding the enzyme system that catalyzes the formation of cyclic 3′,5′-AMP. Adenosine triphosphate is converted to cyclic 3′,5′-AMP in the presence of the enzyme system and magnesium ions; inorganic pyrophosphate also appears during the reaction (7). For simplicity, the term adenyl cyclase or cyclase will be used to describe the enzyme system until other evidence regarding enzymatic mechanism is available. It is realized that more than one enzyme may be involved in the cyclization reaction and that the terms adenyl and cyclase may be too restrictive.

The relation of adenyl cyclase to the action of several hormones has been reviewed recently (8); in brief summary, several hormones or neurohormones influence the activity of adenyl cyclase. Some additional observations are included in accompanying papers (9, 10). The action of the cyclic nucleotide on enzymes was reviewed even more recently (11) and will not be discussed in detail in this series.

In this paper the distribution, preparation, and properties of adenyl cyclase are described. The capability of forming cyclic 3′,5′-AMP exists in many diverse animals and animal tissues. The enzyme catalyzing the formation of the cyclic nucleotide is associated with particulate material, the cellular origin of which will be discussed in a later paper. Preparation of adenyl cyclase in a purified or greatly simplified state has been difficult because of its association with particulate material and because of its ready inactivation by various procedures.

EXPERIMENTAL PROCEDURE

Preparation of Homogenates—Dogs were anesthetized to the surgical stage with sodium secobarbital, the carotid vessels were severed, and the thorax and abdomen were opened. Tissues were removed as rapidly as possible and chilled in cold 0.9% sodium chloride solution or occasionally in sucrose solutions. When the whole liver was used, it was perfused briefly with cold isotonic solutions. Small animals were killed and bled by decapitation. Material from the slaughterhouse was chilled in ice for transportation and then was rinsed with cold 0.9% NaCl or 0.25 M sucrose before homogenization.

Small samples of tissue were homogenized with 3 to 9 volumes of buffer in all-glass homogenizers of several designs. The Kontes Duall type was frequently used in tissue distribution studies in which some tissues were resistant to homogenization. Larger samples of tissue were usually homogenized in a Waring Blendor for 20 seconds or longer.

Fractionation Procedures—All steps from homogenization through fractionation to testing were carried out in the cold at a temperature near 3°, unless otherwise specified. International Centrifuges, models 3 and 2, were useful for fractionation by centrifugation when low gravitational forces were employed. The Lourdes centrifuge (model SL, volume rotor) or the Spinco preparative ultracentrifuges were used for other centrifugation steps.

Chromatography with cellulose resins was conducted in a cold room, and frequently slight pressure was exerted to increase the flow rate. This pressure, when applied, was generally equivalent to approximately 1 meter of water or less.

Assay of Adenyl Cyclase—The amounts of cyclic 3′,5′-AMP formed by adenyl cyclase preparations were not always proportional to the amount of enzyme added, even though very small amounts of cyclic 3′,5′-AMP were formed. In many cases, this lack of proportionality appeared related to the presence of large amounts of ATPase in the preparations. In any event, it was found necessary to prepare and assay several dilutions of enzyme, especially cruder enzymes, in order to assay in a range in which the formation of cyclic 3′,5′-AMP was proportional to the amount of enzyme added. Enzyme preparations were usually diluted in a glycylglycine buffer, 0.002 M, pH 7.4, containing 0.001 M $MgSO_4$; large tip pipettes were used for particulate fractions and careful mixing was employed.

In the majority of cases, 1.0 ml of cold enzyme was added to 1.4 ml of buffer and additives at room temperature; the standard reaction mixture contained, as final concentrations after enzyme, 2×10^{-3} M ATP, 3.5×10^{-3} M $MgSO_4$, 6.67×10^{-3} M caffeine, 1×10^{-2} M NaF, and 4×10^{-2} M Tris, pH 7.4. Incubation was carried out at 30°, with shaking in a Dubnoff metabolic incubator for varying periods of time, usually for 15 minutes, which is the time chosen as the standard for calculation of enzyme activity. To terminate incubation, samples were transferred from the

* This investigation was supported in part by grants from the National Institutes of Health of the United States Public Health Service (H-2745) and from Eli Lilly and Company.

† Research Fellow in Pharmacology. Present address, C/10 Karnatak Building, Moghul Lane, Mahim, Bombay 16, India.

incubating reaction mixtures to culture tubes, heated for 3 minutes in boiling water, and then chilled. Insoluble material was removed by centrifugation, and the heated extracts were assayed for cyclic 3′,5′-AMP content. The heated extracts could be stored at −20° for later assay if immediate assay was not desired.

The assay of cyclic 3′,5′-AMP has been described previously in this journal (3); a few modifications have been described more recently (12). In brief summary, cyclic 3′,5′-AMP samples were incubated at 25° with ATP, magnesium ions, and liver dephosphophosphorylase in the presence of fractions from liver homogenates containing dephosphophosphorylase kinase. After this incubation, the active phosphorylase, formed from the inactive or dephosphoenzyme, was measured. The stimulation of kinase activity was compared with that resulting from addition of known amounts of crystalline cyclic 3′,5′-AMP to similar reaction mixtures. A unit of adenyl cyclase is defined as that amount of enzyme that will catalyze the formation of 1.0 μmole of cyclic 3′,5′-AMP in 15 minutes in the reaction mixture described in the previous paragraph. Specific activity is expressed as units per mg of protein.

Assay of Other Enzymes—Pyrophosphatase was estimated by incubating enzyme samples in reaction mixtures containing 4×10^{-2} M Tris, 2.5×10^{-3} M $MgSO_4$, and 1×10^{-3} M sodium pyrophosphate as final concentrations at pH 7.4. After 15 minutes of incubation at 30°, aliquots were removed and fixed in trichloroacetic acid for determination of orthophosphate formation from the pyrophosphate. Since the pyrophosphatase was usually activated by freezing or by addition of Triton X-100 (Rohm and Haas Company), estimation of recovery in various steps required use of correction factors.

Adenosine triphosphatase was estimated in a similar fashion, except that the substrate concentration (ATP) was 2×10^{-3} M. ATPase activity was also estimated at times in the reaction mixture used for adenyl cyclase determination; *i.e.* 1×10^{-2} M NaF and 6.67×10^{-3} M caffeine were present in the reaction mixture.

Phosphodiesterase[1] activity leading to inactivation of cyclic 3′,5′-AMP was determined as previously described (4). The standard substrate concentration (cyclic 3′,5′-AMP) was 1×10^{-4} M. At times, substrate concentration was lowered as noted to approximate the concentration of cyclic 3′,5′-AMP that would be expected to accumulate during an individual experiment.

Cytochrome oxidase was determined by the method of Cooperstein and Lazarow (13). Inorganic phosphate was determined by the method of Fiske and SubbaRow (14) and protein, by a micro method described previously (15), modified according to the procedure of Lowry *et al.* (16).

With rare exception created by technical difficulty, glass-distilled (redistilled) water was used throughout the preparative procedures. Diethylaminoethyl cellulose (DEAE-cellulose) was obtained from Brown and Company. In most cases, a reagent grade (type 20) was used (0.88 meq per g). The cellulose was adjusted to the desired pH and, after preliminary batch washing with buffer, salt, and Triton, was placed in glass chromatography columns for final washing.

[1] The name cyclic 3′,5′-AMP phosphodiesterase frequently has been applied to this enzyme. Studies regarding the preparation and properties of this enzyme are in preparation for publication by R. W. Butcher and E. W. Sutherland.

RESULTS

Distribution of Adenyl Cyclase in Various Animals and Animal Tissue—Estimations of adenyl cyclase activity in animal homogenates were subject to several possible errors. In some tissues the substrate, ATP, was found to be almost completely hydrolyzed before completion of the incubation time used in the standard assay of adenyl cyclase. Inclusion of fluoride during incubation decreased the ATPase activity, and this action of fluoride, coupled with suitable dilution of enzyme, minimized this possible error. Fluoride appeared to stimulate the over-all cyclase reaction in addition to its effect on ATPase, and, when studied, the activity of cyclase in the presence of 0.01 M NaF has not been increased by addition of hormone. It was assumed that 0.01 M NaF allowed full expression of cyclase activity, although it was realized that this had not been proved for all tissues. Variable amounts of cyclic 3′,5′-AMP phosphodiesterase were present in the various tissues; the action of this enzyme was minimized by inclusion of caffeine in the reaction mixture.

In most instances biological assay alone has been used, both as evidence for the formation of cyclic 3′,5′-AMP and for estimation of the extent of formation of the cyclic nucleotide. As reported previously, cyclic 3′,5′-AMP has been isolated from reaction mixtures containing preparations from liver, heart, skeletal muscle, brain (4), beef adrenal cortex (17), or *Fasciola hepatica* (18), with recovery in various fractions being in reasonable agreement with estimates made from biological assay of the unfractionated or crude reaction mixtures. (Such agreement was not found when whole tissue extracts were studied. As noted previously (19), a number of inhibitors of the assay are present in extracts of whole tissues or organs. Methods for accurate measurement of cyclic 3′,5′-AMP in such extracts are currently under investigation.)

Results of estimatation of adenyl cyclase activity in 11 different tissues from a single dog are summarized in Table I. Adenyl cyclase activity of the whole cerebral cortex was consistently

TABLE I

Distribution of adenyl cyclase in tissues of dog

A large spayed female dog was given intravenous sodium secobarbital to the stage of light surgical anesthesia and was bled. Tissue samples were removed rapidly, chilled in cold NaCl, and then homogenized with 9 volumes of 0.001 M $MgSO_4$ and 0.002 M glycylglycine, pH 7.4, in a Kontes Duall homogenizer. The homogenates were diluted in the homogenization medium and assayed for adenyl cyclase activity as described in the text. Three or more dilutions of each homogenate were assayed.

Tissue	Units per 100 g (wet weight)	Relative activity (wet weight)	Relative activity (protein basis)
Brain (cortex)	91	8	11
Spleen	34	3	2
Skeletal muscle	32	3	2
Heart (ventricle)	28	2.5	2
Lung	23	2	1.5
Kidney (cortex)	13	1	1
Liver	11	1	0.5
Aorta	11	1	1
Intestinal muscle	11	1	1
Femoral artery	5.0	0.5	1.5
Adipose (omental)	0.53	0.05	1

TABLE II

Distribution of cyclase in various animals and animal tissues

The units listed are expressed as units per 100 g of wet weight. Tissues in the intermediate range contained from 5 to 25 units per 100 g; those in the lower range, 5 or fewer units per 100 g.

Animal	Tissue	Cyclase distribution*
		units/100 g
High range:		
Beef	Cerebral cortex	80
Calf	Cerebral cortex	60
	Cerebellum	38
Sheep	Cerebral cortex	37
	Cerebellum	24
Pig	Cerebral cortex	41
Fasciola hepatica		80+
Lumbricus terrestris	Anterior portion	50
Intermediate range:		
Dog	Testis	
	Uterus	
	Intestinal mucosa	
Cat	Liver	
Rabbit	Skeletal muscle	
Rat	Liver	
Guinea pig	Uterus	
	Intestinal mucosa	
Chicken	Liver	
Pigeon	Skeletal muscle	
	Blood cells	
Low range:		
Rat	Epididymal fat pad†	
Chicken	Blood cells	
Fly larva	Whole animal	
Minnow	Whole animal	

* These are approximate values.

† Considerably more activity than noted with omental fat of dog.

much higher than that of other tissues, either on a wet weight or on a protein basis. A sample of gray matter of the cerebral cortex of ox, mixed with some white matter, was found to have several times more activity than a sample of white matter from the cortex.

Results of studies with other tissues or animals are summarized in Table II. It can be seen that the tissues with relatively high cyclase activity were from the central nervous system or from flat or segmented worms. The estimation of cyclase activity in *Fasciola hepatica* was derived from published data (18) based on incubation conditions which probably were suboptimal.

Distribution in Broken Cell Preparations—Most of the adenyl cyclase activity of broken cell preparations has been found in particulate fractions collected with low gravitational forces, for example, 600 × g or 2000 × g for 17 minutes. Homogenates prepared in isotonic or hypotonic solutions yielded similar results. The precipitate collected with such forces could be washed several times with centrifugation after each wash, and the activity could be recovered in high yield. With some tissues, it was found necessary to include some salt, such as 0.02 M NaCl, in the wash solutions.

In early experiments it was apparent that microsomal fractions and nonsedimenting fractions of tissues contained relatively small amounts of cyclase activity. For example, in Table III it may be seen that the total activity of the homogenate was recovered in the nuclear and mitochondrial fractions. Microscopic examinations of various fractions obtained by centrifugation showed that the distribution of particulate elements was as expected from the centrifugal forces used. In addition, inorganic pyrophosphatase activity of liver was measured, since this enzyme in liver has been reported to be concentrated in microsomes. The inorganic pyrophosphatase activity of liver was not correlated with cyclase activity.

Since varying amounts of mitochondria may be present in the "nuclear" fractions, where most cyclase activity was found, it seemed advisable to investigate whether the cyclase activity was associated with the mitochondria, especially heavy mitochondria. Rat liver homogenates were used in this study because of the numerous published experiments on the separation of cell particles by centrifugation of such preparations. The homogenization method of de Duve *et al.* (20) was chosen to minimize damage to mitochondria. Guides to fractionation were gravitational forces employed, microscopic examination, and determination of cytochrome oxidase, which is reported to be localized in mitochondria. The results of such an experiment are summarized in Table III. It is clear that the distribution of cyclase activity was not correlated with the distribution of cytochrome oxidase activity. In agreement with this finding was the observation that cyclase activity was not correlated with mitochondria content as judged by microscopic examination. On a protein basis, the specific activity of cyclase was highest in the nuclear fraction. In some experiments, the bulk of the cyclase activity in the frozen "nuclear" fraction required collec-

TABLE III

Fractionation of liver homogenate by centrifugation

Livers from rats that had been deprived of food were perfused briefly with 0.25 M sucrose and then were homogenized with 3 volumes of 0.25 M sucrose and passed through gauze. The precipitate collected at 600 × g for 10 minutes was rehomogenized and centrifuged twice more; the three supernatant fractions were pooled. From these pooled supernatant fluids the 2,200 × g fraction and the 2,200 to 10,000 × g fraction were collected by centrifugation for 10 minutes at room temperatures near 3°. The "nuclear" fraction (600 × g) was diluted with an equal volume of 0.25 M sucrose and then was frozen rapidly and stored at −70°. Storage at −70° for 4 days had reduced the production of cyclic 3′,5′-AMP by 50%. After thawing, the fraction was diluted with 8 volumes of glass-distilled water and separated into the fractions noted below by centrifugation for 10 minutes. All results are expressed as per cent of the total activity in each subfraction.

Fraction	Adenyl cyclase activity	Cytochrome oxidase activity
	%	%
"Nuclear" fraction (600 × g)	77	24
"Heavy" mitochondria (2200 × g)	23	36
"Light" mitochondria (10,000 × g)	0.4	40
Subfractionation of frozen "nuclear" fraction		
200 × g	81	6
600 × g	18	27
10,000 × g	1	67

tion at forces of $600 \times g$, rather than at the $200 \times g$ that was effective in the experiment reported in Table III.

From such experiments, it can be concluded that the particles associated with adenyl cyclase sediment readily and, at least in liver, are not mitochondria or microsomes. The "nuclear" fraction has not been well defined and has been reported to contain nuclei and cell membranes, as will be discussed later.

Preparation of Adenyl Cyclase

Separation of a particulate portion of homogenates still remains as the first step in the preparation of adenyl cyclase from various tissues. The precipitates were washed in hypotonic solutions, frozen, and washed again in hypotonic solution, then in hypertonic or other salt solutions and refrozen. Adenyl cyclase was then dispersed in a 1.8% Triton solution after having been previously washed with 0.1% Triton solution. In some cases, further fractionation on DEAE-cellulose columns has been helpful. Numerous other protein fractionation procedures have not been useful to date.

Tissues that have been fractionated by the following procedure or by variations of this procedure include cerebral cortex, heart, skeletal muscle, and liver. The over-all increase in specific activity on a protein basis was slightly greater when liver, heart, and skeletal muscle were used, but since cerebral cortex had a much higher initial specific activity, preparations from this tissue have been used most frequently. In all cases the increases of specific activity on a protein basis have been relatively small.

1. Preparation of Third Precipitate—Tissues were sliced and homogenized for 20 to 45 seconds in a Waring Blendor with 9 volumes of homogenizing medium containing 2×10^{-3} M glycylglycine, pH 7.4, 1×10^{-3} M $MgSO_4$, 1×10^{-2} M NaCl, and 1×10^{-2} M KCl. The homogenate was strained through a single layer of grade 60 cheesecloth to remove fibrous or other larger pieces of tissue. Only small losses of activity were noted if filtration was used at this stage; large losses of cyclase activity were incurred if filtration was carried out at any later steps before solubilization. (Occasionally filtration was omitted.) The filtrate was centrifuged at $2000 \times g$ for 20 minutes, and the supernatant fluid was decanted, usually by aspiration, and discarded. The first precipitate was suspended with the homogenizing medium to a volume of approximately one-half of the filtrate volume and was homogenized for 10 seconds in a Waring Blendor. This second homogenization in the hypotonic buffer solution was found to disrupt effectively cells that had not broken in the initial homogenization. The suspension was diluted with the buffer solution to the filtrate volume and was centrifuged again at $2000 \times g$ for 20 minutes. The second precipitate was suspended to the filtrate volume in the homogenizing medium and was centrifuged as before. The third precipitate was suspended in a volume of the homogenizing medium equal to the precipitate volume, frozen in a Cellosolve-dry ice bath as rapidly as possible in several portions, and stored at −70°. The cyclase activity of the third precipitate was not decreased by freezing under these conditions, and the preparation could be stored at −70°, or lower, for at least several weeks without loss of activity. As noted in Table III, sucrose preparations from liver were much more labile. It will be shown in a subsequent section that the presence of higher concentrations of salt may increase the lability to freezing and storage.

2. Preparation of Salt-washed Precipitate—The frozen third precipitate was thawed as rapidly as possible with swirling first in a 45° water bath, then in a 37° bath, and was placed in an ice water bath as thawing approached completion. The suspension was diluted with the homogenizing medium solution to a volume equal to that of the original filtrate. The fourth precipitate was collected by centrifugation at $2000 \times g$ for 20 minutes, and the supernatant fluid was decanted and discarded.

The fourth precipitate was suspended in the homogenizing medium solution to a volume equal to one-half of the filtrate volume, and to this suspension was added an equal volume of 0.38 M KCl containing 0.04 M glycylglycine, pH 7.4, and 0.001 M $MgSO_4$. The suspension, containing a final concentration of 0.2 M KCl, was stirred for 3 minutes and centrifuged at $2000 \times g$ for 20 minutes. The resulting precipitate (0.2 M KCl) was suspended in an equal volume of a solution 0.02 M glycylglycine and 0.001 M $MgSO_4$ and then was diluted to one-half the filtrate volume by addition of a solution containing 0.10 M KCl, 0.02 M glycylglycine, pH 7.4, and 0.001 M $MgSO_4$. The precipitate (0.1 M KCl precipitate) was collected by centrifugation and was suspended in 1.5 volumes of the solution containing 0.02 M glycylglycine and 0.001 M $MgSO_4$. The suspension was frozen rapidly and stored at −70°. The packed volume of the precipitate at this stage was often equal to or somewhat greater than the volume of tissue used as starting material. Therefore, the volume of the suspension containing this precipitate was often approximately 3 times that of the tissue from which it was derived. It was noted that the washing of the precipitate with salt concentrations higher than 0.2 M KCl could be employed but had no obvious advantages. Certain properties of salt-washed preparations will be discussed in a later section.

3. Preparation of Soluble Enzyme—The frozen precipitate, containing 0.04 M KCl, 0.02 M glycylglycine, and 0.001 M $MgSO_4$, was thawed as rapidly as possible. To this suspension was added a small amount of 27% Triton, volume for volume in H_2O, sufficient to bring the final concentration to 0.1% Triton. The resulting mixture was stirred thoroughly and then was centrifuged for 30 minutes at $40{,}000 \times g$. The supernatant fluid was discarded. The resulting precipitate was suspended with a solution containing 0.04 M KCl, 0.02 M glycylglycine, pH 7.4, and 0.001 M $MgSO_4$ to a volume equal to that of the thawed suspension before addition of Triton. (In some instances KCl was omitted from the suspending solution.) To this suspension was added 27% Triton in an amount sufficient to bring the final concentration of Triton to 1.8%. The suspension was stirred for several minutes and then was centrifuged at $40{,}000 \times g$ for 30 minutes. Usually the major portion of the cyclase activity recovered appeared in the supernatant fluid. In all studies to date, addition of 1.8% Triton has caused some loss of cyclase activity. This is evident in the example presented in Table IV. The following percentages of the original activity were found in the indicated fraction when the final concentrations of Triton were the values designated in the parentheses: 1.0%, supernatant fluid (0.1%); 31%, supernatant fluid; 21%, precipitate (1.8%); 44%, supernatant fluid (1.8%). In the second of the two experiments, the specific activity of the supernatant fluid was 0.021; the activity of the precipitate was not measured.

Soluble preparations have been made in an identical or similar manner from heart, skeletal muscle, and liver. The fractionation system was identical with that just described, with the 0.04 M KCl being omitted from the solution used when the precipitate obtained by addition of 0.1% Triton was collected and suspended.

TABLE IV

Fractionation of cyclase from beef cerebral cortex

Three beef brains were obtained from the slaughterhouse, chilled in cracked ice during transportation, and later were rinsed with a cold 0.9% NaCl solution. Portions of the cortex were sliced and homogenized for 30 seconds in a Waring Blendor with 9 volumes of buffer solution containing 0.002 M glycylglycine, pH 7.4, 0.001 M $MgSO_4$, 0.01 M NaCl, and 0.01 M KCl (60 g of sliced cortex + 540 ml of buffer solution for each homogenization). The homogenate was strained through one layer of grade 60 cheesecloth. Six 900-ml portions, representing the filtrate from 540 g of tissue, were centrifuged for 20 minutes at 2000 × *g*. Fractionation of the entire preparation was carried out as described in the text through freezing of the 0.1 M KCl precipitate. From this stage on, aliquots were used for further fractionation and results were calculated as if the entire preparation had been used.

Fraction	Cyclase activity	Total protein	Specific activity	Recovery of cyclase	Recovery by steps	Recovery of other enzymes: Pyrophosphatase	3′,5′-AMP diesterase	ATPase
	units	*g*	× 10^{-3}	%	%	%	%	%
Homogenate	340	38.0	9.0	100	100	100	100	100
First supernatant	60	11.5	5.0	18	18	80	63	21
Third precipitate	199	20.0	10.0	59	59	3	21	53
Third precipitate					100	100	100	100
Fourth supernatant	0	1.3	0	0	0	55	17	2
Fourth precipitate	208	20.4	10.2	61	104	35	67	84
0.1 M KCl precipitate	195	19.0	10.3	57	98	23	68	58
0.1 M KCl precipitate					100	100	100	100
0.1% Triton supernatant	2	1.2	1.7	0.6	1	54	17	2
1.8% Triton supernatant A	60	3.2	17.8	18	31	21	25	20
1.8% Triton precipitate	41	16.0	2.6	12	21	8	37*	16
1.8% Triton supernatant B	86	4.14	21	25	44			

*Inexact value since substrate was utilized to the extent of 98%.

When cyclase was fractionated from heart ventricle muscle of beef, the recovery of cyclase activity was similar in pattern to the data shown in Table IV for brain, with an over-all increase in specific activity of approximately 4-fold (from 0.9 × 10^{-3} to 3.5 × 10^{-3}). The specific activities of dog liver preparations likewise were increased approximately 4-fold by these procedures, *i.e.* from approximately 0.7 × 10^{-3} to 3.0 × 10^{-3}. Such soluble preparations from muscle and liver were, of course, less active than those from brain; however, for use in certain experiments they were fractionated further as described in the following section.

4. Fractionation of Soluble Enzyme on DEAE-Cellulose—Preliminary experiments showed that cyclase activity could be removed from supernatant fractions containing 1.8% Triton by passage over DEAE-cellulose gel columns in columns ranging from pH 6 to 7.5, but it was not removed by passage over carboxymethyl cellulose columns in this pH range. Considerable variation was encountered in the ability to adsorb and to elute cyclase activity from various DEAE-cellulose gels. Variations have been reduced by careful attention to the usual factors, such as pH, ionic strength, type of gel, etc., but have not been eliminated completely. Rapid increases in salt concentration, even if rather small, have been most effective in promoting elution of cyclase from columns. Prolongation of washing times when cyclase is adsorbed on the gels decreases the ability to elute the enzyme subsequently. To date, it has not been possible to elute cyclase unless Triton was included in the eluting fluid.

The DEAE-cellulose gels were given a final wash in the chromatography columns, usually at a pH of approximately 6.8, with imidazole (0.01 M) used as buffer instead of glycylglycine. The KCl concentration of the soluble enzyme fraction applied to the gel was usually 0.04 M or 0.08 M. The gel wash included 1.8% Triton, buffer and the appropriate amount of KCl.

An example of fractionation by this procedure is summarized in Table V. Here a soluble preparation from dog skeletal muscle was concentrated several fold in Fraction 4a with an increase of specific activity from 2.5 to 7.1. This fraction was frozen and used in an experiment described in an accompanying paper (7). Recovery of cyclase in Fraction 4a varied from 10% to 50%, with values of approximately 30% being observed frequently. These values are not corrected for the loss of activity incurred because of the lability of the preparation, the activity of which, on standing in an ice bath (without fractionation), would frequently decrease 25% during the course of the column fractionation.

Fractionation on DEAE-cellulose gels was helpful in reducing the ATPase content of the preparation. In the example given in Table V, 20% of the total ATPase activity (tested in the presence of 0.01 M NaF) was present in Fraction 4a.

This partial separation from ATPase activity was also ob-

TABLE V

Fractionation of soluble muscle cyclase on DEAE-cellulose

A fraction from skeletal muscle of dog (in 1.8% Triton) was prepared as described in the text and Table IV and contained 0.04 M KCl, 0.01 M glycylglycine, and 0.001 M $MgSO_4$. This preparation (310 ml) was applied to a DEAE-cellulose column 2.5 cm in diameter and 8.7 cm high. The DEAE-cellulose had been washed batchwise and then as a column with a solution containing 1.8% Triton, 0.04 M KCl, and 0.01 M imidazole, pH 6.7. After application of the 310 ml of preparation, 125 ml of the 1.8% Triton, 0.04 M KCl, and 0.01 M imidazole solution were added, followed by 100 ml of 1.8% Triton, 0.15 M KCl, and 0.01 M imidazole, pH 6.7; this solution served as the eluting agent. The flow rate was 3.4 ml per minute. A band was visible as the eluting fluid proceeded down the column and was collected into tube 4a.

Fraction	Volume of fraction	Cyclase activity	Total protein	Recovery	Specific activity*
	ml	*units*	*g*	%	× 10^{-3}
1.8% Triton supernatant	310	2.08	0.820	100	2.5
2a	230	0.18	0.120	9	1.5
2b	100	0.09	0.154	4	0.6
3a	40	0.03	0.049	1	0.6
3bc	100	0.07	0.048	3	1.5
4a	25	1.05	0.163	51	7.1
4b	23	0.17	0.036	8	4.7

* Specific activity of fraction 4a was 0.0071.

served when soluble liver preparations were fractionated. In these fractionation procedures, much of the ATPase preceded the cyclase (*i.e.* was less readily adsorbed or more readily washed off the gel). In one experiment, 37% of the cyclase was recovered from a soluble liver preparation in a fraction similar to Fraction 4a of Table V. This fraction had a specific activity of 12.6, which was 4.5 times greater than that of the soluble preparation applied to the column. The recovery of ATPase in this fraction was approximately 20%. Although greater separation than this has been achieved, ATPase activity has not been completely separated from cyclase activity in any preparation.

Properties of Cyclase Preparations

Stability on Storage—Washed particulate preparations containing cyclase activity were relatively stable, could be frozen and stored, and, in general, were as responsive to hormone addition as were any cruder or fresher preparations. Several factors influenced the ability to freeze and to store these preparations with retention of activity. First, the composition of the suspending medium was of prime importance; various preparations could be frozen and stored in hypotonic solutions with good retention of activity, whereas activity was lost in the presence of added salt, *e.g.* KCl. Stability of various preparations to freezing and storage in salt solutions varied greatly, generally salt concentrations from 0.01 M to 0.04 M were used to minimize loss, although occasional preparations could be stored in higher salt concentrations. Sucrose preparations were often unstable when frozen and stored.

Second, the speed of freezing and the temperature of storage were also important. In general, rapid freezing was better, although occasional preparations containing very little salt retained activity as well, or better, if allowed to freeze slowly on standing at $-20°$ or $-70°$. In almost all cases, storage at $-70°$ was equal to, or better than, storage at $-20°$. With most preparations, other than from brain, the temperature of storage was very critical, and therefore storage at $-70°$ was adopted as a routine procedure.

Third, it was also found that addition of glycylglycine at a final concentration of 0.2 M, pH 7.4, protected most preparations during the freezing-storage process, even when the salt concentration was relatively high. It was rarely necessary, however, to use glycylglycine in the preparative procedure.

In addition to these factors, there was a large variation in sensitivity to freezing that depended on the source of the preparation. For example, preparations from brain have been relatively stable.

As mentioned previously, the soluble fractions were much more labile than the particulate fractions. The factors described above were also important in the freezing and storage of soluble fractions, except that glycylglycine was often ineffective in protecting the soluble preparations during freezing and storage.

Other Stability Properties—The washed particulate preparations were relatively stable to storage at 2° and could be washed and centrifuged over a period of hours with little or no loss of activity. The washed particulate preparations have been mixed vigorously for 1 or 2 minutes in a Waring Blendor under nitrogen gas with little or no loss of activity. As noted previously, addition of 1.8% Triton to these preparations caused a variable loss of activity often amounting to 50% or more.

Particulate preparations from the several tissues studied could be lyophilized, and the dried powder stored in a vacuum desiccator in the cold, with full retention of activity. In a single experiment, the dried powder from beef brain (0.1 M KCl precipitate washed so as to contain 0.01 M KCl and lyophilized) was stored for 3 days at room temperature without loss of activity. The lyophilized preparations retained their responsiveness to hormone addition. Solubilized preparations (in Triton) were lyophilized with variable retention of activity; in one experiment most of the activity of an extract of brain (1.8% Triton extract) remained after lyophilization. Variations in samples from a single fraction were encountered after lyophilization, an indication that small differences in the freeze-dry technique were important. The dried powders from particulate preparations from heart and brain could be extracted with ether in the cold with little or no loss of activity. Extraction with cold acetone resulted in small or moderate losses in activity. Extraction with ethanol-ether (1:1) or with chloroform-methanol (1:1) produced a large or complete loss of activity.

Cyclase activity of a particulate fraction from skeletal muscle was lost on incubation with trypsin or chymotrypsin and was not lost on incubation with ribonuclease or deoxyribonuclease.

Solubility Properties—In Table IV an example is shown in which considerable amounts of cyclase activity appeared in 1.8% Triton solutions. Solubilization of the enzyme[2] has not been achieved with agents other than Triton. Various other detergents have not been fully explored; however, in preliminary trials a 1.8% solution of Tween 80 detergent did not solubilize cyclase. In a single experiment, deoxycholate (1.6%) was found to destroy or to suppress cyclase activity nearly completely. Little or no cyclase activity was extracted into aqueous solutions from lyophilized powders that had been extracted with ether.

Soluble preparations containing Triton were not soluble when separated from Triton. Cyclase activity could not be eluted with DEAE-cellulose gels in the absence of Triton. Whereas soluble cyclase preparations could be adsorbed on calcium phosphate gels, cyclase activity could not be eluted in the absence of Triton. (Cyclase activity could be eluted from calcium phosphate gels in the presence of Triton, but to date without significant increase in specific activity.)

Attempts to separate Triton from the soluble cyclase have been unsuccessful. Dialysis was not effective since the Triton diffused through Visking casing very slowly. Addition of ammonium sulfate caused formation of an insoluble mixture of Triton and protein. The resulting insoluble Triton-protein mixture in ammonium sulfate rose to the surface on centrifugation, and although considerable cyclase activity remained in the floating pad, the specific activity was the same or lower than that of the starting material.

pH Optimum, Fluoride, Metal, and O_2 Requirements—The pH optimum for the formation of cyclic 3′,5′-AMP was broad, being relatively flat between pH 7.2 and 8.2. Cyclase activity fell at pH values above or below this range. For example, at pH 6.5 the activity was approximately 40% of that found at pH 7.4. When tested in the absence of catecholamines and fluoride, heart

[2] The term solubilization may be inexact. The extracts were almost clear, although with slight and variable opalescence. The adenyl cyclase in 1.8% Triton extracts did not sediment or rise when subjected to gravitational forces of approximately 50,000 × *g*.

particulate preparations appeared to have a higher pH optimum than in the presence of either fluoride or catecholamines.

Previous reports have discussed the requirements for ATP and Mg^{++} (3, 21) in the formation of cyclic 3′,5′-AMP and have mentioned the stimulatory effects of certain hormones (3, 8) and fluoride (3) on this formation or accumulation. The influence of hormones will be described in greater detail in Papers III and IV of this series (9, 10). Fluoride at a concentration of 0.01 M stimulated the formation of cyclic 3′,5′-AMP; the hormone effects were largely or entirely erased. The stimulatory effect of fluoride was less prominent with preparations from brain, although even with these preparations, a doubling or tripling of accumulation of product was noted. The mechanism of the fluoride stimulation has not yet been clarified.

It was found that manganese ions could replace magnesium ions when brain preparations were employed, and manganese ions were partially effective in replacing magnesium ions when heart preparations were studied. Calcium, cobaltous, and zinc ions did not appear to replace magnesium ions appreciably. Small amounts of barium ions mixed in the usual reaction mixture had little or no effect on the reaction, whereas small amounts of zinc ions mixed in the usual reaction mixture were strongly inhibitory to the cyclase reaction. An example of this inhibition may be seen in Paper II of this series (7), where small amounts of zinc ions (1 to 2 $\times$ 10^{-4} M) are shown to inhibit cyclase activity by approximately 80%. With preparations from brain it was found that the accumulation of cyclic 3′,5′-AMP proceeded as well in the presence of solutions containing sodium ions and no potassium ions as in the presence of potassium ions without the presence of sodium ions. It was also noted that the accumulation of cyclic 3′,5′-AMP in washed particulate preparations proceeded as well in the presence of nitrogen gas as in the presence of oxygen.

DISCUSSION

The capability of forming cyclic 3′,5′-AMP is found in tissues of all animals tested to date in four phyla, with the possible exception of red blood cells from dogs. One or two preliminary attempts to demonstrate adenyl cyclase in plant tissue have been negative, but it is felt that these preliminary experiments with plant tissue have not yet provided an adequate basis for assuming an absence of cyclase in the plants tested. Further study of the distribution of this enzyme activity in plants and microorganisms will be of interest. The distribution of cyclase activity within the cell also appears worthy of further exploration, since the nature of the particulate material associated with the cyclase activity is unknown at the present time. The particulate material containing the enzyme activity sediments readily and may be separated from mitochondria or microsomes. Nuclei and cell membranes are currently the two most likely materials with which the cyclase activity might be associated. This question may be clarified by fractionation of avian erythrocytes in which separation of cell membranes from nuclei may be possible.

The distribution of adenyl cyclase within the tissues of an animal or in the several phyla studied tempts speculation that the function of this enzyme, or the product(s) of its activity, may have broader significance in biological events than the stimulation of phosphorylase activation alone. Mansour and Menard (22) have reported that the cyclic nucleotide or serotonin stimulates the activity of phosphofructokinase in homogenates of the flatworm, *Fasciola hepatica*. More recently, Mansour, Le Rouge, and Mansour (23) have reported that cyclic 3′,5′-AMP activates phosphofructokinase in a soluble fraction derived from such homogenates; serotonin was inactive in this soluble fraction. Previously, Berthet (24) had reported a stimulatory effect of cyclic 3′,5′-AMP on ketone body formation from acetate in liver slices. Later, Pryor and Berthet found that cyclic 3′,5′-AMP, like glucagon, inhibits the incorporation of amino acids into liver proteins (25). Belocopitow (26) has reported that cyclic 3′,5′-AMP decreases the activity of glycogen synthetase in extracts of rat muscle, while simultaneously increasing phosphorylase activity. These observations indicate that cyclic 3′,5′-AMP may serve in a regulatory role related to a number of enzyme activities.

The preparation of adenyl cyclase in a simplified and purified form has been hampered by the association with particulate material in the "nuclear" fraction, by its lability, and by its close association with the detergent Triton after solubilization. It is hoped that the enzyme may be solubilized by other procedures and that more stable soluble preparations can be achieved.

The current preparative procedures reduce the concentration of enzymes that catalyze the hydrolysis of components of the cyclase reaction. The ATPase activity of muscle was reduced considerably during the solubilization step, a reduction that was not prominent in the detailed example of fractionation of cerebral cortex shown in Table IV. Pyrophosphatase and cyclic 3′,5′-AMP phosphodiesterase activity was decreased by most procedures, especially the washing in hypotonic solutions. However, to date no cyclase preparations have been entirely free from these other enzymes. Increases of specific activity on a protein basis have been relatively small, *e.g.* only 2- or 3-fold for brain preparations and 15-fold or so for liver preparations. Surprisingly large amounts of protein are present in the 1.8% Triton extract, and this protein remains in close association with both the detergent and the cyclase.

The lability to freezing and the dispersal or solubilization in high concentrations of Triton have indicated that adenyl cyclase may be a lipoprotein; however, additional information is required before reaching this conclusion.

SUMMARY

1. The enzyme or enzyme system catalyzing the formation of adenosine 3′,5′-phosphate (adenyl cyclase) was found in all animals or animal tissues studied in four phyla, with the possible exception of dog red blood cells. Highest activity was found in the gray matter of brain and in *Fasciola hepatica.*

2. Adenyl cyclase was associated with particulate material in broken cell preparations, the origin of which was not mitochondria or microsomes. The cellular source of the particulate material is unknown, but it may be derived from cell membranes or from nuclei.

3. Preparations containing adenyl cyclase were washed in hypotonic solutions, frozen, washed again with hypotonic and salt solutions, and then refrozen with good recovery of activity. After a wash with 0.1% Triton solution, much of the activity was solubilized or dispersed in a 1.8% Triton solution. Concentrations of adenosine triphosphatase, pyrophosphatase, and adenosine 3′,5′-phosphate phosphodiesterase relative to that of adenyl cyclase were lowered by these procedures.

4. Washed particulate or soluble preparations were lyophilized without and with loss of activity. The dried powders could be

extracted with certain organic solvents, such as diethyl ether, without loss of activity.

5. The pH optimum for the formation of adenosine 3′,5′-phosphate was broad, being relatively flat between pH 7.2 and 8.2 when brain and heart preparations were tested in the presence of epinephrine or fluoride. Manganese ions could replace magnesium ions in brain preparations and could partially replace magnesium ions in heart preparations.

Acknowledgment—The authors wish to thank Mrs. Miriam Dixon for technical assistance in these studies.

REFERENCES

1. Rall, T. W., Sutherland, E. W., and Berthet, J., *J. Biol. Chem.*, **224**, 463 (1957).
2. Sutherland, E. W., and Rall, T. W., *J. Am. Chem. Soc.*, **79**, 3608 (1957).
3. Rall, T. W., and Sutherland, E. W., *J. Biol. Chem.*, **232**, 1065 (1958).
4. Sutherland, E. W., and Rall, T. W., *J. Biol. Chem.*, **232**, 1077 (1958).
5. Cook, W. H., Lipkin, D., and Markham, R., *J. Am. Chem. Soc.*, **79**, 3607 (1957).
6. Lipkin, D., Cook, W. H., and Markham, R., *J. Am. Chem. Soc.*, **81**, 6198 (1959).
7. Rall, T. W., and Sutherland, E. W., *J. Biol. Chem.*, **237**, 1228 (1962).
8. Sutherland, E. W., and Rall, T. W., *Pharmacol. Revs.*, **12**, 265 (1960).
9. Murad, F., Chi, Y.-M., Rall, T. W., and Sutherland, E. W., *J. Biol. Chem.*, **237**, 1233 (1962).
10. Klainer, L. M., Chi, Y.-M., Freidberg, S. L., Rall, T. W., and Sutherland, E. W., *J. Biol. Chem.*, **237**, 1239 (1962).
11. Rall, T. W., and Sutherland, E. W., *Cold Spring Harbor Symposia on Quantitative Biology, XXVI, Cellular regulatory mechanisms*, Long Island Biological Association, Cold Spring Harbor, Long Island, N. Y., 1961, in press.
12. Rall, T. W., and Sutherland, E. W., in S. P. Colowick and N. O. Kaplan (Editors), *Methods in enzymology, Vol. V*, Academic Press, Inc., New York, 1961, p. 377.
13. Cooperstein, S. J., and Lazarow, A., *J. Biol. Chem.*, **189**, 665 (1951).
14. Fiske, C. H., and SubbaRow, Y., *J. Biol. Chem.*, **66**, 275 (1925).
15. Sutherland, E. W., Cori, C. F., Haynes, R., and Olsen, N. S., *J. Biol. Chem.*, **180**, 825 (1949).
16. Lowry, O. H., Rosebrough, N. J., Farr, A. L., and Randall, R. J., *J. Biol. Chem.*, **193**, 265 (1951).
17. Haynes, R. C., Jr., *J. Biol. Chem.*, **233**, 1220 (1958).
18. Mansour, T. E., Sutherland, E. W., Rall, T. W., and Bueding, E., *J. Biol. Chem.*, **235**, 466 (1960).
19. Butcher, R. W., Jr., Sutherland, E. W., and Rall, T. W., *Pharmacologist*, **2**, 66 (1960).
20. de Duve, C., Pressman, B. C., Gianetto, R., Watteau, B., and Appelmans, F., *Biochem. J.*, **60**, 604 (1955).
21. Berthet, J., Sutherland, E. W., and Rall, T. W., *J. Biol. Chem.*, **229**, 351 (1957).
22. Mansour, T. E., and Menard, J. S., *Federation Proc.*, **19**, 50 (1960).
23. Mansour, T. E., Le Rouge, N. A., and Mansour, J. M., *Federation Proc.*, **20**, 226 (1961).
24. Berthet, J., *Fourth International Congress of Biochemistry, Vienna, 1958*, Pergamon Press, Inc., New York, 1960.
25. Pryor, J., and Berthet, J., *Biochim. et Biophys. Acta*, **43**, 556 (1960).
26. Belocopitow, E., *Arch. Biochem. Biophys.*, **93**, 457 (1961).

Walsh et al. (1968) J. Biol. Chem. **243**, 3763–3765

An Adenosine 3′,5′-Monophosphate-dependant Protein Kinase from Rabbit Skeletal Muscle*

(Received for publication, April 2, 1968)

D. A. Walsh,‡ John P. Perkins,§ and Edwin G. Krebs

From the Department of Biochemistry, University of Washington, Seattle, Washington 98105

SUMMARY

A protein kinase that catalyzes an adenosine 3′,5′-monophosphate (cyclic AMP)-dependent phosphorylation of casein and protamine has been purified from rabbit skeletal muscle. The K_m values of cyclic AMP for these reactions are 1 × 10^{-7} and 6 × 10^{-8} M, respectively. The protein kinase markedly increases the rate of the cyclic AMP-dependent activation and phosphorylation of phosphorylase kinase by ATP.

Significant evidence has been obtained by Sutherland and Rall (1), Sutherland, Øye, and Butcher (2), and others (3–6) which indicates that adenosine 3′,5′-monophosphate (cyclic AMP) is an intracellular mediator of numerous hormonal responses. The direct interactions by which this nucleotide effects its role as a "second messenger" (2) for each of these hormones remains to be established in all cases. Studies in this laboratory by Posner, Stern, and Krebs (7) have demonstrated that epinephrine administration to either rats or frogs results in a 2- to 3-fold increase in the levels of cyclic AMP in skeletal muscle and the activation of phosphorylase kinase. The ATP-dependent activation of this enzyme *in vitro* is enhanced by cyclic AMP (8), but no significant binding of the cyclic nucleotide to the purified phosphorylase kinase could be demonstrated (9). A different protein kinase which exhibits a complete dependence on cyclic AMP for its activity has now been partially purified from rabbit skeletal muscle. This enzyme catalyzes the phosphorylation of casein and protamine and augments the rate of activation and phosphorylation of phosphorylase kinase by ATP.

Purification of Protein Kinase—The initial fractionation was performed utilizing a semiquantitative assay based on the acceleration of the rate of phosphorylation of phosphorylase kinase. Later, fractions were assayed more quantitatively utilizing casein as the substrate. The supernatant solution from the pH 6.1 acid precipitation step of a standard phosphorylase kinase purification (8) served as the source of the enzyme. This solution obtained from a preparation utilizing 2.4 kg of rabbit skeletal muscle was adjusted to pH 5.5 and the precipitate was removed by centrifugation. The supernatant solution was adjusted to pH 6.8 with 1 M potassium phosphate buffer, pH 7.2, and fractionated by the addition of 32.5 g of ammonium sulfate per 100 ml of protein solution. The precipitate, collected by centrifugation, was dissolved in 150 ml of 0.005 M potassium phosphate buffer, pH 7.0, containing 0.002 M EDTA, and dialyzed extensively against the same buffer. The protein solution was centrifuged at 78,000 × *g* for 1 hour and the precipitate was discarded. All buffers used in the succeeding steps of the purification contained 0.002 M EDTA. The enzyme was adsorbed on a column (24 × 4.5 cm) of DEAE-cellulose (Sigma), that had been equilibrated with 0.005 M potassium phosphate buffer, pH 7.0, and the cellulose was then washed with 800 ml of the same buffer. The protein kinase was eluted with 0.03 M potassium phosphate buffer, pH 7.0. The enzyme solution was equilibrated with 0.005 M Tris-chloride buffer, pH 7.5, and the enzyme was adsorbed on a Whatman No. DE52 cellulose column (24 × 1.25 cm), equilibrated with the same buffer. It was eluted from the cellulose with a linear gradient of Tris-chloride, pH 7.5, in the fractions between 0.025 and 0.05 M buffer. The protein was precipitated from the solution by the addition of ammonium sulfate (0.4 g per ml), redissolved in a minimal volume of 0.005 M potassium phosphate buffer and chromatographed on Sephadex G-200. Fractions obtained slightly after the breakthrough peak which contained the enzyme were pooled. The enzyme was precipitated by the addition of ammonium sulfate (0.4 g per ml), redissolved in a minimal volume of 0.05 M glycerol phosphate buffer, and the protein solution was equilibrated with the same buffer by chromatography on Sephadex G-25. This fraction, which can be stored at −15°, was used for all the studies reported here. Purification resulted in approximately a 300-fold increase in specific activity and a 7% yield based on the initial enzymatic activity of the pH 5.5 supernatant. The protein kinase fraction contained no phosphorylase kinase activity.

Properties of Protein Kinase Isolated from Rabbit Skeletal Muscle—It was found that the partially purified enzyme catalyzed the transfer of ^{32}P to casein from $\gamma^{32}P$-ATP (Fig. 1) in a reaction which exhibited a complete dependence on the presence of cyclic AMP. This requirement for the presence of cyclic AMP could not be replaced by any of the following nucleotides

* This investigation was supported in part by Grant AM 07873 from the National Institutes of Health, United States Public Health Service.

‡ Postdoctoral Fellow of the American Cancer Society.
§ Postdoctoral Fellow of the National Institutes of Health, United States Public Health Service.

used at a concentration of 1×10^{-4} M: 5′-AMP, 3′-AMP, ADP, GMP, GDP, GTP, CMP, UMP, and IMP. The reaction (Fig. 1) was linear over a wide range of enzyme concentrations and could be used as a convenient assay system for the protein kinase. The phosphorylation of casein was markedly sensitive to low concentrations of cyclic AMP (Fig. 2). Under the conditions used for the assay (pH 6.0; ATP, 1.2×10^{-3} M; magnesium, 3.6×10^{-3} M), the apparent K_m value for cyclic AMP was 1×10^{-7} M.

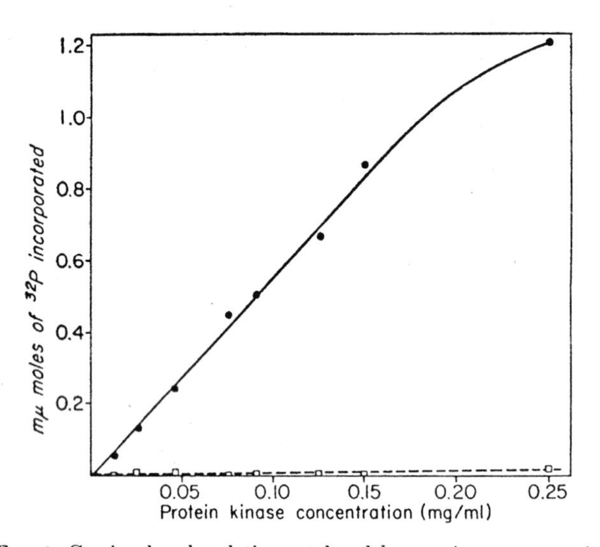

FIG. 1. Casein phosphorylation catalyzed by varying concentrations of protein kinase in the presence (●——●) or absence (□---□) of cyclic AMP. The uptake of phosphate by casein was measured in an incubation mixture containing: sodium glycerol phosphate, pH 6.0, 5 μmoles; casein, pH 6.0, 0.6 mg; γ^{32}P-ATP, 0.12 μmole; magnesium acetate, 0.36 μmole; sodium fluoride, 2 μmoles; theophylline, 0.2 μmole; ethylene glycol bis(β-aminoethyl ether)-N,N'-tetraacetic acid, 0.03 μmole; cyclic AMP (where added), 10 mμmole; and varying aliquots of purified protein kinase in a total volume of 0.1 ml. The incubation was performed at 30° for 10 min. The reaction was initiated by the addition of the solution of γ^{32}P-ATP and magnesium acetate and terminated by the addition of 0.2 ml of bovine serum albumin (6.25 mg per ml) and 0.5 ml of 10% trichloracetic acid. The protein-bound ^{32}P was determined essentially by the method described previously (9).

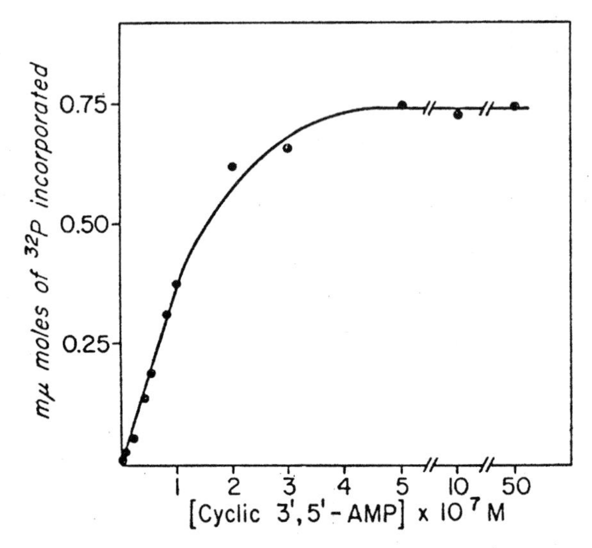

FIG. 2. The variation in casein phosphorylation with the concentration of cyclic AMP. The experiment was performed under identical conditions with those described in the legend of Fig. 1 utilizing a protein kinase concentration of 0.113 mg per ml.

TABLE I

Effect of protein kinase on activation and phosphorylation of phosphorylase kinase

Nonactivated phosphorylase kinase, 0.3 mg per ml, was incubated for 2 min at 30° with 0.2 mM γ^{32}P-ATP, 0.6 mM magnesium acetate in reaction mixtures containing 7 mM glycerol phosphate, 20 mM β-mercaptoethanol, 0.2 mM EDTA buffer, pH 6.8, and additions (as shown) of protein kinase, 9 μg per ml, and 0.01 mM cyclic AMP. Aliquots were removed at 2 min for determination of phosphorylase kinase activity and protein-bound ^{32}P utilizing the methods described previously (8, 9). Control experiments carried out with the protein kinase alone showed that it bound a negligible quantity of ^{32}P.

Tube	Additions		Increase[a] in phosphorylase kinase activity at pH 6.8	Protein-bound ^{32}P
	Protein kinase	Cyclic AMP		
			units/mg	*mole/10⁵ g*
1	−	+	3000	0.097
2	−	−	0	0.023
3	+	+	7300	0.240
4	+	−	0	0.024

[a] Phosphorylase kinase activity at pH 6.8 before activation = 870 units per mg.

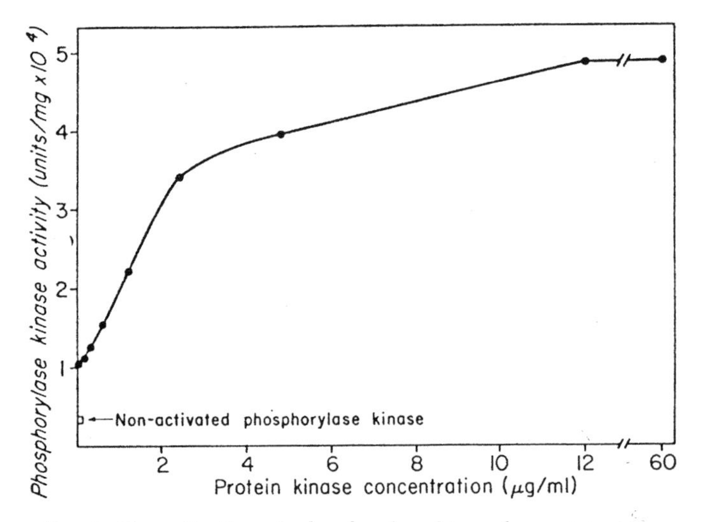

FIG. 3 The activation of phosphorylase kinase by protein kinase. Nonactivated phosphorylase kinase, 0.3 mg per ml, was incubated for 1 min at 30° in reaction mixtures containing: sodium glycerol phosphate, pH 6.8, 7 μmoles; β-mercaptoethanol, 20 μmoles; ATP, 0.18 μmole; magnesium acetate, 0.6 μmole; cyclic AMP, 0.1 μmole; EDTA, 0.2 μmole; and varying aliquots of purified protein kinase in a total volume of 1 ml. The reactions were initiated by the addition of purified phosphorylase kinase (9) and terminated by a 1:21 dilution of the incubation mixtures into 0.01 M glycerol phosphate buffer containing 0.1 M β-mercaptoethanol maintained at 0°. The phosphorylase kinase activity at pH 6.8 of each of these solutions was determined by the method described previously (8). The phosphorylase kinase activity at pH 6.8 before activation = 3450 units per mg.

It was shown that the purified kinase preparation also catalyzed the phosphorylation of protamine. In experiments performed in collaboration with Drs. Gordon H. Dixon and Bengt Jergil, University of British Columbia, it was demonstrated that the uptake of phosphate by protamine occurred at a rate 5 to 10 times that observed with casein as the substrate. Phosphorylation of protamine in the absence of cyclic AMP occurred at a rate of less than 8% of that occurring in the presence of the nucleotide. The apparent K_m value of cyclic AMP under the conditions of the protamine phosphorylation reaction (pH 7.5; NaCl, 0.3 M; ATP, 2.5×10^{-4} M; magnesium, 1×10^{-2} M) was 6×10^{-8} M.

Effect of Protein Kinase on Phosphorylase Kinase Activation—Phosphorylase kinase as extracted and purified from skeletal muscle is essentially inactive at pH 6.8 (10). Incubation of the purified kinase with ATP in the presence of Mg^{++} ions activates the enzyme; this process is accompanied by phosphorylation of the protein (9). The new protein was shown to enhance the activation and the concomitant phosphorylation of phosphorylase kinase (Table I). Under the conditions used in the experiment of Table I both the activation and phosphorylation of phosphorylase kinase that occurred in either the presence or absence of added protein kinase required the presence of cyclic AMP. The effect of varying concentrations of the protein kinase on phosphorylase kinase activation rate is shown in Fig. 3. It will be noted that 300 μg of phosphorylase kinase were almost completely activated in 1 min in the presence of 5 μg of the protein kinase.

DISCUSSION

The protein kinase purified from skeletal muscle exhibits a complete dependence on cyclic AMP for activity. It is attractive to think that this enzyme serves as the link between the epinephrine stimulation of adenyl cyclase (11) and the activation of phosphorylase kinase that has been demonstrated *in vivo* by Posner *et al.* (7). If this is the biological function of the newly isolated protein kinase, then it should be called phosphorylase kinase kinase. It is probable that the cyclic AMP enhanced activation of purified phosphorylase kinase that occurs in the absence of added protein kinase *in vitro* (8) is catalyzed by traces of the new protein kinase in the preparation. This explanation is supported by the observation that there is a good agreement between the apparent K_m values of cyclic AMP determined previously (9) for the activation of phosphorylase kinase in the absence of added protein kinase (7×10^{-8} M), with those values established in this study for the phosphorylation of casein (1×10^{-7} M) and protamine (6×10^{-8} M) by the isolated enzyme. It can be calculated from the data presented in Fig. 3 that the protein kinase need only be present in phosphorylase kinase at a level of less than 0.5% on a weight basis to account fully for the observed rate of activation. The phosphorylation and activation of phosphorylase kinase that has been previously demonstrated to occur at a slow rate in the absence of cyclic AMP at higher ATP concentrations than those used in this study (9) is probably an autocatalytic process.

REFERENCES

1. Sutherland, E. W., and Rall, T. W., *Pharmacol. Rev.*, **12,** 265 (1960).
2. Sutherland, E. W., Øye, I., and Butcher, R. W., *Recent Progr. Hormone Res.*, **21,** 623 (1965).
3. Orloff, J., and Handler, J., *Amer. J. Med.*, **42,** 757 (1967).
4. Gilman, A. G., and Rall, T. W., *Fed. Proc.*, **25,** 617 (1966).
5. Jungas, R. L., *Proc. Nat. Acad. Sci. U. S. A.*, **56,** 757 (1966).
6. Rizack, M. A., *J. Biol. Chem.*, **239,** 392 (1964).
7. Posner, J. B., Stern, R., and Krebs, E. G., *J. Biol. Chem.*, **240,** 982 (1965).
8. Krebs, E. G., Love, D. S., Bratvold, G. E., Trayser, K. A., Meyer, W. L., and Fischer, E. H., *Biochemistry*, **3,** 1022 (1964).
9. DeLange, R. J., Kemp, R. G., Riley, W. D., Cooper, R. A., and Krebs, E. G., *J. Biol. Chem.*, **243,** 2200 (1968).
10. Krebs, E. G., Graves, D. J., and Fischer, E. H., *J. Biol. Chem.*, **234,** 2867 (1959).
11. Robinson, G. A., Butcher, R. W., and Sutherland, E. W., *Ann. N. Y. Acad. Sci.*, **139,** 703 (1967).

Identification and characterization of G-proteins

2

The study of the mechanisms by which plasma membrane receptors couple to adenyl cyclase to regulate intracellular levels of cyclic AMP (see Section 1) was facilitated by the fact that receptor coupling remains intact in isolated membrane fragments. The study of cyclic AMP generation in such 'cell-free' preparations allowed the determination of the intracellular components that are required for adenyl cyclase activation. In 1971, Martin Rodbell and co-workers were able to demonstrate a key role for guanyl nucleotides, particularly GTP. First, they showed that binding of the hormone glucagon to its receptor was modified by guanyl nucleotides [1]. The second significant observation presented in the paper which we have chosen [2], was that guanyl nucleotides were an essential requirement for glucagon-stimulated cyclic AMP synthesis in isolated liver plasma membranes. This key observation eventually led to the discovery of the GTP-binding ('G'-) proteins which link receptors to adenyl cyclase activation, and which were subsequently found to regulate many target enzymes in cell signalling pathways and effectors such as ion channels. These G-proteins therefore play a central role in many aspects of cell regulation.

Following from the demonstration that guanyl nucleotides were an essential requirement was the finding that, in addition to the receptor, two other distinct components were required for adenyl cyclase activity [3,4]. One activity was heat-labile (the catalytic subunit), whereas the second component was relatively heat-stable but inactivated by *N*-ethylmaleimide and by trypsin, demonstrating that this factor was a protein. The experimental demonstration of these two distinct components was only possible due to the fact that the guanyl-nucleotide regulation of adenyl cyclase activity could be reconstituted, following inactivation of one component, by addition of the second in detergent extracts of plasma membrane fractions. A second significant experimental factor was the availability of a mutant lymphoma cell line, the S49 AC^- or cyc^- mutant [5], in which the adenyl cyclase catalytic subunit was normal but the factor (the G-protein) required for cyclase activation was inactive. Extracts of membranes from these cells provided a source of catalytic subunit in reconstitution studies [3,4]. A year or so earlier a guanyl nucleotide-binding protein was identified and suggested to be involved in adenyl cyclase activation [6]. In addition, Cassel and Selinger [7] found that occupancy of catecholamine receptors increased a GTPase activity apparently linked to adenyl cyclase activation. These various studies raised the question of whether the GTP-binding protein, the GTPase and the second heat-stable protein required for adenyl cyclase activity were one and the same protein. This question required purification of the G-protein for its resolution.

An influential review by Martin Rodbell in 1980 [8] discussed in detail the concept of a GTP-binding protein requirement for the coupling of receptors to adenyl cyclase. The review considered the possible roles of two GTP-binding proteins, termed N_s and N_i, linking stimulatory or inhibitory receptors respectively to adenyl cyclase. N_s and N_i (later to be known as G_s and G_i) quickly became accepted as *bone fide* entities following their purification and characterization. G_s was purified to homogeneity in the same year using reconstitution assays with cyc^- S49 lymphoma cells, and this is described in the second of the chosen papers from the laboratory of Gilman [9]. The significance of this paper is that it led to the detailed molecular characterization of the regulatory factor for the stimulation of adenyl cyclase as a GTP-binding protein consisting of three (α-, β-, γ-) subunits with the α-subunit possessing the GTP-binding and GTPase activities. This is the now well known heterotrimeric G-protein G_s.

The paper by Northup et al. showed that, following an extensive 2000-fold purification of G_s, three major polypeptides of relative molecular mass 52000, 45000 and 35000 were present. Earlier studies [10,11] had shown that one of the G-protein subunits was a target for cholera toxin. This toxin possesses ADP-ribosyl transferase activity and can lead to activation of cyclic AMP synthesis. The purified 52000 M_r and 45000 M_r polypeptides were both found to be ADP-ribosylated by cholera toxin [9]. Subsequently it became clear that both of these polypeptides are α_s variants generated by alternative RNA splicing [12] and the 35000 M_r polypeptide is Gβ. In the paper by Northup et al. the very small Gγ subunit (8000 Da) was not visualized on the polyacrylamide gels used to analyse the composition of the purified G_s, but was later detected as a stoichiometric component of G_s.

As noted above, the α-subunits of G_s are substrates for ADP-ribosylation by cholera toxin. A second toxin, pertussis toxin, allowed the identification of the G-protein (G_i) involved in inhibitory regulation of adenylate cyclase. Subsequent studies led to the current understanding of the GTPase cycle (Figure 2.1)

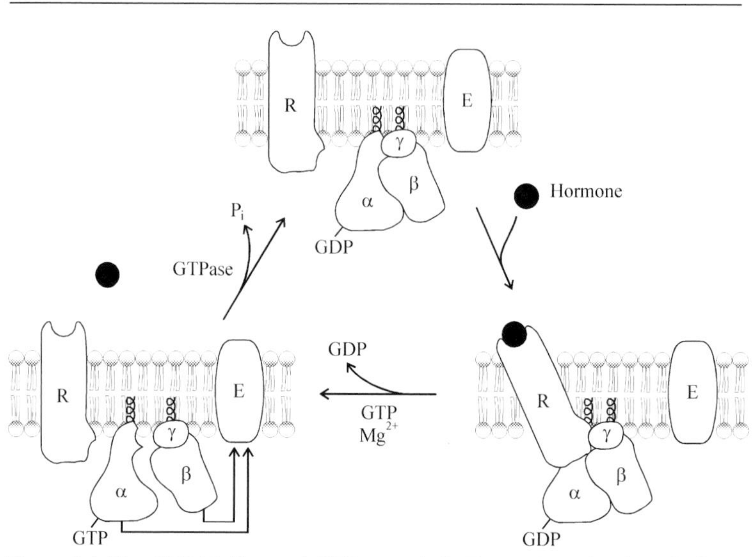

Figure 2.1 The GTP-binding and GTPase cycle in G-protein activation of effectors. In the resting state (top) the α-subunit has bound GDP and the heterotrimeric complex of α-, β- and γ-subunits is not associated with the effector (E). Hormone binding to its receptor (R) leads to a conformational change in the receptor (bottom right). The interaction of the occupied receptor with the G-protein then promotes exchange of GTP for GDP on the α-subunit and subsequent dissociation of the α-subunit from the heterotrimeric complex (bottom left). Both GTP-bound α-subunit and free $\beta\gamma$ can interact with and regulate effector molecules. GTP hydrolysis by the α-subunit switches off the activation and the system returns to the resting state. Modified from [14] with permission.

involved in G-protein activation of effector molecules and timed inactivation of the G-protein. Further work resulted in the discovery of a wide range of G-proteins controlling many different effectors [13]. Initial models proposed that control of the effector was mediated by the α-subunit in its GTP-bound form after dissociation from the heterotrimeric complex (Figure 2.1). It is now known that the $\beta\gamma$ subunits can also exert direct control on a variety of effectors so that dissociation of the complex activates a bifurcating signalling pathway [14].

Many G-protein subunits have been fully characterized by molecular cloning and sequencing, and multiple forms of α-, β- and γ-subunits are now known. The exact functions of all of these forms are yet to be understood. However, a detailed molecular picture of the α-subunit is emerging following the elucidation of the three-dimensional structure of the α-subunit of the retinal rod cell G-protein transducin [15] and $G_{i\alpha 1}$ [16] by X-ray crystallography, and the determination of the conformational switch that occurs after exchange of GTP for GDP and activation of the α-subunit. The structures of two complete G-protein heterotrimers have also been determined [17,18].

The most recent advance in this field has been the discovery of a family of proteins known as the 'regulator of G-protein signalling' or RGS proteins. They interact directly with G-proteins to stimulate their GTPase activity [19] and are likely to be important in the control of G-protein inactivation.

References

1. Rodbell, M., Krans, H.M., Pohl, S.L. and Birnbaumer, L. (1971) *J. Biol. Chem.* **246**, 1861–1871
2. Rodbell, M., Birnbaumer, L., Pohl, S.L. and Krans, M.J. (1971) *J. Biol. Chem.* **246**, 1877–1882
3. Ross, E.M. and Gilman, A.G. (1977) *J. Biol. Chem.* **252**, 6966–6969
4. Ross, E.M., Howlett, A.C., Ferguson, K.M. and Gilman, A.G. (1978) *J. Biol. Chem.* **253**, 6401–6412
5. Haga, T., Ross, E.M., Anderson, H.J. and Gilman, A.G. (1977) *Proc. Natl. Acad. Sci. U.S.A.* **74**, 2016–2020
6. Pfeuffer, T. and Helmrich, E.J.M. (1975) *J. Biol. Chem.* **250**, 867–876
7. Cassel, D. and Selinger, Z. (1976) *Biochim. Biophys. Acta.* **452**, 538–551
8. Rodbell, M. (1980) *Nature (London)* **284**, 17–22
9. Northup, J.K., Sternweis, P.C., Smigel, M.D., Schleifer, L.S., Ross, E.M. and Gilman, A.G. (1980) *Proc. Natl. Acad. Sci. U.S.A.* **77**, 6516–6520
10. Cassel, D. and Pfeuffer, T. (1978) *Proc. Natl. Acad. Sci. U.S.A.* **75**, 2669–2673
11. Gill, D.M. and Meren, R. (1978) *Proc. Natl. Acad. Sci. U.S.A.* **75**, 3050–3054
12. Robishaw, J.D., Smigel, M.D. and Gilman, A.G. (1986) *J. Biol. Chem.* **261**, 9587–9590
13. Casey, P.J. and Gilman, A.G. (1988) *J. Biol. Chem.* **263**, 2577–2580
14. Clapham, D.E. and Neer, E.J. (1993) *Nature (London)* **365**, 403–406
15. Lambright, D.G., Noel, J.P., Hamm, H.E. and Sigler, P.B. (1994) *Nature (London)* **369**, 621–628
16. Coleman, D.E., Berguis, A.M., Lee, E., Lindu, M.E., Gilman, A.G. and Sprang, S.R. (1994) *Science* **265**, 1405–1412
17. Lambright, D.G., Sondek, J., Bohm, A., Skiba, N.P., Hamm, H.E. and Sigler, P.B. (1996) *Nature (London)* **379**, 311–319
18. Wall, M.A., Coleman, D.E., Lee, E., Iniguez-Lluhi, J.A., Posner, B.A., Gilman, A.G. and Sprange, S.R. (1995) *Cell* **83**, 1047–1058
19. Koelle, M.R. (1997) *Current Biol.* **9**, 143–147

Rodbell et al. (1971) J. Biol. Chem. **246**, 1877–1882

The Glucagon-sensitive Adenyl Cyclase System in Plasma Membranes of Rat Liver

V. AN OBLIGATORY ROLE OF GUANYL NUCLEOTIDES IN GLUCAGON ACTION

(Received for publication, October 7, 1970)

MARTIN RODBELL, LUTZ BIRNBAUMER, STEPHEN L. POHL, AND H. MICHIEL J. KRANS*

From the Section on Membrane Regulation, Laboratory of Nutrition and Endocrinology, National Institute of Arthritis and Metabolic Diseases, Bethesda, Maryland 20014

SUMMARY

A method is described for the enzymatic synthesis of 5′-adenylyl imidodiphosphate labeled with ^{32}P at the α position (AMP-PNP-α-^{32}P), an analogue of ATP containing nitrogen substituted for oxygen between the terminal phosphates. The nucleotide is only slowly hydrolyzed during incubation with rat liver plasma membranes and is a substrate for adenyl cyclase in these membranes.

In the presence of 0.2 mM AMP-PNP, glucagon and fluoride ion stimulate adenyl cyclase activity; linear rates are maintained for at least 10 min of incubation at 30°. GTP enhanced the initial rate of basal and glucagon-stimulated adenyl cyclase activity. Reduction in concentration of Mg^{++} in the assay medium or incubation of liver membranes for 5 min at 30° prior to addition of glucagon results in loss of response of adenyl cyclase to glucagon and in reduction in the effects of GTP on basal activity. Under these conditions GTP, GDP, or GMP-PCP are required for glucagon stimulation of the enzyme even though the specific binding sites for glucagon are saturated with hormone; as little as 10 nM GTP or GDP is required. UTP and CTP exert smaller effects than the guanyl nucleotides and act only at concentrations higher than 0.1 mM.

The guanyl nucleotides inhibited the response of adenyl cyclase to fluoride ion (10 mM) over the same concentration range over which they stimulate the response of the enzyme to glucagon. This action of the nucleotides is observed in plasma membranes treated with phospholipase A under conditions that result in loss of glucagon binding and of hormonal response.

It is concluded that guanyl nucleotides play a specific and obligatory role in the activation of adenyl cyclase by glucagon. The nucleotides bind at sites, distinct from the glucagon binding sites, that appear to regulate both the response of adenyl cyclase to glucagon, and, possibly by a related mechanism, the actions of fluoride ion on this system.

* Recipient of a fellowship from the Netherlands Organization for the Advancement of Pure Research (Z.W.O. 1969–1970).

In the previous study (1), it was shown that guanyl nucleotides (GTP or GDP) equally and at concentrations as low as 0.05 μM, alter the properties of the specific binding sites for glucagon in rat liver plasma membranes. GMP-PCP,[1] a nonphosphorylating analogue of GTP, mimicked the actions of the natural nucleotides on glucagon binding, suggesting that the nucleotides act by binding to sites, as yet undefined, that influence the structure of the glucagon binding sites. ATP, ADP, UTP, and CTP similarly affected binding of glucagon but only at concentrations above 0.1 mM.

It was of obvious interest to determine whether the actions of guanyl nucleotides on the glucagon binding sites, which appear to be related to the adenyl cyclase system (2), have their correlate on the response of adenyl cyclase to glucagon. Studies reported elsewhere (3) in preliminary form, showed that guanyl nucleotides enhanced the response of adenyl cyclase to glucagon but only at concentrations considerably higher than those required for their actions on binding of glucagon. The preliminary studies were carried out with concentrations of ATP, the substrate for adenyl cyclase, that affected glucagon binding in the same manner as the guanyl nucleotides. Reduction of ATP concentration to low levels (0.2 mM or less) resulted in rapid hydrolysis of the nucleotide even in the presence of ATP-regenerating systems, creating difficulties in interpretation of kinetic data.

In this study, the enzymatic synthesis of ^{32}P-labeled AMP-PNP is described. This analogue of ATP contains nitrogen in place of oxygen between the terminal phosphates and was found to be resistant to hydrolysis by ATPases in liver membranes. It will be shown that it is a substrate for adenyl cyclase. It will also be shown that guanyl nucleotides, at concentrations as low as 10 nM, play an obligatory role in the activation of adenyl cyclase by glucagon but inhibit, possibly by a related mechanism, the response of the enzyme to fluoride ion.

EXPERIMENTAL PROCEDURE

Only those materials and methods not described in the pre-

[1] The abbreviations used are: GMP-PCP, 5′-guanylyl-diphosphonate; AMP-PNP, 5′-adenylyl-imidodiphosphate; PNP, diphosphoimide; ADP-NH_2, 5′-adenylyl-phosphoamide; cyclic AMP, cyclic-3′,5′-AMP.

ceding papers (1, 2, 4, 5) are documented in the present communication.

Materials—AMP-PNP(Na_4) and PNP(Na_4) were kindly supplied by Dr. Ralph Yount (Washington State University, Pullman). AMP-PNP was contaminated with 15% ADP-NH_2.

Escherichia coli extract was prepared from *E. coli* (strain B) according to the procedure of Nirenberg (6) and represents the 105,000 × *g* supernatant (S-100 fraction) obtained in Step 4 of this procedure.

Preparation and Purification of AMP-PNP-α-^{32}P: Zamecnik and Stephenson have shown (7) that methylene diphosphonate is incorporated into ATP by purified *E. coli* lysyl-tRNA synthetase (EC 6.1.1.60) according to the following reactions:

$$\text{ATP} + \text{Lys} + \text{enzyme}_{\text{Lys}} \rightleftarrows (\text{AMP}\cdot\text{Lys}\cdot\text{enzyme}_{\text{Lys}}) + \text{PP}$$

$$\text{PCP} + (\text{AMP}\cdot\text{Lys}\cdot\text{enzyme}_{\text{Lys}}) \rightarrow \text{AMP-PCP} + \text{enzyme}_{\text{Lys}} + \text{Lys}$$

which in the presence of excess PCP essentially go to completion, particularly if pyrophosphatase is present during incubation. We were informed by Dr. Zamecnik that AMP-PNP can be prepared by the same reaction. Instead of *E. coli* lysyl-tRNA synthetase we used a crude *E. coli* extract containing a number of amino acid tRNA synthetases and which is rich in pyrophosphatase activity. A mixture of 19 naturally occurring amino acids (except leucine), each at 1 mM, was used as substrates for the reaction. The following procedure was used to obtain AMP-PNP-α-^{32}P of high specific activity. Solutions containing 0.15 to 0.3 μmole of ATP-α-^{32}P (3 to 5 Ci per mmole) were evaporated to dryness at room temperature under a stream of nitrogen. The following solutions were added to the residue: 15 μl of 1 M Tris-HCl, pH 7.6, 15 μl of 0.1 M $MgCl_2$, 30 μl of amino acid mixture (see above), 20 μl of 50 mM PNP, and 40 μl of *E. coli* extract. The reaction mixture was incubated for 1 hour at 30°, followed by the addition of 10 μl of a suspension of partially purified rat liver plasma membranes (4) containing 10 mg of membrane protein per ml and further incubation for 15 min at 30°. The latter step served to degrade residual labeled ATP by nucleotidases present in liver membranes (8); AMP-PNP is only slightly hydrolyzed in the presence of liver membranes (see "Results").

All subsequent procedures were carried out at 5°. The above reaction mixture was applied directly to a column (0.4 × 4.0 cm) of DEAE-cellulose (Whatman DE-52) that had been previously washed with 10 ml of 2 M ammonium formate and then washed with 20 ml of distilled water. The column was eluted with a linear gradient formed from 15 ml of water and 15 ml of 0.5 M formate, pH 7.4, at a flow rate of 0.5 ml per min; 0.5-ml fractions were collected. Labeled AMP-PNP appeared in Fractions 42 to 47 which were pooled and treated as follows to remove salt. Washed analytical grade Dowex 50 (H^+ form), 0.5 g (wet weight), was added to the pooled fractions. After stirring for 2 min in an ice bath, the mixture was filtered, the residue was washed twice with 10 ml of cold water, and the combined filtrates were lyophilized. Labeled AMP-PNP was dissolved in 0.5 ml of 25 mM Tris-HCl, pH 7.6, and stored at −10°. Radio purity of AMP-PNP-α-^{32}P was determined by thin layer chromatography (described below) and ranged from 96 to 98%; the major contaminant is ADP-NH_2. Based on the amount of radioactive ATP added to the incubation mixture, yields of labeled AMP-PNP ranged from 60 to 75%.

Separation of Nucleotides by Thin Layer Chromatography—Determinations of changes in concentration of labeled ATP or AMP-PNP during incubation with liver membranes were carried out by thin layer chromatography (ascending) on precoated, aluminum backed sheets of PEI-cellulose F (Brinkmann 6820310-4). A solution of 0.5 N LiCl-2 M formic acid was used as developing solvent. With this solvent system the R_F value for AMP-PNP is 0.68; ATP, 0.51; ADP-NH_2, 0.78. ADP does not separate from AMP-PNP in this system but readily separates on PEI-cellulose chromatography using 1 M Tris-HCl, pH 7.6. Labeled nucleotides were cochromatographed with unlabeled nucleotides which were detected under ultraviolet light. The areas containing the nucleotides were cut out and placed in counting vials containing 10 ml of Bray's scintillation fluid (9). Radioactivity was determined in a liquid scintillation counter.

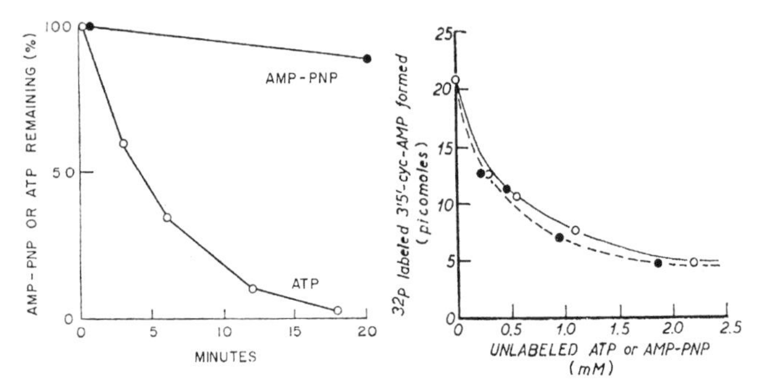

FIG. 1 (*left*). Effect of incubation of liver plasma membranes, under adenyl cyclase assay conditions, on levels of ATP and AMP-PNP. *a*, liver plasma membranes (0.7 mg of protein per ml) were incubated at 30° for varying periods of time in medium containing 0.2 mM ATP-α-^{32}P (140 cpm per pmole), 5 mM $MgCl_2$, 25 mM Tris-HCl, pH 7.6, and an ATP-regenerating system consisting of 20 mM creatine phosphate and 1 mg per ml of creatine kinase. *b*, liver plasma membranes (0.4 mg of protein per ml) were incubated at 30° for 20 min in medium containing 0.25 mM AMP-PNP-α-^{32}P (205 cpm per pmole), 4.0 mM $MgCl_2$, 1 mM EDTA, 1 mM cyclic AMP, 0.2% albumin, and 25 mM Tris-HCl, pH 7.6. Reactions were terminated by addition of 1 volume of 4 M formic acid. Zero times were prepared by addition of liver plasma membranes after formic acid. The percentage of ATP or AMP-PNP remaining was determined by chromatography on thin layer sheets of PEI-cellulose F (see "Experimental Procedures").

FIG. 2 (*right*). Effect of addition of unlabeled ATP or AMP-PNP on formation of cyclic AMP-^{32}P from ATP-α-^{32}P. Liver plasma membranes (0.4 mg of protein per ml) were incubated for 10 min at 30° in 0.05 ml of medium containing 0.53 mM ATP-α-^{32}P (81 cpm per pmole), 5.0 mM $MgCl_2$, 12 μM glucagon, 1.0 mM EDTA, 25 mM Tris-HCl, pH 7.6, an ATP-regenerating system consisting of 20 mM creatine phosphate and 1 mg per ml of creatine kinase, and the indicated concentrations of unlabeled ATP (○——○) or AMP-PNP (●- - -●). The reactions were terminated and the cyclic AMP-^{32}P formed was determined as described in Table I.

RESULTS

In previous studies (4), measurements of adenyl cyclase activity in rat liver plasma membranes were carried out with 3.2 mM ATP as substrate. At this concentration, ATP affects binding of glucagon to about the same extent as 1.0 μM GTP (1). Reduction of ATP concentration to 0.2 mM as a means of minimizing its effects on glucagon binding resulted in rapid hydrolysis of the nucleotide despite the presence of an ATP-regenerating system, as illustrated in Fig. 1. AMP-PNP-α-^{32}P, at 0.2 mM, was hydrolyzed by liver membranes to a slight extent (about 15%) during 20 min of incubation with liver membranes in the absence of an ATP-regenerating system. The slight hydrolysis observed

TABLE I

Effect of glucagon and fluoride ion in absence and in presence of GTP on formation of cyclic AMP from AMP-PNP

Liver membranes (0.4 mg of protein per ml) were incubated for 10 min at 30° in medium containing 0.2 mM AMP-PNP-α-^{32}P (193 cpm per pmole), 4.0 mM $MgCl_2$, 1.0 mM EDTA, 1.0 mM cyclic AMP, 0.2% albumin, 25 mM Tris-HCl, pH 7.6, and the indicated additions. Final volume was 0.05 ml. The reactions were terminated by the addition of 0.10 ml of solution containing 40 mM ATP, 12.5 mM cyclic AMP-^{3}H (approximately 30,000 cpm), and 1% sodium dodecyl sulfate, followed by immediate boiling for 3½ min. Cyclic AMP-^{32}P formed was isolated according to Krishna, Weiss, and Brodie (11) as described previously (12, 13).

Addition	Cyclic AMP formed		
	Control	5 μM glucagon	10 mM NaF
	pmoles/mg protein		
None	125	248	890
GTP (0.2 mM)	207	835	565

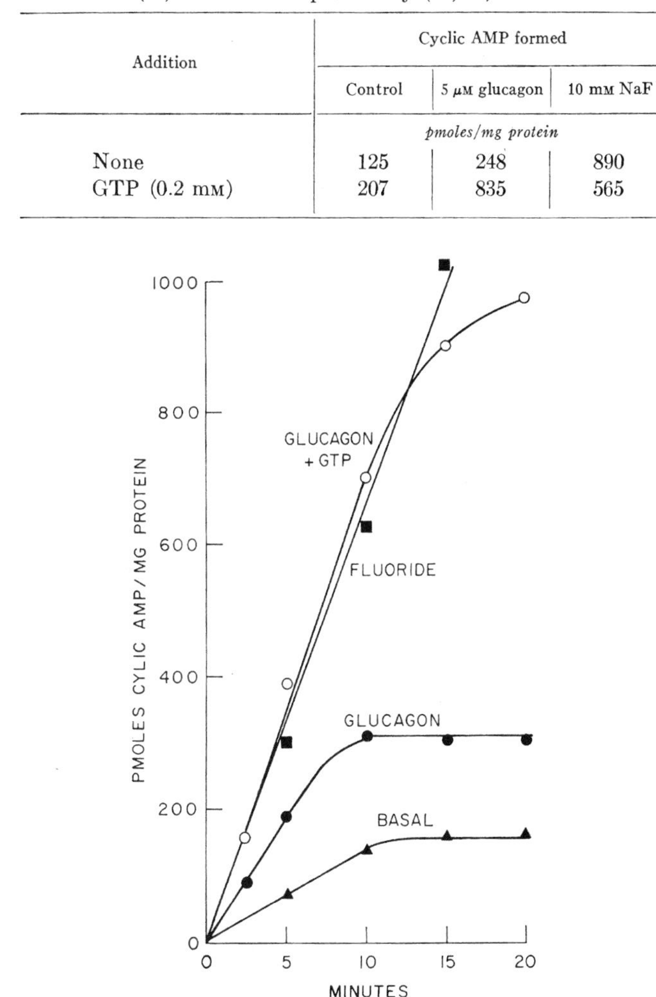

FIG. 3. Effect of fluoride ion, glucagon, and combination of glucagon and GTP on formation of cyclic AMP from AMP-PNP. Liver plasma membranes (0.38 mg of protein per ml) were incubated at 30° in 0.05 ml of medium containing 0.2 mM AMP-PNP-α-^{32}P (190 cpm per pmole), 4.0 mM $MgCl_2$, 1 mM EDTA, 1 mM cyclic AMP, 0.2% albumin, 25 mM Tris-HCl, pH 7.6, and either 10 mM NaF plus 5 μM glucagon or 5 μM glucagon plus 0.01 mM GTP. The reactions were terminated and the cyclic AMP-^{32}P formed was determined as described in Table I.

may have been due to nucleotide triphosphate pyrophosphohydrolases reported in liver membranes (10).

Addition of varying concentrations of AMP-PNP and ATP to the assay medium containing ATP-α-^{32}P (0.5 mM) resulted in proportional reduction in the amount of labeled cyclic 3′,5′-AMP formed by adenyl cyclase (Fig. 2). These findings suggested

TABLE II

Specificity of action of nucleotides on glucagon-stimulated adenyl cyclase activity

Liver plasma membranes (0.36 mg of protein per ml) were incubated in 0.05 ml of medium containing 0.24 mM AMP-PNP-α-^{32}P (230 cpm per pmole), 2.5 mM $MgCl_2$, 1 mM EDTA, 1 mM cyclic AMP, 0.2% albumin, 25 mM Tris-HCl, pH 7.6, 5 μM glucagon, and the indicated additions. Incubations were for 10 min at 30°. The reactions were terminated and cyclic AMP-^{32}P formed was determined as described in Table I.

Addition	Cyclic AMP formed
	pmoles/mg protein
None	21
GTP (0.01 mM)	235
GDP (0.01 mM)	237
GMP-PCP (0.1 mM)	201
UTP (0.1 mM)	75
CTP (0.1 mM)	53

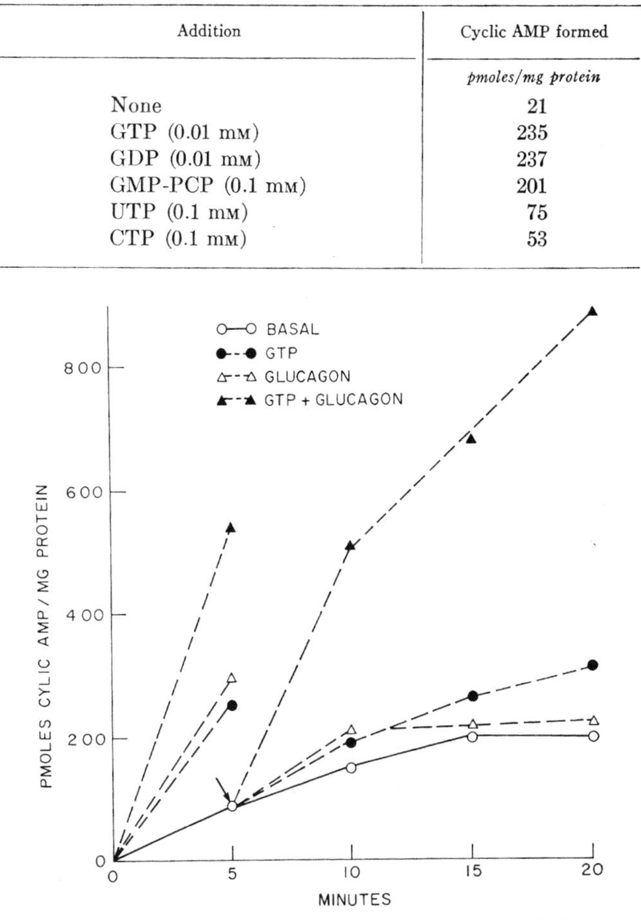

FIG. 4. Effect of time of addition of glucagon and GTP on formation of cyclic AMP from AMP-PNP. Liver plasma membranes (0.4 mg of protein per ml) were incubated at 30° in 0.05 ml of medium containing 0.24 mM AMP-PNP-α-^{32}P (180 cpm per pmole), 4.0 mM $MgCl_2$, 1 mM EDTA, 1 mM cyclic AMP, 0.2% albumin, and 25 mM Tris-HCl, pH 7.6. The following additions were made at either zero time or after 5 min of incubation: 5 μM glucagon, 0.01 mM GTP, and 5 μM glucagon plus 0.01 mM GTP. The reactions were terminated at the indicated times and the cyclic AMP-^{32}P formed was determined as described in Table I.

that ATP and AMP-PNP behaved identically as substrate for adenyl cyclase.

Incubation of AMP-PNP-α-^{32}P with liver membranes resulted in the formation of labeled cyclic 3′,5′-AMP which was stimulated by glucagon and fluoride ion (Table I). This provided direct evidence that AMP-PNP serves as substrate for adenyl cyclase. It will be noted also that GTP (0.2 mM) stimulated basal activity by 65% and caused a 3.5-fold stimulation of the response of the enzyme system to glucagon. The guanyl nucleotide inhibited the fluoride response by 37%.

Glucagon (5.0 μM) and fluoride ion (10 mM) stimulated the

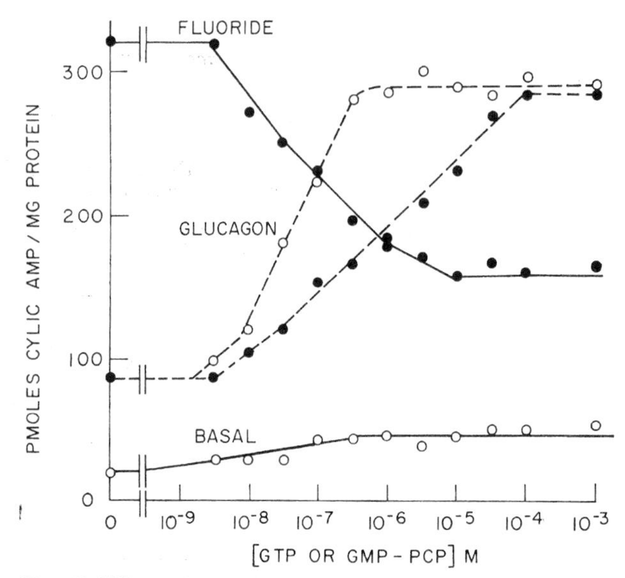

FIG. 5. Effect of varying concentrations of GTP and GMP-PCP on adenyl cyclase activity determined in the absence and in the presence of glucagon and fluoride ion. Liver membranes were incubated in the presence of 5 μM glucagon (*broken lines*) with GTP (*open circles*) or GMP-PCP (*closed circles*). Adenyl cyclase activity in response to NaF (10 mM) was investigated only in the presence of GMP-PCP. Basal activity was studied in the presence of GTP or GMP-PCP, both of which gave identical activities. Other incubation conditions consisted of 0.25 mM AMP-PNP-α-^{32}P, 2.0 mM $MgCl_2$, 1 mM EDTA, 1 mM cyclic AMP, 0.2% albumin, and 25 mM Tris-HCl, pH 7.6.

production of labeled cyclic 3′,5′-AMP at a linear rate for at least 10 min at 30° (Fig. 3); GTP enhanced the initial rate of glucagon-stimulated activity.

GTP or GDP, equally at 100 μM, stimulated the response of adenyl cyclase to glucagon by at least 10-fold when the Mg^{++} concentration was reduced to 2.5 mM. This is shown in Table II (*cf.* Table I where the Mg^{++} concentration was 4.0 mM). The role of magnesium ion in the process of activation of the enzyme by glucagon and guanyl nucleotides is under current investigation. We have shown in a previous study (4) that the Mg^{++} concentration is critical for the response of the enzyme to glucagon. UTP and CTP, at 100 μM exerted smaller stimulatory effects on the response of adenyl cyclase to glucagon.

As shown in Fig. 4, GTP (or not shown, the other guanyl nucleotides) also stimulated basal adenyl cyclase activity (absence of glucagon or fluoride ion) in the presence of 4.0 mM Mg^{++}; under these conditions basal activity in the presence of GTP was about equal to that observed with glucagon (5 μM) alone. As shown, incubation of the membranes for 5 min at 30° prior to addition of GTP or glucagon resulted in a marked decrease in the effect of GTP on basal activity and in essentially no effect of glucagon alone; addition of GTP was required to obtain a glucagon response. The loss of response of adenyl cyclase to glucagon after the 5-min incubation period was not due to a change in binding of glucagon; uptake of ^{125}I-glucagon (4.5 nM) was found to be identical before and after incubation of the membranes under the conditions described in Fig. 4.

The basis of the loss of response of adenyl cyclase to glucagon or the diminished effect of GTP on basal activity after incubation is not known but may be related, respectively, to destruction of endogenous, membrane-bound guanyl nucleotides by nucleotidases and of membrane-bound glucagon by glucagon-inactivating process or processes; both destructive activities are present in liver membranes (2).

In the previous study (1), it was shown that GTP and GDP alter binding and dissociation of bound glucagon over a concentration range of 0.05 μM to 50 μM; GMP-PCP also altered these processes but at higher concentrations. Maximal effects of the analogue were observed at 100 μM. Investigations of the response of the adenyl cyclase system to glucagon, illustrated in Fig. 5, revealed that guanyl nucleotides exert their effects on this system at somewhat lower concentrations than were observed in the binding studies. Significant effects of GTP and GMP-PCP were observed at 0.01 μM and 0.05 μM, respectively. As was observed in the binding studies, maximal effects of GMP-PCP on glucagon response were seen at 100 μM.

Since GTP inhibits the response of the enzyme system to fluoride ion (Table I), it was of interest to determine the effects of GMP-PCP on this process. As shown in Fig. 5, the analogue of GTP inhibited the fluoride response at concentrations similar to those at which it stimulated the response of the enzyme to glucagon.

Treatment of liver membranes with phospholipase A results in complete loss of glucagon binding or activation of adenyl cyclase but in retention of the fluoride response (2, 5). Under conditions in which the glucagon response of adenyl cyclase was abolished and the fluoride response was reduced by 50%, phospholipase A treatment did not alter the inhibitory effects of GMP-PCP on the response of adenyl cyclase to fluoride ion (Table III).

TABLE III

Effect of treatment of liver plasma membranes with phospholipase A on inhibitory action of GMP-PCP on fluoride-stimulated adenyl cyclase activity

Liver plasma membranes (1.2 mg of protein per ml) were incubated in the absence and presence of 150 units per ml of phospholipase A in medium containing 1 mM $CaCl_2$ and 25 mM Tris-HCl, pH 7.6. Incubation was for 5 min at 30°. Treatment was terminated by addition of 0.1 volume of 25 mM EDTA and immediate cooling to 0°. Adenyl cyclase activity was subsequently determined on 10-μl aliquots in the absence or in the presence of 0.1 mM GMP-PCP in 0.05 ml of medium containing 0.24 mM AMP-PNP-α-^{32}P (195 cpm per pmole), 2.5 mM $MgCl_2$, 1 mM EDTA, 1 mM cyclic AMP, 0.2% albumin, 25 mM Tris-HCl, pH 7.6, and 10 mM NaF. Incubations were for 10 min at 30°. The reactions were terminated and cyclic AMP-^{32}P formed was determined as described in Table I.

Addition during treatment	Cyclic AMP formed		Inhibition due to GMP-PCP
	Control	GMP-PCP	
	pmoles/mg protein		%
None	173	113	35
Phospholipase A	86	58	33

Studies were also carried out to determine whether guanyl nucleotides alter the concentration of glucagon required for half-maximal response of adenyl cyclase, which was previously shown to be 0.004 μM (4). In the presence of 0.1 mM GTP, adenyl cyclase activity in response to 0.005 μM glucagon was 66% of that given by glucagon at 5 μM, indicating that GTP does not influence the apparent affinity of adenyl cyclase system for the hormone. It can be concluded from these findings that guanyl nucleotides and glucagon interact with the adenyl cyclase system in a noncompetitive fashion.

DISCUSSION

The purpose of the present study was to ascertain the effects of guanyl nucleotides on the response of adenyl cyclase to glucagon. The use of AMP-PNP as substrate for adenyl cyclase facilitated this study. Compared to ATP, AMP-PNP was only slowly hydrolyzed during incubation with liver membranes, thus circumventing the usual problems encountered in maintaining substrate concentration during kinetic analysis of adenyl cyclase activity. For this reason, addition of ATP-regenerating systems, which may complicate analysis because of their possible influence on the adenyl cyclase system, can be avoided. Initial rates of adenyl cyclase activity were maintained for at least 10 min; changes in rate that occurred can be attributed to factors other than changes in substrate concentration. Indeed, the main advantages of using AMP-PNP in this study was that effects of nucleotides on adenyl cyclase activity could be assessed, at low concentrations of substrate, with certainty that the effects were not due to changes in substrate concentration.

A major finding was that guanyl nucleotides (GTP, GDP, and GMP-PCP) stimulated, by an immediate action, the response of adenyl cyclase to glucagon. Hormonal response was dependent upon the presence of guanyl nucleotides either when the magnesium ion concentration was lowered from 4.0 to 2.0 mM or after incubation of the liver membranes. The basis for these conditions leading to dependence of hormone response on guanyl nucleotides is under current investigation. An important point is that the response of adenyl cyclase to glucagon required as little as 10 nM GTP. Only the guanyl nucleotides stimulated the hormonal response of adenyl cyclase at concentrations less than 0.1 mM, indicating that the hormone-activation process is relatively specific for guanyl nucleotides. The low concentrations required, the equivalent effects of GTP and GDP, and the similar action of GMP-PCP, a nonphosphorylating analogue of GTP, on the response of adenyl cyclase to glucagon suggest that the guanyl nucleotides regulate this process through binding but not through phosphorylation.

The present studies were initiated because of the previous findings (1) that guanyl nucleotides affected the binding sites for glucagon in a manner that results in enhanced dissociation of bound hormone, and in decreases in both the apparent affinity of the binding sites for glucagon and in the amount of glucagon bound. The guanyl nucleotides stimulated release of the total quantity of glucagon bound, indicating that all of the binding sites were affected by nucleotides. It was also shown that the binding sites were specific for glucagon and that these sites have the same apparent affinity for glucagon as does the process involved in activation of adenyl cyclase (2). It was also found that inhibition of binding by such agents as urea, detergents, and phospholipase A leads to inhibition of activation of adenyl cyclase by glucagon (2). Taken *in toto*, the previous studies indicate that all of the binding sites share characteristics with the initial process through which glucagon activates adenyl cyclase and can be appropriately termed "discriminator," as defined previously to be that material which is involved in the primary action of glucagon (2). It remains unknown how binding of glucagon to its discriminator, which appears to be a separate entity from adenyl cyclase (5), is potentially translated into activation of adenyl cyclase.

The findings that guanyl nucleotides alter both the binding of glucagon and the response of the enzyme to the hormone offer another means of relating binding to the hormone-activation process. There appears to be some relationship between the effects of guanyl nucleotides on glucagon binding and hormone activation as evidenced by the following correlations. (*a*) Both processes are relatively specific for guanyl nucleotides; UTP, CTP, and ATP act only at concentrations greater than 0.1 mM; (*b*) GTP and GDP are equally effective, GMP-PCP is relatively less effective on hormone binding and activation; and (*c*) the nucleotides exert their effects on both processes independently of the concentration of glucagon in the medium. Such correlates do not establish how these effects of the guanyl nucleotides are interrelated, but indicate that both processes are affected specifically by guanyl nucleotides through sites that are independent of the sites reacting with glucagon. Since, under appropriate incubation conditions, both glucagon and guanyl nucleotides are required for activation of adenyl cyclase, it appears that there are two regulatory sites, requiring the binding, respectively, of glucagon and guanyl nucleotides for activation of the enzyme.

In a previous study (5), it was shown that glucagon and fluoride ion activate adenyl cyclase in liver membranes through processes that have different characteristics, indicating that fluoride ion does not operate through the discriminator. It was of interest, therefore, to find that guanyl nucleotides, specifically and at low concentrations, inhibited the response of adenyl cyclase to fluoride ion. GMP-PCP inhibited the fluoride response over the same range of concentrations as it stimulated the response of adenyl cyclase to glucagon. Furthermore, GMP-PCP inhibited the fluoride response in liver membranes treated with phospholipase A under conditions that abolished both glucagon binding and activation of adenyl cyclase by the hormone. The specificity of action of guanyl nucleotides and the similar concentration ranges over which the nucleotides alter glucagon binding, hormonal activation of adenyl cyclase, and inhibition of fluoride response suggest that the effects are related, possibly by a common action. However, the site and mode of action of guanyl nucleotides on this complex adenyl cyclase system remains obscure.

These studies were initiated on the premise, widely held, that hormone receptors receive and transmit information imparted by the hormone to its target cell depending only upon the circulating levels of the hormone. This premise seems untenable for glucagon in view of the finding that guanyl nucleotides play an obligatory role in regulating the response of liver adenyl cyclase to glucagon. Further studies of the actions of the guanyl nucleotides on the liver and other adenyl cyclase systems may provide new insights into not only how hormones regulate target cell metabolism at the receptor level but also how a target cell metabolite regulates the initial response to a hormone.

REFERENCES

1. Rodbell, M., Krans, H. M. J., Pohl, S. L., and Birnbaumer, L., *J. Biol. Chem.*, **246**, 1861 (1971).
2. Rodbell, M., Krans, H. M. J., Pohl, S. L., and Birnbaumer, L., *J. Biol. Chem.*, **246**, 1872 (1971).
3. Rodbell, M., Birnbaumer, L., Pohl, S. L., and Krans, H. M. J., *Acta Diabetol. Lat.*, **7** (Suppl. 1), 9 (1970).
4. Pohl, S. L., Birnbaumer, L., and Rodbell, M., *J. Biol. Chem.*, **246**, 1849 (1971).
5. Birnbaumer, L., Pohl, S. L., and Rodbell, M., *J. Biol. Chem.*, **246**, 1857 (1971).
6. Nirenberg, M. W., in S. P. Colowick and N. O. Kaplan (Editors), *Methods in enzymology, Vol. VI*, Academic Press, New York, 1963, p. 17.
7. Zamecnik, P. C., and Stephenson, M. L., in H. M. Kalckar, H. Kenow, A. Munch-Peterson, M. Ottesen, and Y. H.

TAYSEN (Editors), *The role of nucleotides for the function and conformation of enzymes*, Academic Press, New York, 1969, p. 276.
8. BENEDETTI, E. L., AND EMMELOT, P., in A. J. DALTON AND F. HAGAMAN (Editors), *Ultrastructure in biological systems, Vol. 4*, Academic Press, New York, 1968, p. 33.
9. BRAY, G. A., *Anal. Biochem.*, **1**, 279 (1960).
10. LIEBERMAN, I., LANSING, A. I., AND LYNCH, W. E., *J. Biol. Chem.*, **242**, 736 (1967).
11. KRISHNA, G., WEISS, B., AND BRODIE, B. B., *J. Pharmacol. Exp. Ther.*, **163**, 379 (1968).
12. RODBELL, M., *J. Biol. Chem.*, **242**, 5744 (1967).
13. KRISHNA, G., AND BIRNBAUMER, L., *Anal. Biochem.*, **35**, 393 (1970).

Northup et al. (1980) Proc. Natl. Acad. Sci. U.S.A. 77, 6516–6520

Purification of the regulatory component of adenylate cyclase

(adenosine 3′,5′-cyclic monophosphate/hormone receptors/reconstitution/guanine nucleotide)

JOHN K. NORTHUP, PAUL C. STERNWEIS, MURRAY D. SMIGEL, LEONARD S. SCHLEIFER, ELLIOTT M. ROSS, AND ALFRED G. GILMAN

Department of Pharmacology, University of Virginia School of Medicine, Charlottesville, Virginia 22908

Communicated by Theodore T. Puck, August 8, 1980

ABSTRACT The regulatory component (G/F) of adenylate cyclase [ATP pyrophosphate-lyase (cyclizing), EC 4.6.1.1.] from rabbit liver plasma membranes has been purified essentially to homogeneity. The purification was accomplished by three chromatographic procedures in sodium cholate-containing solutions, followed by three steps in Lubrol-containing solutions. The specific activity of G/F was enriched 2000-fold from extracts of membranes to 3–4 μmol·min^{-1}·mg^{-1} (reconstituted adenylate cyclase activity). Purified G/F reconstitutes guanine nucleotide-, fluoride-, and hormone-stimulated adenylate cyclase activity in the adenylate cyclase-deficient variant of S49 murine lymphoma cells. G/F also recouples hormonal stimulation of the enzyme in the uncoupled variant of S49. Preparations of pure G/F contain three polypeptides with approximate molecular weights of 52,000, 45,000, and 35,000. The active G/F protein behaves as a multisubunit complex of these polypeptides. Treatment of G/F with [^{32}P]NAD^{+} and cholera toxin covalently labels the molecular weight 52,000 and 45,000 polypeptides with ^{32}P.

The purification of adenylate cyclase [ATP pyrophosphate-lyase (cyclizing), EC 4.6.1.1] has been an elusive goal for more than 20 years since the description of this important enzyme (1). Although the activity can readily be extracted from plasma membranes with nonionic detergents, attempts to fractionate these preparations have accomplished little. Either the enzyme has behaved as a polydisperse species or its activity has been lost.

Insight toward resolution of this problem was provided by experiments that demonstrated the multicomponent nature of adenylate cyclase. Pfeuffer (2) demonstrated that a guanine nucleotide binding protein could be partially resolved from the putative catalytic subunit of adenylate cyclase by affinity chromatography with GTP-Sepharose. Such chromatography resulted in a partial loss of guanine nucleotide- and fluoride-stimulated enzymatic activity. These activities could be restored by the combination of material that flowed through the column with fractions that were eluted by the addition of a guanine nucleotide.

Study of the requirements for reconstitution of adenylate cyclase activity in genetic variants of the murine S49 lymphoma cell line then indicated that adenylate cyclase consists of a labile catalytic moiety (C) and a relatively stable regulatory component (G/F), which confers upon the catalyst the ability to utilize its physiological substrate, MgATP, and the ability to be activated by fluoride and guanine nucleotides (3–5). The activity of G/F is absent in the adenylate cyclase-deficient cyc^{-} variant of S49, and reconstitution of adenylate cyclase activity in cyc^{-} membranes thus becomes a method for the assay of the regulatory component. Because G/F is considerably more stable than the catalytic moiety of adenylate cyclase, purification of the regulatory component was initiated.

METHODS AND MATERIALS

Assays. The regulatory component of adenylate cyclase was usually measured by its ability to reconstitute fluoride-stimulated MgATP-dependent adenylate cyclase activity in membranes of the cyc^{-} S49 murine lymphoma cell (5, 6). Specific activities of G/F were determined under conditions in which reconstituted activity was a linear function of the quantity of protein (G/F) added (see *Results*); the specific activity is thus defined as the amount of reconstituted adenylate cyclase activity observed per amount of G/F added to the reconstitution. Because the catalytic subunit is kept in excess, the specific activity of the cyc^{-} membranes used has relatively little impact on the value of the specific activity calculated for G/F. Generally, 15 μl of G/F in a buffered solution containing 0.8% cholate or 0.1% Lubrol was mixed with 25 μl of cyc^{-} membranes (1.5–2 mg/ml) and then with 20 μl of 150 mM NaHepes, pH 8/15 mM $MgCl_2$/1.25 mM ATP/150 μM GTP/30 mM NaF/30 μg of pyruvate kinase per ml/9 mM potassium phospho*enol*pyruvate/0.3 mg of bovine serum albumin per ml. After incubation at 30°C for 10 min, 40 μl of a mixture containing 0.62 mM ATP, ≈10^6 cpm of [α-^{32}P]ATP, 125 mM NaHepes at pH 8, 70 mM $MgCl_2$, 25 μg of pyruvate kinase per ml, 7.5 mM potassium phospho*enol*pyruvate, 0.25 mM 4-(3-butoxy-4-methoxybenzyl)-2-imidazolidinone (RO20-1724), and 2.5 mM EDTA was added and the incubation was continued for 30 min at 30°C. Reactions were terminated and cyclic [^{32}P]AMP was isolated by the method of Salomon *et al.* (7).

Protein was quantified as described by Lowry *et al.* (8) or by staining with amido black, as described by Schaffner and Weissman (9).

Membrane Preparations. Plasma membranes of S49 cells were prepared as described (10).

Enriched fractions of plasma membranes from rabbit liver were prepared as follows. Approximately 500 g of frozen rabbit liver (Pel-Freez) was coarsely pulverized (hammered) and then thawed at room temperature in 1 liter of buffer A (10 mM Tris·HCl, pH 8/1 mM EDTA/0.1% 2-mercaptoacetic acid) plus 2 mM $MgCl_2$ and 10% (wt/vol) sucrose. All further operations were performed at 0–4°C. The thawed liver was ground in an electric meat grinder, diluted with an equal volume of buffer A with 2 mM $MgCl_2$, and homogenized for 30 sec in 800-ml portions with a Brinkmann Polytron model PT20 at a setting of 3. The homogenized mixture was further diluted to a total volume of 6 liters and centrifuged at 5000 rpm for 20 min in a Beckman JA10 rotor. The pellets were suspended in buffer A with 2 mM $MgCl_2$ with a Potter–Elvehjem homogenizer (approximate final volume 500 ml); the sucrose concentration was then adjusted to 47% by addition of a saturated solution of

The publication costs of this article were defrayed in part by page charge payment. This article must therefore be hereby marked "*advertisement*" in accordance with 18 U. S. C. §1734 solely to indicate this fact.

Abbreviations: G/F, regulatory component of adenylate cyclase; cyc^{-}, adenylate cyclase-deficient; UNC, uncoupled; TED, Tris/EDTA/dithiothreitol; AMF, ATP/$MgCl_2$/NaF; GTP[γS], guanosine 5′-[γ-thio]triphosphate.

sucrose in buffer A with 2 mM $MgCl_2$. Portions (100 ml) of this suspension were poured into 250-ml bottles, overlaid with 80 ml of 43% sucrose solution in buffer A, and centrifuged at 14,000 rpm for 120 min in a Beckman JA14 rotor. The membranes floating over the layer of 43% sucrose were collected and homogenized, diluted to approximately 2.5 liters with buffer A, and sedimented at 14,000 rpm for 60 min in a JA14 rotor. The pellets from this and subsequent centrifugations were composed of three distinct layers. An upper pink layer was decanted with the supernatant after it was loosened with a spatula. The light-brown middle layer, which contains a higher proportion of plasma membranes, was then separated from the dark bottom pellet by agitation, and the former was suspended in buffer A. This procedure was repeated twice additionally. The final pellet (2 g of protein) was homogenized in 50 ml of buffer A and frozen quickly at −80°C.

Purification Procedures. Purification of G/F was performed in a common buffer solution (TED), consisting of 20 mM Tris·HCl at pH 8.0, 1 mM EDTA, and 1 mM dithiothreitol. In several steps, solutions also contained 1 mM ATP, 6 mM $MgCl_2$, and 10 mM NaF (AMF), a condition that stabilizes G/F activity and activates the protein* (11). All steps were carried out at 4°C unless otherwise stated.

Frozen hepatic membranes (about 10 g of protein) were thawed and washed with 2 liters of TED/500 mM NaCl prior to extraction. After collection by centrifugation, the membranes were suspended to a final volume of 1 liter of TED containing 50 mM NaCl and 1.0% sodium cholate and extracted for 60 min by stirring at 0°C. Membranes were then removed by sedimentation for 75 min at 35,000 rpm in a Beckman 35 rotor. The clear supernatant was removed carefully from beneath a turbid floating layer and was diluted with 2 vol of TED/AMF/0.9% cholate. The diluted supernatant (2.5 liters) was applied to a column of DEAE-Sephacel (Pharmacia) (5 × 60 cm), which had been equilibrated with 3 liters of TED/AMF/0.9% sodium cholate. After application of the extract, the column was eluted with a 2-liter linear gradient of NaCl (0–250 mM) in TED/AMF/0.9% sodium cholate. G/F eluted as a symmetrical peak in the middle of the gradient. Peak fractions were pooled and concentrated to approximately 30 ml by filtration with an Amicon PM-30 membrane in a stirred cell.

The concentrated pool was next fractionated on a column (5 × 60 cm) of Ultrogel AcA 34 (LKB), which was equilibrated and eluted with TED/AMF/0.9% sodium cholate/100 mM NaCl. A peak of G/F was obtained at a K_d of approximately 0.45. Fractions with peak activity (50–70 ml) were pooled and diluted with 1.5 vol of TED/AMF/100 mM NaCl.

The diluted Ultrogel AcA 34 pool was applied to a 50-ml column (1.5 × 27 cm) of heptylamine-Sepharose, which had been equilibrated with TED/AMF/0.4% sodium cholate/100 mM NaCl. The column was washed successively with 50 ml of TED/AMF/0.4% sodium cholate/100 mM NaCl and with 50 ml of TED/AMF/0.3% sodium cholate/500 mM NaCl. The gel was then eluted with a 250-ml linear gradient of TED/AMF/0.3% sodium cholate/250 mM NaCl and TED/AMF/1.3% sodium cholate/100 mM NaCl. The increasing concentration of cholate eluted a peak of G/F in the middle of the gradient. At this stage and beyond, fractions containing G/F were collected in silicone-treated tubes (Siliclad).

* As previously shown (11), G/F may be activated in the absence of the catalytic subunit by incubation with ATP, Mg^{2+}, and F^- or with certain guanine nucleotide analogs. cyc^- membranes reconstituted with G/F so treated require no activators in the assay to express fully stimulated adenylate cyclase activity. As opposed to membrane-bound adenylate cyclase, G/F activated in solution reverses (half-time of minutes at 20°C) to the basal state (deactivates) upon removal of fluoride.

Peak fractions from the heptylamine-Sepharose column were pooled and applied directly to a 5-ml column (0.9 × 8 cm) of hydroxyapatite (Bio-Rad HTP), which had been equilibrated in TED/0.1% Lubrol/100 mM NaCl. During hydroxyapatite chromatography, the concentration of EDTA in TED was reduced to 0.1 mM. After application of the protein, the column was washed with 20 ml of TED/0.1% Lubrol/100 mM NaCl. In general, the sample was allowed to remain bound to the column overnight after the initial wash with the Lubrol-containing buffer. Minimally, several hours were allowed to elapse before elution to allow deactivation of G/F after removal of AMF (11). The column was then washed with 10 ml of TED/0.1% Lubrol/100 mM NaCl/30 mM potassium phosphate, followed by 20 ml of TED/0.1% Lubrol/100 mM NaCl/300 mM potassium phosphate. Most of the G/F activity eluted with the 300 mM phosphate wash; this fraction was desalted by passage through a column of Sephadex G-25.

Protein was next bound to 1 ml of GTP-Sepharose by incubation in the presence of TED/0.3% Lubrol/2.5 mM $MnCl_2$ for 30 min at 30°C. The gel was poured into a column and washed at room temperature with 7 ml of TED/0.5% Lubrol/2.5 mM $MnCl_2$ and 7 ml of TED/0.5% Lubrol/3.5 mM $MnCl_2$/1 mM GTP; G/F was eluted (by removal of free Mn^{2+}) with 10 ml of TED/0.5% Lubrol.

The peak fractions from the GTP-Sepharose column were activated by addition of AMF and applied to a column of DEAE-Sephacel (0.5 × 7 cm) that had been equilibrated with TED/AMF/0.5% Lubrol. After washing with 5 ml of TED/AMF/0.5% Lubrol, the gel was eluted with a linear gradient (40 ml) of NaCl from 50 to 300 mM in TED/AMF/0.5% Lubrol. Peak fractions of G/F activity were eluted at approximately 150 mM NaCl.

Materials. Heptylamine-Sepharose was prepared as described by Shaltiel (12). However, Sepharose CL-4B (Pharmacia) was used as the support, and only 1 mol of heptylamine was used per mol of cyanogen bromide. As a precaution, the derivatized gel was incubated with 1 M ethanolamine for 2 hr prior to storage in H_2O at 4°C.

GTP was immobilized on Sepharose CL-4B through its γ phosphate as described by Pfeuffer (2). A modification of the published procedure was the use of only 50% of the specified amount of γ-aminobutyric acid. The product contained 0.65 μmol of nucleotide per ml of packed gel.

[α-^{32}P]ATP (13) and [^{32}P]NAD (^{32}P in the α position of the ADP moiety) (14) were synthesized as described; sodium cholate was purified by chromatography on DEAE-cellulose as described (6). Lubrol 12A9 was obtained from Imperial Chemical Industries and was deionized with a mixture of Dowex 1 and 50. Cells were cultured and harvested as described (11).

RESULTS

Purification of G/F was the logical outgrowth of the ability to assay its activity by reconstitution of a functional adenylate cyclase complex. Because the reconstitution of G/F depends on its prior extraction from membranes with detergents, such extracts represent the first point at which the specific activity of the protein can be measured. The source of G/F chosen for purification was rabbit liver, because large quantities of partially purified plasma membranes can be obtained with moderate ease and the specific activity of G/F is comparable to or better than that observed from several other sources. Detergent extracts of membranes prepared as described under *Methods and Materials* exhibit a 5-fold greater specific activity of G/F than do those obtained from the total particulate fraction of a liver homogenate.

The purification of G/F is documented in Table 1. The

Table 1. Purification of the regulatory component (G/F) of adenylate cyclase from rabbit liver

Step	Protein, mg	Total units, nmol/min	Recovery, %	Specific activity, nmol·min^{-1}·mg^{-1}
Cholate extract of membranes	2020	4380	100	2.2
DEAE-Sephacel	114	2930	67	26
Ultrogel AcA 34	9.8	1820	42	190
Heptylamine-Sepharose	0.70	740	17	1060
Hydroxyapatite	0.30	350	8.0	1200
GTP-Sepharose	0.18	290	6.6	1600
DEAE-Sephacel	0.039	150	3.5	3800

Rabbit liver plasma membranes (8.6 g of protein) were extracted and purified. Activity was measured by the reconstitution of fluoride-activated adenylate cyclase activity in cyc$^-$ membranes. Specific activity is defined as the amount of reconstituted activity per amount of G/F added to the reconstitution. Recoveries are cumulative through the preparation.

method allows a 2000-fold purification of the activity in the extract, with a 3–4% yield. The six chromatographic steps employed can be divided into two stages. The first stage, consisting of the first three steps, has proven to be very reliable and consistently yields G/F with a specific activity greater than 1000 nmol·min^{-1}·mg^{-1}. The specific activity of the initial extract shown here is unusually high; a specific activity of 1–1.5 nmol·min^{-1}·mg^{-1} is usual and purification of 1000-fold is normally achieved with these three steps. At this stage of the purification, G/F is about 30% pure and the recovery of activity is reasonably high. Overall recovery of activity can be closer to 30% at this point without significant sacrifice of specific activity if a greater percentage of the heptylamine-Sepharose eluate is pooled (see *Discussion*). The last three steps of the purification scheme are inefficient but, to date, have been necessary to remove the specific contaminants that remain in the preparation. Because G/F at this stage of purification is unstable to further manipulation in cholate and does not bind to GTP-Sepharose in this detergent, further purification proceeds in Lubrol 12A9. Hydroxyapatite provides an excellent medium to effect this exchange of detergents, because neither cholate nor Lubrol binds significantly to the gel, whereas G/F binds quantitatively. If the protein is then allowed to deactivate on the gel by the removal of AMF, some purification (often 2-fold) can be achieved by the elution procedure described. G/F can then be quantitatively adsorbed to GTP-Sepharose. Prior to such binding, gel filtration of the peak from hydroxyapatite was necessary, because salt, especially phosphate, prevents adsorption of G/F to the derivatized Sepharose. Activation of the G/F with guanosine 5′-[β,γ-imido]triphosphate, guanosine 5′-[γ-thio]triphosphate (GTP[γS]), or AMF also prevents adsorption; however, these ligands were ineffective in eluting G/F from the matrix once bound. The G/F can be recovered efficiently by elimination of divalent cations from the eluting buffer. The binding capacity appears to exceed 20 μg/ml of packed gel. A final step, chromatography on DEAE-Sephacel (in Lubrol), yields nearly homogeneous G/F. This is an effective procedure if performed in the presence of AMF and a relatively high concentration of detergent.

Analysis of the purified preparation of G/F in NaDodSO$_4$/polyacrylamide gels is presented in Fig. 1A. Two major bands of protein are observed, with molecular weights of approximately 45,000 and 35,000. A third protein, with a molecular weight of 52,000, is also present in much smaller quantities and is thought to be a component of G/F (see *Discussion*). Scanning

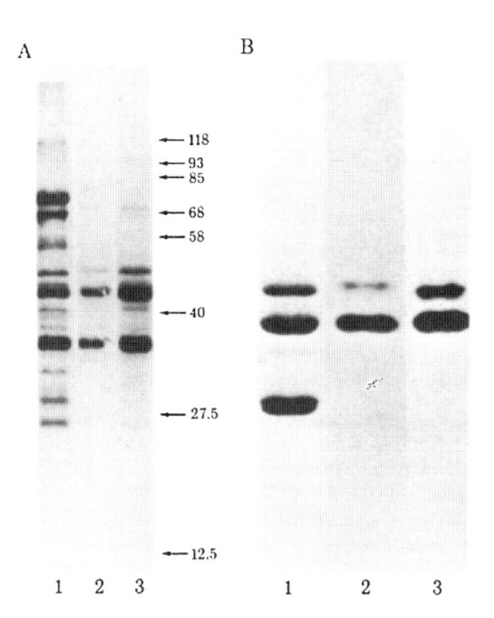

FIG. 1. (*A*) Polyacrylamide gel electrophoresis in NaDodSO$_4$ of purified fractions of G/F. Samples were run in 11% acrylamide gels by the method of Laemmli (15). Samples were prepared for electrophoresis by precipitation with 15% trichloroacetic acid in the presence of 2% NaDodSO$_4$ in order to remove cholate or Lubrol from the proteins. The pellets were rinsed with 1 ml of diethyl ether to extract residual trichloroacetic acid and were then dissolved in the sample buffer. Samples were applied to the gel as follows: lane 1, 7.5 μg of heptylamine-Sepharose peak; lane 2, 3 μg of DEAE-Sephacel peak; lane 3, 8 μg of DEAE-Sephacel peak. The arrows indicate the migration of calibrating proteins, with masses in kilodaltons of: β-galactosidase, 118; phosphorylase b, 93; glycogen synthase, 85; bovine serum albumin, 68; catalase, 58; aldolase, 50; α-chymotrypsinogen, 27.5; and cytochrome *c*, 12.5.

(*B*) Labeling of purified G/F with [^{32}P]NAD$^+$ and cholera toxin. The cholera toxin labeling pattern of G/F was obtained by reconstitution of cyc$^-$ with purified G/F and incubation with [^{32}P]NAD$^+$ and cholera toxin as described (16). After a 60-min incubation the reaction was terminated by addition of NaDodSO$_4$ sample buffer, and the sample was applied to an 11% polyacrylamide gel. Lane 1 shows the Coomassie blue staining pattern of the purified G/F, run as described in *A*. Lanes 2 and 3 are autoradiograms of the cholera toxin-labeled G/F developed for 16 and 48 hr, respectively. Under the conditions utilized, there is essentially no labeling of cyc$^-$ membranes in the absence of prior reconstitution with G/F (see figure 1 of ref. 16).

densitometry of the Coomassie blue-stained gel indicates that greater than 95% of the protein is contained in these three bands. The relative proportion of stain bound to these polypeptides was 1:5:4 for the 52,000-, 45,000-, and 35,000-dalton bands, respectively. Lane 1 of Fig. 1A shows G/F that had been purified through the first three steps of the procedure; the 35,000- and 45,000-dalton peptides are clearly distinguished at this point. The last three steps then remove the remaining impurities, including an excess of the 35,000-dalton polypeptide (see *Discussion*).

Fig. 1B compares a stained gel containing purified protein with an autoradiogram of a NaDodSO$_4$ gel containing G/F that had been labeled (presumably ADP-ribosylated) with [^{32}P]-NAD$^+$ and cholera toxin. In this experiment, cyc$^-$ membranes were reconstituted with purified G/F, and the reconstituted membranes were labeled as described (16).† Both the 45,000- and the 52,000-dalton polypeptides are preferred substrates for covalent modification by the toxin. The 35,000-dalton poly-

† Purified G/F is not a good substrate for cholera toxin when incubated with activated toxin, NAD$^+$, and GTP. Labeling of G/F is greatly stimulated if the protein is reconstituted into cyc$^-$ membranes or if a factor(s) provided by a cholate extract of cyc$^-$ membranes is included in the reaction (unpublished data).

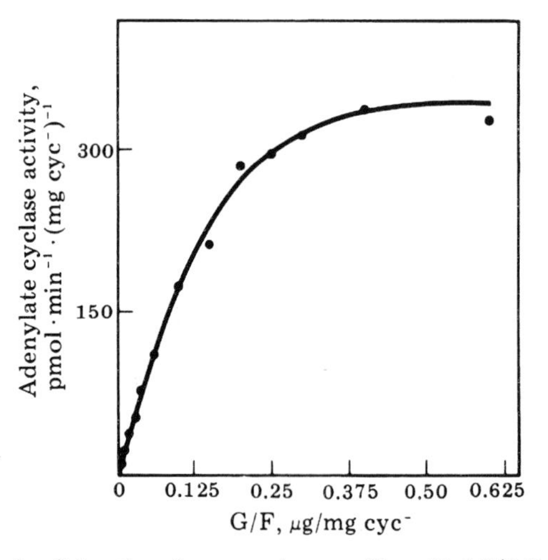

FIG. 2. Saturation of cyc$^-$ membranes with purified G/F. Purified G/F was diluted into 20 mM Hepes, pH 8/1 mM EDTA/0.1% Lubrol. Then 15 μl of the dilutions containing the indicated amounts of G/F was added to 25 μl of cyc$^-$ membranes (1.6 mg/ml). The samples were assayed for NaF-stimulated adenylate cyclase activity.

peptide was not labeled in this experiment and has never been labeled, even with prolonged incubation with high concentrations of NAD^+ and toxin. Although not visible in Fig. 1*B*, two very minor bands in the G/F preparation (38,000 and 42,000 daltons) can also be labeled and are discernible with prolonged exposure of autoradiograms or, more readily, in less pure preparations.

The regulatory component of adenylate cyclase is assayed by its reconstitution with the catalytic subunit in cyc$^-$ membranes. The saturation of this catalytic component by purified G/F is shown in Fig. 2. Reconstituted activity increases in a hyperbolic fashion with added G/F. Adenylate cyclase activity, stimulated by fluoride, was restored to a specific activity of 350 pmol·min^{-1}·(mg membrane protein)$^{-1}$; this activity is comparable to that observed in membranes from wild-type S49 cells (10). Higher specific activities [up to 1000 pmol·min^{-1}·(mg cyc$^-$ protein)$^{-1}$] have been observed with better preparations of cyc$^-$ membranes. The reconstituted activity in cyc$^-$ membranes is essentially linear with small amounts of G/F (up to $\approx$150 pmol·min^{-1}·mg^{-1}); this linear relationship constitutes a quantitative assay for G/F. Apparent half-maximal saturation of cyc$^-$ membranes is obtained with 100 ng of G/F per mg of cyc$^-$ membranes. This value is entirely consistent with the degree of purification achieved. Similar saturation profiles for reconstitution of cyc$^-$ membranes are also observed for GTP[γS]-activated adenylate cyclase activity. The specific activity of G/F, calculated for GTP[γS]-activated adenylate cyclase, is approximately 2 μmol·min^{-1}·mg^{-1}.

The purified G/F reconstitutes adenylate cyclase activatable by hormone. In Table 2, cyc$^-$ and uncoupled (UNC) membranes were reconstituted with purified G/F in cholate (6). Clearly, the restoration of adenylate cyclase activity in cyc$^-$ is accompanied by the capability of stimulating this activity with isoproterenol. The stimulation in reconstituted UNC membranes is, however, less than that previously reported for membranes reconstituted with cholate extracts of wild-type membranes (6). It is not known if this reflects some substantial difference in the properties of purified G/F from liver.

There is also a quantitative difference between the reconstituted activities shown in Fig. 2 and those in Table 2. The procedure utilized in Table 2, which involves recovery of reconstituted membranes by centrifugation prior to assay, requires considerably greater concentrations of G/F to reconstitute the same level of fluoride-stimulated enzymatic activity.

Table 2. Reconstitution of hormone-stimulable adenylate cyclase with purified G/F

Reconstituted membrane*		Adenylate cyclase activity,† pmol·min^{-1}·(mg membrane)$^{-1}$		
Membrane	μg of G/F per mg of membrane	GTP	GTP and isoproterenol	NaF
cyc$^-$	0	3.8	3.5	3.9
cyc$^-$	0.33	3.2	27	60
cyc$^-$	0.82	5.7	68	150
cyc$^-$	3.3	13	190	290
UNC	0	23	22	360
UNC	0.26	30	52	420
UNC	0.65	36	71	470
UNC	2.6	33	87	320

* cyc$^-$ or UNC membranes (250 μl at 1.5 mg/ml or 2.0 mg/ml, respectively) were mixed with 150 μl of purified G/F in TED/100 mM NaCl/0.9% sodium cholate. After a 20-min incubation on ice, samples were diluted with 200 μl of 150 mM NaHepes, pH 8/15 mM $MgCl_2$/300 μg of bovine serum albumin per ml/9 mM phospho*enol*pyruvate/30 μg of pyruvate kinase per ml/1.25 mM ATP/100 μM GTP and incubated for 10 min at 30°C. Reconstituted membranes were then collected at 0°C by centrifugation at 140,000 × *g* for 20 min and resuspended for assay in 300 μl of 50 mM NaHepes, pH 8/2 mM $MgCl_2$/1 mM EDTA.

† Reconstituted membranes (50 μl) were assayed for adenylate cyclase as described (10) in the presence of the indicated effectors: GTP (100 μM), isoproterenol (4 μM), NaF (10 mM).

This is explained in part by the increased thermal lability of both G/F and the catalytic moiety in the presence of cholate. In addition, activation of soluble G/F by fluoride, which can occur in the single-step reconstitution procedure used in Fig. 2, facilitates incorporation of G/F into membranes (11).

DISCUSSION

Previous experiments have demonstrated that adenylate cyclase requires at least two separable components for the expression of MgATP-dependent enzymatic activity (2–5). We have described herein a procedure for the purification of one of these proteins, the regulatory component (G/F), to near homogeneity. The protein has been purified approximately 2000-fold from detergent extracts. This can be extrapolated to represent a purification of 5000- to 10,000-fold from plasma membranes or nearly 100,000-fold from total cellular protein. Upon reconstitution with the catalytic subunit and activation by fluoride, 1 mg of G/F can stimulate the synthesis of 3–4 μmol of cyclic AMP per minute. If an approximate molecular weight of 130,000 is assumed for G/F (17), a molar turnover number of 7 sec^{-1} is obtained for this component of adenylate cyclase.

Analysis of the purified preparation of G/F by $NaDodSO_4$/polyacrylamide gel electrophoresis identified the presence of three predominant polypeptide bands with molecular weights of 52,000, 45,000, and 35,000. From the molecular weight of native G/F (130,000), calculated from hydrodynamic properties (17), we predict that G/F has a multisubunit structure of one or more of these polypeptides. Evidence indicates that the three polypeptides in the purified preparation of G/F are all relevant to the activity of the protein. They are not resolved by the several chromatographic procedures described here, during centrifugation in sucrose gradients, or by several other fractionation procedures that allow preservation of activity.

A powerful method for identification of the components of the regulatory protein is by their labeling with [^{32}P]NAD^+ and cholera toxin. Cholera toxin is believed to ADP-ribosylate G/F specifically; this appears to constitute the mechanism by which the toxin activates adenylate cyclase (14, 18). Prior studies of labeling with cholera toxin in the wild-type and variant clones of S49 have shown cholera toxin-specific incorporation of ^{32}P into two bands with molecular weights of about 52,000 and 42,000 (19). Such labeling is absent in the cyc$^-$ variant, which lacks G/F activity. We have confirmed these findings, and have also demonstrated that both the 45,000-dalton polypeptide and the 52,000-dalton polypeptide have altered isoelectric points in the UNC S49 cell variant (16).‡ The 45,000- and 52,000-dalton bands of purified G/F can both be labeled with [^{32}P]-NAD^+ and cholera toxin, and they migrate identically with the corresponding bands in S49 cell membranes. The data, taken together, strongly suggest the relevance of both of these polypeptides to G/F. Previously, Cassel and Pfeuffer (14) observed identically migrating bands at 42,000 daltons when pigeon erythrocytes were labeled with either cholera toxin or a photoaffinity analog of GTP (2). The single toxin-labeled band from turkey erythrocytes migrates identically with the 45,000-dalton band of pure G/F (unpublished data).

An apparent excess of the 35,000-dalton polypeptide is removed during the purification of G/F. During elution of the hydroxyapatite column with 30 mM potassium phosphate, 35,000-dalton protein is removed as an essentially pure species, such that the stoichiometric ratio of the 35,000 band to the two other bands is near 1:1 in the peak of G/F activity. Rechromatography of the G/F peak under the same conditions does not remove more of the 35,000-dalton protein. The isolated 35,000-dalton protein has no identifiable activity. It and G/F were examined by two-dimensional electrophoresis (electrofocusing and $NaDodSO_4$ electrophoresis). Identical patterns of stained protein in the 35,000-dalton region were obtained with both preparations, confirming the suspected identity of the two 35,000-dalton polypeptides. When purified G/F is inactivated (by incubation at 4°C in detergent solution) and is then rechromatographed over hydroxyapatite, all of the 35,000-dalton protein elutes during the low phosphate wash and it is resolved from the 45,000- and 52,000-dalton species. One interpretation of this result is that the observed inactivation is coincident with dissociation of the subunits of G/F. Furthermore, it is suggested that the apparent excess of the 35,000-dalton protein is a result of inactivation of G/F during purification. In support of this contention, inactivation also leads to partial resolution of the 35,000- and 45,000-dalton peptides during chromatography on DEAE-Sephacel or GTP-Sepharose. Additional evidence for the identity of the 35,000-dalton protein as a subunit of G/F has been obtained by its purification from turkey erythrocytes. G/F that has been purified to near homogeneity from this source contains both the 35,000-dalton and the 45,000-dalton proteins in a ratio of approximately unity (unpublished data). Definitive proof of the contribution of all three species to the activity of G/F will require their resolution and reconstitution; unfortunately, combination of resolved 35,000- and 45,000-dalton polypeptides has not yet resulted in restoration of G/F activity.

We have obtained some evidence for the existence of complexes of G/F with different subunit composition. There is a partial resolution of the 52,000- and 45,000-dalton polypeptides during chromatography on heptylamine-Sepharose. The 52,000-dalton band is more prominent in the front of this peak, while the 45,000-dalton species dominates the back. The 35,000-dalton protein distributes across the entire peak. The 45,000- and 52,000-dalton species are not completely resolved from each other, and no further separation is observed upon rechromatography of portions of this peak. In the preparation documented above, the peak from the heptylamine-Sepharose was taken such that most of the 52,000-dalton protein was eliminated; this contributed to a lower cumulative recovery than can be achieved when a wider peak is taken.

The 50-μg yield of pure G/F from this six-step procedure permits detailed study of the biochemical properties of this protein. G/F has been postulated to hydrolyze GTP (20) and to alter the affinity of hormone receptors for agonist ligand (5). These and other hypotheses concerning the biochemical activities of G/F may now be tested directly. In this regard it may be noted that the GTPase activity of the preparation described herein is less than 1 nmol$\cdot$min$^{-1}\cdot$mg^{-1} (less than 0.001 of the specific activity of reconstituted adenylate cyclase). If G/F can in fact catalyze the hydrolysis of GTP, it seems probable that other proteins (e.g., hormone receptors or the catalytic subunit) will be necessary for the observation of a reasonable specific activity. Finally, it is hoped that portions of this procedure can be modified to allow preparation of greater quantities of material for physical and chemical analysis. Membrane preparation is currently rate-limiting, and, as noted, the final three steps of the procedure are inefficient.

We express special thanks to Thomas Rall for endless hours of membrane preparation. We also thank Debra Lewis and Pamela Van Arsdale for excellent technical assistance, Hannah Anderson and Kim Thompson for production of cultured cells, and Wendy Deaner and Jane Rall for editorial assistance. This work was supported by Grants NS10193 and AM17042 from the U.S. Public Health Service and by Grant BC240 from the American Cancer Society. Other support has included U.S. Public Health Service Fellowships AM05934 (to J.K.N.), NS05957 (to P.C.S.), and AM05930 (to M.D.S.) and a Research Career Development Award (to A.G.G.).

1. Rall, T. W., Sutherland, E. W. & Berthet, J. (1957) *J. Biol. Chem.* **224,** 463–475.
2. Pfeuffer, T. (1977) *J. Biol. Chem.* **252,** 7224–7234.
3. Ross, E. M. & Gilman, A. G. (1977) *Proc. Natl. Acad. Sci. USA* **74,** 3715–3719.
4. Ross, E. M. & Gilman, A. G. (1977) *J. Biol. Chem.* **252,** 6966–6970.
5. Ross, E. M., Howlett, A. C., Ferguson, K. M. & Gilman, A. G. (1978) *J. Biol. Chem.* **253,** 6401–6412.
6. Sternweis, P. C. & Gilman, A. G. (1979) *J. Biol. Chem.* **254,** 3333–3340.
7. Salomon, Y., Londos, C. & Rodbell, M. (1974) *Anal. Biochem.* **58,** 541–548.
8. Lowry, O. H., Rosebrough, N. J., Farr, A. L. & Randall, R. J. (1951) *J. Biol. Chem.* **193,** 265–275.
9. Schaffner, W. & Weissman, C. (1973) *Anal. Biochem.* **56,** 502–504.
10. Ross, E. M., Maguire, M. E., Sturgill, T. W., Biltonen, R. L. & Gilman, A. G. (1977) *J. Biol. Chem.* **252,** 5761–5775.
11. Howlett, A. C., Sternweis, P. C., Macik, B. A., Van Arsdale, P. M. & Gilman, A. G. (1979) *J. Biol. Chem.* **254,** 2287–2295.
12. Shaltiel, S. (1974) *Methods Enzymol.* **34,** 126–140.
13. Johnson, R. A. & Walseth, T. F. (1979) *Adv. Cyclic Nucleotide Res.* **10,** 135–168.
14. Cassel, D. & Pfeuffer, T. (1978) *Proc. Natl. Acad. Sci. USA* **75,** 2669–2673.
15. Laemmli, U. K. (1970) *Nature (London)* **227,** 680–689.
16. Schleifer, L. S., Garrison, J. C., Sternweis, P. C., Northup, J. K. & Gilman, A. G. (1980) *J. Biol. Chem.* **255,** 2641–2644.
17. Howlett, A. C. & Gilman, A. G. (1980) *J. Biol. Chem.* **255,** 2861–2866.
18. Gill, D. M. & Meren, R. (1978) *Proc. Natl. Acad. Sci. USA* **75,** 3050–3054.
19. Johnson, G. L., Kaslow, H. R. & Bourne, H. R. (1978) *J. Biol. Chem.* **253,** 7120–7123.
20. Cassel, D. & Selinger, Z. (1976) *Biochim. Biophys. Acta* **452,** 538–551.

‡ The predominant toxin-labeled bands from S49 cells, rabbit liver, and turkey erythrocytes migrate identically during $NaDodSO_4$/polyacrylamide gel electrophoresis. We have estimated the molecular weight of this protein to be 45,000. This is the protein band estimated by others (e.g., ref. 19) to have a molecular weight of 42,000.

Structure of the β-adrenergic receptor

3

The plasma membrane receptors that are coupled to their effectors via G-proteins were characterized extensively during the 1970s by the use of pharmacological approaches (for further information on G-proteins, see Section 2). These techniques, using radiolabelled ligands in binding assays, allowed the determination of receptor density and tissue distribution, and the identification of pharmacologically distinguishable subtypes of receptors responding to related agonists. The further characterization of these receptors required determination of their primary amino acid sequences. This information enabled the investigation of structure–function relationships, the identification of further subtypes of receptors and comparison of the structures of the various G-protein-coupled receptors. The chosen paper [1] describes the amino acid sequence of the β_2-adrenergic receptor, which was found to demonstrate a high level of similarity to both the amino acid sequence and likely transmembrane orientation of rhodopsin which is required for phototransduction in the retina. This work initiated the development of the idea that all G-protein-coupled receptors are part of a superfamily of related proteins with a common seven-transmembrane-segment structure (Figure 3.1).

Progress towards the molecular characterization of the G-protein-coupled receptors required their purification for amino acid analysis. The β-adrenergic receptors had been studied extensively by pharmacological approaches. These receptors are coupled to cyclic AMP production by adenyl cyclase, and two forms, β_1 and β_2, were identified on the basis of agonist/antagonist sensitivity [2]. β-Adrenergic receptors regulate many different cellular functions, and, while they are expressed at low density, they are detectable in almost all mammalian tissues. Methods were developed to allow purification of these receptors by up to 100 000-fold using specific affinity-purification techniques [3]. The purified receptors were shown to be active using reconstitution techniques with the receptor and G-protein incorporated together into lipid vesicles [4].

The β_2-adrenergic receptor was purified from hamster lung [3] and partial amino acid sequences were generated from peptide fragments [1]. This then allowed the preparation of complementary oligonucleotides, based on one of the peptide sequences, for use in the screening of a hamster genomic DNA library and the isolation of DNA clones spanning the coding sequence for the receptor. Unexpectedly, the β_2-adrenergic receptor gene is intronless, allowing the protein sequence to be deduced directly from the gene sequence. This was confirmed subsequently by isolation of cDNA. Analysis of the complete amino acid sequence revealed significant sequence similarity between the β_2-adrenergic receptor and bovine opsin [5] and showed that both proteins would be predicted to possess seven transmembrane domains based on the distribution of their hydrophobic amino acids (Figure 3.1).

Subsequent molecular cloning revealed the existence of three β-adrenergic receptor subtypes, β_1, β_2 and β_3, and six closely related α-adrenergic receptor subtypes, as well as many other similar seven-transmembrane-segment receptors including the muscarinic acetylcholine receptor. It is now known that the family includes many hundreds of mammalian receptor proteins that couple to G-proteins [6].

The availability of the cloned β_2-adrenergic receptor allowed the use of this receptor as a model in mutagenesis studies to determine several aspects of receptor function [6,7]. The production of chimaeric receptors, in which the domains of β_2-adrenergic and α-adrenergic receptors were interchanged, helped identify regions (within the transmembrane domains) potentially involved in agonist and antagonist binding. Further studies on these chimaeric

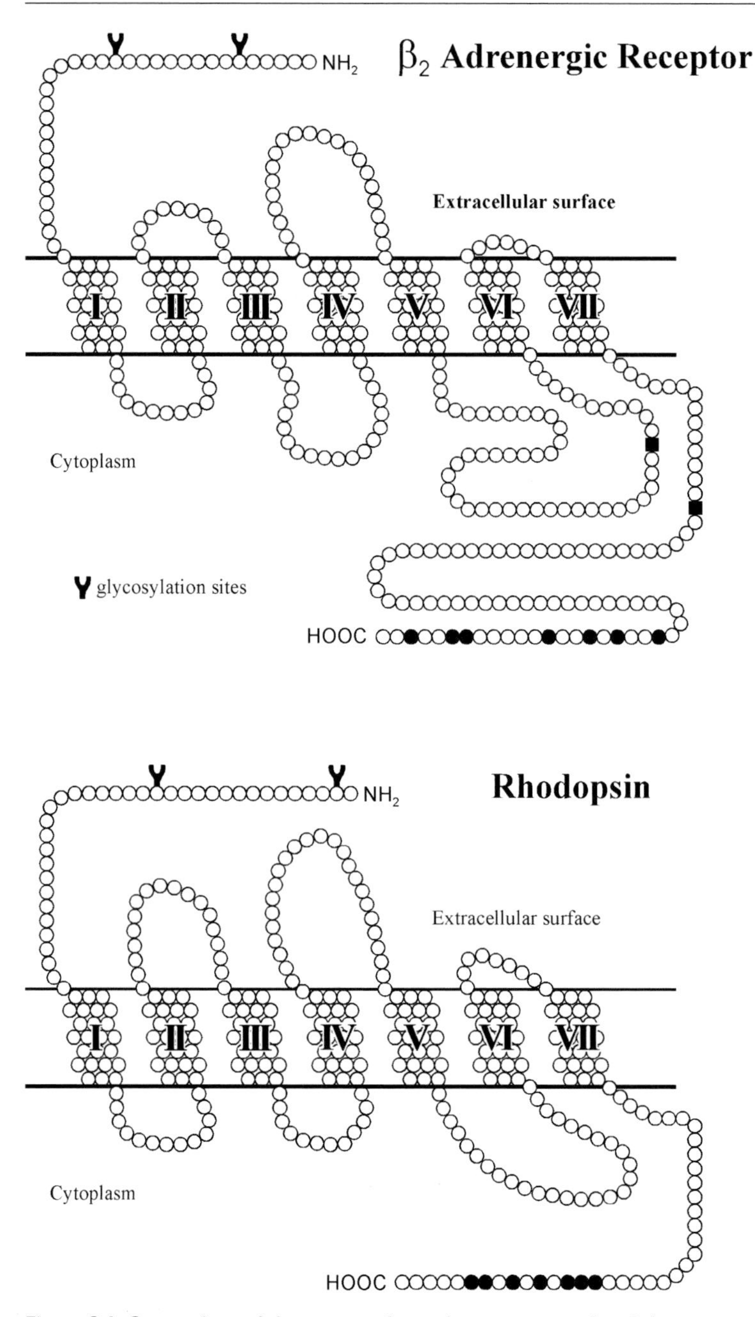

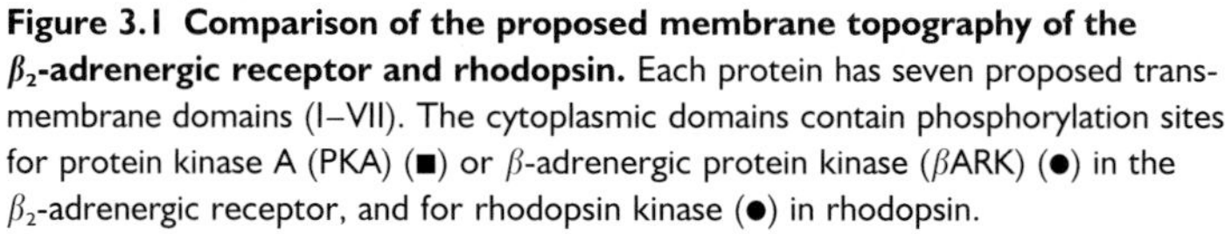

Figure 3.1 Comparison of the proposed membrane topography of the β_2-adrenergic receptor and rhodopsin. Each protein has seven proposed transmembrane domains (I–VII). The cytoplasmic domains contain phosphorylation sites for protein kinase A (PKA) (■) or β-adrenergic protein kinase (βARK) (●) in the β_2-adrenergic receptor, and for rhodopsin kinase (●) in rhodopsin.

receptors indicated that the third intracellular loop between transmembrane domains V and VI is required for, and provides specificity of, coupling to the G-proteins. This loop determines whether expressed receptors couple to adenyl cyclase or phosphoinositide production [7]. Site-directed mutagenesis has also been used to examine the role of protein phosphorylation in the desensitization of the β_2-adrenergic receptor.

In common with many other receptors, the β_2-adrenergic receptor shows desensitization following prolonged exposure to agonists. Long-term desensitization is caused by a reduction in cell-surface receptor density. Short-term desensitization, which occurs within minutes, is due to uncoupling of the receptor from G-proteins caused by phosphorylation of the receptors. This can be brought about by two separate protein kinases. The β_2-adrenergic receptor was found to possess two potential sites for phosphorylation of serines by cyclic AMP-dependent protein kinase (PKA) in the third intracellular loop and in the C-terminal domain (Figure 3.1). Site-directed mutagenesis to change each of these serines showed that the presence of the serine in the third intracellular loop was required and sufficient for PKA-mediated desensitization by exposure to low levels of agonist [8,9]. It may be significant that it is this intracellular loop that is involved in G-protein coupling. Exposure to high levels of agonist leads to phosphorylation of the β-adrenergic receptor by a receptor-specific protein kinase (the β-adrenergic receptor kinase; βARK) which phosphorylates the occupied receptor on several sites in the C-terminal domain [10]. PKA phosphorylation of the third loop results directly in uncoupling, but phosphorylation by βARK leads to the binding, to the phosphorylated receptor, of the protein β-arrestin. It is this process that leads to uncoupling of the receptor from G-proteins. Rhodopsin also possesses phosphorylation sites in its C-terminal domain that are substrates for a specific rhodopsin kinase [11]. Phosphorylation by this kinase reduces the ability of the receptor to couple to the retinal G-protein transducin by an analogous mechanism.

The elucidation of the primary amino acid sequence of the β_2-adrenergic receptor gave rise to a now widely accepted model for the membrane orientation of G-protein-coupled receptors. Subsequent molecular analysis of the β_2-adrenergic receptor has revealed significant structure–function relationships in the nature and regulation of receptor G-protein coupling. These relationships have been found to be relevant to the many other members of the seven-transmembrane-segment receptor family.

References

1. Dixon, R.A.F., Kobilka, B.K., Strader, D.J., Benovic, J.L., Dohlman, H.G., Frielle, T., Bolanowski, M.A., Bennett, C.D., Rands, E., Diehl, R.D., Mumford, R.A., Slater, E.E., Sigal, I.S., Caron, M.G., Lefkowitz, R.J. and Strader, C.D. (1986) *Nature (London)* **321**, 75–79
2. Lefkowitz, R.J., Stadel, J.M. and Caron, M.G. (1983) *Annu. Rev. Biochem.* **52**, 159–186
3. Benovic, J.L., Shoor, R.G.L., Caron, M.G. and Lefkowitz, R.J. (1984) *Biochemistry* **23**, 4510–4518
4. Cerione, R.A., Sibley, D.R., Codina, J., Benovic, J.L., Winslow, J., Neer, E.J., Birnbaumer, L., Caron, M.G. and Lefkowitz, R.J.. (1984) *J. Biol. Chem.* **259**, 9979–9982
5. Nathans, J. and Hogness, D.S (1983) *Cell* **34**, 807–814
6. Dohlman, H.G., Thorner, J., Caron, M.G. and Lefkowitz, R.J. (1991) *Annu. Rev. Biochem.* **60**, 653–688

7. Ostrowski, J., Kjelsberg, M.A., Caron, M.G. and Lefkowitz, R.J. (1992) *Annu. Rev. Pharmacol. Toxicol.* **32**, 167–183
8. Hausdorff, W.P., Bouvier, M., O'Dowd, B.F., Irons, G.P., Caron, M.G. and Lefkowitz, R.J. (1989) *J. Biol. Chem.* **264**, 12657–12663
9. Clark, R.B., Friedman, J., Dixon, R.A.F. and Strader, C.D. (1989) *Mol. Pharmacol.* **36**, 343–348
10. Benovic, J.L., Strasser, R.H., Caron, M.G. and Lefkowitz, R.J. (1986) *Proc. Natl. Acad. Sci. U.S.A.* **83**, 2797–2801
11. Thompson, P. and Findlay, J.B.C. (1984) *Biochem. J.* **220**, 773–780

Dixon et al. (1986) Nature (London) **321**, 75–79

Cloning of the gene and cDNA for mammalian β-adrenergic receptor and homology with rhodopsin

Richard A. F. Dixon*, Brian K. Kobilka†, David J. Strader‡, Jeffrey L. Benovic†, Henrik G. Dohlman†, Thomas Frielle†, Mark A. Bolanowski†, Carl D. Bennett§, Elaine Rands*, Ronald E. Diehl*, Richard A. Mumford‡, Eve E. Slater‡, Irving S. Sigal*, Marc G. Caron†, Robert J. Lefkowitz† & Catherine D. Strader‡

Departments of *Virus and Cell Biology Research and §Medicinal Chemistry, Merck Sharp and Dohme Research Laboratories, West Point, Pennsylvania 19486, USA
† Howard Hughes Medical Institute, Department of Medicine, Biochemistry and Physiology, Duke University Medical Center, Durham, North Carolina 27710, USA
‡ Department of Biochemistry and Molecular Biology, Merck Sharp and Dohme Research Laboratories, Rahway, New Jersey 07065, USA

The adenylate cyclase system, which consists of a catalytic moiety and regulatory guanine nucleotide-binding proteins, provides the effector mechanism for the intracellular actions of many hormones and drugs[1]. The tissue specificity of the system is determined by the particular receptors that a cell expresses. Of the many receptors known to modulate adenylate cyclase activity, the best characterized and one of the most pharmacologically important is the β-adrenergic receptor (βAR). The pharmacologically distinguishable subtypes of the β-adrenergic receptor, β_1 and β_2 receptors, stimulate adenylate cyclase on binding specific catecholamines[1]. Recently, the avian erythrocyte β_1, the amphibian erythrocyte β_2 and the mammalian lung β_2 receptors have been purified to homogeneity and demonstrated to retain binding activity in detergent-solubilized form[1–5]. Moreover, the β-adrenergic receptor has been reconstituted with the other components of the adenylate cyclase system *in vitro*[6], thus making this hormone receptor particularly attractive for studies of the mechanism of receptor action. This situation is in contrast to that for the receptors for growth factors and insulin, where the primary biochemical effectors of receptor action are unknown. Here, we report the cloning of the gene and cDNA for the mammalian β_2AR. Analysis of the amino-acid sequence predicted for the βAR indicates significant amino-acid homology with bovine rhodopsin and suggests that, like rhodopsin[7], βAR possesses multiple membrane-spanning regions.

Hamster lung βAR was purified to homogeneity by sequential affinity chromatography and molecular-sieve HPLC as described previously[2,5]. The purified receptor bound ligand with theoretical specific activity and migrated on SDS-polyacrylamide gel electrophoresis as a single broad band at a relative molecular mass (M_r) of 64,000 (64K). Initial attempts to obtain N-terminal sequence data on intact βAR failed, presumably because the N-terminus of this protein was blocked. Therefore, peptide fragments generated by CNBr cleavage of pure βAR were isolated by reverse-phase HPLC.

Figure 1*a* (solid line) shows a peptide map generated from 1 nmol of pure receptor; the broken line shows the HPLC profile resulting from CNBr treatment of the detergent alone. The βAR-derived peptides produced at least nine specific absorbance peaks which were reproducibly observed in five separate βAR preparations. The most prominent of these peptides (marked with arrows in Fig. 1) were subjected to N-terminal sequence analysis, yielding the amino-acid sequences in Fig. 1*b*.

To confirm that the determined amino-acid sequences were those of the βAR polypeptide, we raised anti-peptide antibodies against peptide 7. Peptide 7 was expressed in *Escherichia coli* as a C-terminal peptide fused to the N-terminal domain of the

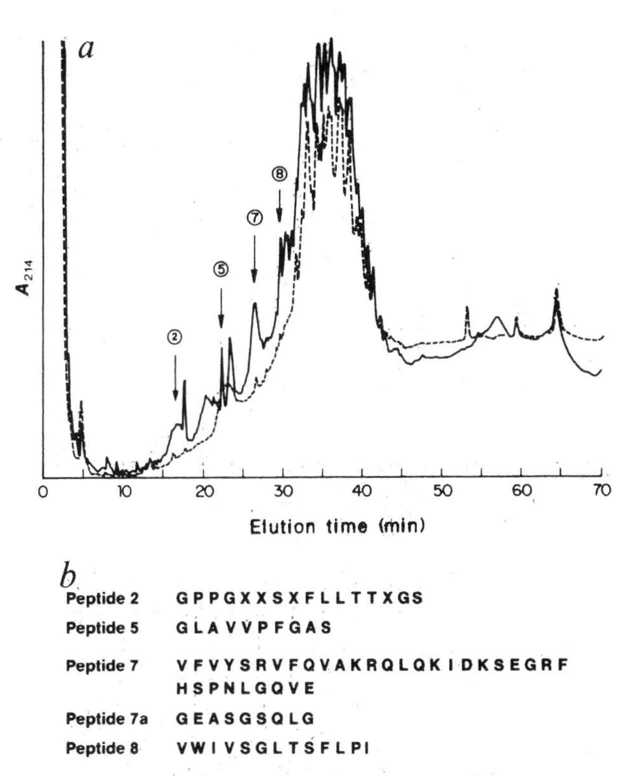

Fig. 1 Amino-acid sequence of peptides derived from CNBr-treated β-adrenergic receptor. *a*, Absorbance profiles represent CNBr treatment of pure βAR (solid line) or digitonin (dashed line). The arrows indicate the peptides that were sequenced. *b*, Amino-acid sequences identified by HPLC following each cycle of the sequenator. Two of the four blank cycles (X) in the amino-acid sequence for peptide 2 are presumed to be due to *N*-linked glycosylation. Peptides 7 and 7a were located within the same peak.
Methods. βAR was purified to homogeneity from hamster lung membranes by the method of Benovic *et al.*[5], using affinity chromatography followed by molecular-sieve HPLC. Binding of ^{125}I-CYP to intact cells or to solubilized βAR was determined according to Caron and Lefkowitz[22]. For peptide preparation, ~1 nmol of pure βAR was treated with CNBr (0.4 mM) in 70% formic acid at 23 °C for 20 h. After lyophilization, the sample was resuspended in 20 mM trifluoroacetic acid (TFA) and the peptides separated by reverse-phase HPLC on a Synchropak C-4 column, eluted with a 10–70% acetonitrile gradient containing 20 mM TFA. The N-terminal sequence analysis was performed by the method of Hewick *et al.*[23], using a gas-phase sequenator (Applied Biosystems). The phenylthiohydantoin (PTH) amino acids produced at each step were separated and quantitated by HPLC[24].

yeast *RAS*^{sc1} protein SC1N (ref. 8). Rabbits injected with the isolated fusion protein produced antibodies which reacted with ^{125}I-labelled immunogen as well as pure ^{125}I-labelled βAR or pure ^{125}I-cyanopindolol-labelled (^{125}I-CYP) βAR. Figure 2*a* shows an immunoprecipitation titration curve of this antibody against ^{125}I-CYP-labelled solubilized βAR from hamster lung, hamster heart, A431 epidermoid carcinoma cells and turkey erythrocytes. No immunoprecipitation of counts above background was observed in control experiments when ^{125}I-CYP was incubated with the antibody in the absence of receptor.

Antibody to the hamster lung βAR-derived peptide was capable of recognizing the human βAR from the A431 line, albeit with a slightly lower sensitivity (50%) (Fig. 2*a*). The antibody also cross-reacted slightly with hamster heart βAR, a tissue containing a β_1 subtype of receptor, but did not immunoprecipitate the β_1AR of turkey erythrocytes. These differences in antibody sensitivity could reflect differences in either primary sequence or the conformation of this region of the protein within the various receptor subtypes and receptors from different species. To confirm that the antibody was recog-

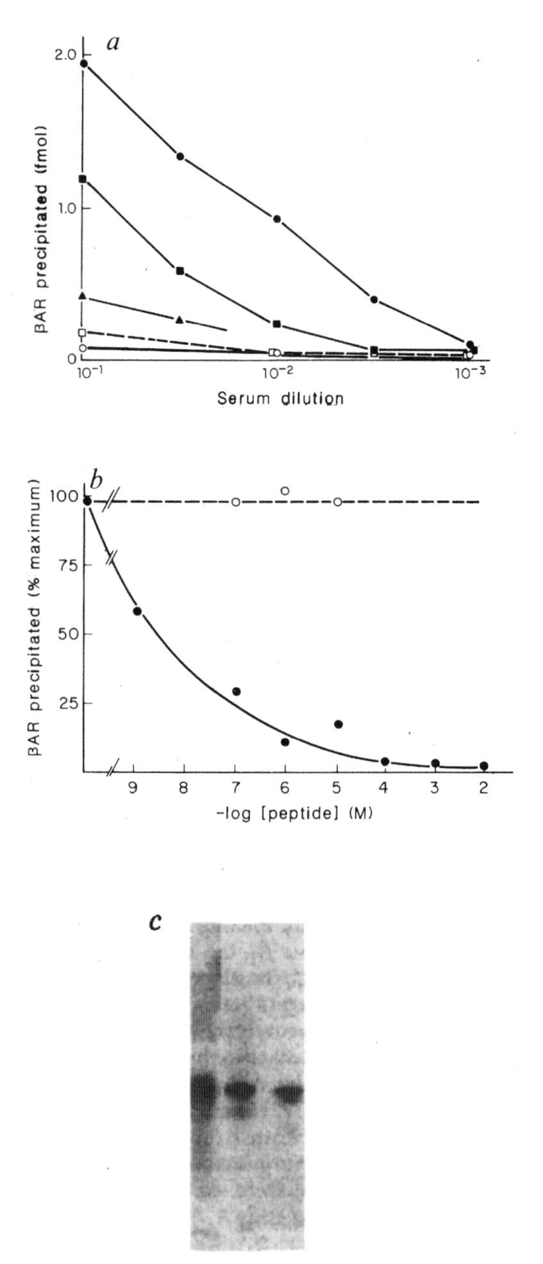

Fig. 2 Immunoreactivity of β-adrenergic receptor. *a*, Immunoprecipitation of ^{125}I-CYP-βAR from hamster lung (●), hamster heart (▲), A431 cells (■) or turkey erythrocytes (○) by serum from rabbits immunized with βAR peptide 7. □, Immunoprecipitation of hamster lung receptor by preimmune serum. *b*, Immunoprecipitation of ^{125}I-CYP-labelled hamster lung βAR with anti-βAR peptide 7 antibody following preincubation with a synthetic peptide containing a portion of the sequence for peptide 7 (●), or with an unrelated peptide, atrial natriuretic factor (○). *c*, Protein immunoblotting of the βAR. Protein samples were separated on a 10% polyacrylamide gel[25], transferred to nitrocellulose and treated sequentially with antibodies and ^{125}I-protein A (10^6 c.p.m. per 25 ml) as described elsewhere[26,27]. Lane 1, pure hamster lung βAR (5 pmol); lane 2, A431 cell lysate from 4×10^5 cells; lane 3, RPMI 1846 lysate from 10^5 cells.

Methods. The immunogen for induction of anti-peptide antibodies was expressed in *E. coli.* Two oligonucleotides encoding the 19 amino acids of peptide 7 (QVAKRQLQKIDKSEGRFHS) were synthesized (see Fig. 1 legend), annealed and ligated[28] into the *Acc*I and *Hin*dIII sites of plasmid pSC1N[8] to give the plasmid pβP1. *E. coli* transformed with pβP1 overexpresses a protein of apparent M_r 23K, while *E coli* containing pSC1N overexpresses a protein of apparent M_r 21K (data not shown). This observed difference in relative molecular mass of the two proteins is consistent with the encoded fusion protein containing 19 additional amino acids. To prepare antigen, plasmid-containing cells were grown in L-broth containing ampicillin and isopropylthiogalactoside at 37 °C for 16 h, harvested by centrifugation, lysed by sonication and the soluble proteins removed by centrifugation at 40,000*g*. The cell pellet was seqentially extracted with 1 M NaCl, 1% Triton X-100 and 1.75 M guanidinium-HCl. The SC1N βAR fusion protein was extracted from the cells with 3.25 M guanidinium-HCl, dialysed against phosphate-buffered saline and used directly as an immunogen. Approximately 100 mg of fusion protein of 90% purity was obtained from 1 litre of starting culture. Antibodies were detected in serum from injected rabbits by incubation of the serum with ^{125}I-CYP-labelled soluble βAR in 10 mM Tris-HCl, 0.1 M NaCl, 0.1% digitonin, 0.5% bovine serum albumin (BSA) *p*H 7.4. After 2 h at 25 °C, the antibody was precipitated by addition of either $(NH_4)_2SO_4$ to 50% or *Staphylococcus aureus* protein A, followed by incubation in ice for 30 min. The precipitated protein was collected by centrifugation, and the radioactivity contained in the antibody pellet measured. For the peptide blocking experiment, the peptide YAKRQLQKIDKSEGR was synthesized[29] using a SAM II peptide synthesizer (Biosearch), and purified on a Whatman C-18 Magnum column in a H_2O/acetonitrile gradient containing 0.2% TFA. The resulting product was judged to be pure by amino-acid sequencing and mass spectral analysis. Increasing concentrations of this peptide were added to a 1:100 dilution of anti-peptide 7 antiserum and incubated for 2 h at 23 °C. The treated antiserum was then mixed with ^{125}I-CYP-labelled βAR and assayed as above.

nizing the amino-acid sequence of peptide 7, a chemically synthesized peptide was used as a specific inhibitor of antibody–receptor interactions. At concentrations $\geq$100 μM, this synthetic peptide completely prevented the immunoprecipitation of ^{125}I-CYP-labelled βAR by the antibody (Fig. 2*b*). An unrelated peptide, atrial natriuretic factor[9], had no effect on the immunoprecipitation.

The specificity of the antibody for the βAR was demonstrated further by protein immunoblotting. As shown in Fig. 2*c* (lane 1), the antibody reacted specifically with pure hamster lung βAR; a single protein of the same relative molecular mass (64K) was also observed in human A431 and hamster melanoma RPMI 1846 cells (Fig. 2*c*, lanes 2, 3), both of which were found to contain βAR on the basis of ^{125}I-CYP binding[10] (data not shown). This specific immunoreactive band was not observed on prior treatment of the antibody with the synthetic peptide and was not present when normal rabbit serum was substituted for the anti-peptide 7 antibody (data not shown).

To facilitate cloning of the βAR gene, oligonucleotides complementary to the DNA encoding the amino-acid sequence of peptide 7 were synthesized for use as hybridization probes (see Fig. 3 legend). In hybridization experiments performed at high stringency on blots of hamster genomic DNA, a single hybridizing band of 5.2 kilobases (kb) was observed in *Eco*RI digests and a band of 1.3 kb was observed in *Hin*dIII digests (data not shown). When a complete hamster genomic library was screened under the same conditions, five clones were isolated. Restriction analysis of the phage DNA revealed that all these clones contained a 1.3-kb *Hin*dIII and a 5.2-kb *Eco*RI fragment which hybridized to the probes (data not shown). Mapping of the phage DNA indicated that these clones overlap to give a total of 30 kb of contiguous genomic DNA. Figure 3 shows the restriction map of the genomic DNA containing the βAR-related sequences. Sequencing of the 1.3-kb *Hin*dIII fragment revealed a continuous open reading frame encoding 435 amino acids; the sequences of all the CNBr peptides shown in Fig. 1 were contained within this putative polypeptide.

Using the 1.3-kb *Hin*dIII gene fragment as a probe, seven clones were obtained from an unamplified hamster cDNA library (2×10^6 recombinants). Two of these cDNAs hybridized to oligonucleotide probes specific for the N-terminal, middle and C-terminal portions of the βAR gene. The nucleotide sequence of these two cDNA clones (Fig. 4) extends from 210 nucleotides (nt) 5′ to the open reading frame encoding the βAR

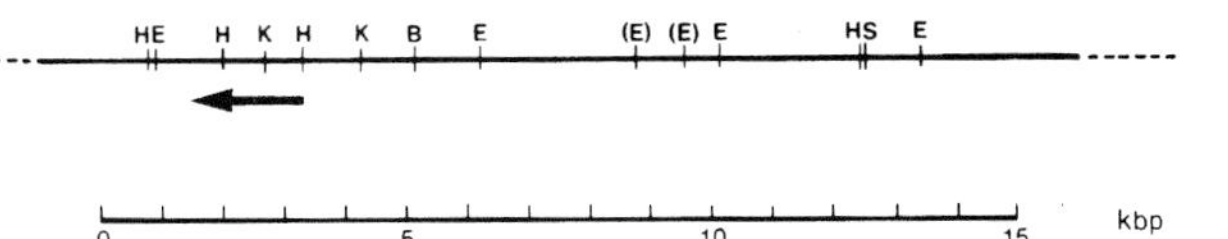

Fig. 3 Restriction map of the hamster βAR gene. A portion of the hamster DNA is shown. The 1.3-kb *Hin*dIII fragment which hybridizes to oligonucleotides specific for peptide 7, is underlined with an arrow indicating the direction of transcription of the βAR gene. The restriction enzyme sites are: B, *Bam*HI; E, *Eco*RI; H, *Hin*dIII; K, *Kpn*I; S, *Sal*I. Those sites shown in parentheses have not been unequivocally ordered.

Methods. All restriction enzymes, *E. coli* DNA polymerase I, T_4 DNA ligase and T_4 polynucleotide kinase were purchased from New England Biolabs. Radiolabelled nucleotides were purchased from Amersham. λ EMBL3A phage arms[30] and λ *in vitro* DNA packaging extracts were purchased from Vector Cloning Systems. Standard recombinant DNA and microbiological procedures were used throughout[28]. Genomic libraries were constructed using high-M_r genomic DNA isolated from hamster lung cells in the vector λ EMBL3 (refs 6, 30). Probes for peptide 7 coding sequences, oligonucleotides ON225(5′ pCTCCACCTGGCCCAGGTTGGG-AGAGTGGAACCTGCCCTCAGACTTGTCGAT) and ON229(5′ pAGGCAGCTGCAGAAGATCGACAAGTCTGAG) and ON168 (5′ pTTCCAGGTGGCCAAGCGGCAGCTGCAGAAGATCGA-CAA) and ON169(5′ pATGGTCTTTGTCTACTCCCGGGTCT-TCCAGGTGGCCAA), were synthesized on an Applied Biosystems Model 3A DNA synthesizer. The oligonucleotides were labelled by either a fill-in reaction (ON225, ON229) using Klenow DNA polymerase and all four [α-^{32}P]dNTPs[31] or by phosphorylation using T_4 polynucleotide kinase and [γ-^{32}P]ATP[28]. Phage libraries were screened by the method of Benton and Davis[32] using the hybridization conditions of Ullrich *et al.*[33]. DNA was isolated from CsCl-banded phage as described elsewhere[28]. For restriction analysis, DNA was digested with the appropriate enzyme and electrophoresed on 0.8% agarose gels. DNA was transferred to nitrocellulose by the procedure of Southern[34] and hybridized as above.

Fig. 4 (Right) Nucleotide and deduced amino-acid sequence of the βAR cDNA. The nucleotides are numbered on the right-hand side of each line beginning with the first nucleotide of the most 5′ cDNA clone. The translated amino-acid sequence is shown beneath the corresponding nucleotide sequence and is numbered to the left of each line. Underlined amino acids represent the CNBr peptides whose sequences are given in Fig. 1. All predicted amino acids agree with those determined by peptide sequencing, with the single exception of a cysteine for serine substitution in peptide 7a. All derived peptide sequences are preceded by a methionine, consistent with CNBr cleavage. The underlined nucleotides preceding the βAR sequences denote the first methionine codon and an in-frame termination codon. The boxed nucleotides at the 3′ end of the sequence represent the polyadenylation signal. Postulated glycosylation sites (Asn-X-Ser/Thr) are indicated by asterisks. Putative protein kinase A phosphorylation sites are boxed.

Methods. Reverse transcriptase was purchased from Seikagaku-America. RNase H was from Pharmacia and λ gt10 arms were from Vector Cloning. Total cellular RNA was isolated from growing cultures of DDT1-MF2 cells by the guanidinium isothiocyanate-CsCl method[47]. Poly(A)$^+$ RNA was purified by chromatography on oligo(dT)-cellulose. Double-stranded cDNA was synthesized by oligo(dT)-primed reverse transcription of the poly(A)$^+$ RNA, followed by treatment with *E. coli* DNA polymerase and RNase[48]. The ends of the cDNA were blunted with T_4 DNA polymerase[28]. Following protection of the *Eco*RI sites with *Eco*RI methylase, *Eco*RI linkers were added[28]. *Eco*RI-digested cDNA was size-fractionated by agarose gel electrophoresis to obtain cDNAs between 2 and 7 kb long. The cDNA was ligated to the vector λ gt10 (ref. 49) and packaged *in vitro*. The resulting library was screened unamplified as described in Fig. 3 legend. The 1.3-kb *Hin*dIII fragment was labelled using [α-^{32}P]dCTP by nick-translation for use as a hybridization probe[28]. The cDNA inserts contained in the positive phage were subcloned into pUC13 or M13 mp19 for DNA sequence analysis[35-37]. Both strands of the clones were sequenced with no discrepancies.

```
                                                                                     70
CAGCGTTCAA GCTGCTGTTA GCAGGCACCG CGAGCCCCGG GCACCCCACG AGCTGAGTGT GCAGGACGCG
                                                                                    140
CCCCCAGCAC AGCCACCTAC AGCCGCTGAA TGAAGCTTCC AGGAGTCTGC CTCCGGCCGG CTGCGCCCCG
                                                                                    210
TCGGAGGTGC ACCCGCTGAG AGCGCCAGGG CACCAGAAAG CCGGTGCGCT CACCTGCTCG TCTGCCAGCG
                                                                                    264
ATG GGG CCA CCC GGG AAC GAC AGT GAC TTC TTG CTG ACA ACC AAC GGA AGC CAT
MET Gly Pro Pro Gly Asn Asp Ser Asp Phe Leu Leu Thr Thr Asn Gly Ser His
 1           2      ***                                 ***
                                                                                    318
GTG CCA GAC CAC GAT GTC ACT GAG GAA CGG GAC GAA GCA TGG GTG GTA GGC ATG
Val Pro Asp His Asp Val Thr Glu Glu Arg Asp Glu Ala Trp Val Val Gly MET
 19
                                                                                    372
GCC ATC CTT ATG TCG GTT ATC GTC CTG GCC ATC GTG TTT GGC AAC GTG CTG GTC
Ala Ile Leu MET Ser Val Ile Val Leu Ala Ile Val Phe Gly Asn Val Leu Val
 37
                                                                                    426
ATC ACA GCC ATT GCC AAG TTC GAG AGG CTA CAG ACT GTC ACC AAC TAC TTC ATA
Ile Thr Ala Ile Ala Lys Phe Glu Arg Leu Gln Thr Val Thr Asn Tyr Phe Ile
 55
                                                                                    480
ACC TCC TTG GCG TGT GCT GAT CTA GTC ATG GGC CTA GCG GTG GTG CCG TTT GGG
Thr Ser Leu Ala Cys Ala Asp Leu Val MET Gly Leu Ala Val Val Pro Phe Gly
 73                                             5
                                                                                    534
GCC AGT CAC ATC CTT ATG AAA ATG TGG AAT TTT GGC AAC TTC TGG TGC GAG TTC
Ala Ser His Ile Leu MET Lys MET Trp Asn Phe Gly Asn Phe Trp Cys Glu Phe
 91
                                                                                    588
TGG ACT TCC ATT GAT GTG TTA TGC GTC ACA GCC AGC ATT GAG ACC CTG TGC GTG
Trp Thr Ser Ile Asp Val Leu Cys Val Thr Ala Ser Ile Glu Thr Leu Cys Val
109
                                                                                    642
ATA GCA GTG GAT CGC TAC ATT GCT ATC ACA TCG CCA TTC AAG TAC CAG AGC CTG
Ile Ala Val Asp Arg Tyr Ile Ala Ile Thr Ser Pro Phe Lys Tyr Gln Ser Leu
127
                                                                                    696
CTG ACC AAG AAT AAG GCC CGA ATG GTC ATC CTA ATG GTG TGG ATT GTA TCC GGC
Leu Thr Lys Asn Lys Ala Arg MET Val Ile Leu MET Val Trp Ile Val Ser Gly
145                                                       8
                                                                                    750
CTT ACC TCC TTC TTG CCC ATT CAG ATG CAC TGG TAC CGT GCC ACC CAC CAG AAA
Leu Thr Ser Phe Leu Pro Ile Gln MET His Trp Tyr Arg Ala Thr His Gln Lys
163
                                                                                    804
GCC ATC GAC TGC TAT CAC AAG GAG ACT TGC TGC GAC TTC TTC ACG AAC CAG GCC
Ala Ile Asp Cys Tyr His Lys Glu Thr Cys Cys Asp Phe Phe Thr Asn Gln Ala
181
                                                                                    858
TAC GCC ATT GCT TCC TCC ATT GTA TCT TTC TAC GTG CCT CTA GTG GTC ATG GTC
Tyr Ala Ile Ala Ser Ser Ile Val Ser Phe Tyr Val Pro Leu Val Val MET Val
199
                                                                                    912
TTT GTC TAT TCC AGG GTC TTC CAG GTG GCC AAA AGG CAG CTC CAG AAG ATA GAC
Phe Val Tyr Ser Arg Val Phe Gln Val Ala Lys Arg Gln Leu Gln Lys Ile Asp
217     7
                                                                                    966
AAA TCT GAG GGA AGA TTC CAC TCC CCA AAC CTC GGC CAG GTG GAG CAG GAT GGG
Lys Ser Glu Gly Arg Phe His Ser Pro Asn Leu Gly Gln Val Glu Gln Asp Gly
235
                                                                                   1020
CGG AGT GGG CAC GGA CTC CGA AGG TCC TCC AAG TTC TGC TTG AAG GAG CAC AAA
Arg Ser Gly His Gly Leu Arg Arg [Ser] [Ser] Lys Phe Cys Leu Lys Glu His Lys
253
                                                                                   1074
GCC CTC AAG ACT TTA GGC ATC ATC ATG GGC ACA TTC ACC CTC TGC TGG CTG CCC
Ala Leu Lys Thr Leu Gly Ile Ile MET Gly Thr Phe Thr Leu Cys Trp Leu Pro
271
                                                                                   1128
TTC TTC ATT GTC AAC ATC GTG CAC GTG ATC CAG GAC AAC CTC ATC CCT AAG GAA
Phe Phe Ile Val Asn Ile Val His Val Ile Gln Asp Asn Leu Ile Pro Lys Glu
289
                                                                                   1182
GTT TAC ATC CTC CTT AAC TGG TTG GGC TAT GTC AAT TCT GCT TTC AAT CCC CTC
Val Tyr Ile Leu Leu Asn Trp Leu Gly Tyr Val Asn Ser Ala Phe Asn Pro Leu
307
                                                                                   1236
ATC TAC TGT CGG AGT CCA GAT TTC AGG ATT GCC TTC CAG GAG CTT CTA TGC CTC
Ile Tyr Cys Arg Ser Pro Asp Phe Arg Ile Ala Phe Gln Glu Leu Leu Cys Leu
325
                                                                                   1290
CGC AGG TCT TCT TCA AAA GCC TAT GGG AAC GGC TAC TCC AGC AAC AGT AAT GGC
Arg Arg [Ser] [Ser] [Ser] Lys Ala Tyr Gly Asn Gly Tyr Ser Ser Asn Ser Asn Gly
343
                                                                                   1344
AAA ACA GAC TAC ATG GGG GAG GCG AGT GGA TGT CAG CTG GGG CAG GAA AAA GAA
Lys Thr Asp Tyr MET Gly Glu Ala Ser Gly Cys Gln Leu Gly Gln Glu Lys Glu
361                           7a
                                                                                   1398
AGT GAA CGG CTG TGT GAG GAC CCC CCA GGC ACG GAA AGC TTT GTG AAC TGT CAA
Ser Glu Arg Leu Cys Glu Asp Pro Pro Gly Thr Glu Ser Phe Val Asn Cys Gln
379
                                                                                   1452
GGT ACT GTG CCT AGC CTT AGC CTT GAT TCC CAA GGG AGG AAC TGT AGT ACA AAT
Gly Thr Val Pro Ser Leu Ser Leu Asp Ser Gln Gly Arg Asn Cys Ser Thr Asn
397                                                 ***             ***
                                                                                   1517
GAC TCA CCG CTG TAA TGCAGGCTTT CTGCTTTTTA AGACCCCTCC CTGACAGGAC ACTAACCAGA
Asp Ser Pro Leu
415
                                                                                   1587
CTATTTAACT TGAGTGTAAT AACTTTAGAA TAAAACTGTA TAGAGATTTG CAGAAGGGGA GCATCCTTCT
                                                                                   1657
GCCCTTTTTT ATTTTATTTT TTTAAGCCGC AAAAATAGAG AGGGAGAGAA ACTGTACTTG AGTGCTTGTT
                                                                                   1727
TGTTTCTTGT GCAATTCAGT TCCTCTTTGC GTGGAACTTA AAAGTTTCTG TCTGAAGTAT GTTGGGTTCT
                                                                                   1797
AGAGGACTGT CTGTATGTTT AGATGATTTT CCATGCATCT ACCTCACTCG TCAAGTGTTA GGGGATACGC
                                                                                   1867
TGCTAGTAAT TTGTACCTGA AGGAAATTTT CCTTCCTGTA CCCTTACACT TGTCAATCCT GTGTCTTGGA
                                                                                   1937
CCTTTCTGCT GTGAATATAT ACTCTCTCCC GCTCCACTTA TTTGCTCAAA TGGAGTGTGT AGACAGGGAT
                                                                                   2007
CTTGAGGGAC AGCTTCAGTT GGTTTTTTTT TTTTTTTTGA GCAAAGTCTA AAGTTTACAG TAAATAAATT
                   2026
GTTTGACCAC GAAAAAAAAA
```

peptides to 560 nt 3′ to the termination codon. A hexanucleotide AATAAA occurs near the 3′ end, followed by a poly(A) stretch. The nucleotide sequence of the genomic clones is identical to that of the cDNA clones up to but not including the poly(A) tail. These data demonstrate an absence of introns within the coding and 3′-untranslated regions of the βAR gene but the possibility of introns in the untranslated region 5′ to the sequence isolated cannot be ruled out. While the lack of introns is unusual, it is not unprecedented in that sea urchin histone genes[38] and mammalian α- and β-interferon genes[39,40] are also uninterrupted. Previous studies with simian virus 40 and β-globin indicate that introns are necessary for efficient expression of those genes[41,42], so the lack of introns within the hamster βAR gene may account in part for its low level of expression.

Whereas most eukaryotic genes are translated using the first AUG encountered in the messenger RNA[43], the open reading frame (ORF) encoding the βAR peptides begins at the second AUG. The first AUG is followed by a termination codon after only 19 amino acids. This open reading frame is in-frame with the AUG codon beginning the βAR polypeptide. A similar situation has recently been reported for the oestrogen receptor, where the reading frame of that receptor is preceded by a short 20-amino-acid ORF[44]. Note that the DNA sequence around the AUG codon for the second ORF agrees well with the consensus eukaryotic translation initiation sequence[43] (CAGCGAUGG compared with CCGCCAUGG), whereas the sequence surrounding the first AUG does not.

Translation beginning with the second AUG as the initiation site would produce a polypeptide of 418 amino acids, with a M_r (46K) in close agreement with the apparent M_r of the deglycosylated βAR (49K)[11]. The protein sequence immediately following the initiator methionine residue is identical to the sequence determined for peptide 2. Apparently, like several other integral membrane proteins—for example, bovine opsin[45]—the βAR does not contain a cleavable signal sequence and may use internal signals for the insertion of the protein into the membrane. The receptor has been shown to have two sites of *N*-linked glycosylation (R.J.L. *et al.*, in preparation). Both sites are present in peptide 2, consistent with the results of peptide sequencing. The presence of two consensus protein kinase A and C sites[46] within the coding region (Fig. 4) agrees with *in vitro* phosphorylation results[20].

Hydropathicity profiles of the predicted βAR amino-acid sequence were produced using the analyses of Hopp and Woods[12] and of Kyte and Doolittle[13], with similar results. As shown in Fig. 5*a*, the βAR sequence should encode a largely hydrophobic polypeptide, with the N-terminal region of the receptor being predominantly hydrophobic and the C-terminal region of the molecule being hydrophilic. The βAR hydropathicity profile is remarkably similar to that of the rhodopsins (Fig. 5*b*), of which bacteriorhodopsin[7] is known to contain seven membrane-spanning helices. Not only does a similar pattern of repeating hydrophobic sequences 20–25 residues long occur in the predicted βAR sequence, but also the amino-acid composition of these postulated helices is similar to that of the rhodopsins, having a high proportion of proline and aromatic amino acids. The exact number of transmembrane regions remains to be determined. Amino-acid homology was apparent when the sequences corresponding to the postulated helices V, VI and VII of bovine opsin[14], which comprise the retinal binding site[15], were aligned with those for the analogous regions of βAR (Fig. 5*c*).

The sequence homology between βAR and rhodopsin parallels similarities in their function: both rhodopsin and βAR are involved in signal transduction mechanisms that involve interaction with the guanine nucleotide regulatory proteins transducin[16] and G_s (ref. 17), respectively. Moreover, it seems that phosphorylation has an important role in the regulation of both rhodopsin and βAR[18,19]. Rhodopsin is multiply phosphorylated at its C-terminus, which contains seven serine

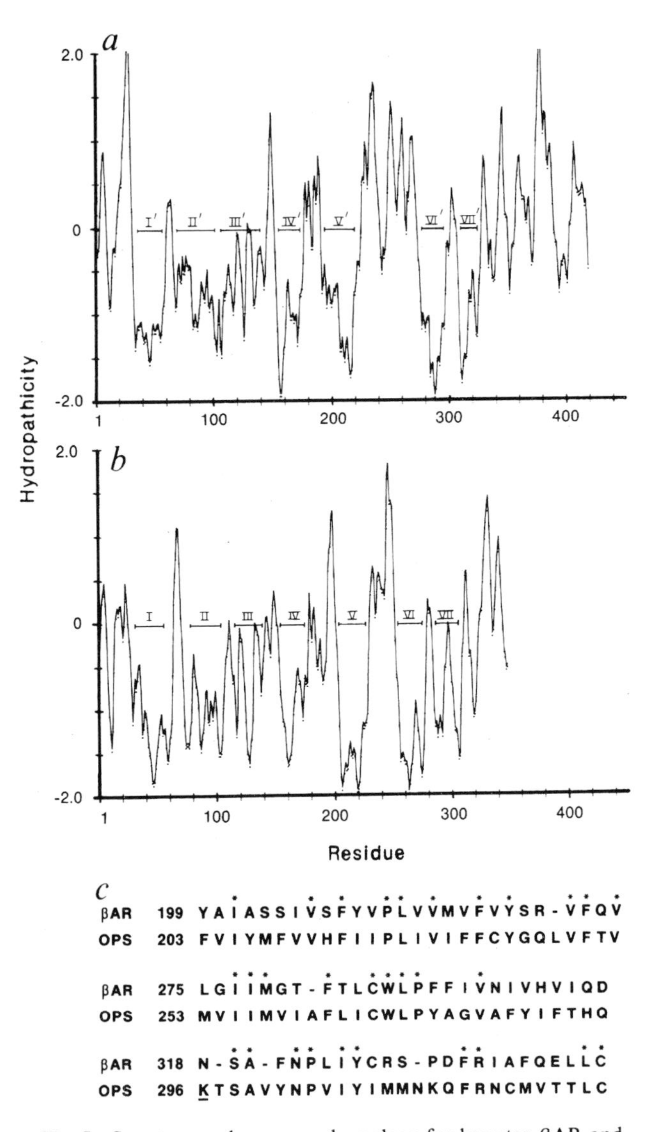

Fig. 5 Structure and sequence homology for hamster βAR and bovine opsin . Hydropathicity profiles are shown for hamster βAR (*a*) and bovine opsin[14] (*b*). Values were calculated by the method of Hopp and Woods[12]. Hydrophobicity increases with decreasing values. Horizontal lines indicate hydrophobic peptide regions 20–25 residues long. The putative transmembrane helices of bovine opsin are numbered as in ref. 14. We have designated the proposed transmembrane helices of βAR based on the analogous hydropathicity patterns of βAR and opsin. *c*, Amino-acid sequence homologies between the hamster βAR and bovine opsin (OPS). Positions of amino-acid identities are designated by asterisks. The underlined lysine residue (296) of bovine rhodopsin is involved in a Schiff base with retinal[15].

and threonine residues[18]. The C-terminus of the postulated sequence for βAR contains several serine and threonine residues that could serve as sites for phosphorylation[19].

The similarity in both the amino-acid sequence and proposed structure for the retinal binding site of rhodopsin and the analogous region in βAR suggests a specific mode of action for the ligands that modulate βAR function. We propose that these compounds interact with the βAR in a manner similar to that by which retinal interacts with opsin[21]; that is, they intercalate among the hydrophobic transmembrane helices and thereby determine whether the receptor is in its active or inactive conformation. According to this model, agonists and antagonists would mimic the action of photoactivated and native retinal. The generality of this hypothesis will be tested once the genes for similar membrane-bound receptors, such as those for leukotrienes, prostaglandins, dopamine and histamine, have been isolated. Our proposed model for the structure of βAR and its interaction with pharmacologically important ligands should, together with the biochemical and genetic studies now possible, provide a rational basis for a new approach to the development of more selective drugs.

We thank Lenora Davis, Jill D'Alonzo, Ester Hou, Tong Sun Kobilka, John Rodkey, Claudia Staniszewski, H. Vincent Strout Jr and Barbara Zemcik for technical assistance, Jeanne White and Lynn Tilley for manuscript preparation, and Drs Eugene Cordes, Ronald Ellis, Russell Kaufman, Paul Keller, Pete Kniskern, Richard Lebovitz, Mark Riemen and Michael Rosenblatt for helpful discussions. We particularly thank Dr Edward Scolnick for continued support and direction.

Received 10 March; accepted 11 April 1986.

1. Lefkowitz, R. J., Stadel, J. M. & Caron, M. G. *A. Rev. Biochem.* **52,** 159-186 (1983).
2. Shorr, R. G., Lefkowitz, R. J. & Caron, M. G. *J. biol. Chem.* **256,** 5820-5826 (1981).
3. Shorr, R. G., Strohsacker, M. W., Lavin, T. N., Lefkowitz, R. J. & Caron, M. G. *J. biol. Chem.* **257,** 12341-12350 (1982).
4. Homcy, C. J., Rockson, S. G., Countaway, J. & Egan, D. A. *Biochemistry* **22,** 660-668 (1983).
5. Benovic, J. L., Shorr, R. G. L., Caron, M. G. & Lefkowitz, R. J. *Biochemistry* **23,** 4510-4518 (1984).
6. Cerione, R. A. *et al. J. biol. Chem.* **259,** 9979-9982 (1984).
7. Engelman, D. M., Goldman, A. & Steitz, T. A. *Meth. Enzym.* **88,** 81-88 (1982).
8. Temeles, G. L., Gibbs, J. B., D'Alonzo, J. S., Sigal, I. S. & Scolnick, E. M. *Nature* **313,** 700-703 (1985).
9. Seidah, N. G. *et al. Proc. natn. Acad. Sci. U.S.A.* **81,** 2640-2644 (1984).
10. Delavier-Klutchko, C., Hocheke, J. & Strasberg, A. D. *FEBS Lett.* **169,** 151-155 (1984).
11. Stiles, G. L., Benovic, J. L., Caron, M. G. & Lefkowitz, R. J. *J. biol. Chem.* **259,** 8655-8663 (1984).
12. Hopp, T. P. & Woods, K. R. *Proc. natn. Acad. Sci. U.S.A.* **78,** 3824-3828 (1981).
13. Kyte, J. & Doolittle, R. F. *J. molec. Biol.* **157,** 105-132 (1982).
14. Nathans, J. & Hogness, D. S. *Cell* **34,** 807-814 (1983).
15. Hubbell, W. L. & Bownds, M. D. *A. Rev. Neurosci.* **2,** 17-34 (1979).
16. Stryer, L. *A. Rev. Neurosci.* **9,** 87-119 (1986).
17. Gilman, A. G. *Cell* **36,** 577-579 (1984).
18. Hargrave, P. A. *Prog. Retinal Res.* **1,** 1-51 (1982).
19. Benovic, J. L., Strasser, R. H., Caron, M. G. & Lefkowitz, R. J. *Proc. natn. Acad. Sci. U.S.A.* (in the press).
20. Benovic, J. L. *et al. J. biol. Chem.* **260,** 7094-7101 (1985).
21. Thomas, D. D. & Stryer, L. *J. molec. Biol.* **154,** 145-157 (1982).
22. Caron, M. G. & Lefkowitz, R. J. *J. biol. Chem.* **251,** 2374-2384 (1976).
23. Hewick, R. M., Hunkapiller, M. W., Hood, L. E. & Dreyer, W. J. *J. biol. Chem.* **256,** 7790-7797 (1981).
24. Spiess, J., Rivier, J. E., Rodkey, J. A., Bennett, C. D. & Vale, W. *Proc. natn. Acad. Sci. U.S.A.* **76,** 2974-2978 (1979).
25. Laemmli, U. K. *Nature* **117,** 680-685 (1970).
26. Towbin, H., Staehelin, T. & Gordon, J. *Proc. natn. Acad. Sci. U.S.A.* **76,** 4350-4356 (1979).
27. Johnson, D. A. & Elder, J. H. *J. exp. Med.* **159,** 1751-1756 (1983).
28. Maniatis, T., Fritsch, E. F. & Sambrook, J. *Molecular Cloning, A Laboratory Manual* (Cold Spring Harbor Laboratory, New York, 1982).
29. Merrifield, R. B. *J. Am. chem. Soc.* **85,** 2149-2154 (1963).
30. Frischauf, A. M., Lehrach, H., Paustka, A. & Murray, N. *J. molec. Biol.* **170,** 827-842 (1983).
31. Lauffer, L. *et al. Nature* **318,** 334-338 (1985).
32. Benton, W. D. & Davis, R. W. *Science* **196,** 180-182 (1977).
33. Ullrich, A., Berman, C. H., Dull, T. J., Gray, A. & Lee, J. M. *EMBO J.* **3,** 361-364 (1984).
34. Southern, E. M. *J. molec. Biol.* **98,** 503-517 (1975).
35. Maxam, A. M. & Gilbert, W. *Proc. natn. Acad. Sci. U.S.A.* **74,** 560-565 (1977).
36. Vieira, J. & Messing, J. *Gene* **19,** 259-268 (1982).
37. Sanger, F., Nicklen, S. & Coulson, A. R. *Proc. natn. Acad. Sci. U.S.A.* **74,** 5463-5467 (1977).
38. Schaffner, W. *et al. Cell* **14,** 655-671 (1978).
39. Nagata, S., Mantei, N. & Weissman, C. *Nature* **287,** 401-408 (1980).
40. Lawn, R. M. *et al. Nucleic Acids Res.* **9,** 1045-1052 (1981).
41. Volckaert, G., Feunteun, J., Crawford, L. V., Berg, P. & Fiers, W. J. *J. Virol.* **30,** 674-682 (1979).
42. Hamer, D. H. & Leder, P. *Cell* **18,** 1299-1302 (1979).
43. Kozak, M. *Nucleic Acids Res.* **12,** 857-872 (1984).
44. Green, S. *et al. Nature* **320,** 134-139 (1986).
45. Friedlander, M. & Blobel, G. *Nature* **318,** 338-343 (1985).
46. Feramisco, J. L., Glass, D. B. & Krebs, E. G. *J. biol. Chem.* **255,** 4240-4245 (1980).
47. Chirgwin, J. M., Przybyla, A. E., MacDonald, R. J. & Rutter, W. J. *Biochemistry* **18,** 5294-5299 (1979).
48. Gubler, U. & Hoffman, B. J. *Gene* **25,** 263-269 (1983).
49. Huynh, T. V., Young, R. A. & Davis, R. W. in *DNA Cloning a Practical Approach* Vol. 1, 49-78 (IRL, Oxford, 1985).

The patch-clamp technique

Electrophysiological investigations employing intracellular electrodes treat the plasma membrane as a black box. Measurements are made of the transmembrane electrical potential difference or of transmembrane current flow. The influences of neurotransmitters or of electrical stimuli are described in terms of changes in membrane conductance or membrane permeability to various ions. While this approach has been enormously successful in defining and accounting for the macroscopic transmembrane ionic currents associated with, for example, the nerve action potential [1], it is by its nature incapable of giving direct information about the underlying microscopic events.

Electrical currents in cells are mediated mainly by a class of proteins in the plasma membrane called ion channels or pores. The first direct observation of current flowing through single ion channels came from studies on bimolecular lipid membranes in the presence of antibiotics [2]. In intact cells the problem is to detect single-channel currents in the presence of background electrical noise. Conventional intracellular microelectrode methods for current measurement are associated with a background noise of at least 100 pA, whereas the current flowing when a single channel opens is only a small fraction of this background noise.

Neher and Sakmann [3] solved this problem through the development of the patch-clamp method. Instead of inserting a microelectrode into a cell they pressed a microelectrode tip on to the surface, effectively isolating a patch of membrane. The intrinsic noise increases with the area of the membrane under study, and when a small area (1–10 μm^2) is isolated the extraneous noise levels are so low that the picoampere currents flowing through single channels can be measured directly.

The seal between the tip of the microelectrode and the outer surface of the cell membrane has, under suitable conditions (fire-polished and clean micropipette tip, together with clean membrane surface), a high electrical resistance [in the order of giga (10^9) Ω] and is surprisingly mechanically stable. The discovery in 1980 of this high-resistance seal by Sigworth and Neher [4] turned out to be very important as it made possible entirely new types of experiments [4,5]. In the paper we have selected [6], Hamill et al. described the new patch-clamp techniques in such detail and with such clarity that they surprisingly quickly became *the* techniques for investigating the electrical properties of cell membranes.

It was expected in the early 1980s that patch-clamp single-channel current recording would quickly reveal the key properties of the most important ion channels in electrically excitable tissues. This indeed turned out to be the case, and this book contains some examples of other important landmark papers based on the patch-clamp technique (see Sections 5, 6 and 7). What perhaps came as a surprise was how the whole focus of electrophysiology changed and how the new experimental techniques came to be preferred.

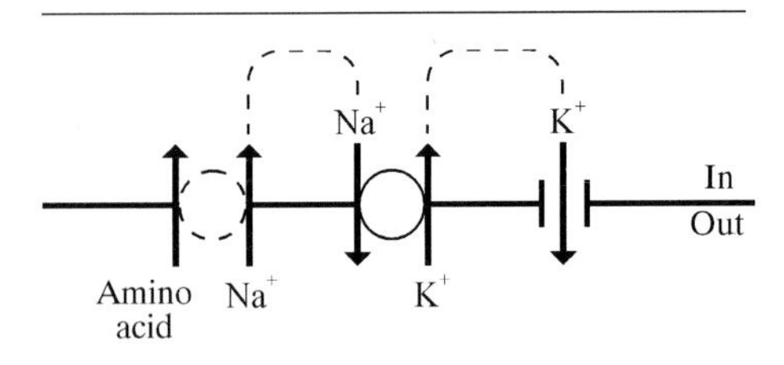

Figure 4.1 Simplified diagram illustrating the three different types of ion transporters present in cell membranes: carriers (example Na^+–amino acid co-transporter), active pump (example Na^+–K^+ ATPase) and channel (example K^+ channel).

Although the main excitement in the field of membrane transport has been due to progress in our understanding of ion channels and their functions, it should not be forgotten that these pores represent just one class of membrane-transport proteins dealing with ion movements. Figure 4.1 shows a simple minimal membrane model that illustrates what could be regarded as the most fundamental transport systems occurring in animal cell membranes. It shows an example of each of the three main types of membrane ion transporters: channels, carriers and pumps. The resting membrane potential in virtually all cells (inside negative with respect to outside) is due to the large transmembrane K^+ gradient and the presence of resting K^+ channels. The transmembrane K^+ gradient is established and maintained by the ATP-driven Na^+,K^+-pump. The transmembrane Na^+ gradient, which results from the operation of the Na^+,K^+-pump together with the existence of the membrane potential, is used as the driving force for the up-hill movement of many different substrates (for example a number of neutral amino acids) as well as ions. In the steady state, K^+ recirculates via the Na^+,K^+-pump and K^+ channel, whereas Na^+ recirculates via the pump and, in the example chosen in Figure 4.1, the Na^+–amino acid carrier (co-transporter). Figure 4.1 illustrates the functional linkage between these three different types of transporter. Although the most spectacular progress has been in the characterization of channel properties, the patch-clamp technique (whole-cell configuration) has also allowed careful studies of electrogenic carriers such as the Na^+–L-alanine co-transporter [7] or the Na^+–Ca^{2+} exchanger [8].

Two very important extensions of the patch-clamp technique were not described in the original methods paper [6]. The first is the measurement of electrical capacitance in the whole-cell recording configuration. By monitoring membrane capacitance, which is proportional to the cell surface area, it is possible to follow the time courses of membrane insertion into the cell surface by exocytosis or retrieval of membrane from the surface by endocytosis [9]. This has become a very useful technique for investigations of secretion by exocytosis [10].

The other extension of the patch-clamp technique is to work with cells *in situ*. Since the success of the patch-clamp technique depends on a high seal-resistance between the micropipette and the surface cell membrane, it was initially applied almost exclusively to enzymically cleaved single cells or cell clusters or to cells in culture. Several years later a tissue-slice preparation for patch-clamp recordings from neurons of the mammalian central nervous system was described by Sakmann and his collaborators [11]. By examining thin slices, individual neurons could be identified and then exposed by a stream of physiological saline blown through a cleaning pipette. This technique has become very important for the characterization of synaptic events in the nervous system.

Patch-clamp methods have changed radically our views on the preferred cell types for electrophysiological studies. In the 1950s the success of voltage-clamp experiments, as reviewed by Hodgkin on the squid axon [1], initiated an era in which the study of large cells provided by far the most useful information. Since the 1980s the patch-clamp experiments have been best suited to small round cells and studies on mammalian gland cells have been particularly successful in clarifying aspects of cellular physiology that could not have been investigated with classic intracellular microelectrode techniques [12]. Early experiences with patch-clamp techniques were summarized in the first edition of *Single-Channel Recording* edited by Sakmann and Neher [13]. More recent reviews on patch-clamp techniques can be found in the second edition of this book [14] and in a large multi-author volume edited by Rudy and Iverson [15].

The whole-cell recording technique has become much more important for cell physiological investigations than originally envisaged [16]. One example is provided by the ability to combine patch-clamp whole-cell current recording with simultaneous measurement of the free intracellular Ca^{2+} concentration, using a Ca^{2+}-sensitive fluorescent probe introduced into the cytoplasm via the patch pipette [17]. This approach has proven immensely useful in numerous studies on Ca^{2+} signalling mechanisms on very many different systems.

References

1. Hodgkin, A.L. (1976) *J. Physiol. (London)* **263**, 1–21
2. Hladky, S.B. and Haydon, D.A. (1970) *Nature (London)* **225**, 451–453
3. Neher, E. and Sakmann, B. (1976) *Nature (London)* **260**, 799–802
4. Sigworth, F.J. and Neher, E. (1980) *Nature (London)* **287**, 447–449
5. Horn, R. and Patlak, J. (1980) *Proc. Natl. Acad. Sci. U.S.A.* **77**, 6930–6934
6. Hamill, O.P., Marty, A., Neher, E., Sakmann, B. and Sigworth, F.J. (1981) *Pflügers Arch.* **391**, 85–100
7. Jauch, P., Petersen, O.H. and Lauger, P. (1986) *J. Membr. Biol.* **94**, 99–115
8. Hilgemann, D.W., Matsuoka, S., Nagel, G.A. and Collins, C. (1992) *J. Gen. Physiol.* **100**, 905–932
9. Neher, E. and Marty, A. (1982) *Proc. Natl. Acad. Sci. U.S.A.* **79**, 6712–6717
10. Fernandez, J.M., Neher, E. and Gomperts, B.D. (1984) *Nature (London)* **312**, 453–455
11. Edwards, F.A., Konnerth, A., Sakmann, B. and Takahashi, T. (1989) *Pflügers Arch.* **414**, 600–612
12. Petersen, O.H. (1992) *J. Physiol. (London)* **448**, 1–51
13. Sakmann, B. and Neher, E. (eds.) (1983) *Single-Channel Recording*, Plenum, New York
14. Sakmann, B. and Neher, E. (eds.) (1995) *Single-Channel Recording*, 2nd edn., Plenum, New York
15. Rudy, B. and Iverson, L.E. (eds.) (1992) *Methods Enzymol.* **207**, 1–917
16. Sigworth, F.J. (1986) *Fed. Proc.* **45**, 2673–2677
17. Osipchuk, Y.V., Wakui, M., Yule, D.I., Gallacher, D.V. and Petersen, O.H. (1990) *EMBO J.* **9**, 697–704

Hamill et al. (1981) Pflügers Arch. **391**, 85–100

Improved Patch-Clamp Techniques for High-Resolution Current Recording from Cells and Cell-Free Membrane Patches

O. P. Hamill, A. Marty, E. Neher, B. Sakmann, and F. J. Sigworth

Max-Planck-Institut für biophysikalische Chemie, Postfach 968, Am Fassberg, D-3400 Göttingen, Federal Republic of Germany

Abstract. 1. The extracellular patch clamp method, which first allowed the detection of single channel currents in biological membranes, has been further refined to enable higher current resolution, direct membrane patch potential control, and physical isolation of membrane patches.

2. A description of a convenient method for the fabrication of patch recording pipettes is given together with procedures followed to achieve giga-seals i.e. pipette-membrane seals with resistances of 10^9-10^{11} Ω.

3. The basic patch clamp recording circuit, and designs for improved frequency response are described along with the present limitations in recording the currents from single channels.

4. Procedures for preparation and recording from three representative cell types are given. Some properties of single acetylcholine-activated channels in muscle membrane are described to illustrate the improved current and time resolution achieved with giga-seals.

5. A description is given of the various ways that patches of membrane can be physically isolated from cells. This isolation enables the recording of single channel currents with well-defined solutions on both sides of the membrane. Two types of isolated cell-free patch configurations can be formed: an inside-out patch with its cytoplasmic membrane face exposed to the bath solution, and an outside-out patch with its extracellular membrane face exposed to the bath solution.

6. The application of the method for the recording of ionic currents and internal dialysis of small cells is considered. Single channel resolution can be achieved when recording from whole cells, if the cell diameter is small (<20 μm).

7. The wide range of cell types amenable to giga-seal formation is discussed.

Key words: Voltage-clamp – Membrane currents – Single channel recording – Ionic channels

Introduction

The extracellular patch clamp technique has allowed, for the first time, the currents in single ionic channels to be observed (Neher and Sakmann 1976). In this technique a small heat-polished glass pipette is pressed against the cell membrane, forming an electrical seal with a resistance of the order of 50 MΩ (Neher et al. 1978). The high resistance of this seal ensures that most of the currents originating in a small patch of membrane flow into the pipette, and from there into current-measurement circuitry. The resistance of the seal is important also because it determines the level of background noise in the recordings.

Recently it was observed that tight pipette-membrane seals, with resistances of 10–100 GΩ, can be obtained when precautions are taken to keep the pipette surface clean, and when suction is applied to the pipette interior (Neher 1981). We will call these seals "giga-seals" to distinguish them from the conventional, megaohm seals. The high resistance of a "giga-seal" reduces the background noise of the recording by an order of magnitude, and allows a patch of membrane to be voltage-clamped without the use of microelectrodes (Sigworth and Neher 1980).

Giga-seals are also mechanically stable. Following withdrawal from the cell membrane a membrane vesicle forms occluding the pipette tip (Hamill and Sakmann 1981; Neher 1981). The vesicle can be partly disrupted without destroying the giga-seal, leaving a cell-free membrane patch that spans the opening of the pipette tip. This allows single channel current recordings from isolated membrane patches in defined media, as well as solution changes during the measurements (Horn and Patlak 1980; Hamill and Sakmann 1981). Alternatively, after giga-seal formation, the membrane patch can be disrupted keeping the pipette cell-attached. This provides a direct low resistance access to the cell interior which allows potential recording and voltage clamping of small cells.

These improvements of the patch clamp technique make it applicable to a wide variety of electrophysiological problems. We have obtained giga-seals on nearly every cell type we have tried. It should be noted, however, that enzymatic treatment of the cell surface is required in many cases, either as part of the plating procedure for cultured cells, or as part of the preparation of single cells from adult tissues.

In this paper we describe the special equipment, the fabrication of pipettes, and the various cell-attached and cell-free recording configurations we have used. To illustrate the capabilities of the techniques we show recordings of AChR-channel currents in frog muscle fibres and rat myoballs, as well as Na currents and ACh-induced currents in bovine chromaffin cells.

Part I

Techniques and Preparation

Giga-seals are obtained most easily if particular types of pipettes are used and if certain measures of cleanliness are

Send offprint requests to B. Sakmann at the above address

taken. The improved resolution requires a more careful design of the electronic apparatus for lowest possible background noise. These experimental details will be described in this section.

1. Pipette Fabrication and Mechanical Setup

Pipette Fabrication. Patch pipettes are made in a three-stage process: pulling a pipette, coating of its shank with Sylgard, and the final heat polishing of the pipette tip.

First step-pulling: Patch pipettes can be pulled from flint glass or borosilicate glass. Flint glass has a lower melting point, is easier to handle, and forms more stable seals than borosilicate glass, which however has better electrical properties (see below). We routinely use commercially available flint capillaries made for hemocytometric purposes (Cee-Bee hemostat capillaries), or melting point determination capillaries. The borosilicate (Pyrex) glass is in the form of standard microelectrode capillaries (Jencons, H 15/10). The pipettes are pulled in two stages using a vertical microelectrode puller (David Kopf Instruments, Tujunga, CA, USA, Model 700C) and standard Nichrome heating coils supplied with it. In the first (pre-)pull the capillary is thinned over a length of 7 – 10 mm to obtain a minimum diameter of 200 µm. The capillary is then recentered with respect to the heating coil and in the second pull the thinned part breaks, producing two pipettes. To obtain large numbers of pipettes of similar properties it is advisable to use a fixed pulling length and fixed settings for the two stages. For example with Cee-Bee capillaries and the David Kopf puller we use the following settings. The prepull is made at 19A with a pulling length of 8 mm. The thinned part of the capillary is then recentered by a shift of approximately 5.5 mm. The final pull is made at a critical heat setting around 12 A. Slight variations of the heat setting around this value produce tip openings between fractions of a µm and several µm. We aim at openings between 1 and 2 µm. These pipettes, then, have steep tapers at the very tip (see for example Fig. 10C). The Pyrex capillaries require higher heat settings of 24 and 15 A for the two stages; the resulting pipettes have thicker walls at the tip, and often the tips break unevenly in pulling.

Second step-coating: In order to reduce the pipette-bath capacitance and to form a hydrophobic surface, pipette shanks are coated with Sylgard to within about 50 µm from the tip. Already-mixed Sylgard can be stored for several weeks at $-20°$ C. It is applied to the pipette using a small glass hook taking care that the very tip remains uncoated. We apply the Sylgard while the pipette is mounted in a microforge and cure it by bringing the heated filament close to the pipette for a few seconds. The Sylgard coating is not required for giga-seal formation; it only serves to improve background noise.

Third step – heat polishing: Polishing of the glass wall at the pipette tip is done on a microforge shortly after Sylgard coating. We observe this step at 16×35 magnification using a compound microscope with a long-distance objective. The heat is supplied by a V-shaped platinum-iridium filament bearing a glass ball of $\simeq 0.5$ mm diameter. The filament is heated to a dull red glow and a stream of air is directed towards the glass ball, restricting the heat to its immediate vicinity. The tip of the pipette is brought to within 10 – 20 µm of the ball for a few seconds; darkening of the tip walls indicates polishing of the tip rim. If the pipettes are coated with Sylgard, it is preferable to heat-polish them within an

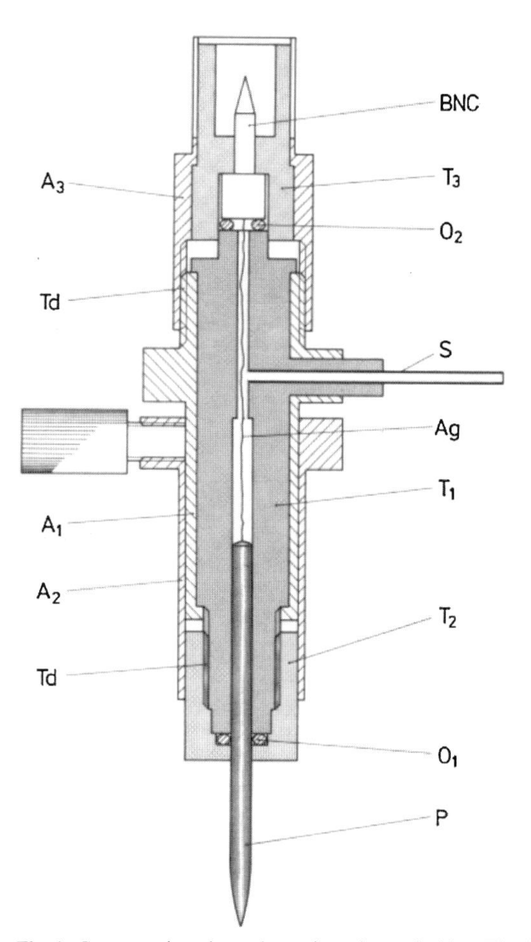

Fig. 1. Cross section through suction pipette holder. The holder serves two basic functions, firstly to provide electrical connection between the patch pipette solution and the pin of a BNC connector, and secondly to allow suction or pressure to be applied to the pipette interior. The holder has a Teflon body T_1 with a central bore for tight fitting of a patch pipette P and a chlorided silver wire Ag which is soldered to the pin of a BNC connector. The BNC pin is held by Teflon piece T_3. The pipette is tightened by a screw cap T_2. Outlet S connects to Silicone rubber tubing for application of suction or pressure to the inner compartment, which is made airtight by the O-rings O_1 and O_2. A_1 and A_3 are aluminium shields to the body; A_2 is a sliding shield to the pipette. *Td* indicates screw threads. The unit (without pipette) is 55 mm long

hour after coating; after this time, it is difficult to obtain a steep taper at the pipette tip. When pipettes have to be stored more than a few hours they should be cleaned before use by immersion in methanol while a positive pressure is applied to their interior.

Sylgard-coated patch pipettes usually do not fill by capillary forces when their tip is immersed into solution. They can be filled quickly by first sucking in a small amount of pipette solution and then back-filling. All the solutions used for filling should be filtered using effective pore sizes smaller than 0.5 µm. We use pipettes with resistance values in the range 2 – 5 MΩ. These have opening diameters between 0.5 and 1 µm.

Mechanical Setup. The patch pipettes are mounted on a suction pipette holder shown schematically in Fig. 1. It consists of inner parts made of Dynal or Teflon T_1, T_2, T_3) and is shielded by metal caps (A_1, A_2, A_3). The outlet S is connected to silicone rubber tubing through which suction is applied, usually by mouth. It is critical that the O-rings, O_1 and O_2 fit tightly. Otherwise the pipette tip can move slightly

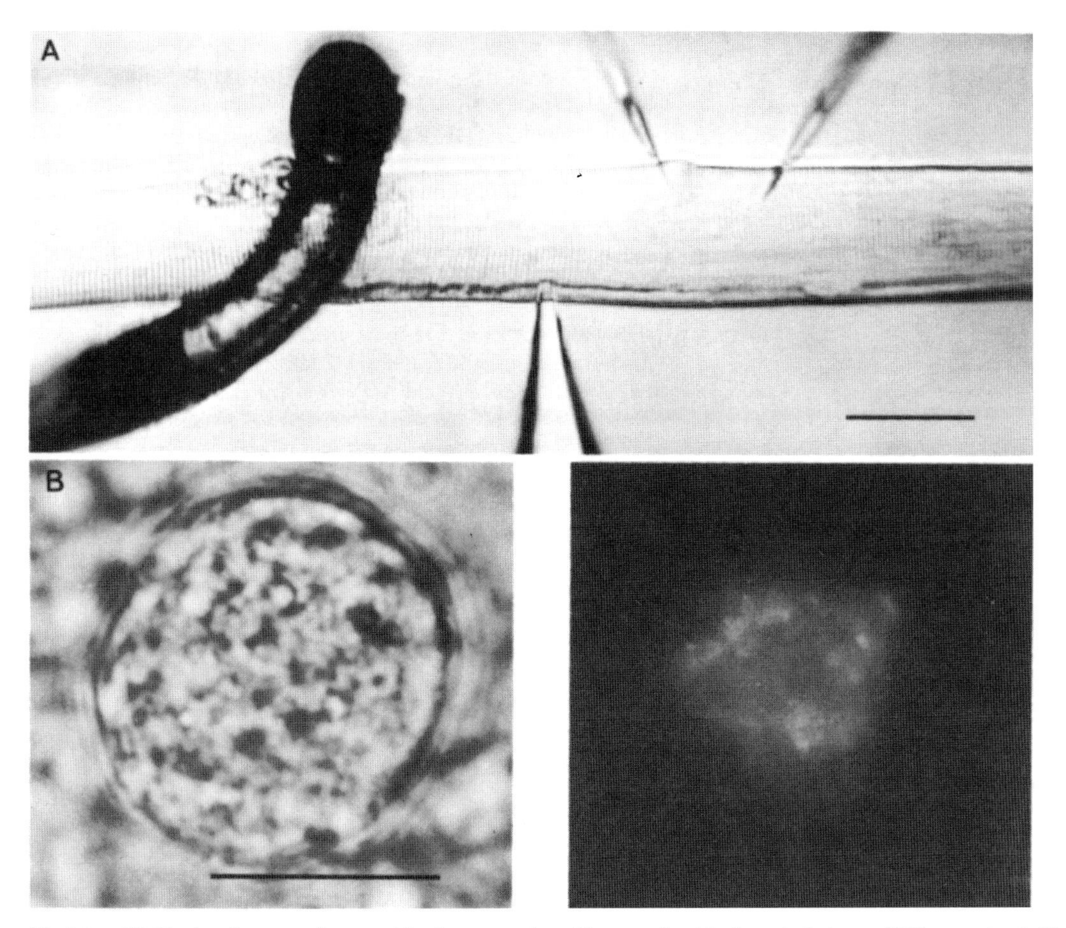

Fig. 2A and B. Single cell preparations used for demonstration of improved patch clamp techniques. (**A**) Enzyme treated frog (*R. temporaria*) cutaneous pectoris muscle fibre. The end-plate region of this fibre is viewed by Normarski optics. The fibre is supported by a glass hook. The fibre is stripped of its nerve terminal. The patch pipette is seen in contact with the synaptic trough. Two intracellular glass microelectrodes are used here to voltage clamp the fibre locally. Alternatively, the measurement can be performed at the natural resting potential without intracellular electrodes. (**B**) Primary culture of rat myoball. The same myoball is viewed in bright field optics on the left side and, using fluorescence microscopy, on the right side after labelling with fluorescent Rhodamine-conjugated α-BuTX. The fluorescence pattern illustrates the "patchy"/distribution of AChR's in this preparation. Calibration bars: 50 μm (upper), and 25 μm (lower)

during suction, tearing off a membrane patch from the cell. The pipette holder connects to a BNC connector of the amplifier head stage which is mounted on a Narashige MO-103 hydraulic micromanipulator. This, in turn, is mounted onto another manipulator for coarse movements (Narishige MM 33). The pipette holder should be repeatedly cleaned by methanol and a jet of nitrogen.

2. Preparations

The development of giga-seals requires a "clean" plasma membrane; that is, no sign of a surface coat should be detectable in conventionally-stained EM-sections. This requirement is met in many tissue-cultured cells, for example myotubes, spinal cord cells and dorsal root ganglion cells. In adult tissue however individual cells are covered with surface coats and enzymatic cleaning of the cell surface must precede the experiment. The exact protocol of enzymatic cleaning varies from tissue to tissue (see Neher 1981). Here we describe a treatment procedure adequate for frog skeletal muscle fibres. We also briefly describe the preparation of rat myoballs. These cells, as well as the chromaffin cells, require no enzyme treatment before use.

a) End-Plate Region of Frog Muscle Fibres. From innervated muscle a useable preparation can be obtained within 2–3 h using the following procedure. The whole cutaneous pectoris muscle is bathed for an hour at room temperature in normal frog Ringer solution containing 1 mg/ml collagenase (Sigma type I). At this point overlying fibre layers can be easily cut away, such that a monolayer of fibres remains. The muscle endplate region is subsequently superfused with Ringer solution containing 0.07 mg/ml Protease (Sigma, type VII) for 20–40 min. The tendinous insertions of the muscle fibres are protected by small 3–7 mm guides made from Perspex to restrict the flow of protease containing solution to the endplate region of the muscle (Neher et al. 1978). This procedure results in a preparation of ≈20 fibres with ends firmly attached to skin and sternum. When a single fibre is viewed the bare synaptic trough can be easily seen with a ×16 objective (Zeiss 0.32) and ×16 eyepieces using Nomarski interference contrast optics (Fig. 2A). Although currents can be recorded from the synaptic area, the perisynaptic AChR density within 10–50 μm of the synaptic trough is high enough in most preparations to allow recording of ACh activated single channel currents at low ACh concentrations (<1 μM). Preparations kept in phosphate-buffered Ringer solution remain viable and can be used for up to 48 h when kept at <10°C. All bath solutions contain 10^{-8} M Tetrodotoxin to avoid muscle contraction during the dissection.

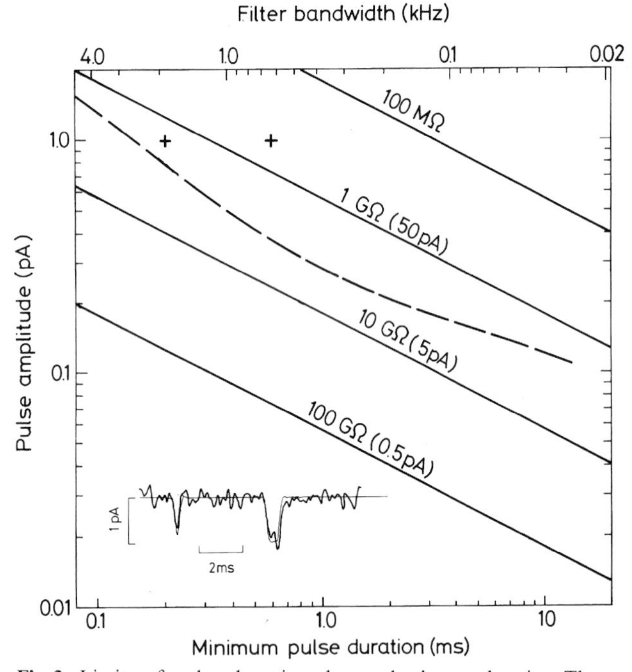

Fig. 3. Limits of pulse detection due to background noise. The relationship between filter settings (*top scale*) or minimum detectable rectangular pulse durations (bottom scale) and the pulse amplitudes (*vertical scale*) is shown for various background noise levels. The solid lines represent theoretical limits imposed by Johnson noise and shot noise sources and are discussed below. The dashed curve represents the background noise in an actual recording situation (50 GΩ seal on a myoball) and was computed from the spectrum in Fig. 5. Using this curve and the top scale, an appropriate low-pass filter setting can be found for observing single-channel currents whose amplitude is given by the vertical scale. The filter bandwidths (−3dB frequency, Gaussian or Bessel response) are chosen to make the standard deviation of the background noise $^1/_8$ of the given amplitude. When the bottom scale is used the curve shows whether or not a current pulse of given amplitude and duration can be recognized in the presence of the background noise. Each combination of amplitude and duration corresponds to a point in the figure. All points lying above the curve represent pulses that can be detected reliably. For example, 1 pA channel currents can be detected if the duration is at least 0.15 ms; pulses of 0.2 pA amplitude can be detected only if the duration is greater than 2.2 ms. Comparison of the top and bottom scales in the figure then gives the filter setting yielding the best signal-to-noise ratio (peak signal to rms noise) for a pulse of the given minimum duration. However, at this filter setting, the time-course of such a minimum-width pulse is distorted. This is illustrated in the *inset*, which shows the simulated response of our recording system at 2 kHz to two rectangular pulses, both with original amplitudes of 1 pA but with durations of 0.2 and 0.6 ms. The thin trace shows the response in the absence of noise; in the thicker trace a recording of background noise (same data as in Fig. 5) has been added. The parameters of the two pulses are indicated as crosses in the figure. The shorter pulse represents the movement of 1250 elementary charges; it is clearly detectable, but the *a. posteriori*, simultaneous estimation of pulse amplitude and duration is impossible. The rectangular form of the longer pulse is just recognizable, however, allowing these parameters to be estimated. It can be seen from this example that for kinetic analysis of channel gating, the shortest mean event duration should be considerably longer (preferably by an order of magnitude or more) than the minimum given in the figure. The solid lines show the ultimate theoretical detection limits imposed by Johnson noise in the seal and membrane resistance, computed from Eq. (1). Since seal resistances above 100 GΩ are often observed, it should in principle be possible to resolve much smaller pulses than is presently possible. These lines also represent the resolution limits that would be imposed by shot noise (Eq. 2) in channels carrying the indicated current levels in the case that shot noise is the predominant noise source. For the calculation of the detection limits in this figure a filter with Gaussian frequency response was assumed; the filter bandwidth was chosen to give a risetime t_r equal to 0.9 times the minimum pulse duration. The minimum detectable pulse amplitude was taken to be 8 times the standard deviation σ of the background noise. A minimum-width pulse is attenuated to 6.7 σ by the filter; with a detection threshold of 4.7 σ the probability of missing an event is less than 0.02, while the probability per unit time of a background fluctuation being mistaken for a pulse is less than $3 \times 10^{-6}/t_r$

b) Myoballs from Embryonic Rat Muscle. The procedure to obtain spherical "myoballs" is essentially the same as that used by other laboratories (Horn and Brodwick 1980). The growth medium (DMEM + 10% fetal calf serum) is changed on day 3 and on day 6 after plating of the cells on 18 mm cover slips placed into culture dishes. For 2 days starting on day 8, medium containing 10^{-7} M colchicine is used. Thereafter normal growth medium is used again and changed every third day. This procedure results in ≃ 100 spherical myoballs of 30 − 80 μm diameter (Fig. 2B) per cover slip. A single cover slip can be cracked with a scapel blade into 8 − 10 small pieces which can be transferred individually into the experimental chamber. The culture medium is exchanged for normal bath solution before the experiment. This solution has the following composition (in mM): 150 NaCl, 3 KCl, 1 $MgCl_2$, 1 $CaCl_2$, 10 HEPES, pH adjusted to 7.2 by NaOH.

As visualized with fluorescent α-bungarotoxin, the ACh receptors are unevenly distributed in myoballs (Fig. 2B); however in virtually all patches single channel currents could be recorded with low ACh concentrations (< 1 μM). For experiments with ACh-activated channels it is advisable to work within 2 − 5 days following colchicine treatment. During a later period 5 − 10 days following colchicine treatment myoballs are a suitable preparation for investigating properties of electrically excitable Na and K channels as well as Ca-dependent K channels.

c) Chromaffin Cells. As an example of cells obtained by enzymatic dispersion of an adult organ we use bovine chromaffin cells. These cells are dispersed by perfusion with collagenase of the adrenal gland (Fenwick et al. 1978), and are subsequently kept in short term culture for up to 8 days (Medium 199, supplemented with 10% fetal calf serum and 1 mg/ml BSA).

3. Background Noise and Design of Recording Electronics

One of the main advantages of the giga-seal recording technique is the improvement, by roughly an order of magnitude, in the resolution of current recordings. The resolution is limited by background noise from the membrane, pipette and recording electronics.

Theoretical Limits. Apart from noise sources in the instrumentation there are inherent limits on the resolution of the patch clamp due to the conductances of the patch membrane and the seal. One noise source is the Johnson noise of the membrane-seal combination, which has a one-sided current spectral density

$$S_I(f) = 4\,kT\,Re\,\{Y(f)\} \tag{1}$$

where $4kT = 1.6 \times 10^{-20}$ Joule at room temperature, and $Re\,\{Y(f)\}$ is the real part of the admittance, which depends, in general, on the frequency f. If the membrane-seal parallel combination is modelled as a simple parallel $R - C$ circuit, then $Re\,\{Y(f)\} = 1/R$. Integrating the resulting (in this case constant) spectral density over the frequency range of interest gives the noise variance, which decreases with increasing patch resistance. From the variance we have calculated the size of the smallest detectable current pulses, which is plotted in Fig. 3 against the minimum pulse durations for various values of R.

Another background noise source is the "shot noise" expected from ions crossing the membrane, for example through leakage channels or pumps. Although the size and spectrum of this noise depends on details in the ion translocation process, a rough estimate of the spectral density can be made assuming that an ion crosses the membrane rapidly (Stevens 1972; Läuger 1975),

$$S_I = 2\ Iq \tag{2}$$

where q is the effective charge of the current carrier (we assume a unit charge $q_e = 1.6 \times 10^{-19}$ Coulomb) and I is the unidirectional current. The shot noise at $I = 0.5$ pA is nearly the same as the Johnson noise with $R = 100\ \mathrm{G}\Omega$. If R is determined mainly by "leakage channels" in the membrane patch, the shot noise may be comparable to the Johnson noise in size.

Intrinsic Noise in the Pipette. As can be seen in Fig. 3, the background noise in our present recording system is several times larger than the limit imposed by the patch resistance. The excess results from roughly equal contributions from noise sources in the pipette and sources in the current-to-voltage converter. We are aware of three main sources of Johnson noise in the pipette, each of which can be roughly modelled by a series $R-C$ circuit. The current noise spectral density in such a circuit is given by (1) with

$$Re\,\{Y(f)\} = \frac{\alpha^2}{R(I+\alpha^2)}, \tag{3}$$

where $\alpha = 2\,\pi f RC$. In the high frequency limit (α large) this approaches $1/R$; in the low frequency limit $Re\,\{Y\} = (2\,\pi fC)^2 R$, which increases with frequency.

The potentially most serious noise source arises from a thin film of solution that creeps up the outer wall of an uncoated pipette. Evidence for the presence of this film is that, when a small voltage step is applied to the pipette, a slow capacitive transient is observed whose size and time constant are influenced by air currents near the pipette. The film apparently has a distributed resistance R of the order of 100 MΩ, and a distributed wall capacitance $C \approx 3$ pF. In the high frequency regime the noise (like that in a 100 MΩ resistor) is very large. A Sylgard coating applied to the pipette reduces the noise considerably: the hydrophobic surface prevents the formation of a film, and the thickness of the coating reduces C.

Secondly, we find that the bulk conductivity of the pipette glass can be significant. The Cee-Bee capillaries, for example, show substantial conductivity above 100 Hz, as evidenced by capacitance transients and noise spectra from pipettes with closed tips. Coating the pipette helps, but even in a Sylgard-coated pipette the effective values of R and C are roughly 2 GΩ and 2 pF. Pyrex electrode glass (Jencons H15/10) has at least an order of magnitude lower conductivity. However, it is more difficult to make pipettes with this hard glass because of its higher melting point.

Finally, the pipette access resistance R_{acc} (in the range 2–5 MΩ) and the capacitance of the tip of the pipette C_{tip} (of the order of 0.3 pF) constitute a noise source. Since the time constant is short, the low-frequency limit of (3) holds. The resulting spectral density increases as f^2, becoming comparable to the 1 GΩ noise level around 10 kHz. This noise could be reduced in pipettes having steeper tapers near the tip, reducing R_{acc}, or having the coating extend closer to the tip, reducing C_{tip}.

Noise in the Current-to-Voltage Converter. Figure 4A shows a simplified diagram of the recording electronics. The pipette current is measured as the voltage drop across the high-valued

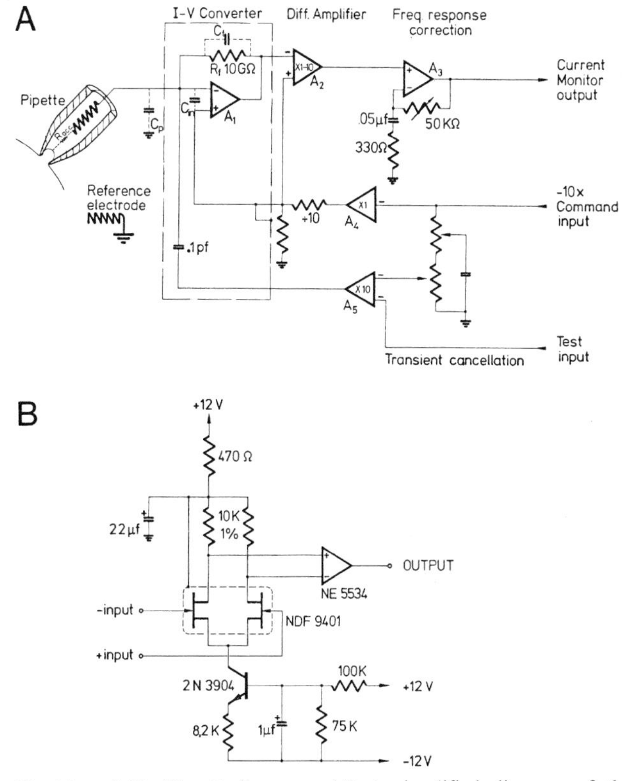

Fig. 4A and B. Circuit diagrams. (**A**) A simplified diagram of the recording system. The current-to-voltage converter is mounted on a micromanipulator, and the pipette holder (Fig. 1) plugs directly into it. Important stray capacitances (indicated by dotted lines) are the feedback capacitance $C_f \approx 0.1$ pf and the total pipette and holder capacitance $C_p = 4-7$ pF. C_{in} represents the input capacitance of amplifier A_1, which is either a Burr Brown 3523J or the circuit shown in **B**. With the values shown, the frequency response correction circuit compensates for time constants $R_f C_f$ up to 2.5 ms and extends the bandwidth to 10 kHz. The transient-cancellation amplifier A_5 sums two filtered signals with time constants variable in the ranges 0.5–10 µs and 0.1–5 ms; only one filter network is shown here. The test input allows the transient response of the system to be tested: a triangle wave applied to this input should result in a square wave at the output. Amplifiers A_2, A_4 and A_5 are operational amplifiers with associated resistor networks. The op amps for A_2 and A_4 (NE 5534, Signetics or LF 356, National Semiconductor) are chosen for low voltage noise, especially above 1 kHz; more critical for A_3 and A_5 (LF 357 and LF 356) are slew rate and bandwidth. For potential recording from whole cells (see part IV), a feedback amplifier is introduced between current monitor output and the voltage command input. (**B**) Circuit of a low-noise operational amplifier for the $I-V$ converter with a selected NDF 9401, dual FET (National Semiconductor) and the following approximate parameters: Input bias current, 0.3 pA; input capacitance, 8 pF; voltage noise density at 3 kHz, 5×10^{-17} V²/Hz; and gain-bandwidth product 20 MHz. The corresponding values for the 3523J are 0.01 pA, 4 pF, 4×10^{-16}V²/Hz and 0.6 MHz. The lower voltage noise of this amplifier is apparent in the $I-V$ converter's background noise above 500 Hz. The high gain-bandwidth product of the amplifier results in a loop bandwidth of ≈ 300 kHz in the $I-V$ converter, so that the frequency response in the 5–10 kHz region is negligibly affected by changes in C_p. The loop bandwidth with the 3523 is about 5 kHz

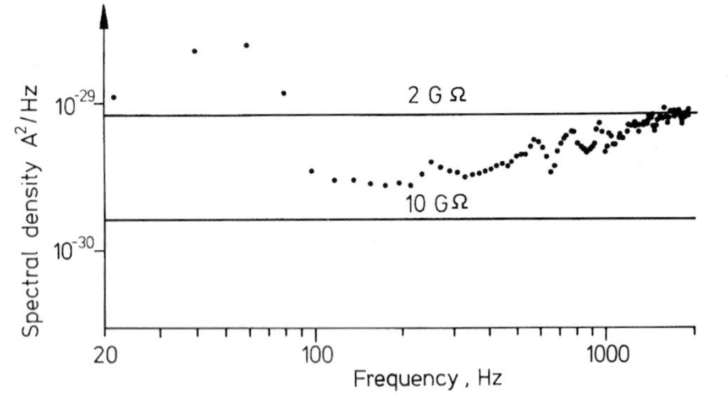

Fig. 5. Power spectrum of the total background noise from a rat myoball membrane patch at resting potential (dots). The amplifier of Fig. 4B was used, with a coated, hard-glass pipette. The patch resistance was 50 GΩ. Lines indicate the lower limit of the noise imposed by the 10 GΩ feedback resistor, and for comparison, the Johnson noise in a 2 GΩ resistor

resistor R_f; the Johnson noise in this resistor is the predominant noise source in the $I-V$ converter below a few hundred Hertz. The substantial shunt capacitance $C_f \approx 0.1$ pF across this resistor affects the frequency response of the $I-V$ converter but makes no contribution to $Re\{Y\}$, and therefore to the noise current, assuming that it is a pure capacitance. This assumption appears to hold for the colloidfilm resistors (Type CX65, Electronic GmbH, Unterhaching/Munich, FRG) we use, since, after correction of the frequency response, we found that the $I-V$ converter's noise spectrum was unchanged when we substituted a homemade tin-oxide resistor having $C_f < 0.01$ pF for the commercial resistor. Other resistor types, including the conductive-glass chip resistors we have previously used (Neher et al. 1978) and colloid-film resistors with higher values do not show the transient response characteristic of a simple $R-C$ combination and therefore probably have a frequency-dependence of $Re\{Y\}$. Correcting the frequency response of these resistors is also more complicated; this is the primary reason why we have not yet used values for R_f above 10 GΩ, even though this might improve the low-frequency noise level.

The other main noise source in the $I-V$ converter is the operational amplifier itself. With both of the amplifiers we use (Burr Brown 3523J, and the circuit of Fig. 4B) the low-frequency (4–100 Hz) spectral density is essentially equal to the value expected from R_f, suggesting that the amplifier current noise is negligible. At higher frequencies however the amplifier voltage noise becomes the dominant noise source. This voltage noise is imposed by the feedback loop on the pipette and the input of the amplifier, causing a fluctuating current to flow through R_f to charge C_p and C_{in} (see Fig. 4). The resulting contribution to the current fluctuations has the spectrum

$$S_I(f) = [2\pi f (C_p + C_{in})]^2 \, S_{V(A)}(f), \quad (4)$$

where $S_{V(A)}$ is the amplifier voltage noise spectral density. The f^2-dependence dominates over the constant or $1/f$ behavior of $S_{V(A)}$ giving an increase of S_I with frequency. This noise source can be reduced by minimizing C_p by using a low solution level and avoiding unnecessary shielding of the pipette and holder. It can also be reduced by choosing an operational amplifier having low values for C_{in} and $S_{V(A)}$; the amplifier of Fig. 4B was designed for these criteria.

Figure 5 shows the spectrum of the background noise during an actual experiment. Because the noise variance is the integral of the spectrum the high-frequency components have much greater importance than is suggested by this logarithmic plot. Below 90 Hz the excess fluctuations mainly come from 50 Hz pickup. In the range 100–500 Hz the spectral density is near the level set by the 10 GΩ feedback resistor. Above 500 Hz, electrode noise sources and the amplifier voltage noise contribute about equally to the rising spectral density.

Capacitance Transient Cancellation. For studying voltage-activated channels voltage jumps can be applied to the pipette. However, a step change in the pipette potential can result in a very large capacitive charging current. For example, charging 5 pF of capacitance to 100 mV in 5 μs requires 100 nA of current, which is 4–5 orders of magnitude larger than typical single channel currents. We use three strategies to reduce this transient to manageable sizes. First, we round the command signal (e. g. with a single time constant of 20 μs) to reduce the peak current in the transient. Second, we try to reduce the capacitance to be charged as much as possible. Metal surfaces near the pipette and holder (excepting the ground electrode in the bath!) are driven with the command signal; this includes the microscope and stage, and the enclosure for the $I-V$ converter. (Alternatively, an inverted command signal could be applied only to the bath electrode.) This measure reduces the capacitance to be charged to 1–2 pF when coated pipettes and low solution levels are used. Notice, however, that while the capacitance to be charged by an imposed voltage change is reduced, C_p is unchanged for the purpose of the noise calculation (Eq. 4).

Third, we use a transient cancellation circuit which injects the proper amount of charge directly into the pipette, so that the $I-V$ converter is required to supply only a small error current during the voltage step. The charge is injected through a small, air-dielectric capacitor (see Fig. 4A) which is driven with an amplified and shaped version of the command voltage. The same capacitor can be used to inject currents for test purposes. With these three measures the transient from a 100 mV step can be reduced to below 10 pA (at 2 kHz bandwidth), which is small enough to allow computer subtraction of the remainder.

Part II

Patch Current Recording with Giga-Seals

1. Development of the Giga-Seal

In the past, seal resistances as high as 200 MΩ could be obtained by pressing a pipette tip against a cell membrane and applying suction. This same procedure can also lead to the formation of a seal in the gigaohm range; the only difference is that precautions must be taken to ensure the cleanliness of the pipette tip (Neher 1981). The main precautions are (1) the use of filtered solutions in the bath as well as in the pipette, and (2) using a fresh pipette for each seal. Further precautions are listed below.

The formation of a giga-seal is a sudden, all-or-nothing increase in seal resistance by as much as 3 orders of magnitude. Figure 6 shows the time-course of the development of a 60 GΩ seal in the perisynaptic region of a frog muscle fibre. When the tip of the pipette was pressed against the enzymatically cleaned muscle surface the seal resistance was 150 MΩ. (The resistance was measured by applying a 0.1 mV voltage pulse in the pipette and monitoring the

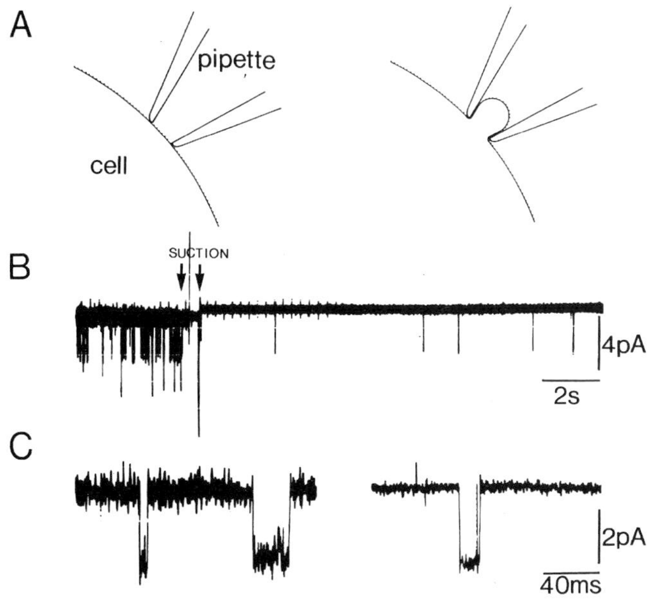

Fig. 6A–C. Giga-seal formation between pipette tip and sarcolemma of frog muscle. (**A**) Schematic diagrams showing a pipette pressed against the cell membrane when the pipette-membrane seal resistance is of the order of 50–100 Megohms (*left*), and after formation of a gigaseal when a small patch of membrane is drawn into the pipette tip (*right*). (**B**) The *upper trace* is a continuous current record before, during and after application of suction. In this experiment a pipette-membrane seal resistance of 150 MΩ was achieved by pressing the pipette against the membrane. Single suberyldicholine-induced channel currents are apparent. During the time indicated by the two arrows slight suction was applied to the pipette interior resulting in the formation of a giga-seal of 60 GΩ resistance. Note reduction in background noise level. The decrease in channel opening frequency presumably resulted from depletion of agonist in the pipette tip during suction. It increased again during the minute following giga-seal formation. The two large current deflections represent artifacts. The *lower traces* show single channel currents at higher resolution before (*left*) and after (*right*) formation of a giga-seal. The single channel current pulse on the right is preceded by capacitive artifacts from a calibration pulse. All records were made at the cell's resting potential of −92 mV and at 11°C. They were low pass filtered at 1 kHz (*upper trace*) or 3 kHz (*lower traces*)

resulting current flow). When a slight negative pressure of 20–30 cm H_2O was applied (arrows) the resistance increased within a few seconds to 60 GΩ. The development of giga-seals usually occurs within several seconds when a negative pressure is applied; seals always remain intact when the suction is subsequently released. In some cases giga-seals develop spontaneously without suction. In other cases suction has to be applied for periods of 10–20 s, or a seal may develop only after suction has been released.

It was previously suggested that upon suction the membrane at the pipette tip is distored and forms and Ω-shaped protrusion (Neher 1981). This is indeed supported by measurements of patch capacitance after giga-seal formation (Sigworth and Neher 1980). The increase in area of glass-membrane contact, going along with such a distortion, probably explains the gradual, 2–4-fold increase of seal resistance which is usually observed during suction shortly before giga-seal formation, and is also seen in cases when giga-seals do not develop. Giga-seal formation, however, is unlikely to be explained solely by such an area increase. Crude estimates of the thickness of a water layer interposed between membrane and glass give values in the range 20–50 Å for values of the seal resistance between 50 and 200 MΩ. These distances are characteristic for equilibrium separations between hydrophilic surfaces in salt solutions (Parsegian et al. 1979; Nir and Bentz 1978).

A seal resistance larger than 10 GΩ, however, is consistent only with glass-membrane separations of the order of 1 Å, i.e. within the distance of chemical bonds. The abrupt change in distance involved may therefore represent the establishment of direct contact between the surfaces, as occurs during transfer of insoluble surface monolayers on to glass substrates (Langmuir 1938; Petrov et al. 1980).

Also in favor of a tight membrane-glass contact is evidence that small molecules do not diffuse through the seal area. After establishment of a giga-seal the application of high ACh concentrations (in the range of 5–10 μM) outside the pipette does not activate single channel currents in the patch, even though the rest of the cell is depolarized by 20–50 mV.

The high-resistance contact area between glass and membrane also seems to be well delineated. We conclude this from the observation that the so-called "rim-channel" currents, which are quite common when using thick-walled pipette tips, are rarely observed after formation of a giga-seal (see below).

Reproducibility of Giga-Seals. The success rate for the establishment of giga-seals varies for different batches of patch pipettes. This variability probably results from a combination of several factors. The following general rules have been found helpful so far.

1. To avoid dirt on the pipette tip, pipettes should always be moved through the air-water interface with a slight positive pressure (10 cm H_2O). Even the first pipette-cell contact should be made with pipette solution streaming outwards. When the pressure is released while the pipette touches the cell the pipette membrane seal resistance should increase by a factor >2.
2. Each pipette should be used only once after positive pressure has been relieved.
3. Following enzyme treatment of muscle preparations the surface of the bathing solution is frequently covered with debris which readily adheres to the pipette tip, preventing giga-seal formation. The water surface can be cleaned by wiping with lens paper or by aspiration.
4. HEPES-buffered pipette solutions should be used when Ca^{2+} is present in the pipette solution. In phosphate buffer small crystals often form at the pipette tip by precipitation.
5. When slightly (10%) hypoosmolar pipette solutions are used the giga-seals develop more frequently. With these precautions, about 80% of all pipettes will develop giga-seals on healthy preparations. However, even after giga-seal formation, irregular bursts of fast current transients are observed on some patches. We interpret these as artifacts due to membrane damage or leakage through the seal.

2. Improved Current Recording After Giga-Seal Formation

When a pipette is sealed tightly onto a cell it separates the total cell surface membrane into two parts: the area covered by the pipette (the patch area) and the rest of the cell. Current entering the cell in the patch area has to leave it somewhere else. Thus, the equivalent circuit of the whole system consists of two membranes arranged in series. This, and the resulting complications will be discussed in a later section (see also Fig. 10). Here we will focus on the simple case that the total cell membrane area is very large with respect to the patch area.

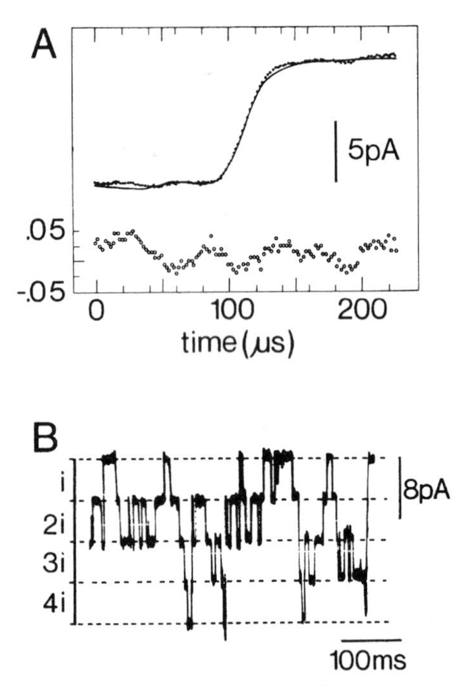

Fig. 7A and B. Demonstration of time resolution and of uniformity in step sizes. (**A**) A digitized record (1.87 μs sample interval) of the opening time course of channels activated by 100 nM SubCh on the perisynaptic region of an adult frog muscle fiber (11° C). The patch was hyperpolarized to approximately −160 mV and a solution containing 100 mM CsCl was used. The single channel currents were −10.5 pA in amplitude. Six individual channel opening events were averaged by superposition after alignment with respect to the midpoint of the transition. The same procedure was followed to obtain the instrument's step-response, using records from capacitively-injected current steps. The step response (continuous line) is superimposed on the channel's opening time course after amplitude scaling. The relative difference between the two curves is plotted below. The amplitude of the fluctuations in the difference record is the same during the transition and during the rest of the record. Based on the size of these fluctuations, an upper limit was estimated for the transition time between the closed and open states of the channel by the following procedure: The transfer function of the electronic apparatus was calculated from the known step response by Fourier transform methods. Then, theoretical responses to open-close transitions of various shapes were calculated and compared to the experimental step response. It was found that the predicted deviations between the two curves were significantly larger than the observed ones only when the open-close transitions were spread out in time over 10 μs or more. (**B**) A current record from a myoball under the following conditions: 1 μM ACh; 18° C; −140 mV holding potential. Individual single channel currents superimpose to form regularly spaced amplitude levels

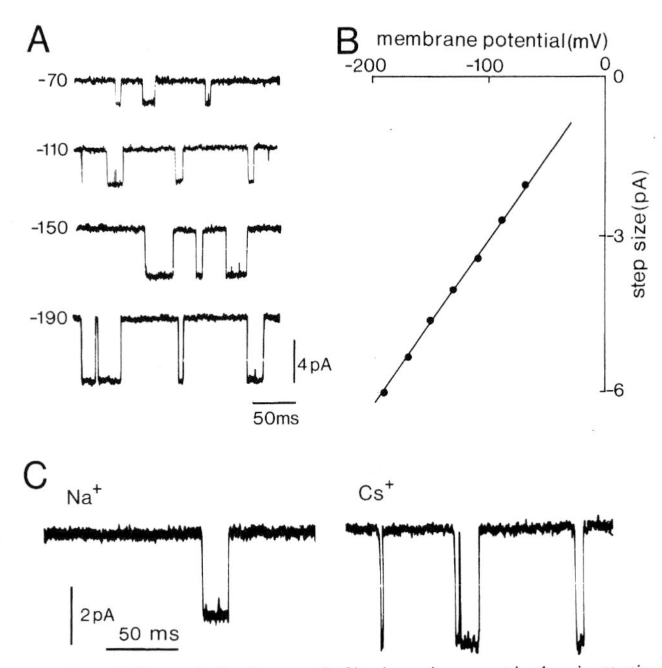

Fig. 8A–C. Control of voltage and of ionic environment in the pipette tip after formation of a giga-seal. (**A**) Single channel current recordings at 50 nM SubCh in the perisynaptic membrane of cutaneous pectoris muscle fibre; 11° C. The membrane potential of this fibre was −89 mV measured by an intracellular microelectrode. The pipette potential was shifted by different amounts to obtain the membrane potential indicated on the left of each trace (in mV). The pipette was filled with Ringer solution in which NaCl concentration was reduced to 100 mM to improve gigaseal formation. (**B**) Current-voltage relationship of channel currents, derived from the experiment shown in **A**. Each point represents the mean current amplitude of ten individual current events. The straight line is drawn by eye and represents a single channel (chord) conductance of 32 pS. (**C**) Single channel current recordings under the conditions of part A at −90 mV membrane potential. The main salt in the pipette solution was 100 mM CsCl (*right*) and 100 mM NaCl (*left*). The average single channel current amplitudes were 2.8 pA and 3.8 pA in Na^+ and Cs^+ solutions respectively

Then, the small patch currents will not noticeably alter the cell's resting potential. For instance, a myoball with 50 MΩ input resistance will be polarized less than 0.5 mV by a patch current of 10 pA. Thus, a patch can be considered "voltage clamped" even without the use of intracellular electrodes. The clamp potential is equal to the difference between the cell potential and the potential in the pipette. Some properties of current recordings done under this type of "voltage clamp" are illustrated here.

a) Increased Amplitude and Time Resolution; the Time Course of Channel Opening is Fast. Due to low background noise and an improved electronic circuit the time course of channel opening and closing can be observed at 5–10 kHz resolution. Figure 7A illustrates the average opening time-course of suberyldicholine-activated channels at the frog endplate region. The large single channel currents and the low background noise level allowed a much higher time resolution to be obtained than in previous recordings. Still, the rising phase of the conductance (dotted line) is seen to be indistinguishable from the step response of the measuring system (continous line). From a comparison of the two time courses it can be estimated that the actual channel opening occurs within a time interval smaller than 10 μs (see legend, Fig. 7).

b) Lack of Rim-Channel Currents. A major problem of the extracellular patch clamp technique has been the occurrence of currents from channels in the membrane area under the rim of the pipette. These currents are not uniform in size; the resulting skewed step-size histograms complicate the estimation of the single channel current amplitudes (Neher et al. 1978). Current recordings with giga-seals, however, show amplitude distributions that are nearly as narrow as expected from noise in the baseline. The improvement reflects a more sharply delineated seal region. A recording is shown in Fig. 7B which demonstrates the regularity in current amplitudes.

The regularity also allows (i) the unequivocal discrimination between channel types of slightly different conductance, e.g. synaptic and extrasynaptic ACh receptor channels, and (ii) analysis of channel activity when the currents of several channels overlap.

c) Voltage Control of the Membrane Patch. Previously the potential inside the pipette had to be balanced to within less than 1 mV of the bath potential; otherwise large, noisy leakage currents flowed through the seal conductance. The high seal resistance now allows the patch membrane potential to be changed. For example, a 100 mV change in pipette potential will drive only 5 pA of current through a 20 GΩ seal. This leakage is comparable in size to a single channel current and is easily manageable by leakage subtraction procedures. The ability to impose changes in membrane potential has been used to activate Na channels in myoballs (Sigworth and Neher 1980). It also allows measurement of single channel currents in adult muscle fibres over a wide range of potential that is not accessible with the conventional two-microelectrode voltage clamp because of local contractions. Figure 8A and B shows representative traces and the current-voltage relationship of single channel currents recorded from a muscle fibre at various patch membrane potentials ranging from −70 to −190 mV.

d) Control of Extracellular Ion Composition. Giga-seals form a lateral diffusion barrier for ions (see above, p. 91). Here we show that the ionic composition on the external side of the patch membrane is that of the pipette solution. Figure 8C illustrates ACh-activated single channel currents from a frog muscle fibre when the major salt in the pipette was CsCl (100 mM) while the bath contained normal Ringer. The open channel conductance was seen to be 1.3 times larger than the conductance in standard Ringer solution, consistent with the larger estimates of conductance that have been made from fluctuation analysis (Gage and Van Helden 1979).

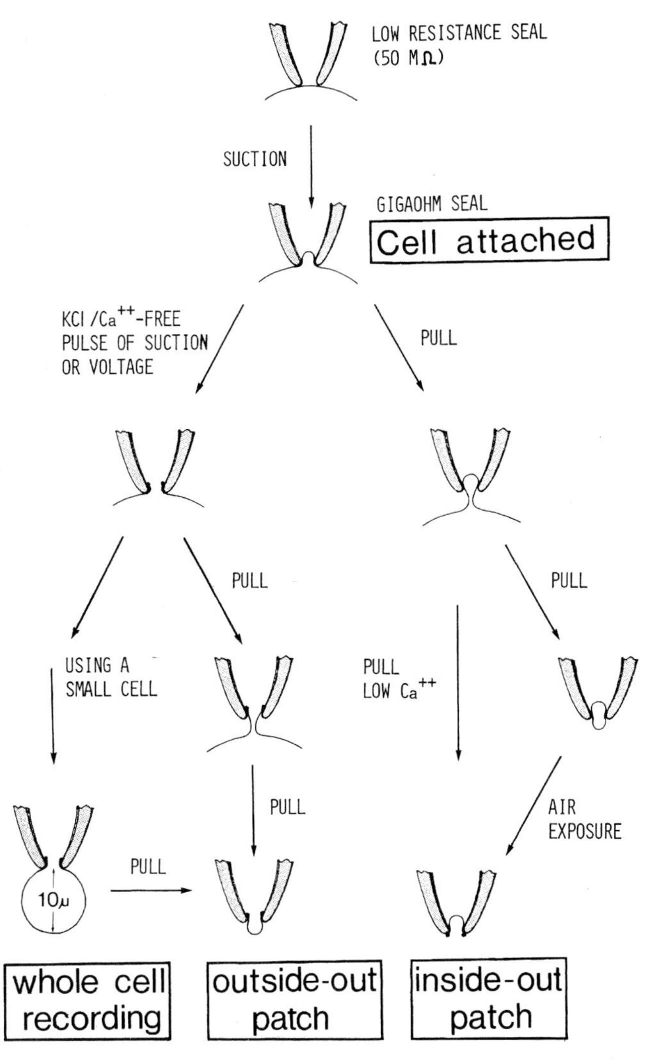

Fig. 9. Schematic representation of the procedures which lead to recording configurations. The four recording configurations, described in this paper are: "cell-attached", "whole-cell recording", "outside-out patch", and "inside-out patch". The upper most frame is the configuration of a pipette in simple mechanical contact with a cell, as has been used in the past for single channel recording ((Neher et al. 1978). Upon slight suction the seal between membrane and pipette increases in resistance by 2 to 3 orders of magnitude, forming what we call a cell-attached patch. This configuration is described in part II of this article. The improved seal allows a 10-fold reduction in background noise. This stage is the starting point for manipulations to isolate membrane patches which lead to two different cell-free recording configurations (the outside-out and inside-out patches described in part III). Alternatively, voltage clamp currents from whole cells can be recorded after disruption of the patch membrane if cells of sufficiently small diameter are used (see part IV of this article). The manipulations include withdrawal of the pipette from the cell (pull), short exposure of the pipette tip to air and short pulses of suction or voltage applied to the pipette interior while cell-attached

Part III

Single Channel Current Recording from "Cell-Free" Membrane Patches

Apparently the contact between cell membrane and glass pipette after formation of a giga-seal is not only electrically tight, but also mechanically very stable. The pipette tip can be drawn away from the cell surface without a decrease in the seal resistance (Hamill and Sakmann 1981; Neher 1981). As will be shown below a tight vesicle sealing off the tip forms, when this is done in normal Ca^{2+}-containing bath solution. Procedures are described by which the resistance of either the inner or the outer part of the vesicle can be made low (< 100 MΩ) without damaging the giga-seal. The remaining intact membrane can then be studied as before. Either "inside-out" or "outside-out" patches can be isolated in this way (see Fig. 9). By varying the composition of the bath solution, the effect of drugs or ion concentration changes on single channel currents can be studied at either the cytoplasmic or the extracellular face of the membrane.

1. Vesicle Formation at the Pipette Tip

The top trace in Fig. 10A shows single channel currents recorded in a frog muscle fibre at its resting potential of −90 mV. The patch pipette was filled with standard Ringer solution plus 50 nM suberyldicholine (SubCh) and was sealed against the surface membrane in the perisynaptic region. When the pipette tip was slowly withdrawn a few μm from the cell surface the shape of single channel currents became rounded and they decreased in size (middle trace). Often a fine cytoplasmic bridge could then be observed between the cell surface and the pipette as shown in Fig. 10D. Upon further removal the cytoplasmic bridge tied off but left the giga-seal intact. Single channel currents further decreased in size and finally disappeared in the background noise within the next 1−2 min. The pipette input resistance remained high (110 GΩ, see bottom trace).

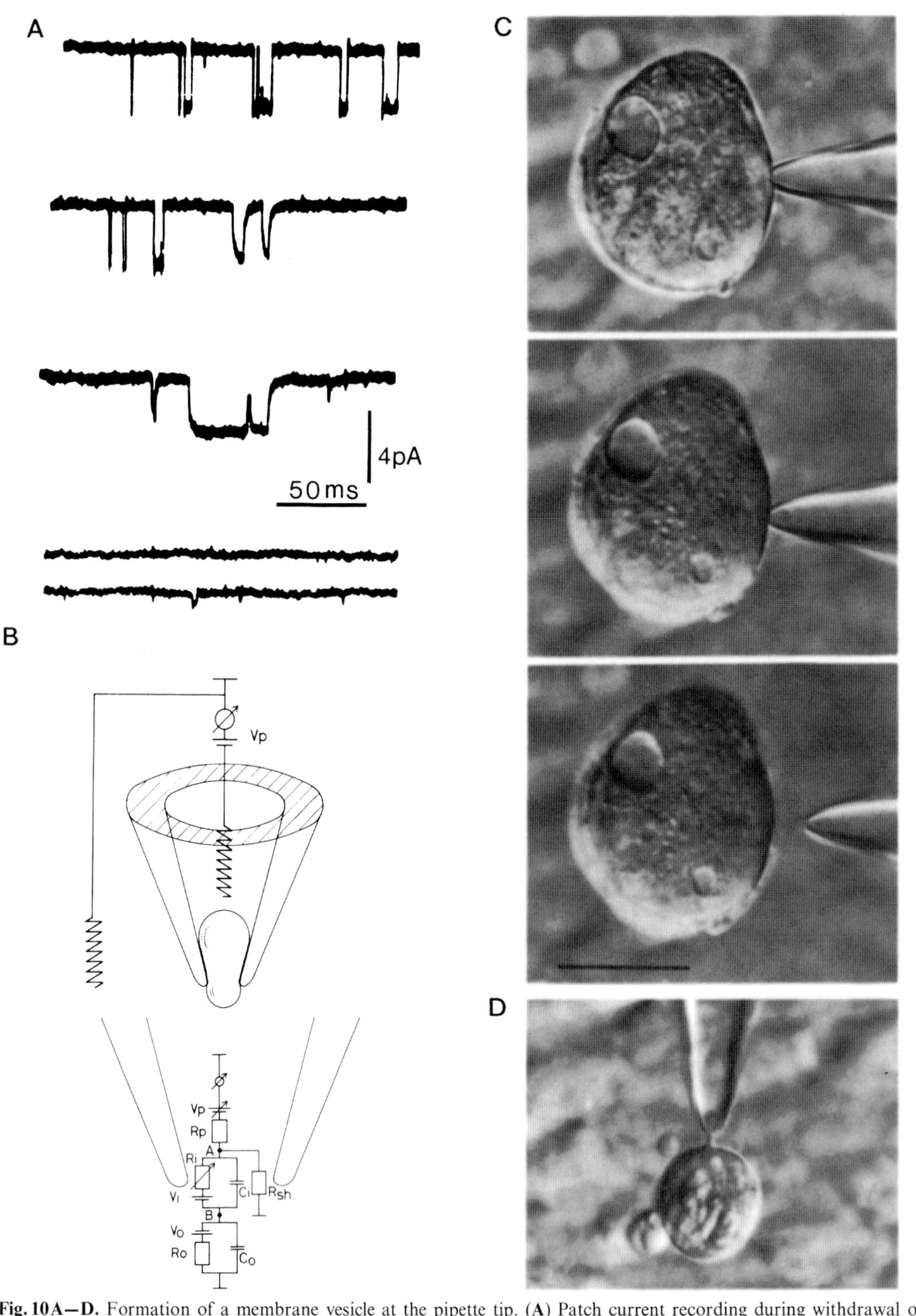

Fig. 10A–D. Formation of a membrane vesicle at the pipette tip. (**A**) Patch current recording during withdrawal of the pipette tip from muscle sacrolemma. The *upper trace* illustrates single channel currents after formation of a gigaseal under conditions similar to those of Fig. 8. During withdrawal of the pipette single channel currents suddenly appeared distorted in shape; currents showed rounded rising and falling time courses and progressively lower amplitudes. In this experiment single channel currents became undetectable within 30 s following the appearance of rounded current events. The pipette-membrane seal resistance remained high, ($>30\,G\Omega$) and current pulses remained undetectable even after hyperpolarizing the membrane by 90 mV. This suggests that the pipette opening became occluded by a membraneous structure, which most probably was a closed vesicle as illustrated in part **B**. (**B**) Schematic diagram of recording situation following formation of a membrane vesicle. The equivalent circuit is modelled by parallel current pathways through the shunt resistance R_{sh} and through the vesicle. The two halves of the vesicle are represented by $R-C$ combinations in series. The inner membrane, exposed to the pipette solution, has resistance R_i and capacitance C_i. The outer membrane, represented by R_o and C_o, is exposed to the bath solution. Point **B** represents the interior of the vesicle. V_o and V_i are the electromotive forces of the outer and inner membranes. R_p is the access resistance of the unsealed pipette. Opening of a channel in the presence of ACh in the pipette solution decreases the resistance of the inner membrane, R_i. Therefore, R_i which in fact is a parallel combination of membrane resistance and open channel resistance R_c is modelled as a variable resistor. (**C**) Visualisation of membrane patch isolation from a rat myoball. The upper micrograph shows the tip of a patch pipette in contact with a myoball. When suction is applied, an Ω-shaped membrane vesicle is pulled into the pipette tip (*middle micrograph*). Following slight withdrawal of the pipette tip from the myoball, a cytoplasmic bridge of sarcolemma between pipette tip and myoball surface was observed (not shown). At this stage of membrane isolation distorted single channel current pulses are recorded like those shown in **A**. Upon further withdrawal the cytoplasmic bridge ruptures leaving a small vesicle protruding from the pipette tip (*lower micrograph*). Note "healing" of the sarcolemmal membrane of the myoball. (**D**) Visualisation of cytoplasmic bridge between myoball and vesicle in the pipette tip in another experiment. Calibration bar: 20 μm. (Normarski interference optics, ×40 water immersion objective)

The most likely explanation for this sequence of events is that during rupture of the cytoplasmic bridge between pipette tip and cell surface a closed vesicle formed at the tip of the pipette. The formation of a vesicle is illustrated in Fig. 10C on a myoball. The vesicle is partly exposed to the bath solution ("outer membrane") and partly to the pipette solution ("inner membrane") in the way shown schematically in Fig. 10B.

Three kinds of observations give indications on the electrical properties of the vesicle. (i) When ACh is added at a low concentration to the bath solution, current pulses of reduced size are sometimes recorded at an applied pipette potential V_p of +90 mV or more. These current pulses reflect an outward current flow through AChR channels on the outer membrane of the vesicle. (ii) Destruction of the barrier properties of the outer membrane (see the next section) results in the reappearance of single channel currents through the inner membrane. (iii) In some experiments single channel currents of decreased amplitude and distored shape could be recorded when the pipette potential was increased to > +70 mV even several minutes following pipette withdrawal. Examples of such distorted current pulses recorded from patches and vesicles are illustrated in Figs. 10A and 11. Single channel currents were of rectangular shape as long as the membrane patch was attached to the cell.

Equivalent circuit. The electrical signals after vesicle formation can be explained by an equivalent circuit shown in Fig. 10B. The two halves of the vesicle form a series combination of two RC-circuits, shunted by the leakage resistance R_{sh}. Assuming a specific conductance of the membrane of 10^{-4} Scm^{-2} and an area of a semi-vesicle in the range of several μm^2, values for the membrane resistances R_i and R_o should be in the range of several hundred GΩ, much larger than the resistance of a single open AChR-channel (about 30 GΩ). Local damage or partial breakdown of the membrane could bring this value down into the range of single open channels. Then, opening of a single channel in either of the two membranes would result in an attenuated current flow through the series combination. Consider the simple case that R_i and R_o are the same and are identical to the resistance of a single channel R_c. Then, upon opening of a single channel the apparent conductance at steady state is attenuated by a factor of 6 with respect to the true conductance.

More generally, the apparent conductance G_{app} of the channel (assumed to open on the inner side of the vesicle) will be

$$G_{app} = R_i^2/((R_i + R_o)\,(R_c R_o + R_c R_i + R_i R_o))\,.$$

The time course will be governed by the time course of the voltage at point B in the equivalent circuit. For a series combination of two RC-elements and a step-like perturbation this will be a single exponential (neglecting R_p of Fig. 10B). The time constant τ of the exponential is equal to

$$\tau = R_c \| R_i \| R_o \cdot (C_i + C_o)\,,$$

where $R_c \| R_i \| R_o$ is the parallel combination of $R_c R_i$ and R_o (see Fig. 10B). For the simple case above $\tau_1 = R_c\,(C_i + C_o)/3$ while the channel is open and $\tau_2 = R_c\,(C_i + C_o)/2$ with the channel closed. The measurement of a time constant, therefore, gives the total membrane capacitance of the vesicle. More generally, determination of G_{app}, τ_1, and τ_2 allows calculation of the three unknowns R_i, R_o and $(C_i + C_o)$ if the conductance of the single channel $(1/R_c)$ is known. This analysis neglects the effects of R_{sh}, which however adds only a constant offset in current (at V_p = constant and $R_p \ll R_i$, R_o).

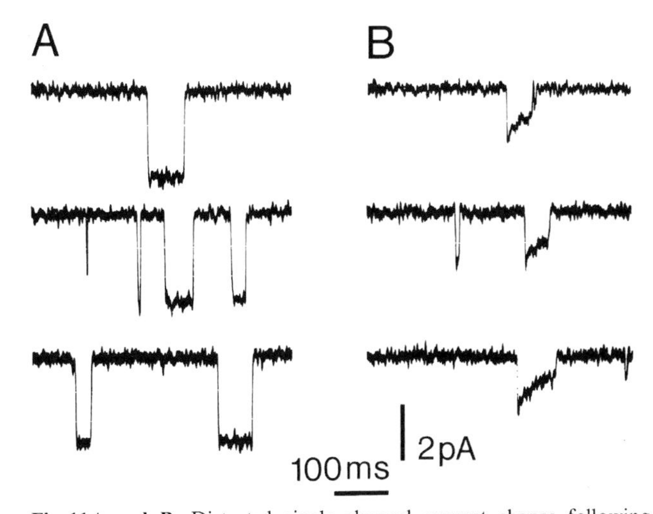

Fig. 11A and B. Distorted single channel current shapes following isolation of a membrane vesicle from myoball sarcolemma. (**A**) Single channel currents recorded from a cell-attached membrane patch. In this experiment the Na^+ in the pipette solution was reduced to 50 mM. Pipette potential was +80 mV. Assuming a resting potential of −70 mV the membrane potential across the patch was −150 mV. The slope conductance was 21 pS. (**B**) Single channel currents after physical isolation of the membrane patch from the myoball. Pipette potential was +80 mV. Single channel currents could be observed at the pipette zero potential indicating that the vesicle had a membrane potential. The apparent slope conductance of the initial peak amplitude of these current events was 12 pS. All records filtered at 0.4 kHz, low pass; 18°C. The duration of the current events is not representative since long duration currents were selected to illustrate differences in the time course in cell-attached and cell-free configurations. Also, the time constant of decay was unusually long in this experiment. In some experiments single channel currents similar to those shown here but of very small amplitude could be observed also without application of a pipette potential. This indicates that the vesicle can have a resting potential

Experimental data gave estimates for $C_i + C_o$ in the range 0.03–0.3 pF. Assuming a specific capacitance of the vesicle membrane of 1 $\mu F/cm^2$ this results in a vesicle area of 3–30 μm^2. The waveform of individual channel responses (Fig. 11) can display either rising or decaying relaxations. The form depends on the specific relation between R_o, R_i, C_o and C_i.

Although closed vesicles form regularly in standard bath solution 1 mM Ca^{2+} and 1 M Mg^{2+}, formation of tight vesicles is infrequently observed when divalent metal cations are left out of the bath solution or if they are chelated by EGTA. Horn and Patlak (1980) used F^--containing bath solution to prevent formation of closed vesicles.

Simultaneous intracellular recording of the membrane potential of the cell under study shows that isolating a membrane vesicle from the cell surface membrane does not damage the cell. In a few cases even the opposite was observed. Cells that depolarized partially while the seal was being formed returned to the normal resting potential after pipette withdrawal.

2. Formation of "Cell-Free" Membrane Patches at the Pipette Tip

Vesicle formation at the tip opening offers the possibility of measuring single channel currents in cell-free patches by selectively disrupting either the inner or the outer membrane of the vesicle. Figure 9 illustrates schematically how this can be done. When the outer membrane of the vesicle is disrupted the cytoplasmic face of the inner membrane is exposed to the bath solution. Its extracellular face is exposed to the pipette solution. We call this an "inside-out" membrane patch. When, on the other hand, the inner membrane of the vesicle is disrupted the cytoplasmic face of the outer membrane of the vesicle is exposed to the pipette solution, its extracellular face to the bath solution. This will be called an "outside-out" patch. The next two sections describe formation of these two configurations of membrane patches.

The "Inside-Out" Membrane Patch. To obtain a membrane whose cytoplasmic face is exposed to the bath solution, either formation of the outer vesicle membrane has to be prevented (Horn and Patlak 1980) or it has to be disrupted once it has been formed (Hamill and Sakmann 1981). This can be done either mechanically or chemically.

A prerequisite for isolated patch formation is a seal of >20 GΩ. Disruption of the outer vesicle membrane is done by passing the pipette tip briefly through the air-water interface of the bath or by touching an air bubble held by a nearby pipette. Brief contact with a drop of hexadecane will sometimes disrupt the vesicle. Figure 12 illustrates the disruption of the barrier properties of the outer vesicle membrane. Initially the pipette tip was sealed against a myoball membrane. The pipette solution contained 0.5 μM ACh. Single channel currents were recorded at the cell's resting potential as shown in the uppermost trace. After withdrawal of the pipette tip from the cell surface single channel currents disappeared. Upon increasing the pipette potential to +70 mV the trace became noisier but single currents were not resolved. By briefly (1–2 s) passing the tip through the air-water interface the outer membrane of the vesicle was disrupted. Upon reimmersion of the tip into the bath solution single channel currents of the expected size were recorded (Fig. 12). Single channel currents, which in most membrane patches of myoballs fall into two classes (Hamill and Sakmann 1981) were similar in their respective amplitudes in the cell-attached and cell-free configuration as shown in Fig. 12B on another patch at −100 mV membrane potential.

The outer membrane of the vesicle can also be made leaky by exposing it to a Ca-free, 150 mM KCl bath solution during isolation, as was originally used by Kostyuk et al. (1976) to disrupt neuronal membranes for internal dialysis. A vesicle tends to reform with this procedure, and mechanical disruption is usually required in addition to open the vesicle completely. The stability of inside-out patches is greatly improved when most of the Cl^- in the bath solution is replaced by SO_4^{2-}. For membrane patches isolated from myoballs an anion mixture of 4 mM Cl^- and 75 mM SO_4^{2-} was found to yield stable recordings for up to several hours.

The "Outside-Out" Membrane Patch. In order to work with the outer membrane of the vesicle, the inner membrane can be made leaky or can be disrupted in very much the same way as described for the case above, i.e. by exposing the inner vesicle membrane to a pipette solution containing 150 mM KCl and only a low ($<10^{-6}$ M) concentration of Ca. Alternatively it can be opened by mechanical rupture of the patch preceding vesicle formation. Isolation of a membrane patch in the outside-out configuration is illustrated in Fig. 13. A pipette containing 150 mM KCl and 3 mM HEPES buffer was used. A few minutes following the establishment of a giga-seal, the background noise increased progressively by several orders of magnitude. This was accompanied by a decrease of the

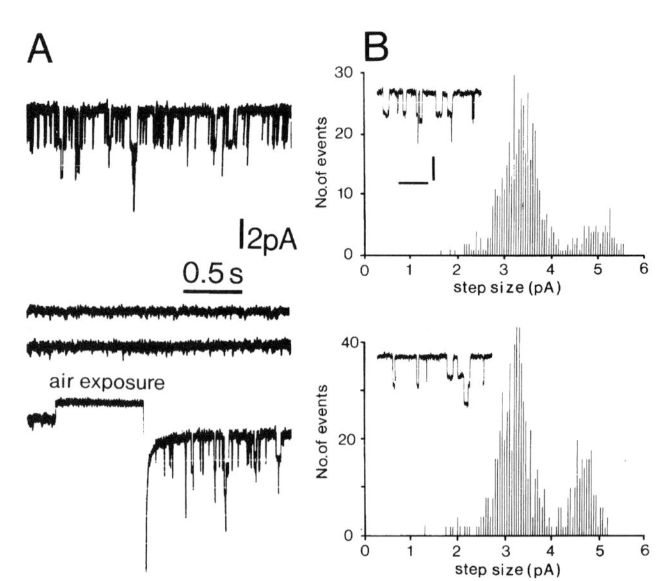

Fig. 12A and B. Isolation of an "inside-out" membrane patch from rat myoball sarcolemma. (**A**) The upper three traces were recorded during the process of vesicle formation and are analogous to those shown in Fig. 10A. Standard solutions were used. The pipette contained 0.5 μM ACh. Single channel currents (*first trace*, recorded at the cell's resting potential of ∼ −70 mV) became undetectable following withdrawal of the pipette tip. Upon shifting the pipette potential from 0 (*second trace*) to +70 mV (*third trace*) the current trace changed by only 0.8 pA, indicating formation of a closed vesicle where $R_o > R_i$ (see Fig. 10B). The pipette tip was then passed briefly (1–2 s) through the air-water interface which disrupted the outer membrane of the vesicle. Upon reimmersion single channel currents were recorded at a pipette potential of +70 mV (*fourth trace*), which were similar to those of the cell-attached case (shown in the first trace). (**B**) Step size distribution of single channel currents recorded from the same patch of membrane in the cell-attached configuration (*upper graph*) and in the cell-free "inside-out" configuration (*lower graph*), both at a membrane potential of −100 mV. The pipette solution contained 0.5 μM ACh. The distribution of step sizes is doubly peaked. It indicates that in this patch two types of AChR channels, junctional and extrajunctional, with slightly different open channel conductances were activated. The larger spread of step size distributions in the cell-attached recording configuration is due to the larger background noise which in this experiment was mostly caused by mechanical instabilities and which disappears following isolation of the patch. In some experiments the average opening frequency of AChR-channels decreased (up to 30%) following isolation of the membrane patch. This is probably due to a decrease of the inner membrane area exposed to the pipette solution. Insets show examples of single channel current events (Calibration bars: 4 pA and 50 ms). A downward deflection of the current trace in this and all other figures indicates cation transfer from the compartment facing the extracellular membrane side to the compartment facing the cytoplasmic side. This is from the pipette to the bath solution for an inside-out patch

pipette-bath resistance to less than 1 GΩ and by the development of a large inward current. Upon withdrawal of the pipette tip from the cell surface the background noise decreased within 1–2 s to the initial low level and the pipette-bath resistance simultaneously increased to a value larger than 10 GΩ.

We attribute the initial decrease of the pipette input resistance to the disruption of the membrane patch and not to an increased leakage of the pipette-membrane seal. This follows from the observation that the inward current inverts at a rather large negative pipette potential presumably equal to the cell resting potential. Also the pipette input capacitance shows an increase corresponding to the cell capacitance (see below, Part IV).

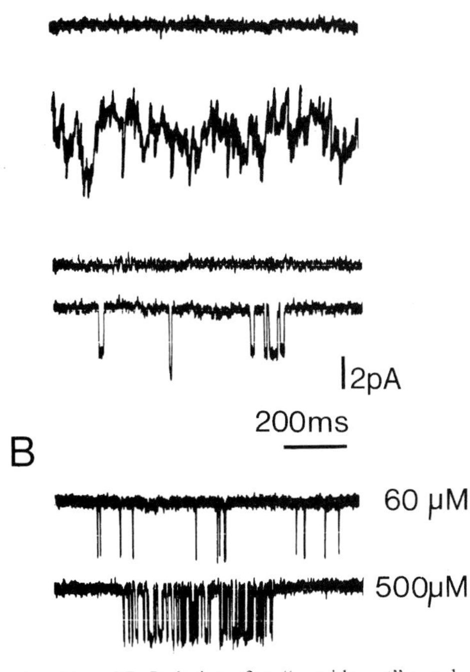

Fig. 13A and B. Isolation of an "outside-out" membrane patch from rat myoball sarcolemma. (**A**) After formation of a "gigaseal" using a pipette containing 150 mM KCl and 3 mM HEPES at pH 7.2 the background noise increased within 2–3 min. The upper two traces represent recordings immediately following gigaseal formation and 3 min later. At this stage a leakage current of > lnA developed which eventually drove the feedback amplifier into saturation. The leakage currents could be reduced or inverted when the pipette potential was shifted to –50 to –70 mV. The pipette access resistance decreased to values < 100 MΩ. Upon withdrawal of the pipette tip the pipette access resistance increased again into the GΩ range, and the background noise decreased (third trace). Addition of 0.5–1 μM ACh to the bath solution induced single channel currents of 2.5 pA amplitude at –70 mV membrane potential (fourth trace). All records are filtered at 0.5 kHz; temp. 18° C. (**B**) Demonstration of equilibration of bath-applied agonists at the extracellular face of an outside-out patch. Single channel currents were recorded at –70 mV membrane potential from the same membrane patch when either 60 μM carbachol (*above*) or 500 μM carbachol (*below*) was added to the bath solution. At the lower concentration single channel current events appeared at random, at the higher concentration current pulses appeared in "bursts". Addition of 10^{-6} M α-BuTx irreversibly blocked agonist activated currents (not shown)

The subsequent sealing observed when pulling away the pipette apparently results from the formation of a new membrane bilayer at the pipette tip, with its external side facing the bath solution (outside-out patch, see Fig. 9). Several observations lead to this conclusion: (i) Air exposure of the pipette tip as described in the previous section results in a decrease of the pipette-bath resistance to values < 100 MΩ. (ii) Addition of low ACh concentrations to the bath solution activates single channel current pulses. At a pipette potential of –70 mV they are similar in their amplitude and average duration to those observed on cell-attached membrane patches (Fig. 13A). (iii) In experiments where only the pipette contained ACh, no single channel currents were recorded at this stage.

The breakdown of the initial membrane patch, which was spontaneous in the experiment illustrated in Fig. 13A, can be accelerated or initiated by applying brief voltage (of up to 200 mV) or negative pressure pulses (matching the pressure of a 100 cm H_2O column) to the pipette interior. In this way, the access resistance can be lowered to a value near the pipette resistance.

The extracellular face of outside-out membrane patches equilibrates rapidly (< 1 s) and reversibly with ACh added to the bath solution by perfusion of the experimental chamber or applied by flow of ACh-containing solution from a nearby pipette of 20–50 μm tip diameter. Figure 13B shows a recording of single channel currents from an outside-out patch when carbachol is applied at 60 μM and 500 μM. At the low carbachol concentration current pulses appear at random intervals, whereas bursts of current pulses are recorded at high agonist concentration as previously reported for cell-attached membrane patches (Sakmann et al. 1980).

3. Equilibration of the Cytoplasmic Face of Cell-Free Patches with Bath or Pipette Solutions

In the previous two sections it was shown that by suitable manipulations either the inner membrane or the outer membrane of the vesicle can be disrupted such that it represents a low series resistance. In order to check whether, using either of the two configurations of cell-free membrane patches, the cytoplasmic face of the patch equilibrates with the experimental solution we have measured I–V relations of ACh-activated channels under various ionic conditions in both inside-out and outside-out configurations (Fig. 14A). The *I–V* relations show the following features which are expected for ionic equilibration between the cytoplasmic membrane face and the bath solutions or pipette solutions: (i) when Na^+ concentration is reduced on one side of the membrane, one branch of the *I–V* relation changes strongly whereas the other branch is minimally affected in its extremes (ii) the two recording configurations result in overlapping *I–V* relationships when ionic compositions on both sides of the membrane are the same (iii) changes in the shape of the *I–V* relations can be reversed and are reproducible from one patch to the next.

It is well established that the disrupted membrane of the vesicle does not represent an appreciable electrical series resistance (see for instance Fig. 12). However it might well contribute to changes in the *I–V* relation if its conductance were high, but ion-selective. In such a case a potential would develop across the disrupted membrane which would produce a parallel shift in the *I–V* relation along the voltage axis. This, however, is contrary to the results shown in Fig. 14A where changes occur in curvature, and neighbouring curves approach each other asymptotically.

Further evidence for ionic equilibration is provided by kinetic studies. It was found that after successful patch isolation the *I–V* relations did not change their properties with time if solutions in the pipette and in the bath were kept constant. Furthermore steady state properties were obtained instantaneously upon a change of environment. This is shown in Fig. 14B.

Both the changes in curvature and the fact that the changes occur instantaneously point towards very efficient exchange of ionic contents between the interior of a vesicle and the compartment neighbouring its disrupted membrane. In addition, ionic exchange across a disrupted membrane between the pipette interior and the cell interior was observed in the whole cell recording configuration (see part IV). These findings, together, make it very likely that true equilibration

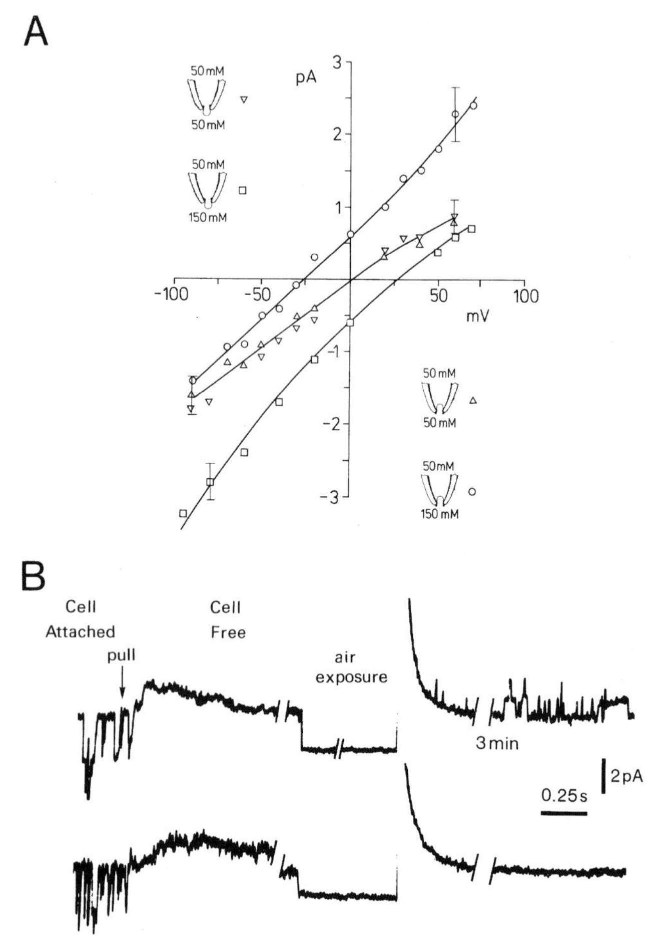

Fig. 14A and B. Equilibration of the cytoplasmic face of cell-free membrane patches with bath or pipette solutions. (**A**) $I-V$ relationships for acetylcholine-activated channels in inside-out and outside-out membrane patches measured under different ionic conditions. For all measurements the pipette solution contained 50 mM NaCl as the predominant salt. The bath solution contained initially 150 mM NaCl, which was changed to a solution containing 50 mM after patch isolation. Note that under symmetrical conditions $I-V$ relationships in both patch configurations overlapped, passed through the pipette zero potential, and showed slight rectification. Under asymmetrical conditions the reversal potentials were shifted by 25 mV. The $I-V$ relationships showed the expected curvature, and, in one branch each, approached the neighbouring symmetrical $I-V$. The symbols ○, △, ▽, □ represent the mean channel currents determined from 16, 3, 2, and 3 separate experiments. SEM's are shown for some averages. All measurements were made at 18°C. Changes in junction potentials caused by solution changes were measured independently, and were corrected for. (**B**) Current records before, during, and after formation of an inside-out patch at zero pipette potential. For both upper and lower traces the pipette was withdrawn from the cell while in normal bath solution. The pipette solutions contained 50 mM NaCl. During withdrawal of the pipette from the myoball, ACh-activated currents, evident while cell-attached, disappeared. For the upper trace the pipette tip was briefly exposed to air, and then returned to the normal bath solution. Inverted currents immediately appeared consistent with the vesicle being opened to the bath solution. The currents recorded did not change their properties over a three-min period. They displayed an $I-V$ relation similar to the inside-out asymmetrical case described in part **A**. For the *lower trace*, after pipette withdrawal from the cell, the bath solution was changed to one containing 50 mM NaCl and the pipette tip was then exposed shortly to air. No ACh-induced currents were evident at the pipette zero potential. They appeared upon polarization and displayed a similar $I-V$ relationship as described for symmetrical cases in part **A**

between the cytoplasmic membrane face and adjacent bulk solutions takes place.

Part IV

Recording of Whole-Cell Voltage Clamp Currents

It was pointed out above that the membrane patch which separates the pipette from the cell interior can be broken without damaging the seal between the pipette rim and the cell membrane. This is the situation occurring at the initial stage during the formation of outside-out patches (see p. 96). Here, we demonstrate the suitability of this configuration for studying the total ionic currents in small cells. The technique to be described can be viewed as a microversion of the internal dialysis techniques originally developed for molluscan giant neurons (Krishtal and Pidoplichko 1975; Kostyuk and Krishtal 1977; Lee et al. 1978) and recently applied to mammalian neurones (Krishtal and Pidoplichko 1980). As it is appropriate only for cells of less than 30 μm in diameter, we take as an example bovine chromaffin cells in short-term tissue culture. These cells have a diameter of 10–20 μm. Their single channel properties will be detailled elsewhere (Fenwick, Marty, and Neher, manuscript in preparation).

The pipette was filled with a solution mimicking the ionic environment of the cell interior (Ca-EGTA buffer, high K^+). After establishment of a giga-seal, the patch membrane was disrupted, usually by suction, as previously described (see above). The measured zero-current potential was typically −50 to −70 mV. This corresponds to the cell resting potential (Brandt et al. 1976). Applying small voltage jumps from this potential revealed a resistance value in the range of 10 GΩ. This resistance is mainly due to the cell membrane since markedly larger resistances (20–50 GΩ) were obtained when using a CsCl solution in the pipette interior and tetrodotoxin in the bath. Small voltage jumps also showed that the disruption of the intial patch is accompanied by a large increase of the input capacitance (Fig. 15). The additional capacitance was about 5 pF, in good agreement with the value expected from the estimated cell surface, assuming a unit capacity of 1 μF/cm². The time constant of the capacity current was of the order of 100 μs, which shows that the series resistance due to the pipette tip is no more than 20 MΩ. However, larger time constants were occasionally observed, indicating an incomplete disruption of the initial membrane patch.

Depolarizing voltage commands elicited Na and K currents which could be well resolved after compensation of the cell capacitance current (Fig. 16). The cell can be considered under excellent voltage clamp since, (i) at the peak inward current, the voltage drop across the series resistance is small (less than 2 mV in the experiment of Fig. 16, assuming a 20 MΩ series resistance), and (ii) the clamp settles within 100 μs as indicated above. The background noise was somewhat larger than that of a patch recording due to the conductance and capacitance of the cell. However, for small cells, resolution was still good enough to record large single channel responses. This is illustrated in Fig. 17 which shows individual ACh-activated channel currents in a chromaffin cell.

The pipette provides a low-resistance access to the cell interior which we used to measure intracellular potentials under current clamp conditions. The recordings showed spontaneous action potentials resembling those published by

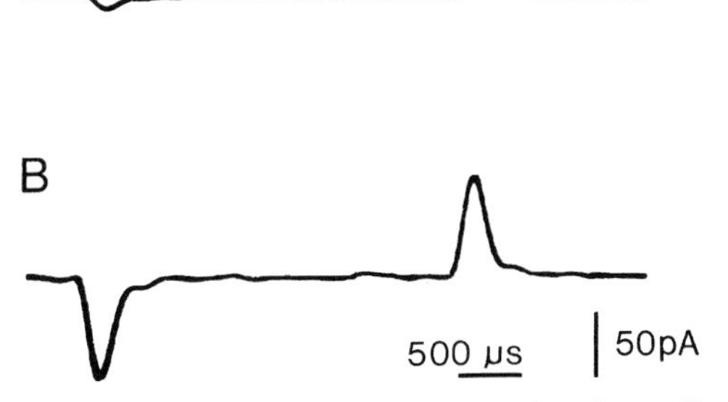

Fig. 15A and B. Capacitive current of a chromaffin cell. (**A**) After establishing a gigaseal on a chromaffin cell, a 20 mV pulse was applied to the pipette interior starting from a holding potential equal to the bath potential. The capacitance of the pipette and of the patch was almost completely compensated (see part I), resulting in very small capacitive artifacts. The pipette input resistance was 20 GΩ. (**B**) After disruption of the patch the response to a 3 mV pulse at a holding potential of -54 mV was measured. The capacitive current was much larger than in **A**, as the cell membrane capacitance had to be charged (compare the amplitudes of the voltage steps). From the integral of the capacitive current a cell membrane capacitance of 5 pF is calculated. The cell had a diameter of 13 µm. Assuming a spherical shape, one obtains a unit capacitance of about 1 µF/cm^2. The DC current was smaller than 2 pA, indicating a cell membrane resistance of several GΩ. The time constant of the capacitive current was less than 0.1 ms, which, together with the value of the cell capacitance, indicates a series resistance smaller than 20 MΩ

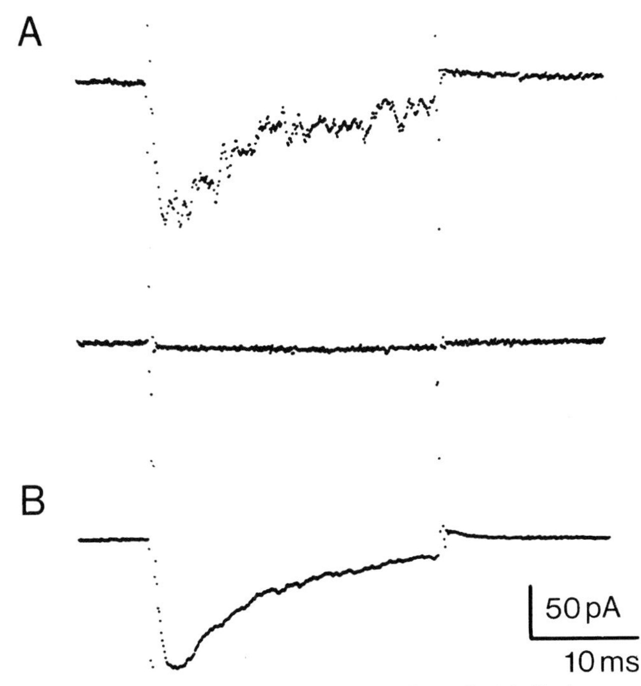

Fig. 16A and B. Na currents in a chromaffin cell. (**A**) Single sweep responses to 34 mV depolarizing (*above*) and hyperpolarizing (*below*) pulses starting from a holding potential of -56 mV. (**B**) Average responses to 25 depolarizing voltage commands, as above. Room temperature; 1 kHz low pass

Brandt et al. (1976). The resting potential, averaging around -60 mV, displayed large fluctuations presumably due to spontaneous opening and closing of ionic channels. Similarly, action potentials and EPSPs could be recorded from small cultured spinal cord neurons.

Fig. 17. Single channel records from a whole chromaffin cell. Normal bath solution with 100 µM ACh added to the chamber. Single ACh-induced currents appeared at varying frequency due to slow changes in ACh concentration and due to desensitization. Two examples at different mean frequency are given. Holding potential was -65 mV; the amplitude of single channel currents was 2.3 pA and their mean open time approximately 30 ms. Filter bandwith 200 Hz

When the pipette was withdrawn from the cell it sealed again, forming an outside-out patch (see Part III,2) of small dimensions. The pipette input capacitance dropped back to a value close to that observed during establishment of the initial giga-seal. Concomitantly background noise was reduced and small channel currents, like those of individual Na channels, could be observed.

In some experiments, the pipette was filled with a Cs-rich solution. The resting potential dropped from a normal value to zero within 10 s after disruption of the initial patch. This observation suggests that the intracellular solution exchanges quickly with the pipette interior. Thus, the method described in this section may be applied not only to record membrane currents, but also to alter the cell's ionic contents. Compared to the nystatin method (Cass and Dalmark 1973) it has the advantage to allow the exchange of divalent ions and of macromolecules.

The present method offers several advantages over the usual recording techniques using glass microelectrodes. It avoids the leakage due to cell penetration with the microelectrode, it allows reliable voltage clamp of small cells, and it offers the possibility of studying macroscopic currents and single channel currents in the same cell. It also allows at least partial control of the ionic milieu of the cell interior.

Conclusions

The methods described here provide several options for voltage- or current-clamp recording on cells or cell-free membrane patches. The size of the cells is not a restriction to the applicability of at least two of these methods. The only requirement is a freely accessible cell surface. This requirement is naturally fulfilled for a number of preparations. It is also fulfilled for most other preparations after enzymatic cleaning. The variety of cell types on which giga-seal formation has been successful is illustrated in Table 1.

The manipulations described here provide free access to either face of the membrane for control of the ionic environment. The giga-seal allows ionic gradients across the membrane to be maintained and high resolution measurements of current through the membrane to be performed.

Table 1. A listing of preparations on which giga-seals have been obtained. Only a fraction of the preparations have been investigated in detail. Bovine chromaffin cells and guinea pig liver cells were plated and kept in short term tissue culture. We acknowledge receiving cells from J. Bormann, T. Jovin, W. D. Krenz, E.-M. Neher, L. Piper, and G. Shaw, Göttingen, FRG; I. Schulz, Frankfurt, FRG; G. Trube, Homburg, FRG, and I. Spector, Bethesda, MD, USA

Cell lines in tissue culture:	Single cells:
Mouse neuroblastoma	Human and avian erythrocytes
Rat basophilic leukaemia cells	Mouse activated macrophages
Primary tissue culture:	Enzymatically dispersed cells:
Rat myotubes and myoballs	Bovine chromaffin cells
Mouse and rabbit spinal cord cells	Guinea pig heart myocytes
Rat cerebellar cells	Guinea pig liver cells
Rat dorsal root ganglion cells	Mouse pancreatic cells
Torpedo electrocytes	Enzyme treated cells:
Rat fibroblasts	Frog skeletal muscle fibres
	Rat skeletal muscle fibres
	Snail ganglion cells

The manipulations are simple and, with some practice, appear to be performed more easily than standard voltage clamp experiments. We expect that they will help to clarify physiological mechanisms in a number of preparations which, so far, have not been amenable to electrophysiological techniques.

Acknowledgements. We thank E. Fenwick for providing dispersed chromaffin cells. We also thank H. Karsten for culturing myoballs and Z. Vogel for a gift of Rhodamine-labelled α-bungarotoxin. O. P. Hamill and F. J. Sigworth were supported by grants from the Humboldt foundation. E. Neher and A. Marty were partially supported by the Deutsche Forschungsgemeinschaft.

References

Brandt BL, Hagiwara S, Kidokoro Y, Miyazaki S. (1976) Action potentials in the rat chromaffin cell and effects of Acetylcholine. J Physiol (Lond) 263:417–439

Cass A, Dalmark M (1973) Equilibrium dialysis of ions in nystatin-treated red cells. Nature (New Biol) 244:47–49

Fenwick EM, Fajdiga PB, Howe NBS, Livett BG (1978) Functional and morphological characterization of isolated bovine adrenal medullary cells. J Cell Biol 76:12–30

Gage PW, Van Helden D (1979) Effects of permeant monovalent cations on end-plate channels. J Physiol (Lond) 288:509–528

Hamill OP, Sakmann B (1981) A cell-free method for recording single channel currents from biological membranes. J Physiol (Lond) 312:41–42P

Horn R, Brodwick MS (1980) Acetylcholine-induced current in perfused rat myoballs. J Gen Physiol 75:297–321

Horn R, Patlak JB (1980) Single channel currents from excised patches of muscle membrane. Proc Natl Acad Sci USA 77:6930–6934

Kostyuk PG, Krishtal OA (1977) Separation of sodium and calcium currents in the somatic membrane of mollusc neurones. J Physiol (Lond) 270:545–568

Kostyuk PG, Krishtal OA, Pidoplichko VI (1976) Effect of internal fluoride and phosphate on membrane currents during intracellular dialysis of nerve cells. Nature 257:691–693

Krishtal OA, Pidoplichko VI (1975) Intracellular perfusion of Helix neurons. Neurophysiol (Kiev) 7:258–259

Krishtal OA, Pidoplichko VI (1980) A receptor for protons in the nerve cell membrane. Neuroscience 5:2325–2327

Läuger P (1975) Shot noise in ion channels. Biochim Biophys Acta 413:1–10

Langmuir I (1938) Overturning and anchoring of monolayers. Science 87:493–500

Lee KS, Akaike N, Brown AM (1978) Properties of internally perfused, voltage clamped, isolated nerve cell bodies. J Gen Physiol 71:489–508

Neher E (1981) Unit conductance studies in biological membranes. In: Baker PF (ed), Techniques in cellular physiology. Elsevier/North-Holland, Amsterdam

Neher E, Sakmann B (1976) Single channel currents recorded from membrane of denervated frog muscle fibres. Nature 260:799–802

Neher E, Sakmann B, Steinbach JH (1978) The extracellular patch clamp: A method for resolving currents through individual open channels in biological membranes. Pflügers Arch 375:219–228

Nir S, Bentz J (1978) On the forces between phospholipid bilayers. J Colloid Interface Sci 65:399–412

Parsegian VA, Fuller N, Rand RP (1979) Measured work of deformation and repulsion of lecithin bilayers. Proc Natl Acad Sci USA 76: 2750–2754

Petrov JG, Kuhn H, Möbius D (1980) Three-Phase Contact Line Motion in the deposition of spread monolayers. J Colloid Interface Sci 73:66–75

Sakmann B, Patlak J, Neher E (1980) Single acetylcholine-activated channels show burst-kinetics in presence of desensitizing concentrations of agonist. Nature 286:71–73

Sigworth FJ, Neher E (1980) Single Na^+ channel currents observed in cultured rat muscle cells. Nature 287:447–449

Stevens CF (1972) Inferences about membrane properties from electrical noise measurements. Biophys J 12:1028–1047

Received March 11 / Accepted May 27, 1981

The nicotinic acetylcholine receptor channel

5

More is known about the neuromuscular junction than about any other synapse, because skeletal muscle is a simple and relatively homogeneous system well suited to biophysical, structural and biochemical studies. The main events in neuromuscular transmission are well known and usually described in some detail at the textbook level [1]. The main steps are as follows: acetylcholine (ACh) is synthesized in the presynaptic nerve terminal from choline and acetate, and concentrated in synaptic vesicles with about 2000–5000 ACh molecules in each vesicle. One presynaptic action potential, via an increase in the free intracellular Ca^{2+} concentration, evokes the release of about 200–500 vesicles, and the ACh concentration in the synaptic cleft increases from virtually nothing to about 1 mM in less than 100 μs. The ACh concentration rise lasts less than 1 ms, as ACh is quickly broken down by ACh esterase. ACh binds to the postsynaptic ACh receptor (two ACh molecules to each receptor), which is a pentameric protein. This causes a conformational change in the receptor, leading to opening of a channel in each receptor that is cation selective. This causes a depolarization of the end-plate membrane that in turn depolarizes the adjacent parts of the muscle surface cell membrane to elicit a propagating action potential [1].

The skeletal muscle nicotinic ACh receptor has been characterized in more detail than any other biological channel. Indeed, the nicotinic ACh receptor channel was the first ion channel in a native biological membrane that was studied directly by single-channel current recording [2]. A combination of biophysical, biochemical and structural (crystallography) analysis has led to the current model of the nicotinic ACh receptor shown in Figure 5.1. In the adult receptor there are five subunits (two α, one β, one δ and one ε), whereas in the fetal form the ε-subunit is replaced by a γ-subunit. There are considerable sequence similarities between the five subunits, all of which span the membrane and form the pore with their M2 segments. The extracellular funnel is also formed by the five subunits, whereas the intracellular protrusion is formed by a tightly linked 43 kDa protein [1].

A very substantial part of our current knowledge concerning the operation of this channel has been the result of the spectacularly successful collaboration between Bert Sakmann and the late Shosaku Numa, in which the patch-clamp technique and the tools of molecular biology were combined [3,4]. In a series of papers [5–12] they explored the properties of the channel and related these to specific structural elements of the receptor.

Site-directed mutagenesis has proved to be an important tool in the elucidation of structure–function relationships for many ion channels. The first ion channel to be investigated by this approach was the nicotinic ACh receptor, and the main strategies now used for studying many different ion channels were first applied to this

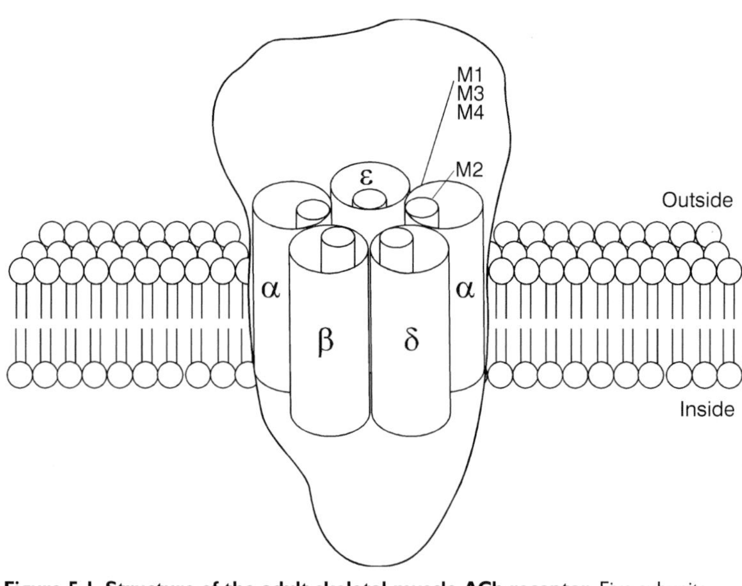

Figure 5.1 Structure of the adult skeletal muscle ACh receptor. Five subunits (two α, one β, one δ and one ε) with considerable sequence similarity contain four hydrophobic segments (M1–M4) that form the membrane-spanning domain. The M2 segment is of particular importance in determining the conductance and selectivity of the channels. Adapted from models shown in [1] and [4]. For further details see text.

receptor [1]. Work with chimaeric subunits constructed from related sequences of bovine subunits and subunits from *Torpedo californica* showed that the hydrophobic M2 segment (Figure 5.1) and its vicinity determined the rate of ion transport through the ACh receptor channel [9]. In the paper that we have chosen from this series [11], the Göttingen and Kyoto groups identified those amino acids that interact with the permeating ions. They introduced various point mutations into the *Torpedo* ACh receptor subunit cDNAs to alter the net charge of the residues around the proposed M2 transmembrane segments. The single-channel conductance properties of these mutants expressed in *Xenopus* oocytes were then evaluated.

There is a conspicuous clustering of charged amino acids bordering the M2 segment in each of the subunits [4], and, by point mutations at these positions, Imoto et al. [11] found that the net number of negative charges determines the conductance irrespective of the subunit into which the mutations had been incorporated. They suggested that the negatively charged amino acids in the five subunits are present in the regions bordering the M2 segment as three ring-like structures at the extra- and intracellular mouths [11,12].

Further work involving mutagenesis within the M2 segment, that forms the narrow part of the channel, was carried out to characterize the selectivity filter. Conductances of mutant channels for Na^+, K^+, Rb^+ and Cs^+ were assessed. These experiments showed that ion selectivity is mainly determined by the amino acid residues located in the M2 segment that are close to the intermediate anionic ring. The channel conductance could be altered by introducing amino acids with different side chain volumes. A side chain with a large volume would reduce the conductance and vice versa. The magnitude of the effect depended on the conducting ion species, with the largest effect seen in the case of Cs^+ and the smallest effect observed with Na^+ [4]. The point mutation analysis identified four positions in each subunit where the amino acid side chains can interact with permeating cations.

The structure–function analysis work, briefly reviewed here, led to a relatively simple model of the ACh receptor channel, in which cations accumulate at the mouths on the extracellular and intracellular sides due to the negative charges of the three anionic rings. The cation selectivity is determined by the narrowest part of the channel inside the M2 segments, by the constriction formed by amino acids with hydroxy-containing side chains [4].

References

1. Ruppersberg, J.P. and Herlitze, S. (1996) in *Comprehensive Human Physiology* (Greger, R. and Windhorst, U., eds.), pp. 307–320, Springer, Berlin
2. Neher, E. and Sakmann, B. (1976) *Nature (London)* **260**, 799–802
3. Numa, S. (1989) *Harvey Lect. Ser.* **83**, 121–165
4. Sakmann, B. (1992) *Neuron* **8**, 613–629
5. Mishina, M., Tobimatsu, T., Imoto, K., Tanaka, K., Fujita, Y., Fukuda, K., Kurasaki, M., Takahashi, H., Morimoto, Y., Hirose, T., Inayama, S., Takahashi, T., Kuno, M. and Numa, S. (1985) *Nature (London)* **313**, 364–369
6. Sakmann, B., Methfessel, C., Mishina, M., Takahashi, T., Takai, T., Kurasaki, M., Fukuda, K. and Numa, S. (1985) *Nature (London)* **318**, 539–543
7. Methfessel, C., Witzemann, V., Takahashi, T., Mishina, M., Numa, S. and Sakmann, B. (1986) *Pflügers Arch.* **407**, 577–588
8. Mishina, M., Takai, T., Imoto, K., Noda, M., Takahashi, T., Numa, S., Methfessel, C. and Sakmann, B. (1986) *Nature (London)* **321**, 406–411
9. Imoto, K., Methfessel, C., Sakmann, B., Mishina, M., Mori, Y., Konno, T., Fukuda, K., Kurasaki, M., Bujo, H., Fujita, Y. and Numa, S. (1986) *Nature (London)* **324**, 670–674
10. Witzemann, V., Barg, B., Nishikawa, Y., Sakmann, B. and Numa, S. (1987) *FEBS Lett.* **223**, 104–112
11. Imoto, K., Busch, C., Sakmann, B., Mishina, M., Konno, T., Nakai, J., Bujo, H., Mori, Y., Fukuda, K. and Numa, S. (1988) *Nature (London)* **335**, 645–648
12. Konno, T., Busch, C., von Kitzing, E., Imoto, K., Wang, F., Nakai, J., Mishina, M., Numa, S. and Sakmann, B. (1991) *Proc. R. Soc. London Ser. B* **244**, 69–79

Imoto et al. (1988) Nature (London) **335**, 645–648

Rings of negatively charged amino acids determine the acetylcholine receptor channel conductance

Keiji Imoto*‡, Christopher Busch*, Bert Sakmann*§, Masayoshi Mishina†, Takashi Konno†, Junichi Nakai†, Hideaki Bujo†, Yasuo Mori†, Kazuhiko Fukuda† & Shosaku Numa†§

* Max-Planck-Institut für biophysikalische Chemie, D-3400 Göttingen, FRG
† Departments of Medical Chemistry and Molecular Genetics, Kyoto University Faculty of Medicine, Kyoto 606, Japan

The structure–function relationship of the nicotinic acetylcholine receptor (AChR) has been effectively studied by the combination of complementary DNA manipulation and single-channel current analysis[1–6]. Previous work with chimaeras between the *Torpedo californica* and bovine AChR δ-subunits has shown that the region comprising the hydrophobic segment M2 and its vicinity contains an important determinant of the rate of ion transport through the AChR channel[5]. It has also been suggested that this region is responsible for the reduction in channel conductance caused by divalent cations[5] and that segment M2 contributes to the binding site of noncompetitive antagonists[7,8]. To identify those amino acid residues that interact with permeating ions, we have introduced various point mutations into the *Torpedo* AChR subunit cDNAs to alter the net charge of the charged or glutamine residues around the proposed transmembrane segments[9–15]. The single-channel conductance properties of these AChR mutants expressed in *Xenopus laevis* oocytes indicate that three clusters of negatively charged and glutamine residues neighbouring segment M2 of the α-, β-, γ- and δ-subunits, probably forming three anionic rings, are major determinants of the rate of ion transport.

‡ Present address: Departments of Medical Chemistry and Molecular Genetics, Kyoto University Faculty of Medicine, Kyoto 606, Japan.
§ To whom correspondence should be addressed.

The regions of the *T. californica* AChR subunits lying between segments M1 and M2 and between segments M2 and M3 (referred to as the M1–M2 portion and the M2–M3 portion, respectively) contain four clusters of negatively charged and polar amino acid residues (Fig. 1). These are located at the α-subunit positions αD238, αE241, αE262 and αS266 (amino acid residues followed by residue numbers[9–11]) and the equivalent positions of the three other subunits. The single-channel current-voltage (*i*–*V*) relation of acetylcholine (ACh)-activated currents (obtained in a nominally divalent-cation-free, symmetrical K^+-rich solution) of the wild-type AChR channel (Fig. 2*a*) is compared with those of the AChR channels with the α-subunit mutation αE262K (Fig. 2*b*) or αD238K (Fig. 2*c*) (amino acid residues in the wild-type and the mutant preceding and following the number of the altered residue, respectively). Both the mutations result in decreased conductance for both inward and outward currents. The *i*–*V* relation of the wild-type channel is nearly symmetrical, whereas those of the mutant channels are asymmetrical, indicating outward (αE262K) or inward (αD238K) rectification of ion transport.

Conductance properties were studied for other AChR mutants in which the residues at the α-subunit position αE262 and at the equivalent positions of the three other subunits were systematically altered. In Fig. 2*d*, the conductances for inward currents of these mutants and of various combinations of them are plotted as a function of the decrease in the total number of net negative charges caused by the mutations. An approximately inverse relationship is found between channel conductance and change in total net negative charge. The data on the α-subunit mutations also indicate that the charge of the side chains is a more important determinant of the rate of ion transport than their size. These results, together with the pseudosymmetrical

```
       238            241 242                                                                               262          266                                          276
α  - Thr [Asp] Ser Gly  - [Glu] [Lys] Met Thr Leu Ser Ile Ser Val Leu Leu Ser Leu Thr Val Phe Leu Leu Val Ile Val [Glu] Leu Ile Pro Ser Thr Ser Ser Ala Val Pro Leu Ile Gly [Lys] Tyr -
       244            247 248                                                                               268   269      272                                        
β  - Pro [Asp] Ala Gly  - [Glu] [Lys] Met Ser Leu Ser Ile Ser Ala Leu Leu Ala Val Thr Val Phe Leu Leu Leu Leu Ala [Asp] [Lys] Val Pro [Glu] Thr Ser Leu Ser Val Pro Ile Ile Ile [Arg] Tyr -
       246            250 251                                                                               271   272      275
γ  - Ala Gln Ala Gly Gly Gln [Lys] Cys Thr Leu Ser Ile Ser Val Leu Leu Ala Gln Thr Ile Phe Leu Phe Leu Ile Ala Gln [Lys] Val Pro [Glu] Thr Ser Leu Asn Val Pro Leu Ile Gly [Lys] Tyr -
       252            255 256                                                                               276   277      280
δ  - Ala [Glu] Ser Gly  - [Glu] [Lys] Met Ser Thr Ala Ile Ser Val Leu Leu Ala Gln Ala Val Phe Leu Leu Leu Thr Ser Gln [Arg] Leu Pro [Glu] Thr Ala Leu Ala Val Pro Leu Ile Gly [Lys] Tyr -
 M1 ┘                   └──────────────────────────────── M2 ─────────────────────────────────┘                                                                    └ M3
```

Fig. 1 Regions surrounding segment M2 of the α-, β-, γ- and δ-subunits of the *T. californica* AChR subjected to point mutations. The relevant amino acid sequences[9–11] are aligned[11,26]. The numbers[9–11] of the relevant amino acid residues and the positions of the hydrophobic segments M1, M2 and M3 (defined in ref. 11) are indicated. Negatively charged residues are boxed with solid lines, and positively charged residues with broken lines.

Methods. cDNAs encoding *T. californica* AChR α-, β-, γ- and δ-subunit mutants were constructed[2,6,27,28] using synthetic oligodeoxyribonucleotides prepared with an automatic DNA synthesizer (Applied Biosystems); some silent mutations were also introduced by this strategy. The plasmids constructed differ from the parental plasmids (given in parentheses) as follows (the substituted nucleotides with residue numbers[9–11] are given, and the plasmids carrying mutated cDNAs linked with the SP6 promoter[29] are named after the mutant specification). pSPα6 (pSPα, ref. 2): A 573, 597, 687; T 594, 648; G627; C630, 651, 690, 691; AGC 676–678. pSPα7 (pSPα): C 717, 748, 750; G 720, 732, 762; A 735, 741, 744, 747; TCG 736–738; T 751; AAGCT 753–757; the junctional sequence between the 3′ end of the α-subunit cDNA and the *Eco*RI site of the vector taken from pSPEα (ref. 2, referred to as junctional sequence). pSPαR209E (pSPα6): G 625; A 626. pSPαR209Q (pSPα6): A 626. pSPαD238R (pSPα6): C 712; G 713. pSPαD238K (pSPα): T 648; C 651, 690, 691; AGC 676–678; A 687, 712; G 714. pSPαD238N (pSPα7): T 648; C 651, 690, 691, 703; AGC 676–678; A 687, 712; G 693, 705. pSPαD238E (pSPα6): G 714. pSPαE241Q (pSPα7): C 721. pSPαE241D (pSPα7): T 720, 723, 732. pSPαK242E (pSPα): T 648, 751, 757; C 651, 690, 691, 748, 750, 756; AGC 676–678; A 687, 726, 735, 741, 744, 747, 753, 754; G 724, 755, 762; TCG 736–738. pSPαK242Q (pSPα7): T 720, 732; C 724; A 726. pSPαE262R (pSPα6): C 784; G 785. pSPαE262K (pSPα): A 784; junctional sequence. pSPαE262Q (pSPα): C 784; junctional sequence. pSPαE262D (pSPα6): C 786. pSPαK276E (pSPα8, see below): G 826. pSPαR301E (pSPα8): GAA 901–903. pSPαD371K (pSPα4, ref. 6): G 1065, 1113; C 1071; A 1111. pSPαE377K (pSPα4): A 1129. pSPαE384K (pSPα4): A 1150. pSPαD389K (pSPα4): A 1162, 1165, 1167; G 1163. pSPαE390K (pSPα4): A 1162, 1168; G 1163. pSPαE391K (pSPα4): A 1162, 1171; G 1163. pSPα(E397K·E398K) (pSPα5, ref. 6): A 1189, 1192; G 1194. pSPαD407K (pSPα5): A 1219, 1221; C 1266. pSPαD407N (pSPα5): A 1219; C 1221, 1266. pSPαE432K (pSPα4): A1294; G 1296. pSPβD244K (pSPβ, ref. 2): A 730; G732. pSPβE247Q (pSPβ): C 739. pSPβK248E (pSPβ): G 742. pSPβD268K (pSPβ): A 802; G 804. pSPβK269E (pSPβ): G 805. pSPβE272K (pSPβ): A 814. pSPγQ246K (pSPγ, ref. 2): A 736. pSPγQ250K (pSPγ): A 748. pSPγK251E (pSPγ): G 751. pSPγQ271K (pSPγ): A 811. pSPγK272E (pSPγ): G814. pSPγE275K (pSPγ): A 823. pSPδE252K (pSPδ, ref. 2): A 754. pSPδE255Q (pSPδ): C 763. pSPδK256E (pSPδ): G 766. pSPδQ276K (pSPδ): A 826. pSPδR277E (pSPδ): G 829; A 830. pSPδE280K (pSPδ): A 838. The plasmid pSPα8 was constructed by ligating the 2.1-kilobase-pair (kb) *Hin*dIII fragment from pSPα6 and the 3.0-kb *Sac*I/*Hin*dIII fragment from pSPα after they had been blunt-ended. Messenger RNAs specific for mutant and wild-type AChR subunits were synthesized *in vitro*[2,5,6], using *Eco*RI- or *Sma*I-cleaved plasmids carrying the respective cDNAs as templates.

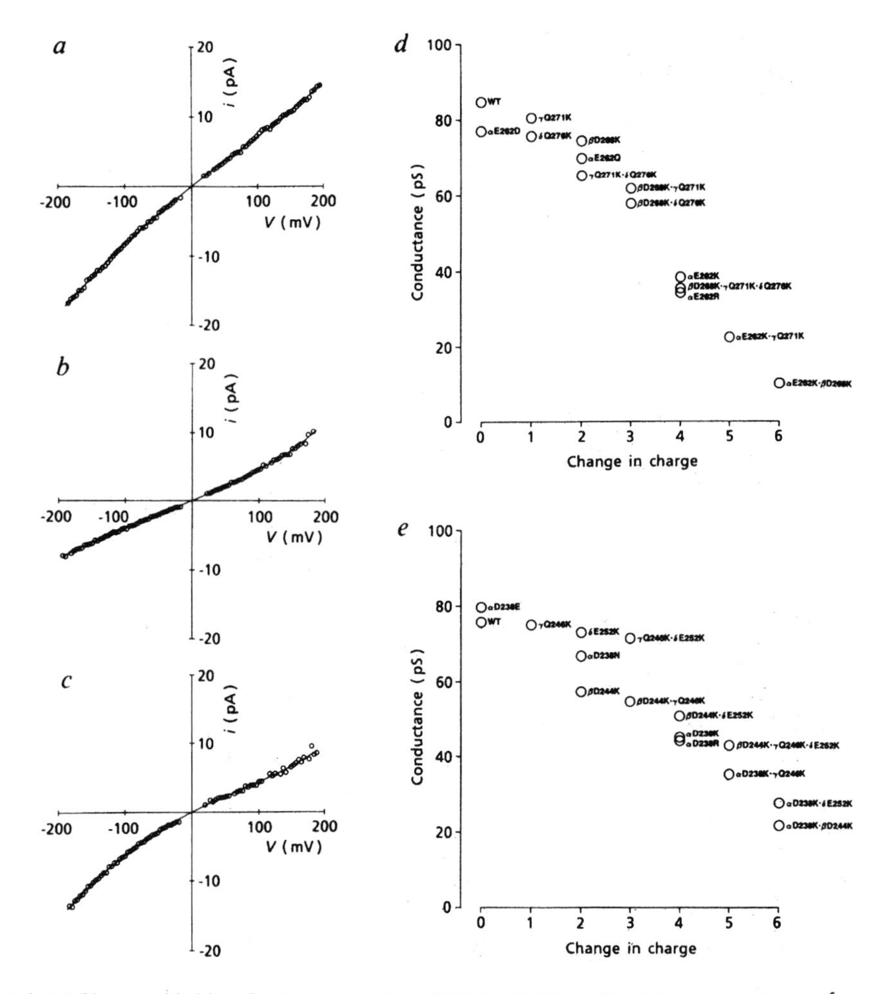

Fig. 2 Effects of point mutations in the M2–M3 and M1–M2 portions. *a–c*, Single-channel *i–V* relations of the wild-type AChR channel (*a*) and the mutant channels with αE262K (*b*) or αD238K (*c*). Each symbol represents a mean of at least four amplitude measurements within 4-mV intervals. Continuous lines represent fitted 5th-order polynomials. *d*, *e*, Changes in single-channel conductance with decreases in the total number of net negative charges. The chord conductances at −100 mV membrane potential of the AChR channels with a mutation or mutations at αE262 and/or the equivalent positions (*d*) or the chord conductances at +100 mV membrane potential of the AChR channels with a mutation or mutations at αD238 and/or the equivalent positions (*e*) are plotted against the reduction in the total number of net negative charges. A change from a negatively-charged to a positively-charged residue is scored as two and a change from a negatively charged to a neutral residue or from a neutral to a positively charged residue as one. Scores are doubled for mutations in the α-subunit because of the $\alpha_2\beta\gamma\delta$ stoichiometry and are summed in case of multiple mutations. Each symbol, followed by the mutant specification, is the mean of 3–5 experiments, the standard deviation being at most 3 pS. WT refers to the wild-type AChR channel.

Methods. *X. laevis* oocytes were injected[2] with the four kinds of mutant or wild-type mRNAs (α-, β-, γ- and δ-subunit-specific mRNAs in a molar ratio of 2:1:1:1) and were incubated[1] for 3–5 days. Single-channel current measurements were made at 12±1 °C in inside-out membrane patches isolated from the oocytes[30]. The bath solution contained 100 mM KCl, 10 mM EGTA and 10 mM HEPES; the solution was adjusted to *p*H 7.2 with KOH. The pipette solution was the same as the bath solution, except that ACh was added in a final concentration of 0.5–4 μM. The *i–V* relations were obtained[5] in the voltage range between −200 mV and +200 mV. In some experiments, the *i–V* relations were shifted by a few millivolts so that the zero current potential was 0 mV.

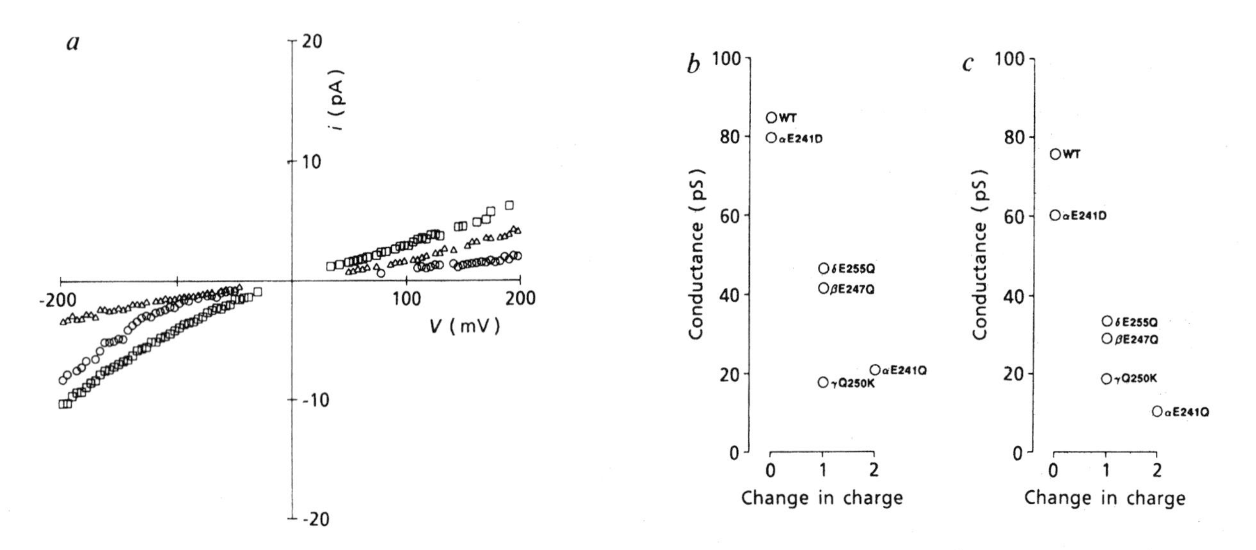

Fig. 3 Single-channel conductances of the AChR channels with mutations in the intermediate ring. *a*, *i–V* relations of the AChR channels with a mutation in the α- (αE241Q, circles), β- (βE247Q, squares) or γ-subunit (γQ250K, triangles); the *i–V* relation of the AChR channel with a mutation in the δ-subunit (δE255Q) is shown in Fig. 4*d*, *h* (squares). *b*, *c*, Changes in single-channel conductance with decreases in the total number of net negative charges. The chord conductances at membrane potentials of −100 mV (*b*) and +100 mV (*c*) are plotted as in Fig. 2*d*, *e*. The experimental conditions were as described in Fig. 2 legend.

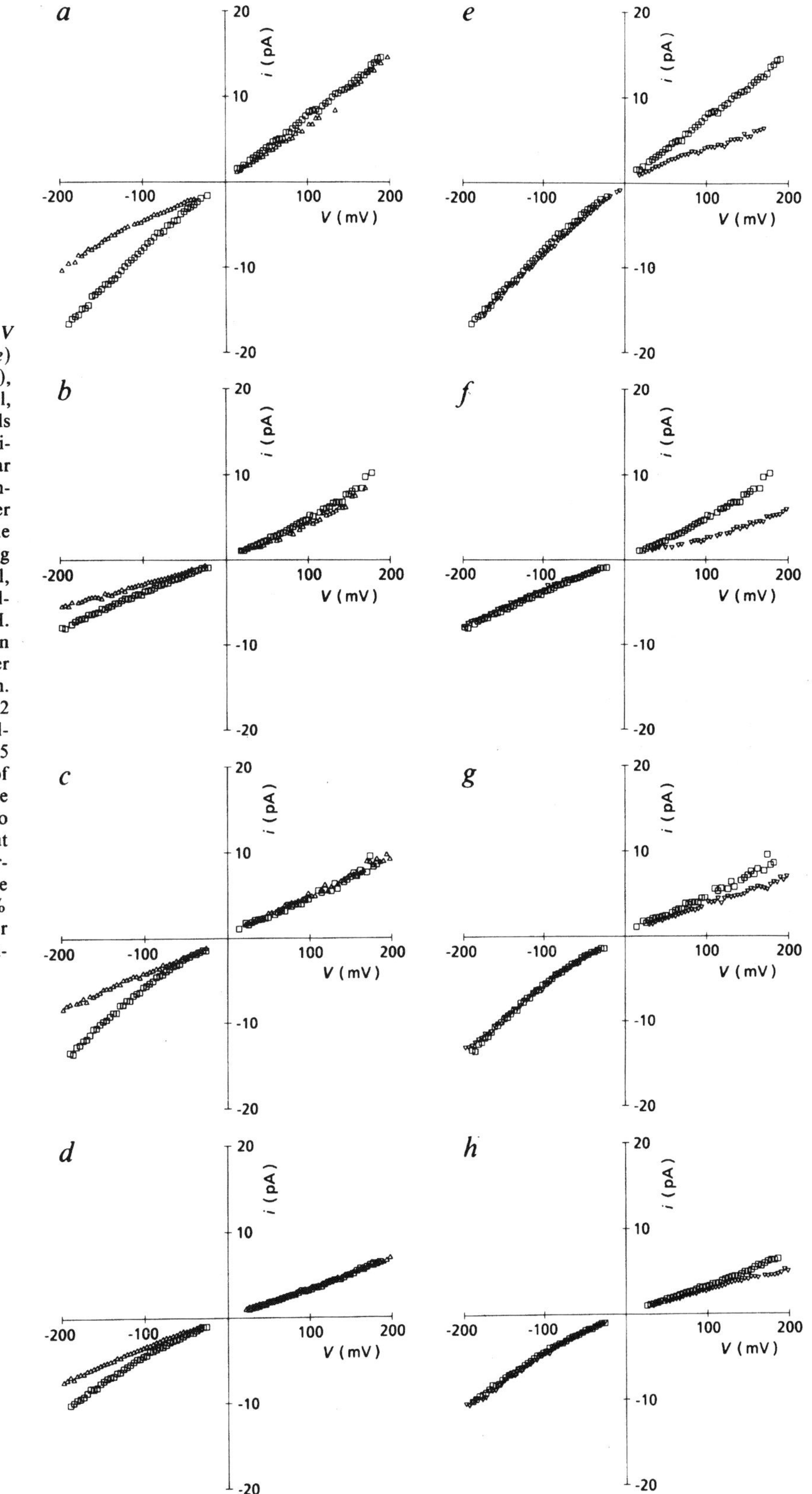

Fig. 4 Sidedness of Mg^{2+} effects on the *i–V* relations of the wild-type AChR channel (*a, e*) and the mutant channels with αE262K (*b, f*), αD238K (*c, g*) or δE255Q (*d, h*). In each panel, two *i–V* relations are shown. Square symbols indicate control experiments under the conditions described in Fig. 2 legend. Triangular symbols indicate experiments under asymmetrical conditions with 0.5 mM Mg^{2+} either on the extracellular side (*a–d*) or on the intracellular side (*e–h*). The Mg^{2+}-containing solution was composed of 100 mM KCl, 0.5 mM $MgCl_2$ and 10 mM HEPES; the solution was adjusted to *p*H 7.2 by adding KOH. The nominally divalent-cation-free solution specified in Fig. 2 legend was used for the other side. ACh was added to the pipette solution. For further experimental details, see Fig. 2 legend. The effects of extracellular and intracellular Mg^{2+} on mean chord conductances (3-5 experiments) at membrane potentials of −150 mV and +150 mV, respectively, are expressed as percentage decreases relative to mean control values (3-7 experiments) without Mg^{2+}. The values for inward and outward currents respectively are 42% and 50% for the wild-type; 29% and 48% for αE262K; 43% and 17% for αD238K; and 31% and 18% for δE255Q. The standard deviation for each conductance measurement was at most 4 pS.

orientation of the AChR subunits assembled in a molar stoichiometry of $\alpha_2\beta\gamma\delta$ (ref. 16), indicate that the negatively charged and glutamine residues in this cluster are nearly equidistant from the channel axis, forming a ring structure. Analogous experiments on AChR channels mutagenized in the cluster at αD238 and the equivalent positions reveal a similar inverse relationship between channel conductance for outward current and change in total net negative charge, indicating that there is another ring of negatively-charged and glutamine residues (Fig. 2*e*). Mutations in the two anionic rings evoke opposite directions of rectification of the *i–V* relation (exemplified in Fig. 2*b, c*), implying that the rings are on opposite sides of the membrane.

Figure 3*a–c* shows the effects of mutating residues in the cluster at αE241 and the equivalent positions. Reducing the net negative charge of the individual residues in this intermediate cluster has a much stronger effect (Fig. 3*b*, *c*) than when the residues of the corresponding subunits in the two other clusters are mutated (Fig. 2*d*, *e*). The intermediate cluster of glutamic acid and glutamine residues, probably also forming a ring-like structure, is therefore more important in determining the rate of ion transport than the two other anionic rings. Notably, the γ-subunit mutation γQ250K causes a greater reduction in conductance than would be expected from the decrease in net negative charge resulting from the mutation. Furthermore, the AChR channel with the γ-subunit mutation shows virtually no rectification of the *i*-*V* relation, whereas inward rectification is evoked by the α- (αE241Q and αE241D), β- (βE247Q) or δ-subunit mutation (δE255Q) (Fig. 3*a*-*c*; see also Fig. 4*d*, *h*). It is possible that the individual residues of the intermediate ring do not occupy strictly equivalent positions in the electrical profile of the channel. The glutamic acid residues in the α-, β- and δ-subunits may be located closer to the anionic ring at αD238 than the glutamine residue in the γ-subunit, which may be closer to the centre of the electrical profile. In this context, it could be relevant that the γ-subunit has an insertion of a glycine in the M1-M2 portion compared with the other subunits (Fig. 1). The difference in rectification and the variable extents of reduction in conductance may indicate that, in the intermediate ring, not only the net negative charges of the constituent residues but also other factors, such as their size, position and interaction, are involved in determining the rate of ion transport. We conclude that the intermediate ring is located between the two other rings, probably nearer to the anionic ring at αD238, and is close to the narrowest part of the channel.

When 0.5 mM Mg^{2+} is added to the extracellular side (Fig. 4*a*-*d*), the inward current of the mutant channel with αE262K (Fig. 4*b*) is found to be less sensitive to Mg^{2+} than that of the wild-type channel (Fig. 4*a*). In contrast, the mutant channel with αD238K (Fig. 4*c*) shows nearly the same relative decrease in conductance for inward current as the wild-type. There is no significant change in the outward current of either channel. When 0.5 mM Mg^{2+} is present on the intracellular side (Fig. 4*e*-*h*), the outward current of the channel with αD238K (Fig. 4*g*) is less affected than that of the wild-type channel (Fig. 4*e*). In contrast, the relative decrease in conductance for outward current is similar in the channel with αE262K (Fig. 4*f*) and in the wild-type channel. The inward current is essentially unaffected in the wild-type and mutant channels. These results indicate that the anionic ring at αE262 and that at αD238 are involved in interactions with Mg^{2+}, suggesting that these rings constitute part of the extracellular and the cytoplasmic portion of the channel, respectively. Figure 4*d*, *h* shows that the inward and the outward current of the channel with δE255Q are less sensitive to extracellular and intracellular Mg^{2+}, respectively, than the currents of the wild-type channel. The intermediate ring is therefore involved in the interaction with Mg^{2+}. The reduced sensitivity to Mg^{2+} on both sides provides further evidence that this ring is located between the extracellular and cytoplasmic rings.

No significant changes in channel conductance result from the following mutations (data not shown): βK248E, γK251E, δK256E, βK269E, γK272E, δR277E, βE272K, γE275K, δE280K, βE272K·γE275K·δE280K, αK276E, αR301E and αE432K (neighbouring segment M2, M3 or M4); αD371K, αE377K, αE384K, αD389K, αE390K, αE391K and α(E397K·E398K) in the amphipathic segment MA (refs 14, 15) and the neighbouring regions. αD407N (neighbouring segment M4) causes a slight decrease in conductance. The data on segment MA are consistent with our previous studies, suggesting that this segment is dispensable for ACh responsiveness, although it is important for efficient expression of AChR in the plasma membrane[6]. Mutations of some charged residues (αR209E, αR209Q, αK242E, αK242Q and αD407K) cause a failure to observe channel openings, which is probably due to insufficient functional expression of AChR.

We have identified three clusters of negatively charged and glutamine residues neighbouring the hydrophobic segment M2 as major determinants of the AChR channel conductance for monovalent cations and as sites of interaction with divalent cations. Our results suggest that these amino acid residues form three anionic rings. The membrane topology of the three rings is based largely on the sidedness of Mg^{2+} effects, which provides evidence that the hydrophobic segment M2 is a transmembrane segment with the M2-M3 portion on the extracellular side and with the M1-M2 portion on the cytoplasmic side. Thus, the extracellular ring is located near the external mouth and the cytoplasmic ring near the internal mouth of the channel. The intermediate ring is positioned between the two other rings and may form a narrow channel constriction. The location of both the intermediate and the cytoplasmic ring in the M1-M2 portion suggests that this portion actually forms part of the transmembrane segment containing segment M2. This transmembrane segment probably constitutes at least part of the channel lining. However, our results do not exclude the possibility that other transmembrane segments also take part in forming the channel lining.

In the regions neighbouring segment M2, the mammalian muscle AChR ε-subunit has two more net negative charges than the γ-subunit, one in each of the extracellular and cytoplasmic rings[17-19]. This may account for the higher channel conductance of the AChR composed of the α-, β-, δ- and ε-subunits compared with the AChR composed of the α-, β-, γ- and δ-subunits[4,19]. Interestingly, the *T. californica* electroplax AChR γ-subunit is equivalent to the ε- rather than to the γ-subunit of the mammalian muscle AChR with respect to the amino acid residues in the three anionic rings[11,12,17-22]. Our conclusion that rings containing charged amino acid residues are major determinants of the rate of ion transport may also be valid for other multisubunit ligand-gated channels, such as the neuronal AChR[23], the $GABA_A$ receptor[24] and the glycine receptor[25].

We thank Drs A. Karlin, F. Conti and E. Neher for reading the manuscript. This investigation was supported in part by research grants from the Ministry of Education, Science and Culture of Japan, the Institute of Physical and Chemical Research, the Mitsubishi Foundation and the Japanese Foundation of Metabolism and Diseases.

Received 27 June; accepted 1 September 1988.

1. Mishina, M. *et al. Nature* **307**, 604-608 (1984).
2. Mishina, M. *et al. Nature* **313**, 364-369 (1985).
3. Sakmann, B. *et al. Nature* **318**, 538-543 (1985).
4. Mishina, M. *et al. Nature* **321**, 406-411 (1986).
5. Imoto, K. *et al. Nature* **324**, 670-674 (1986).
6. Tobimatsu, T. *et al. FEBS Lett.* **222**, 56-62 (1987).
7. Giraudat, J., Dennis, M., Heidmann, T., Chang, J.-Y. & Changeux, J.-P. *Proc. natn. Acad. Sci. U.S.A.* **83**, 2719-2723 (1986).
8. Hucho, F., Oberthür, W. & Lottspeich, F. *FEBS Lett.* **205**, 137-142 (1986).
9. Noda, M. *et al. Nature* **299**, 793-797 (1982).
10. Noda, M. *et al. Nature* **301**, 251-255 (1983).
11. Noda, M. *et al. Nature* **302**, 528-532 (1983).
12. Claudio, T., Ballivet, M., Patrick, J. & Heinemann, S. *Proc. natn. Acad. Sci. U.S.A.* **80**, 1111-1115 (1983).
13. Devillers-Thiery, A., Giraudat, J., Bentaboulet, M. & Changeux, J.-P. *Proc. natn. Acad. Sci. U.S.A.* **80**, 2067-2071 (1983).
14. Guy, H. R. *Biophys. J.* **45**, 249-261 (1984).
15. Finer-Moore, J. & Stroud, R. M. *Proc. natn. Acad. Sci. U.S.A.* **81**, 155-159 (1984).
16. Brisson, A. & Unwin, P. N. T. *Nature* **315**, 474-477 (1985).
17. Takai, T. *et al. Nature* **315**, 761-764 (1985).
18. Takai, T. *et al. Eur. J. Biochem.* **143**, 109-115 (1984).
19. Witzemann, V., Barg, B., Nishikawa, Y., Sakmann, B. & Numa, S. *FEBS Lett.* **223**, 104-112 (1987).
20. Shibahara, S. *et al. Eur. J. Biochem.* **146**, 15-22 (1985).
21. Yu, L., LaPolla, R. J. & Davidson, N. *Nucleic Acids Res.* **14**, 3539-3555 (1986).
22. Boulter, J. *et al. J. Neurosci. Res.* **16**, 37-49 (1986).
23. Boulter, J. *et al. Nature* **319**, 368-374 (1986).
24. Schofield, P. R. *et al. Nature* **328**, 221-227 (1987).
25. Grenningloh, G. *et al. Nature* **328**, 215-220 (1987).
26. Kubo, T. *et al. Eur. J. Biochem.* **149**, 5-13 (1985).
27. Zoller, M. J. & Smith, M. *Meth. Enzym.* **100**, 468-500 (1983).
28. Nakamaye, K. L. & Eckstein, F. *Nucleic Acids Res.* **14**, 9679-9698 (1986).
29. Melton, D. A. *et al. Nucleic Acids Res.* **12**, 7035-7056 (1984).
30. Methfessel, C. *et al. Pflügers Arch. ges. Physiol.* **407**, 577-588 (1986).

The role of cyclic GMP in photoreceptors

Following the publication of the detailed description of the patch-clamp technique (see Section 4) many different types of ion channels were quickly discovered. In addition to the classic voltage-sensitive and agonist-operated ion channels, it had already become clear in the early 1980s that there were also ion channels controlled by hormones or neurotransmitters via the intracellular messengers cyclic AMP [1] and Ca^{2+} [2]. The cyclic AMP effect on the so-called S potassium channel in *Aplysia* sensory neurons [1] is mediated via cyclic AMP-dependent phosphorylation and is therefore relatively slow [3]. Much faster effects could be obtained if binding a cyclic nucleotide to an ion channel was able to control its function directly. In 1985 no less than five papers describing direct actions of guanosine 3′,5′-cyclic monophosphate (cyclic GMP) on ion channels in retinal rod and cone photoreceptors were published [4–8]. A few years later a cyclic nucleotide-gated conductance in olfactory receptor cilia was also discovered [9]. Cyclic GMP-activated cation channels have since been found in numerous places including the heart, aorta, kidney, testis and colon [10].

The paper we have selected [4] reports the results of experiments performed on the light-adapted frog retina. The actions of cyclic GMP were studied on excised inside-out patches from the plasma membrane of the rod outer segment. Fesenko et al. [4] found that cyclic GMP acting on the inner side of the membrane reversibly increased the cationic conductance. Since the cyclic GMP action could occur in the absence of ATP, they concluded that the effect was not mediated by protein phosphorylation, but was a result of a direct action on the membrane. The effect of cyclic GMP was very specific, and even very high cyclic AMP concentrations (mM) were ineffective. The authors noted that, because of similarities between the properties of the light-dependent conductance of the rod outer segment membrane and the cyclic GMP-dependent conductance of the rod membrane patches, it was most likely that cyclic GMP controlled the light-dependent conductance.

The direct action of cyclic GMP on ion channels is advantageous, indeed necessary, for the proper function of photoreceptors where the response time can be less than 1 ms. A model for the light-induced events in rod outer segments is shown in Figure 6.1. It is well known that photoreceptors increase their membrane potential (hyperpolarize) in response to light [11]. Figure 6.1 shows how this works. There is a high intracellular cyclic GMP level in the dark and this keeps a non-selective cation channel in the plasma membrane open, thereby allowing a steady inflow of cations depolarizing the membrane. When rhodopsin absorbs light it activates a heterotrimeric GTP-binding protein, usually referred to as transducin, which in turn activates a phosphodiesterase that breaks down cyclic GMP. The resulting reduction in the cellular cyclic GMP concentration closes the non-selective cation channel, and this allows the K^+ channels to become dominant, thereby hyperpolarizing the membrane [11].

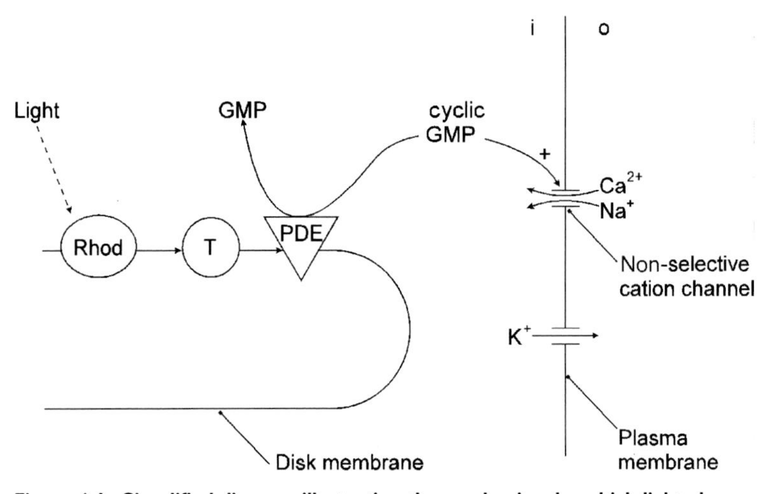

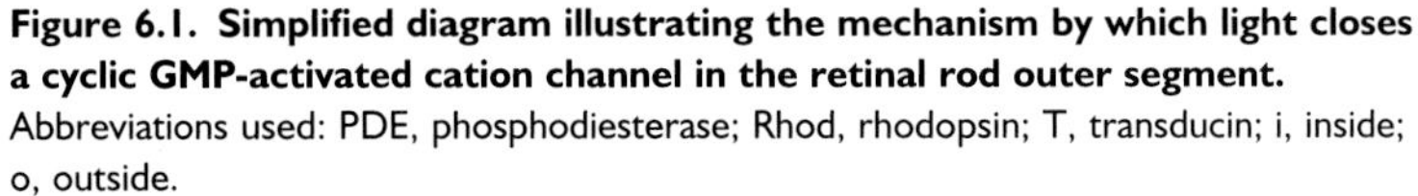

Figure 6.1. Simplified diagram illustrating the mechanism by which light closes a cyclic GMP-activated cation channel in the retinal rod outer segment. Abbreviations used: PDE, phosphodiesterase; Rhod, rhodopsin; T, transducin; i, inside; o, outside.

It is important that the cyclic GMP-activated cation channel is non-selective since, as illustrated in Figure 6.1, this also allows some Ca^{2+} influx. In the dark, Ca^{2+} entering through the channel inhibits guanyl cyclase, the enzyme that catalyses the synthesis of cyclic GMP from GTP. When light, via the steps shown in Figure 6.1, closes the cation channel, there is an increase in the synthesis of cyclic GMP which counteracts the effect of the phosphodiesterase. This means that during continuous light exposure there is a decrease in the electrical photoresponse as the cyclic GMP level partially recovers, allowing some reopening of the cyclic GMP-dependent cation channels. This is the molecular basis for light adaptation [11].

One physiologically very important advantage of the steps between light absorption and channel closure shown in Figure 6.1 concerns signal amplification. One activated phosphodiesterase molecule catalyses the hydrolysis of about 1000 cyclic GMP molecules per second. The overall amplification of the cascade shown in Figure 6.1 can be up to one million, but the adaptation process already mentioned reduces this factor considerably [12].

The rod photoreceptor cyclic GMP-activated cation channel has been purified from bovine retina and the cDNA cloned [12]. The putative folding pattern indicates six transmembrane segments, similar to that of many voltage-activated K^+ channels, although the cyclic GMP-activated channel is not voltage-gated [10]. The cyclic nucleotide binding site is located in the cytoplasmic C-terminus and has structural similarity to the equivalent region on the cyclic GMP-dependent protein kinase [10]. There are at least two different subunits in the channel and it is most likely that the functional channels are tetramers, but the subunit stoichiometry in the native channel is not yet known.

The work of Fesenko et al. [4], closely followed by that of several other groups [5–8], has not only clarified the very important mechanism of photoreception, but also led to studies of an entirely new family of cyclic nucleotide-regulated ion channels with different functions in different tissues [10].

References

1. Siegelbaum, S.A., Camardo, J.S. and Kandel, E.R. (1982) *Nature (London)* **299**, 413–417
2. Maruyama, Y. and Petersen, O.H. (1982) *Nature (London)* **300**, 61–63
3. Siegelbaum, S.A., Belardetti, F., Camardo, J.S. and Shuster, M.J. (1986) *J. Exp. Biol.* **124**, 287–306
4. Fesenko, E.E., Kolesnikov, S.S. and Lyubarsky, A.L. (1985) *Nature (London)* **313**, 310–313
5. Yau, K.-W. and Nakatani, K. (1985) *Nature (London)* **317**, 252–255
6. Koch, K.-W. and Kaupp, U.B. (1985) *J. Biol. Chem.* **260**, 6788–6800
7. Haynes, L.W. and Yau, K.-W. (1985) *Nature (London)* **317**, 61–64
8. Cobbs, W.H., Barkdoll, A.E. and Pugh, E.N. (1985) *Nature (London)* **317**, 64–66
9. Nakamura, T. and Gold, G.H. (1987) *Nature (London)* **325**, 442–444
10. Finn, J.T., Grunwald, M.E. and Yau, K.-W. (1996) *Annu. Rev. Physiol.* **58**, 395–426
11. Dowling, J.E. (1996) in *Comprehensive Human Physiology* (Greger, R. and Windhorst, U., eds.), pp. 773–788, Springer, Heidelberg
12. Kaupp, U.B. and Koch, K.-W. (1992) *Annu. Rev. Physiol.* **54**, 153–175

Induction by cyclic GMP of cationic conductance in plasma membrane of retinal rod outer segment

Evgeniy E. Fesenko, Stanislav S. Kolesnikov & Arkadiy L. Lyubarsky

Institute of Biological Physics, USSR Academy of Sciences, Pushchino, Moscow Region 142292, USSR

Vertebrate rod photoreceptors hyperpolarize when illuminated, due to the closing of cation-selective channels in the plasma membrane. The mechanism controlling the opening and closing of these channels is still unclear, however. Both 3′, 5′-cyclic GMP[1,2] and Ca^{2+} ions[3] have been proposed as intracellular messengers for coupling the light activation of the photopigment rhodopsin to channel activity and thus modulating light-sensitive conductance. We have now studied the effects of possible conductance modulators on excised 'inside-out' patches from the plasma membrane of the rod outer segment (ROS), and have found that cyclic GMP acting from the inner side of the membrane markedly increases the cationic conductance of such patches (EC_{50} 30 μM cyclic GMP) in a reversible manner, while Ca^{2+} is ineffective. The cyclic GMP-induced conductance increase occurs in the absence of nucleoside triphosphates and, hence, is not mediated by protein phosphorylation, but seems rather to result from a direct action of cyclic GMP on the membrane. The effect of cyclic GMP is highly specific; cyclic AMP and 2′, 3′-cyclic GMP are completely ineffective when applied in millimolar concentrations. We were unable to recognize discrete current steps that might represent single-channel openings and closings modulated by cyclic GMP. Analysis of membrane current noise shows the elementary event to be 3 fA with 110 mM Na^+ on both sides of the membrane at a membrane potential of −30 mV. If the initial event is assumed to be the closure of a single cyclic GMP-sensitive channel, this value corresponds to a single-channel conductance of 100 fS. It seems probable that the cyclic GMP-sensitive conductance is responsible for the generation of the rod photoresponse *in vivo*.

Experiments were performed on the light-adapted retina of the frog *Rana temporaria*. Figure 1*a* shows the recording method. A retina treated with trypsin (10 mg ml^{-1}, 30 min, 20 °C) was shaken gently in a small volume of saline and the solution containing detached ROSs and whole photoreceptor cells was then pipetted into a perfusion chamber. Under visual observation, the recording electrode was advanced to make contact with the lateral surface of the outer segment of a whole rod, or more often, an isolated ROS. Brief suction was applied to the pipette to produce a seal of 1–20 GΩ (ref. 4) and a patch of membrane was excised from the cell by a sharp shift of the pipette. This exposed the intracellular surface of the membrane to the bathing solution in the chamber while the extracellular surface was bathed with the solution contained in the pipette. The results reported below were obtained from membrane fragments lacking the high-conductive anion channels described previously[5].

During these experiments, we obtained 136 patches sufficiently stable for the solution in the chamber to be changed several times. Of these, 70 responded to the application of cyclic GMP by an increase in conductance (see Fig. 1*b*), while the current fluctuated in a random manner without any recognizable steps which could represent single-channel activity (see below). The effect of cyclic GMP was completely reversible and could be observed several times in the same patch, with restoration to baseline when the cyclic GMP was washed out. The sensitivity of most patches to cyclic GMP remained constant for 10–15 min after the patch was isolated (during this time the solution in the chamber was changed five to eight times) before it began to decrease.

To our surprise, the cyclic GMP-dependent increase in the conductance of the patch occurred in the absence of nucleoside

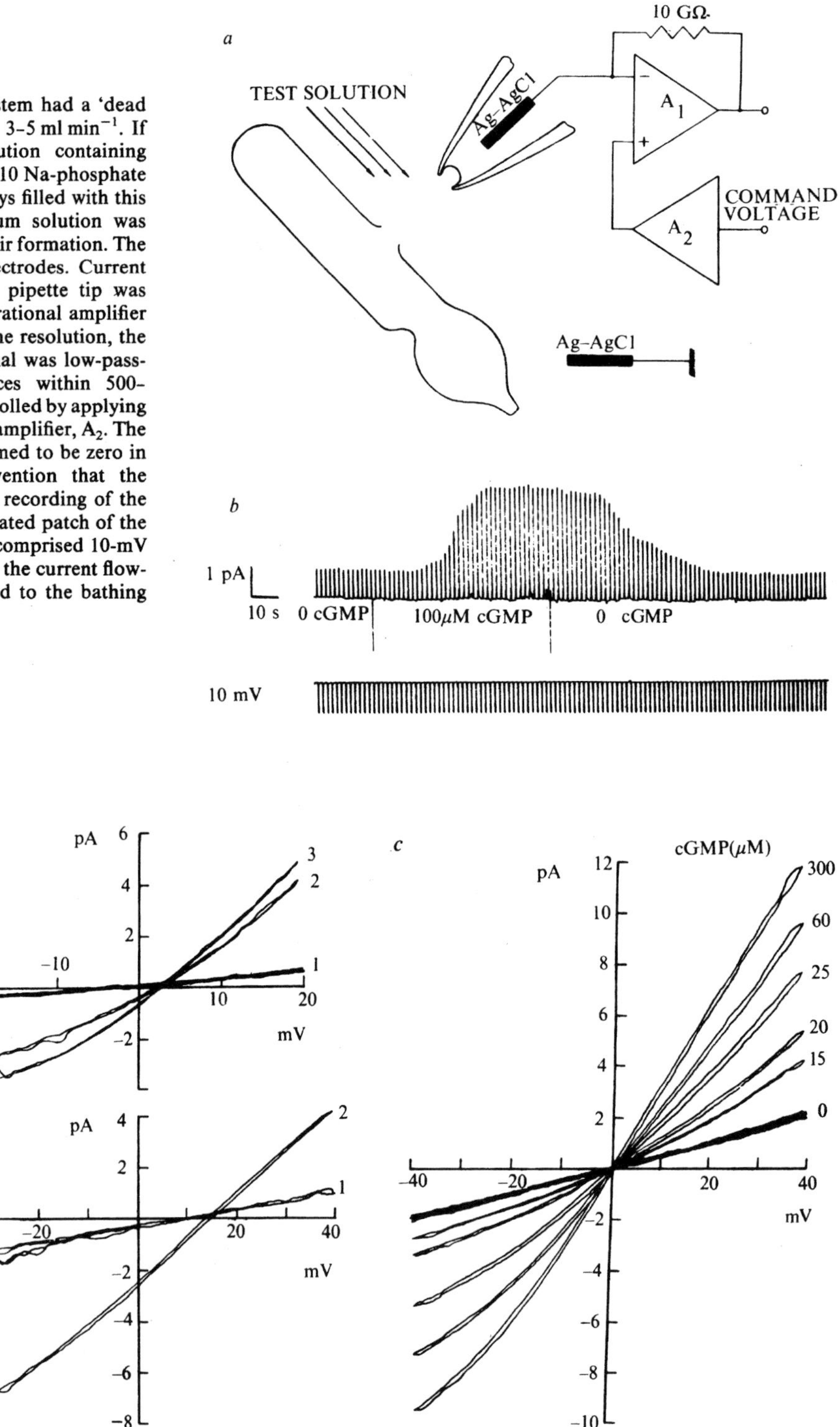

Fig. 1 ***a***, Recording method. The perfusion system had a 'dead volume' of ~2 ml and the flow rate was usually 3-5 ml min^{-1}. If not otherwise stated, a standard Ringer solution containing (in mM) 90 NaCl, 10 KCl, 2 $MgCl_2$, 0.1 $CaCl_2$, 10 Na-phosphate *p*H 7.5 was used. The microelectrodes were always filled with this saline. The yield of vesicles in this low-calcium solution was < 20%[5], so no precautions were taken against their formation. The pipette and bath contained Ag/AgCl pellet electrodes. Current flowing across the patch of membrane at the pipette tip was measured with a virtual ground circuit (A_1, operational amplifier with a 10-GΩ feedback resistor). To improve time resolution, the recovery of high frequencies was used. The signal was low-pass-filtered (6-pole Bessel filter, cut-off frequences within 500-10,000 Hz). The transmembrane voltage was controlled by applying a command voltage through a buffer operational amplifier, A_2. The voltage inside the measuring electrode was assumed to be zero in agreement with the electrophysiological convention that the extracellular space is at zero potential. ***b***, Chart recording of the action of cyclic GMP (cGMP; Sigma) on an isolated patch of the ROS plasma membrane. The command voltage comprised 10-mV pulses (bottom trace). The upper trace represents the current flowing through the patch. Cyclic GMP was applied to the bathing solution.

Fig. 2 Voltage-current relations of patches of the ROS plasma membrane. A saw-toothed voltage (5 mV s^{-1}) was applied to patches and the *x*-input of an *x*-*y* recorder. The output signal of the current-voltage converter was applied to the *y*-input of the recorder. The relations were recorded using one to three cycles of the saw-toothed voltage. The stability of the baseline was controlled after every application of cyclic GMP. ***a***, Effect of Ca^{2+} and cyclic GMP on a patch of the ROS plasma membrane. Curve 1, 10 nM or 1 mM Ca^{2+}, no cyclic GMP; curve 2, 100 μM cyclic GMP, 10 nM Ca^{2+}; curve 3, 100 μM cyclic GMP, 1 mM Ca^{2+}. The composition of the solution was (in mM): 100 KCl, 10 Na-phosphate, 2 $MgCl_2$, 1 $CaCl_2$ in high-calcium solution, 0.1 and 0.384 EGTA in the low-calcium one (the dissociation constants for Ca-EGTA complexes given in ref. 23 were used), *p*H 7.5. ***b***, Voltage-current relation in a low-salt solution containing (mM): 45 NaCl, 5 KCl, 2 $MgCl_2$, 0.1 $CaCl_2$, 5 Na-phosphate, *p*H 7.5. Curve 1, no cyclic GMP; curve 2, 100 μM cyclic GMP. ***c***, Family of voltage-current relations recorded at different cyclic GMP concentrations. Cyclic GMP was applied in standard Ringer. The relation at 500 μM cyclic GMP (not shown) was identical to that at 300 μM cyclic GMP.

triphosphates; the addition of 1 mM ATP and 1 mM GTP was without effect. It is generally considered that the effects of cyclic nucleotides are mediated by protein phosphorylation by specific kinases[6]. Analysis by TLC of the cyclic GMP preparation used shows it to contain $< 1/10^6$ nucleoside di- and triphosphates, that is, a concentration of $< 10^{-10}$ M in 100 μM cyclic GMP solution applied to the membrane, a level apparently too low for kinases to function. Phosphorylation supported by endogenous triphosphates in the membrane also seems unlikely in our experiments, because any triphosphates present during the first application of cyclic GMP would be washed out during the repeated exchange of solution. We conclude that cyclic GMP acts directly on the channels in the membrane.

Because we could not detect current jumps corresponding to single-channel activity, the conductive properties of the ROS plasma membrane patches were determined from their integral voltage-current relations (Fig. 2). Ca^{2+} ions (10^{-8}–10^{-3} M) had no effect on the conductance of the system in the absence of cyclic GMP. The cyclic GMP-dependent component of conductance decreased when the Ca^{2+} concentration was lowered; at 10^{-8} M Ca^{2+} it was 20-30% lower than that at 10^{-3} M Ca^{2+}.

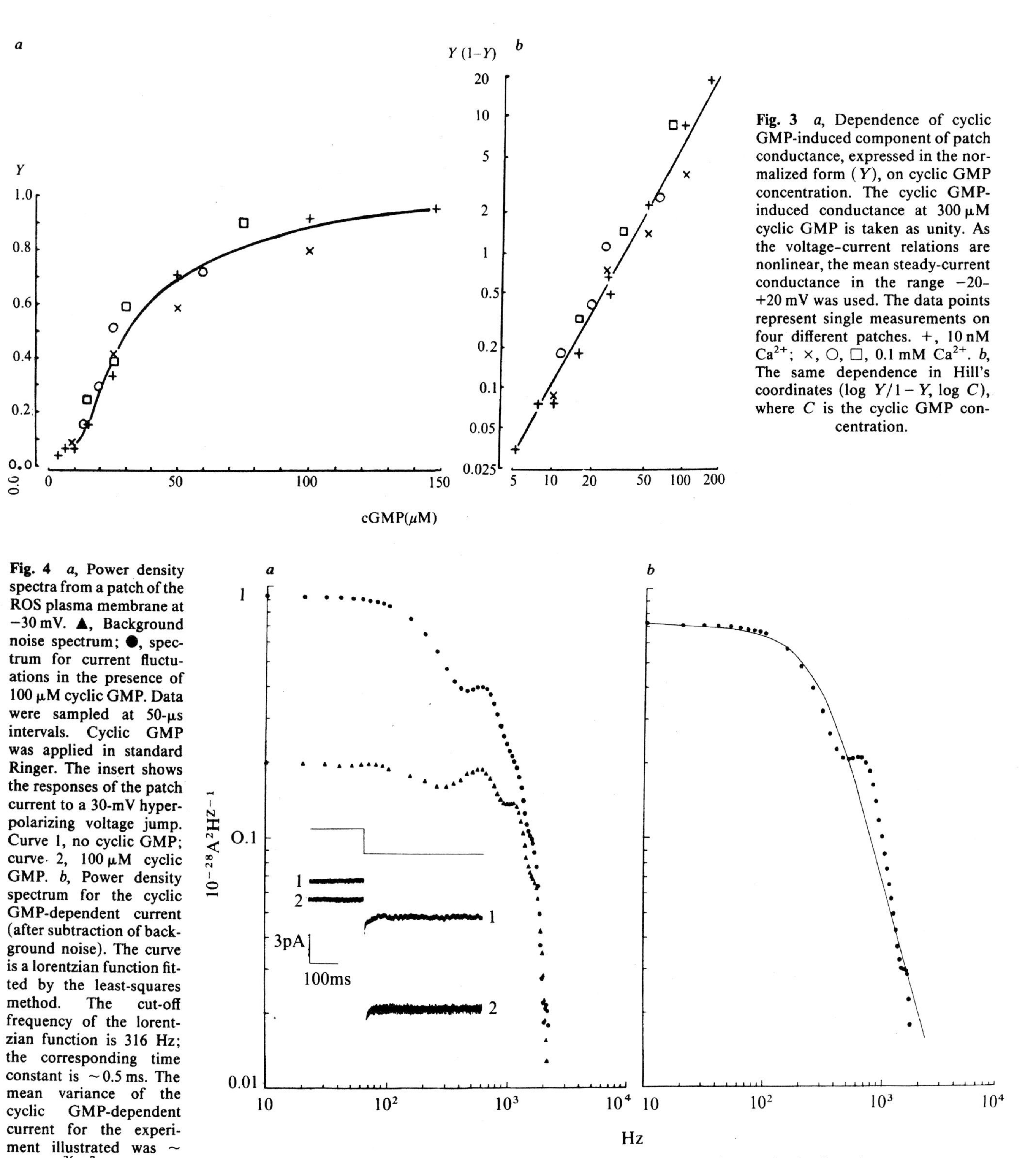

Fig. 3 *a*, Dependence of cyclic GMP-induced component of patch conductance, expressed in the normalized form (Y), on cyclic GMP concentration. The cyclic GMP-induced conductance at 300 μM cyclic GMP is taken as unity. As the voltage-current relations are nonlinear, the mean steady-current conductance in the range −20–+20 mV was used. The data points represent single measurements on four different patches. +, 10 nM Ca^{2+}; ×, ○, □, 0.1 mM Ca^{2+}. *b*, The same dependence in Hill's coordinates ($\log Y/1-Y$, $\log C$), where C is the cyclic GMP concentration.

Fig. 4 *a*, Power density spectra from a patch of the ROS plasma membrane at −30 mV. ▲, Background noise spectrum; ●, spectrum for current fluctuations in the presence of 100 μM cyclic GMP. Data were sampled at 50-μs intervals. Cyclic GMP was applied in standard Ringer. The insert shows the responses of the patch current to a 30-mV hyperpolarizing voltage jump. Curve 1, no cyclic GMP; curve 2, 100 μM cyclic GMP. *b*, Power density spectrum for the cyclic GMP-dependent current (after subtraction of background noise). The curve is a lorentzian function fitted by the least-squares method. The cut-off frequency of the lorentzian function is 316 Hz; the corresponding time constant is ~0.5 ms. The mean variance of the cyclic GMP-dependent current for the experiment illustrated was ~ 3.5×10^{-26} A^2, and the change of the mean current on cyclic GMP application at −30 mV was 10.5 pA. The amplitude of an elementary event is 3.5×10^{-26} $A^2/10.5\times10^{-12}$ A = 3.3 fA.

To test the ion selectivity of cyclic GMP-dependent conductance, we measured the reversal potentials for the cyclic GMP-dependent current when the intracellular surface of a patch was bathed with media of various ionic compositions, thereby enabling us to compare the data obtained from different patches showing different cyclic GMP-induced conductance. Our analysis reveals that cyclic GMP induces cationic conductance, because, when the NaCl gradient was doubled, the reversal potential obtained was equivalent to that of Na^+ ions (16·5 ± 0.3 mV, six determinations on different patches, Fig. 2*b*). Substitution of NaCl in the Ringer solution bathing the intracellular side of the membrane by KCl, LiCl, RbCl or CsCl resulted in respective reversal potentials for the cyclic GMP-dependent current of 2.7 ± 0.7 mV, 2.2 ± 2.1 mV, 5 ± 1 and 6 ± 2.2 mV (8, 8, 4 and 4 determinations, respectively). These values indicate that the channels controlled by cyclic GMP are moderately selective for monovalent cations in the sequence $Na^+ > Li^+$, $K^+ > Rb^+$, $Cs^+ \gg Cl^-$.

The family of voltage-current relations for a single patch, obtained at different cyclic GMP concentrations (Fig. 2*c*), shows that the most marked conductance changes occur in the range of 20-60 μM cyclic GMP and that at 300 μM there is complete saturation of the cyclic GMP-dependent conductance. When studying the dependence of patch conductance on cyclic GMP concentration, we therefore expressed the cyclic GMP-induced conductance in normalized form, taking the conductance at 300 μM cyclic GMP as unity.

The curve showing the dependence of conductance on cyclic GMP concentration was S-shaped (Fig. 3*a*), with a Hill coefficient of ~ 1.8 (Fig. 2*b*) and a 50% effect at 30 μM cyclic GMP. Neither Ca^{2+} ions nor 3′,5′-cyclic AMP, 2′,3′-cyclic GMP or 5′ GMP (1 mM) had any effect on the cyclic GMP dependence of the conductance, confirming that cyclic GMP is a specific agent for inducing the membrane conductance increase.

The magnitude of the events underlying the cyclic GMP-induced conductance was estimated approximately by dividing the change in the current produced by cyclic GMP by the change in mean current[7]. The amplitude of the estimated unit event for three patches was ~3 fA (±10%) in standard Ringer solution at a membrane voltage of −30 mV. If it is assumed that the initial event is the closure of a single cyclic GMP-sensitive channel, then a single channel has a conductance of 100 fS. At this value, the channel responsible for the cyclic GMP-induced conductance could be either a carrier or a pore.

As the cyclic GMP-dependent current is not inactivated, at least within minutes, the power density spectrum of the steady-state current can be deduced. The spectra obtained at −30 mV in the absence and presence of cyclic GMP are shown in Fig. 4*a*. The spectrum for the cyclic GMP-dependent current with the best least-square lorentzian function fit (Fig. 4*b*) indicated a cut-off frequency of ~300 Hz, corresponding to a time constant of ~0.5 ms.

It is known that the extracellular and intracellular application of cyclic GMP depolarizes the rod in the dark[1,2,8], probably by increasing the light-sensitive component of the dark conductance of the plasma membrane[9]. This suggests that the light-controlled conductivity units are activated by cyclic GMP. (We do not know whether the properties of the cyclic GMP-activated conductance we have studied *in vitro* are the same as those of the light-sensitive conductance recorded *in vivo*.)

It has been shown recently that the photocurrent through the rod plasma membrane previously thought to be dependent on Na^{+} ions[10] can also be carried by Li^{+}, K^{+}, Rb^{+}, Cs^{+}, Tl^{+}, Ca^{2+}, Ba^{2+} and Sr^{2+} ions[11]. The selectivity for monovalent cations found by us corresponds to that reported elsewhere[11]. The disappearance of photoresponses in the absence of Na^{+} observed previously results from the rise in the intracellular Ca^{2+} concentration due to inactivation of the Na^{+}-Ca^{2+} exchanger[11-15].

Estimates of the single-event light-sensitive conductance for the rod plasma membrane of the turtle, lizard and salamander are 600, 50 and 5 fS, respectively[16-18], the same order of magnitude as the initial event of the cyclic GMP-dependent conductance we describe. Information on the power density spectrum of the light-dependent current in rods is very limited and only two values for the time constant are available, 200 ms for the turtle rod[16] and 0.8 ms for the salamander[18]. The reason for this discrepancy between the two estimates is unclear. We consider it unlikely that the ensemble of channels with intrinsic time constants of the order of hundreds of milliseconds could produce responses to bright flashes which would develop within milliseconds[19]. Therefore, the lower value, which is consistent with our estimate of 0.5 ms, seems the more reliable.

In view of the similarities between the properties of the light-dependent conductance of the ROS membrane and those of the cyclic GMP-dependent conductance of the rod membrane patches, it is quite feasible that the cyclic GMP-activated conductances we describe are responsible for the light-dependent conductance of the ROS membrane. If this is so, cyclic GMP, not Ca^{2+}, is the messenger modulating the conductance of the plasma membrane. The effect of Ca^{2+} on the rod photoresponse may be through its action on the cyclic GMP metabolism in the photoreceptor. A decrease in Ca^{2+} concentration is known to produce a marked increase in intracellular cyclic GMP concentration in the rod[20,21], which could be explained by a direct effect of Ca^{2+} on ROS cyclic GMP phosphodiesterase[22].

The direct action of a cyclic nucleotide on the conductance of the plasma membrane has not previously been described. The advantage of such a mechanism is its fast response time, essential in photoreceptors, where the response to intense illumination develops within milliseconds and single-channel open times are less than 1 ms. It will be exciting to see whether the same mechanism exists in other cells.

We thank Dr G. Krapivinsky for performing chromatographic controls and V. Samoylov for help in computer programming.

Received 13 November; accepted 7 December 1984.

1. Lipton, S. A., Rasmussen, H. & Dowling, J. E. *J. gen. Physiol.* **70**, 771-791 (1977).
2. Miller, W. H. & Nicol, G. D. *Nature* **280**, 64-66 (1979).
3. Hagins, W. A. *A. Rev. Biophys. Bioengng* **1**, 131-158 (1972).
4. Hamill, O. P., Marty, A., Neher, E., Sakmann, B. & Sigworth, F. S. *Pflügers Arch. ges. Physiol.* **391**, 85-100 (1981).
5. Kolesnikov, S. S., Lyubarsky, A. L. & Fesenko, E. E. *Vision Res.* (in the presss).
6. Rasmussen, H. & Goodman, D. B. *Physiol. Rev.* **57**, 421-509 (1977).
7. Katz, B. & Miledi, R. *J. Physiol., Lond.* **224**, 665-699 (1972).
8. Waloga, G. *J. Physiol., Lond.* **341**, 341-358 (1983).
9. MacLeish, P. R., Schwartz, E. A. & Tachibana, M. *J. Physiol., Lond.* **348**, 645-664 (1984).
10. Sillman, A. I., Ito, H. & Tomita, T. *Vision Res.* **9**, 1443-1451 (1969).
11. Yau, K.-W. & Nakatani, K. *Nature* **309**, 352-354 (1984).
12. Bastian, B. L. & Fain, G. L. *J. Physiol., Lond.* **330**, 331-347 (1982).
13. Woodruff, M. L., Fain, G. L. & Bastian, B. L. *J. gen. Physiol.* **80**, 517-536 (1982).
14. Capovilla, M., Caretta, A., Cervetto, L. & Torre, V. *J. Physiol., Lond.* **343**, 295-310 (1983).
15. Hodgkin, A. L., McNaughton, P. A., Nunn, B. J. & Yau, K.-W. *J. Physiol., Lond.* **350**, 649-680 (1984).
16. Schwartz, E. A. *J. Physiol., Lond.* **272**, 217-246 (1977).
17. Detwiler, P. B., Conner, J. D. & Bodoia, R. D. *Nature* **300**, 59-61 (1982).
18. Attwell, D. & Gray, P. *J. Physiol., Lond.* **351**, 9P (1984).
19. Penn, R. D. & Hagins, W. A. *Biophys. J.* **12**, 1073-1094 (1972).
20. Cohen, A. I., Hall, I. A. & Ferrendelli, J. A. *J. gen. Physiol.* **71**, 595-612 (1978).
21. Polans, A. S., Kawamura, S. & Bownds, M. D. *J. gen. Physiol.* **77**, 41-48 (1981).
22. Robinson, P. R., Kawamura, S., Abramson, B. & Bownds, M. D. *J. gen. Physiol.* **76**, 631-645 (1980).
23. Caldwell, P. C. in *Calcium and Cellular Function* (ed. Cuthbert, A. W.) 11-16 (St Martin's Press, New York, 1970).

Voltage-sensitive Ca^{2+} channels

In 1958 Fatt and Ginsborg [1] demonstrated membrane action potentials that were due to inflow of Ca^{2+} in crustacean muscle fibres. In the mid-1960s Hagiwara's group [2,3] provided further evidence for a Ca^{2+}-influx pathway that is switched on by membrane depolarization (voltage-gated Ca^{2+} channels) and showed that Ca^{2+} inflow could be enhanced by reducing the intracellular Ca^{2+} concentration.

Hagiwara's group was the first to demonstrate the existence of more than one Ca^{2+} channel type. They could distinguish two Ca^{2+} current components on the basis of steady-state inactivation, activation threshold and selectivity. With the advent of the patch-clamp technique (see Section 4), the properties of single calcium channels began to be studied [4,5].

The paper we have selected [6] presented single-channel current as well as whole-cell current recording data to demonstrate three types of neuronal Ca^{2+} channels. Two subtypes had previously been identified by Hagiwara's group [7] in starfish eggs. Work at the Bogomoletz Institute in Kiev, Ukraine had established the existence in neurons of a low-voltage-activated (LVA) Ca^{2+} current, which was activated at relatively negative membrane potentials (-40 to -60 mV), but rapidly inactivated in a potential-dependent way. This current was easily separated from the high-voltage-activated (HVA) Ca^{2+} current that required larger depolarizations and showed relatively less inactivation [8,9].

Nowycky et al. [6] demonstrated in studies on chick embryo dorsal root ganglia that the HVA Ca^{2+} current component could be subdivided into an inactivating and a steady component. They proposed that there were three types of neuronal Ca^{2+} channels and called them type L (long-lasting current, HVA), T (transient current, LVA) and N (neither L nor T, HVA transient current). In single-channel current experiments it was shown that the channels giving rise to the L current conveniently had a *large* unit conductance whereas the channels underlying the T current had a small (*tiny*) conductance. The N channels had an intermediate unit conductance. These three subtypes could also be differentiated in mouse and rat dorsal root ganglia [10].

Although the separation of voltage-activated Ca^{2+} currents into the T, L and N subtypes [6] was a useful step in the emerging classification of Ca^{2+} channel properties, subsequent work has shown that the situation with regard to the HVA Ca^{2+} channels is much more complex. Currently, the HVA channels are subdivided into L, N, P, Q and R types, largely on the basis of pharmacology [11–14].

There has been substantial progress in our understanding of the structure of the voltage-gated Ca^{2+} channels. These channels are composed of five subunits. Of these the α_1-subunit (210–270 kDa) is the largest and contains the ion-conducting pore and binding sites for channel antagonists. There are four membrane-spanning repeats each with six helices [14,15].

Voltage-gated Ca^{2+} channels play many different roles in different cell types. In neurons one of the best studied functions is the control of neurotransmitter release from synaptic terminals. When the synaptic terminal is depolarized during an action potential, voltage-sensitive Ca^{2+} channels open and the neurotransmitter contained in synaptic vesicles is released into the synaptic cleft via exocytosis (see Section 13). For the fast neurotransmitters this happens on a timescale of fractions of 1 ms and with a spatial precision of a fraction of 1 μm in the active zones where the synaptic vesicles, the release machinery and the Ca^{2+} channels are co-localized [16].

In the heart, the voltage-sensitive Ca^{2+} channels are essential for the Ca^{2+} inflow that activates contraction during the systolic period [17]. A physiologically important channel modulation by noradrenaline mediated via cyclic AMP was originally demonstrated by Reuter's group [18]. In elegant patch-clamp experiments on cultured heart cells they showed that the cyclic AMP-dependent phosphorylation of Ca^{2+} channels promotes an increase in the channel open-state probability, leading to enhanced Ca^{2+} influx and thereby stronger activation of contraction.

Although it is outside the scope of this short commentary to discuss the very rich pharmacology of Ca^{2+} channels, it is worth mentioning that nifedipine and related dihydropyridines are potent organic compounds that block L-type Ca^{2+} channels in the heart and smooth muscle, and, as such, have gained widespread clinical use [15].

References

1. Fatt, P. and Ginsborg, B.L. (1958) *J. Physiol. (London)* **142**, 516–543
2. Hagiwara, S. and Naka, K.I. (1964) *J. Gen. Physiol.* **48**, 141–162
3. Hagiwara, S. and Nakajima, S. (1966) *J. Gen. Physiol.* **49**, 807–818
4. Fenwick, E.M., Marty, A. and Neher, E. (1982) *J. Physiol. (London)* **331**, 599–635
5. Reuter, H., Stevens, C.F., Tsien, R.W. and Yellen, G. (1982) *Nature (London)* **297**, 501–504
6. Nowycky, M.C., Fox, A.P. and Tsien, R.W. (1985) *Nature (London)* **316**, 440–443
7. Hagiwara, S., Ozawa, S. and Sand, O. (1975) *J. Gen. Physiol.* **65**, 617–644
8. Veselovsky, N.S. and Fedulova, S.A. (1983) *Dok. Akad. Nauk. S.S.S.R.* **268**, 747–750
9. Fedulova, S.A., Kostyuk, P.G. and Veselovsky, N.S. (1985) *J. Physiol. (London)* **359**, 431–446
10. Kostyuk, P.G., Shuba, Y.M. and Savchenko, A.N. (1988) *Pflügers Arch.* **411**, 661–669
11. Bean, B.P. (1989) *Annu. Rev. Physiol.* **51**, 367–384
12. Swandulla, D., Carbone, E. and Lux, H.D. (1991) *Trends Neurosci.* **14**, 46–51
13. Llinas, R., Sugimori, M., Hillman, D.E. and Cherskey, B. (1992) *Trends Neurosci.* **15**, 351–355
14. Kostyuk, P.G. and Verkhratsky, A.N. (1995) *Calcium Signalling in the Nervous System*, Wiley, Chichester
15. Glossmann, H. and Striessnig, J. (1990) *Rev. Physiol. Biochem. Pharmacol.* **114**, 1–105
16. Matthews, G. and Von Gersdorff, H. (1996) in Neuronal Ca^{2+} Signalling (Kasai, H. and Petersen, O.H., eds.), pp. 329–334 *Seminars in the Neurosciences*, Academic Press, London
17. Reuter, H. (1983) *Nature (London)* **301**, 569–574
18. Cachelin, A.B., DePayer, J.E., Kokobun, S. and Reuter, H. (1983) *Nature (London)* **304**, 462–464

Three types of neuronal calcium channel with different calcium agonist sensitivity

Martha C. Nowycky*‡, Aaron P. Fox†§ & Richard W. Tsien†

* Section of Neuroanatomy and † Department of Physiology, Yale University School of Medicine, New Haven, Connecticut 06510, USA

How many types of calcium channels exist in neurones? This question is fundamental to understanding how calcium entry contributes to diverse neuronal functions such as transmitter release, neurite extension, spike initiation and rhythmic firing[1–3]. There is considerable evidence for the presence of more than one type of Ca conductance in neurones[4–13] and other cells[14–18]. However, little is known about single-channel properties of diverse neuronal Ca channels, or their responsiveness to dihydropyridines, compounds widely used as labels in Ca channel purification[19–21]. Here we report evidence for the coexistence of three types of Ca channel in sensory neurones of the chick dorsal root ganglion. In addition to a large conductance channel that contributes long-lasting current at strong depolarizations (L), and a relatively tiny conductance that underlies a transient current activated at weak depolarizations (T), we find a third type of unitary activity (N) that is neither T nor L. N-type Ca channels require strongly negative potentials for complete removal of inactivation (unlike L) and strong depolarizations for activation (unlike T). The dihydropyridine Ca agonist Bay K 8644 strongly increases the opening probability of L-, but not T- or N-type channels.

Figure 1 shows evidence for three distinct components of Ca channel current in whole-cell recordings obtained under ionic conditions that minimize contamination by other currents. The components were distinguished kinetically by applying depolarizing test pulses at various levels from different holding potentials (h.p.). With h.p. = −40 mV (uppermost traces in each panel), strong depolarizations (−10, +10, +20 mV) are required to activate any inward current; the peak current-voltage relation (Fig. 1*b*, squares) is typical for a single-current component. Because this inward current component decays very slowly ($t_{1/2}$ of hundreds of milliseconds), we designate it 'L' (for long-lasting). Weak depolarizations from h.p. = −100 mV evoke a different Ca channel current (Fig. 1*a*), seen as a prominent shoulder at negative test potentials in the associated peak current-voltage plot (Fig. 1*b*, circles). The additional current becomes noticeable at −60 mV, and is nearly constant in amplitude between −40 and −10 mV, consistent with a very negative range of activation. As this component decays relatively rapidly ($\tau \sim 25$ ms at −30 mV), we term it 'T' (transient). A third component of calcium current appears with strong depolarizations from h.p. = −100 mV (for example, the traces at +10 or +20 mV in Fig. 1*a*). It is distinguished most easily in a plot of the amplitude of decaying current versus test potential (triangles, Fig. 1*c*). The decay amplitude shows a clear plateau between −30 and −10 mV, as expected for component T; however, it grows substantially with stronger test pulses, reaching a peak at +20 mV. We attribute the extra decaying current to a third component that we term 'N' (neither T nor L). Like T, but unlike L, component N contributes phasic current and requires strongly negative holding potentials for complete removal of inactivation. Like L, but unlike T, N current requires strong depolarizations for activation and is relatively sensitive to the inorganic blocker cadmium. For example, in six cells, 50 μM Cd almost completely abolished N and L current, but left 55±4% of T (see ref. 22). All three components are present in most whole-cell recordings, although the relative amplitudes vary considerably between cells and with different ionic conditions. N current is particularly prominent with 55 or 110 mM Ba as the charge carrier; in whole-cell recordings from four cells (average capacitance 30 pF), the maximal amplitude of the decaying N-type current at +20 mV was 1,340±236 pA, roughly equal to the L current at +20 mV and ~8 times greater than the peak T current at −10 mV.

Recordings from cell-attached patches on dorsal root ganglion (DRG) cell bodies show three types of unitary Ca channel current, readily distinguished by different slope conductances in 110 mM Ba (Fig. 2). The different channel types show kinetic properties that correspond well to components T, N and L in whole-cell recordings. Figure 2*a* shows unitary current activity of a channel that we associate with component T. Inactivation of this channel is complete positive to −40 mV, and is fully removed only at holding potentials more negative than −80 mV.

‡ Present address: Department of Anatomy, The Medical College of Pennsylvania, Philadelphia, Pennsylvania 19129, USA

§ To whom reprint requests should be addressed.

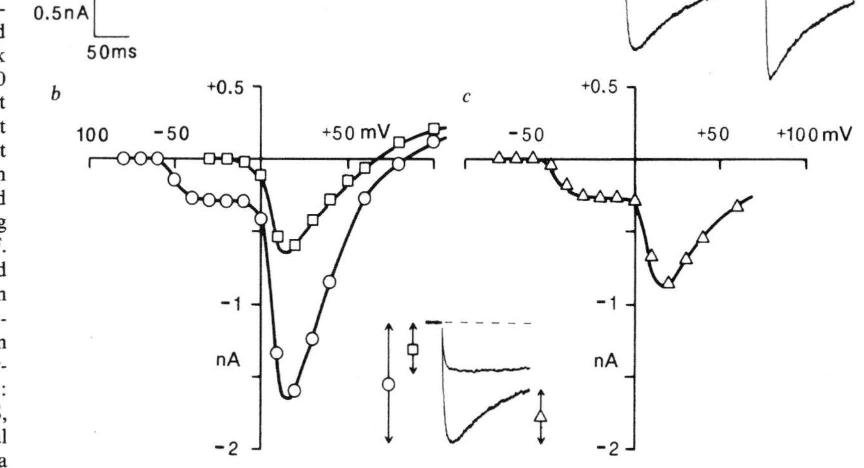

Fig. 1 Three components of Ca channel current in whole-cell recordings with 10 mM extracellular Ca. *a*, Superimposed current traces evoked from h.p. = −40 and h.p. = −100 mV. Test potential indicated above. Linear leak and capacitance have been subtracted with 1 ms blanked at both the on and off of the depolarizing pulse. *b*, Plot of peak current versus test potential from h.p. = −40 (squares) and h.p. = −100 mV (circles). *c*, Plot of magnitude of relaxing current against test potential from h.p. = −100 mV. The current relaxation is the late current subtracted from the peak current. Inset repeats records obtained at +10 mV and shows how peak and relaxing current amplitudes were measured. Cell N82F. **Methods.** Dorsal root ganglia were obtained from 8-12-day-old chick embryos. The isolation and culture procedures followed standard protocols (see ref. 24). Cells were maintained in culture for 2-7 days before recording. The internal solution (pipette) contained (in mM): 100 CsCl, 10 Cs-EGTA, 5 $MgCl_2$, 40 HEPES, 2 ATP, 0.25 cyclic AMP, *p*H 7.3. The external solution was exchanged after formation of a gigaseal from a standard Tyrode solution to one containing (in mM) 10 $CaCl_2$, 135 TEA-Cl, 10 HEPES, and 200 μM tetrodotoxin (TTX), *p*H 7.3. Voltage pulses and current traces were simultaneously generated and sampled online with a PDP 11/23 computer. Whole-cell currents were filtered with an 8-pole Bessel filter (3 dB down at 3 kHz) and sampled at 190-μs intervals. The stimulation rate for both whole-cell and single-channel experiments was 0.25 Hz. All experiments were carried out at room temperature (21-22 °C).

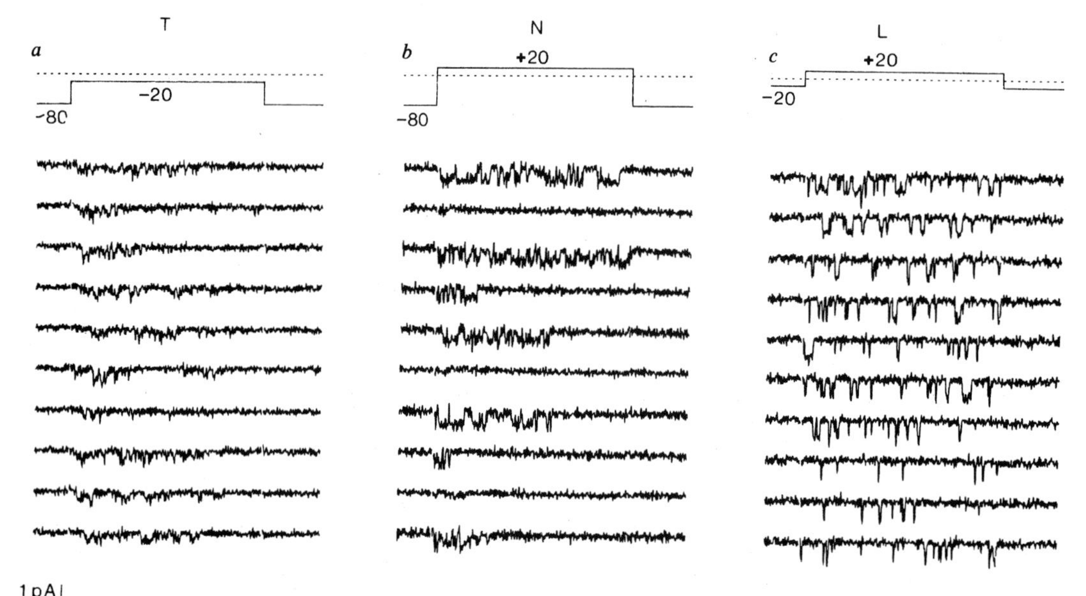

Fig. 2 Three types of unitary Ca channel activity seen in cell-attached patch recordings with Ba as the charge carrier. *a*, T-type channel activity, cell N85A. *b*, N-type channel, cell N75H. *c*, L-type channel, cell N67J. Patch pipettes contained (in mM): 110 $BaCl_2$, 10 HEPES, and 200 nM TTX, *p*H 7.4. To put patch membrane potential on an absolute scale, cell resting potential was zeroed with an external solution containing (mM); 140 K-aspartate, 10 K-EGTA, 10 HEPES, 1 $MgCl_2$, *p*H 7.4. Current signals were filtered at 1 kHz and sampled at 5 kHz. Within each panel traces are consecutive. Values of single-channel amplitude and slope conductances given in the text were obtained from amplitude histograms. The unitary current amplitude of the T-type channel at −20 mV was given by the average of amplitudes from three different patches exhibiting long, well-resolved openings. We have seen each of the three channel types in isolation and in all possible combinations. The number of channels under the patch pipette varied considerably from one to many.

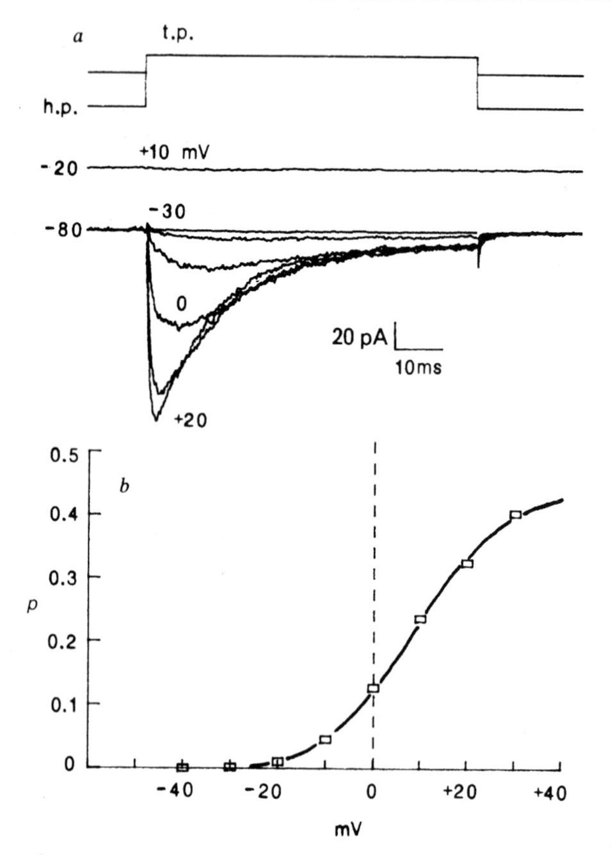

Fig. 3 Voltage-dependent activation of N-type Ca channel activity in a cell-attached patch with ~600 N-type channels. *a*, Voltage-clamp protocols (top); associated current records produced by depolarization from h.p. = −20 to test potential (t.p.) = +10 mV (middle trace) and by depolarizations from h.p. = −80 to test potentials ranging from −30 mV to +20 mV in 10-mV increments (bottom family of traces). Each record was obtained by averaging five individual sweeps after subtraction of linear leak and capacity currents. *b*, Analysis of voltage-dependence of peak opening probability (p) from the same experiment. Values of p were obtained by dividing the peak current ($= N_T \times p \times i$) by the unitary current (i) at each test potential and by an estimate of the total number of channels ($N_T = 599$). Unitary current was given by the equation $i(V) = -1.023\ \text{pA} + (13.2\ \text{pS}) \times (V$
single-channel data from 18 cell-attached patches (correlation coefficient for linear regression was greater than 0.99). N_T was determined by comparison with the single-channel experiment in Fig. 2*b*, which showed that $p = 0.32$ at +20 mV. Patch 84G.

Activation becomes detectable positive to −50 mV. With test depolarizations to −20 or −10 mV, the averaged current decays with a time course very similiar to component T in whole-cell recordings. The T-type channel has a small slope conductance (about 8 pS in 110 mM Ba).

Figure 2*c* illustrates unitary channel activity which corresponds to the L component. It is relatively resistant to inactivation: activity is seen at h.p. = −20 mV (Fig. 2*c*) or even at 0 mV. Activation of the L-type channel usually becomes significant at −10 mV, although openings can sometimes be detected as negative as −30 mV. Averaged currents obtained from many individual traces do not relax at any test potential (see Fig. 4). The channel is characterized by a large slope conductance in 110 Ba (25 pS).

Figure 2*b* illustrates a third type of unitary activity that we term 'N' because its properties account for component N in whole-cell recordings. N-type Ca channels differ from L- and T-type Ca channels in several respects. Their slope conductance with 110 mM Ba is 13 pS, larger than that of T-type channels and smaller than that of L-type channels. The difference in unitary current size is highly significant: for example, at −20 mV, average values (±s.e.m.) for T, N and L were 0.62±0.03 (5 patches), 1.22±0.03 (10 patches) and 2.07±0.09 (6 patches), respectively. At +20 mV, the test potential illustrated in Fig. 2*b* and *c*, the difference between average values for N and L (0.824 and 1.12 pA) is smaller in absolute terms but is still appreciable.

N-type Ca channels can also be distinguished from L-type Ca channels by their inactivation properties. The time-dependence of inactivation of N-type Ca channels appears as a bunching of openings near the beginning of the depolarization and an absence of openings near the end (Fig. 2*b*); the averaged current record (not shown) decays almost completely by the end of a 136-ms test pulse. The steady voltage-dependence of inactivation was studied by varying the holding potential. Unitary N-type activity is completely abolished by holding the patch at −20 mV, a potential at which L-type channels remain largely available for opening (see Fig. 2*c*). N- and T-type Ca channels are similar in their inactivation properties but very different in the voltage-dependence of activation. With isotonic Ba in the pipette, significant activation of N-type channels requires depolarization to −20 mV, whereas opening of T-type channels first becomes detectable beyond −50 mV.

In several experiments, we recorded from cell-attached patches with many N-type Ca channels and no detectable T- or L-type activity. Figure 3 shows kinetic analysis from this kind of recording. The rise and fall of the N-type channel current both become faster as the test pulse depolarization is increased; at strong depolarizations, inactivation is largely complete within ~100 ms (Fig. 3*a*, bottom family of traces). A steady holding potential of −20 mV also produces complete inactivation (Fig. 3*a*, middle trace); 50% inactivation is seen at h.p. = −60 mV (trace not shown). As Fig. 3*b* illustrates, the open probability (*p*) increases over the range between −20 and +30 mV, in good agreement with recordings from whole cells (Fig. 1*c*) and patches with only one or a few N-type channels (Fig. 2*b*).

In addition to differences in kinetics, unitary conductance and sensitivity to Cd, we find that the three channel types differ in their responsiveness to Bay K 8644. This dihydropyridine Ca agonist[23] elevates calcium influx by promoting a pattern of Ca channel gating with very long openings in neurones[24] and heart cells[25-28]. Bay K 8644 strongly enhances averaged L-type channel currents in DRG neurones[11,24] (Fig. 4), but produces no change in the current carried by T-type channels (6 patches) or N-type channels (4 patches). Our results with Bay K 8644 are consistent with findings of dihydropyridine-resistant Ca channels in GH3 cells[18] and cardiac cells[29]. The existence of different dihydropyridine-resistant channels in neurones might help explain the partial or complete lack of effect of dihydropyridines on depolarization-induced Ca entry or transmitter release in many systems[30,31].

Although this is the first report to focus on the coexistence of three types of calcium channels in one preparation, we believe that activity of all three channel types may be found in whole-cell currents reported by others (for example Fig. 1*c* of ref. 8). Single-channel recordings in DRG neurones by Carbone and Lux[9] have shown two types of unitary activity that correspond kinetically to what we call T-type and L-type Ca channels. The relative sizes of the unitary currents with 40 mM Ca as the charge carrier are opposite to what we find with 110 mM Ba; this discrepancy might be explained by the very different ionic selectivity properties of T- and L-type Ca channels[10,17,29].

It is attractive to suppose that the various channels may serve different cellular functions. For example, T-type channels might contribute to threshold behaviour or rhythmic activity and N- or L-type channels may be important for dendritic spikes or neurotransmitter release[3-5,11,17]. Determination of the cellular distribution of the channel types and their sensitivity to neuromodulators will help in the elucidation of their physiological roles.

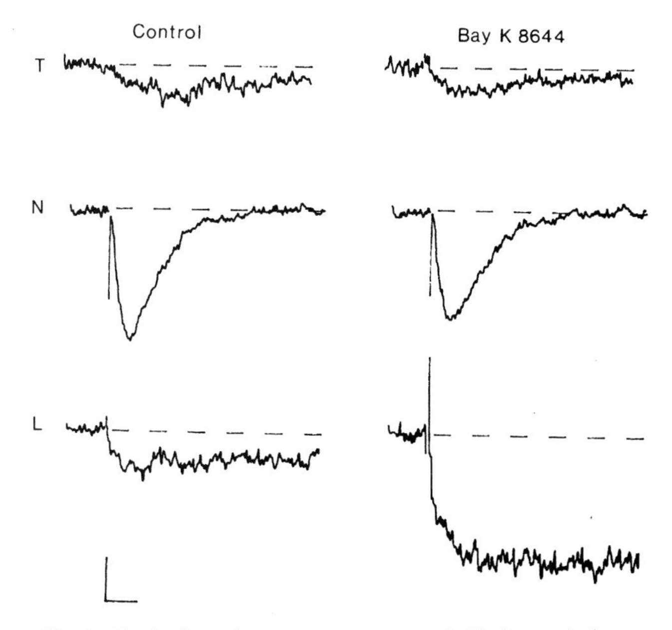

Fig. 4 Single-channel current averages recorded before and after the addition of Bay K 8644. All recording conditions identical to those described in Fig. 2 legend. Averaged currents for three channel types before (left) and after exposure to 5 μM Bay K 8644 (right). Top, T-type activity, voltage step from −80 to −30 mV. Patch 86D. Middle, N-type activity, voltage step from −80 to +10 mV. Patch 84K. Bottom, L-type activity, voltage step from −40 to +10 mV. Patch 89F. Vertical calibration bar represents 1 pA for middle panel and 0.25 for top and bottom panels. Horizontal calibration bar represents 20 ms for top and bottom panels and 10 ms for middle panel.

We thank So-Ching Wong for preparing the cultured neurones and for help with the illustrations, and E. W. McCleskey, P. Hess, B. Nilius and J. B. Lansman for helpful discussion. Support was provided by a Canadian Heart Foundation Fellowship to A.P.F. and USPHS grants to M.C.N. & R.W.T.

Received 7 February; accepted 22 May 1985.

1. Miller, R. J. *Trends Neurosci.* **8**, 45-47 (1985).
2. Hagiwara, S. & Byerly, L. *Trends Neurosci.* **6**, 189-193 (1983).
3. Llinas, R., Steinberg, I. Z. & Walton, K. *Biophys. J.* **33**, 323-352 (1981).
4. Fishman, M. C. & Spector, I. *Proc. natn. Acad. Sci. U.S.A.* **78**, 5245-5249 (1981).
5. Llinas, R. & Sugimori, M. *J. Physiol., Lond.* **305**, 197-213 (1980).
6. Murase, K. & Randic, M. *J. Physiol., Lond.* **344**, 141-153 (1983).
7. Nowycky, M. C., Fox, A. P. & Tsien, R. W. *Biophys. J.* **45**, 36a (1984).
8. Carbone, E. & Lux, H. D. *Biophys. J.* **46**, 413-418 (1984).
9. Carbone, E. & Lux, H. D. *Nature* **310**, 501-502 (1984).
10. Yoshii, M., Tsunoo, A. & Narahashi, T. *Biophys. J.* **47**, 433a (1984).
11. Brown, D. A. *et al. IUPHAR Proc.* (in the press).
12. Fedulova, S. A., Kostyuk, P. G. & Veselovsky, N. S. *J. Physiol., Lond.* **359**, 431-446 (1985).
13. Bossu, J.-L., Feltz, A. & Thomann, J.-M. *Pflügers Arch. ges. Physiol.* (in the press).
14. Hagiwara, S., Ozawa, S. & Sand, O. *J. gen. Physiol.* **65**, 617-644 (1975).
15. Deitmer, J. W. *J. Physiol., Lond.* **355**, 137-159 (1984).
16. Fox, A. P. & Krasne, S. *J. Physiol., Lond.* **356**, 491-505 (1984).
17. Armstrong, C. M. & Matteson, D. R. *Science* **227**, 65-67 (1985).
18. Cohen, C. J. & McCarthy, R. T. *Biophys. J.* **47**, 513a (1985).
19. Curtis, B. M. & Catterall, W. A. *J. biol. Chem.* **258**, 9344-9348 (1983).
20. Norman, R. I., Borsotto, M., Fosset, M. & Lazdunski, M. *Biochem. biophys. Res. Commun.* **111**, 878-883 (1983).
21. Glossman, H., Ferry, D. R., Lubbecke, F., Mewes, R. & Hoffman, F. *Trends pharmac. Sci.* **3**, 431-437 (1982).
22. Nowycky, M. C., Fox, A. P. & Tsien, R. W. *Soc. Neurosci. Abstr.* **10**, 526 (1984).
23. Schramm, M., Thomas, G., Towart, R. & Franckowiak, G. *Nature* **303**, 535-537 (1983).
24. Nowycky, M. C., Fox, A. P. & Tsien, R. W. *Proc. natn. Acad. Sci. U.S.A.* **82**, 2178-2182 (1985).
25. Kokubun, S. & Reuter, H. *Proc. natn. Acad. Sci. U.S.A.* **81**, 4824-4827 (1984).
26. Ochi, R., Hino, N. & Niimi, Y. *Proc. Jap. Acad.* **B60**, 153-156 (1984).
27. Hess, P., Lansman, J. B. & Tsien, R. W. *Nature* **311**, 538-544 (1984).
28. Brown, A. M., Kunze, D. L. & Yatani, A. *Nature* **311**, 570-572 (1984).
29. Nilius, B., Hess, P., Lansman, J. B. & Tsien, R. W. *Nature* **316**, 443-446 (1985).
30. Nachsen, D. A. & Blaustein, M. P. *Molec. Pharmac.* **16**, 579-586 (1979).
31. Turner, T. J. & Goldin, S. M. *J. Neurosci.* **5**, 841-849 (1985).

Ca^{2+} uptake into intracellular stores

In the early 1950s, Setsuro Ebashi, under the guidance of Professor Kumagai, worked on the problem of muscle relaxation. They used a glycerol-extracted muscle (glycerinated muscle fibre, essentially containing the contractile elements), in which ATP could induce contraction. This contraction was not followed by relaxation, but the latter could be evoked by addition of a simple muscle extract (relaxing factor) [1,2]. The relaxing factor [2] turned out to have characteristics in common with a Mg^{2+}-activated ATPase described in 1948 by Kielley and Meyerhof [3]. In Fritz Lipmann's laboratory at the Rockefeller Institute, Ebashi was subsequently able to characterize this muscle-relaxing factor in some considerable detail. In the paper by Ebashi and Lippmann [4] that we have selected, it was shown that there was an ATP-dependent binding of Ca^{2+} to the membrane fraction, and the dependence of the binding on the Ca^{2+} and Mg^{2+} concentrations was determined. ATPase activity and Ca^{2+} binding indicated the existence of a Ca^{2+} transport enzyme now called Ca^{2+},Mg^{2+}-ATPase. George Palade, who was also at the Rockefeller Institute at that time (strangely enough not a co-author of the paper) examined the preparation with the help of the electron microscope, and included in the paper is a picture and a description of the closed vesicles of smooth membrane representing the fragmented sarcoplasmic reticulum. The paper clearly concludes that the relaxing effect is due to Ca^{2+} uptake across the vesicular membrane. The same conclusion was reached independently by Hasselbach and Makinose [5–7]. Thus the active Ca^{2+} accumulation in an intracellular store by a Ca^{2+}-activated ATPase had been discovered. This finding, although primarily made in the context of research into the mechanism of relaxation, is fundamental to our present understanding of cytosolic Ca^{2+} signalling, since it introduced for the first time the concept of an intracellular membrane-bounded Ca^{2+} store.

The Ca^{2+} pumps in the intracellular stores are now generally described under the name SERCA (sarcoplasmic/endoplasmic reticulum Ca^{2+}-ATPase) pump. The SERCA pumps are the products of three different genes known as SERCA 1, 2 and 3. SERCA 1 pumps, first cloned by McLennan et al. [8], are in fast-twitch skeletal muscle, SERCA 2 are in cardiac and slow-twitch muscles, whereas SERCA 3 pumps are expressed in non-muscle tissues [9].

Several years after the discovery of Ca^{2+} pumps in intracellular stores, the plasma membrane Ca^{2+} pump was identified by Schatzmann in human red blood cells [10]. The plasma membrane Ca^{2+}-ATPase (PMCA) pump, although working in the same way as the SERCA pumps and also having 10 putative transmembrane-spanning domains, has surprisingly little sequence similarity with the SERCA pumps [11]. This probably explains the impressive selectivity of the tumour-promoting sesquiterpene lactone, thapsigargin, in inhibiting specifically the SERCA, but not the PMCA, pumps [12]. Thapsigargin has therefore become an

extremely useful experimental tool for depleting intracellular stores of Ca^{2+}.

There are two primary functions of the intracellular Ca^{2+} stores. On the one hand, they provide a rapidly releasable Ca^{2+} pool that can generate a cytosolic Ca^{2+} rise that may cause contraction, secretion or change metabolism according to the cell type. On the other hand, they also provide a powerful mechanism for removing Ca^{2+} from the cytosol, thus restoring the normal resting cytosolic Ca^{2+} concentration. With regard to restoring a normal low cytosolic Ca^{2+} concentration following a Ca^{2+} signal, the SERCA pumps are obviously helped by the PMCA pumps. The SERCA pumps are in general the most important elements in rapidly restoring intracellular Ca^{2+} concentration following cessation of Ca^{2+} signal generation. The ability of the SERCA pumps to remove cytosolic Ca^{2+} is of course dependent on closure of the Ca^{2+}-release channels in the stores [13]. In that situation Ca^{2+} uptake into the endoplasmic reticulum can be considerably more powerful than Ca^{2+} extrusion across the plasma membrane, although the precise relationship between the intensity of these two processes depends, in a complex manner, on the level of intracellular Ca^{2+} [13].

The work on muscle relaxing factor led to the important concept of intracellular Ca^{2+} stores. Because of the importance of cytosolic Ca^{2+} signalling in controlling many different cellular functions, there has in recent years been enormous interest in characterizing and localizing intracellular Ca^{2+} stores and understanding the way they contribute to the generation of cytosolic Ca^{2+} spikes and waves [14,15] (see Section 10).

References

1. Ebashi, S. (1985) in *Structure and Function of Sarcoplasmic Reticulum* (Fleischer, S. and Tonomura, Y., eds.), pp. 1–18, Academic Press, New York
2. Kumagai, H., Ebashi, S. and Takeda, F. (1955) *Nature (London)* **176**, 166
3. Kielley, W.W. and Meyerhof, O. (1948) *J. Biol. Chem.* **176**, 591–601
4. Ebashi, S. and Lippmann, F. (1962) *J. Cell Biol.* **14**, 389–400
5. Hasselbach, W. and Makinose, M. (1960) *Pflügers Arch.* **272**, 45
6. Hasselbach, W. and Makinose, M. (1961) *Biochem. Z.* **333**, 518–528
7. Hasselbach, W. and Makinose, M. (1962) *Biochem. Z.* **339**, 94–111
8. McLennan, D., Brandl, C.J., Korczak, B. and Green, N.M. (1985) *Nature (London)* **316**, 396–400
9. Pozzan, T., Rizzuto, R., Volpe, P. and Meldolesi, J. (1994) *Physiol. Rev.* **74**, 595–636
10. Schatzmann, H.J. (1966) *Experientia* **22**, 364–368
11. Carafoli, E. (1992) *J. Biol. Chem.* **267**, 2115–2118
12. Thastrup, O., Cullen, P.J., Drobak, B.K., Hanley, M.R. and Dawson, A.P. (1990) *Proc. Natl. Acad. Sci. U.S.A.* **87**, 2466–2470
13. Camello, P., Gardner, J., Petersen, O.H. and Tepikin, A.V. (1996) *J. Physiol. (London)* **490**, 585–593
14. Bootman, M.D. and Berridge, M.J. (1995) *Cell* **83**, 675–678
15. Bock, G.R. and Ackrill, K. (eds.) (1995) *Calcium Waves, Gradients and Oscillations*, CIBA Foundation Symposium 188, pp. 1–291, John Wiley and Son, Chichester

Ebashi & Lipmann (1962) J. Cell Biol. **14**, 389–400

ADENOSINE TRIPHOSPHATE-LINKED CONCENTRATION OF CALCIUM IONS IN A PARTICULATE FRACTION OF RABBIT MUSCLE

SETSURO EBASHI, M.D., and FRITZ LIPMANN, M.D.

From The Rockefeller Institute. Dr. Ebashi's present address is Department of Pharmacology, Faculty of Medicine, University of Tokyo, Japan

ABSTRACT

ATPase and ATP-dependent calcium ion concentration was studied with a membrane fraction isolated from homogenized rabbit skeletal muscle by differential centrifugation. Electron micrographs of the fraction indicate that it consists mainly of resealed tubules and vesicles of the endoplasmic reticulum. The up-to-1400-fold concentration of calcium in this fraction might be explained by proposing the existence of an energy-requiring system for the transport of calcium ions into the tubules or vesicles.

Ebashi *et al.* (1, 3, 8) isolated a particulate fraction from muscle which was identified as a component of the relaxation system. This fraction was similar to the magnesium-dependent ATPase-containing particles[1] of Kielley and Meyerhof (6, 7) and, like these, had high ATPase activity. In addition, in preliminary experiments, Ebashi (2, 2a) had observed a retention of calcium ion with this fraction. The metabolic effects found here became even more interesting when it appeared that, by their ultracentrifugal characteristics, what had been termed particles are, indeed, fragments of the endoplasmic, or, in this case, sarcoplasmic reticulum of Porter and Palade (14).

We will describe in this paper the calcium ion concentration function, its dependence on ATP, and its relation to the ATPase effect. Elsewhere, Ebashi (2) has already discussed in part the possible interrelationship between the relaxing activity and the various metabolic effects shown by this fraction. This work was done about 3 years ago but, due to various circumstances, publication has been delayed. In the meantime, Hasselbach and his coworkers (5, 10) have reported on observations similar to the ones described here.

METHODS

Preparation of the Particulate Fraction

The procedures are essentially the same as those described in the previous paper (1) but slight modifications have been made. The hind leg and back muscles of a rabbit were ground in a meat grinder and 200 gm of the minced muscle were added to 600 ml of 0.05 M $KHCO_3$ solution. They were then homogenized in a Waring blendor for 1 minute, centrifuged at 10,400 *g* for 20 minutes in the small Lourdes rotor, after which the supernatant fraction was passed through glass wool to remove the lipid layer, and recentrifuged at 38,000 *g* for 1 hour in the Spinco rotor no. 21. The supernatant was discarded and the pellet dispersed in 100 ml of 0.03 M $KHCO_3$ in a Potter-Elvehjem tissue grinder; the resuspended material was centrifuged at 8,200 *g* for 20 minutes in the small Lourdes rotor and the sediment discarded. This supernatant was again centrifuged for 1 hour at 38,000 *g* in the Spinco rotor no. 30. The precipitate

[1] The following abbreviations are used: AMP, adenosine 5'-monophosphate; ADP, adenosine 5'-diphosphate; ATP, adenosine 5'-triphosphate; PEP, phosphoenolpyruvic acid; PP, pyrophosphate; $\sim$P, energy-rich phosphate.

was suspended in an amount of 0.03 M $KHCO_3$ sufficient to make a 4 to 6 per cent solution by dry weight. Unless otherwise noted, this preparation was used in all experiments.

In cases where further purification is needed the preparation may be diluted 20-fold with 0.03 M $KHCO_3$, centrifuged at 8,200 *g* for 10 minutes, and the supernatant recentrifuged at 38,000 *g* for 45 minutes. The procedure may be repeated to obtain a more purified preparation.

The amount of membrane fraction is given in dry weight. The amount of sediment was generally expressed in wet weight, arbitrarily using five times the dry weight value. Such an approximate weight value was adopted and used for estimating concentrations inside the particulate fraction.

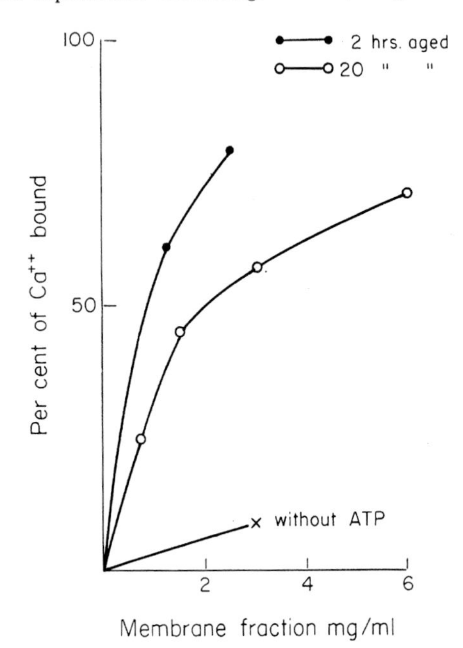

FIGURE 1

ATP-dependent calcium ion binding with various concentrations of normal and aged membrane fraction. See Methods for further details.

Standard Assay for Calcium Ion Binding

Each ml of solution contained, 2 μmoles of ATP, 10 μmoles of $MgCl_2$, 144 μmoles of KCl, 20 μmoles of tris-maleate buffer (pH 6.8), generally 0.01 to 0.1 μmole of $Ca^{45}Cl_2$, and a specified amount of sediment. The total volume was usually 12 ml. Unless otherwise noted, all solutions were kept at below 2°C. The mixture was transferred to Spinco tubes and the reaction was initiated by adding the ATP just before inserting the tube into the centrifuge; for the determination of calcium ion binding, a very fast reaction, the mixture was spun without prior incubation for 15 minutes at 100,000 *g* using rotor no. 40. During spinning, the reaction goes to completion. Both the precipitate and supernatants were then analyzed for Ca^{45} radioactivity.

Assay for ATP-ADP Exchange and ATPase

Each ml of reaction mixture contained: 2.5 μmoles of ATP, 2.5 μmoles of uniformly labeled C^{14}-ADP, 100 μmoles of KCl, 4 μmoles of $MgCl_2$, 30 μmoles of tris-maleate buffer (pH 6.8), and a specified amount of membrane fraction; the final volume was 0.6 ml. The reaction was initiated by adding the ATP-C^{14}-ADP mixture; it was stopped by shaking with 0.1 ml of a 1:1 mixture of chloroform and ether. After removal of the chloroform and ether by aeration, 0.05

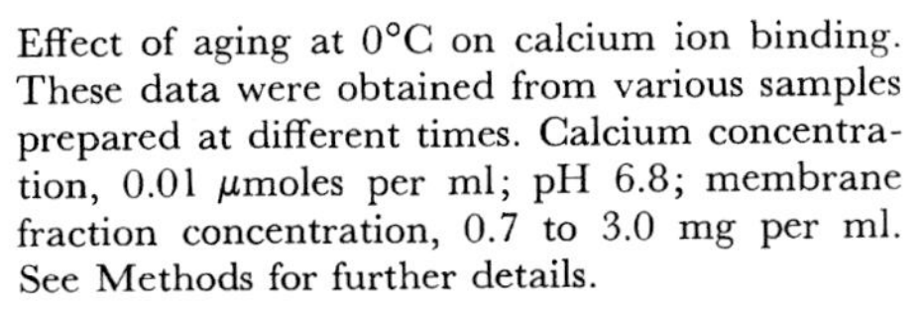

FIGURE 2

Effect of aging at 0°C on calcium ion binding. These data were obtained from various samples prepared at different times. Calcium concentration, 0.01 μmoles per ml; pH 6.8; membrane fraction concentration, 0.7 to 3.0 mg per ml. See Methods for further details.

ml was usually used for assay. The nucleotides were separated by paper electrophoresis according to Markham and Smith (9).

ATPase was determined by measuring inorganic phosphate in the remaining part of the reaction mixture by the method of Fiske and Subbarow (4). Myokinase was determined by the method of Noda and Kuby (11).

RESULTS

Calcium Ion Accumulation in Membrane Fragments

Following up earlier observations by Ebashi (2) on a retention of calcium ion by the membrane fraction, it was discovered that the addition of

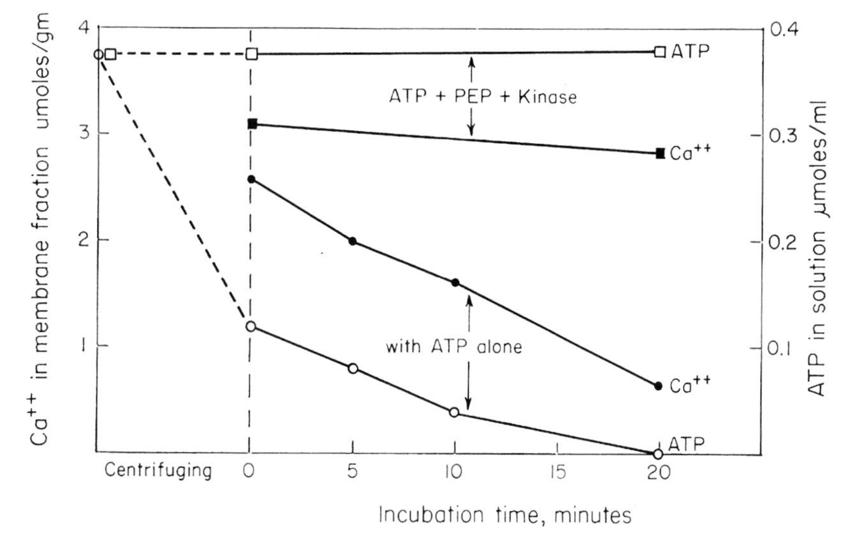

FIGURE 3

Dependence of calcium ion binding on maintenance of ATP concentration. When used to maintain ATP levels, 2 μmoles per ml of PEP and 50 units per ml of pyruvate kinase were added. Reaction mixtures contained 0.8 mg per ml of membrane fraction, were incubated at 15°C for the periods indicated in the figure, and centrifuged for assay of calcium ion binding as described under Methods. On the left of the zero line of the figure indicating the time before centrifugation, the amount of added ATP is noted. During the 15 minutes' centrifugation at 0°C in the Spinco, necessary to separate the membrane fraction for calcium determination, the ATP concentration, if not regenerated by the PEP system, falls to the value noted on the zero line. After 20 minutes, ATP in this series of assays has practically completely disappeared and, correspondingly, calcium has been released, during the incubation, from the membrane fraction. If ATP is maintained, however, by the PEP system, calcium is steadily kept in the membrane-vesicle fraction as shown by the upper two lines.

Assay for ATP Binding

Procedures were essentially the same as those for calcium ion binding, but uniformly labeled C^{14}-ATP was used in place of Ca^{45} as well as 2 μmoles of PEP and about 140 units of pyruvate kinase per ml.

Materials

Myokinase was obtained from Boehringer & Soehne, Mannheim, Germany, and C^{14}-ATP and ADP were obtained from Schwarz Bio-Research Inc., Orangeburg, New York. Ca^{45} was obtained from Oak Ridge National Laboratory.

ATP greatly stimulated this effect. It will appear from the following experiments that calcium ion accumulation is essentially an ATP-linked reaction. Fig. 1 shows the quantitative relation between calcium ion binding and amounts of membrane fraction. With fresh preparations about 80 per cent of the calcium ion in solution was bound maximally. This is equivalent to 1400-fold concentration of calcium ion in the granules as compared with the surrounding fluid. ADP showed a slight effect, which may be due to the presence of myokinase. When ADP was added to

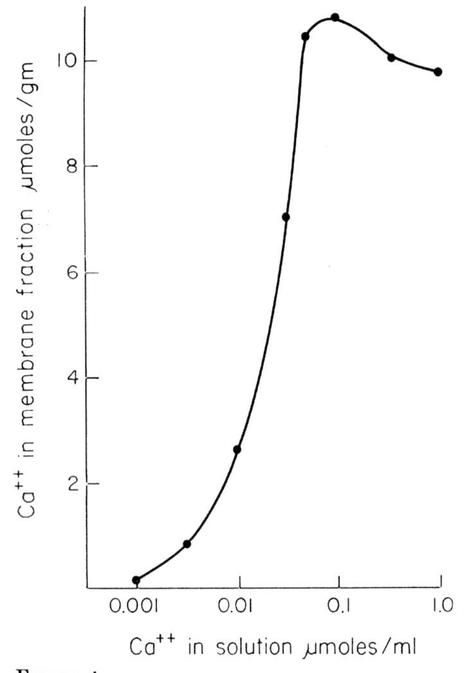

FIGURE 4

Calcium ion binding of membrane fraction as a function of calcium concentration in solution. Each ml of solution contained 2.0 mg of membrane fraction. See Methods for further details.

ATP in equimolar amounts it caused a 30 per cent inhibition. AMP and PP were without effect.

The pH optimum for the calcium ion concentrating reaction is 6.5. As shown by the lower curve in Fig. 1, part of the concentrating ability is lost on aging. This loss appears more clearly in Fig. 2. The relative ease of deterioration parallels that of the relaxing effect, but contrasts with the ATP-splitting reaction which is rather insensitive and may even slightly increase with similar treatment.

In the following experiments a dependence upon a constant supply of ATP for holding calcium ion in the membrane fraction will be demonstrated. As shown in Fig. 3, two parallel sets of tubes were used; in one set ATP was allowed to decompose, in the other its concentration was maintained by PEP + pyruvate kinase + ATP, acting as an energy-rich phosphate feeder system. The tubes were incubated at 15°C for varying periods of time before spinning for 15 minutes in the Spinco at 0°C. Due to reactions taking place during this spin, the samples showed an already marked decrease in ATP at "zero" time. On further incubation, the amount of bound calcium ion decreased parallel with the decrease in ATP concentration in those tubes where the ~P-

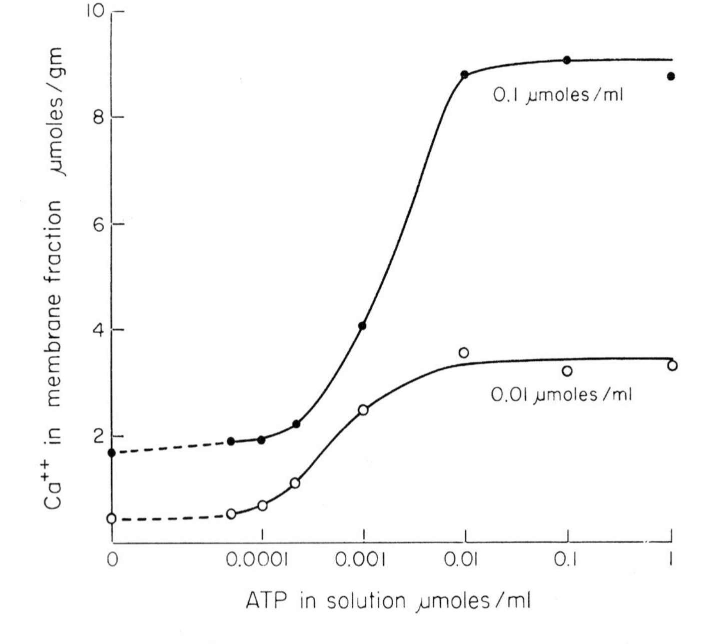

FIGURE 5

Calcium ion binding of membrane fraction as a function of ATP concentration in solution. Each ml of solution contained 2 µmoles of PEP, 140 units of pyruvate kinase, 1.0 mg of membrane fraction and either 0.1 or 0.01 µmoles per ml of calcium ion. See Methods for further details.

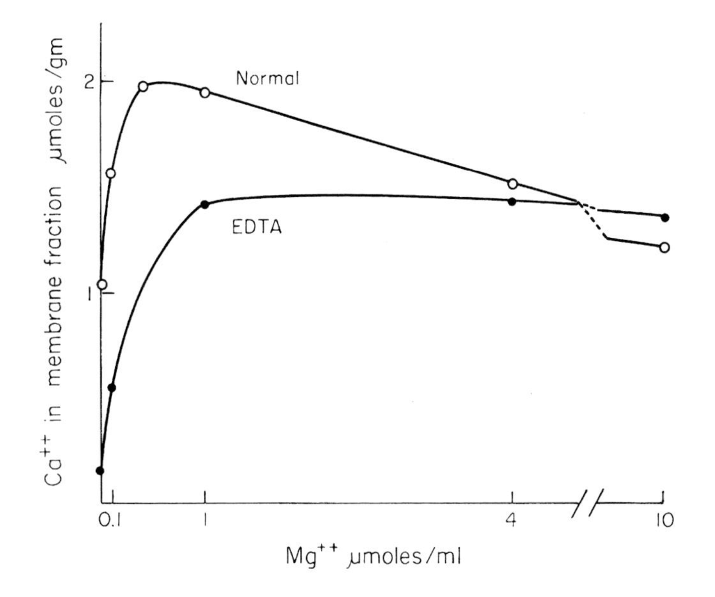

FIGURE 6

Dependence of calcium ion binding on magnesium concentration. In the lower curve the usual preparation was diluted by 0.03 M $KHCO_3$ containing 5 μmoles of EDTA per ml and then centrifuged. Reaction mixtures contained 0.7 mg of membrane fraction per ml in the usual preparation, and 0.8 mg in the EDTA-washed preparation. See Methods for further details.

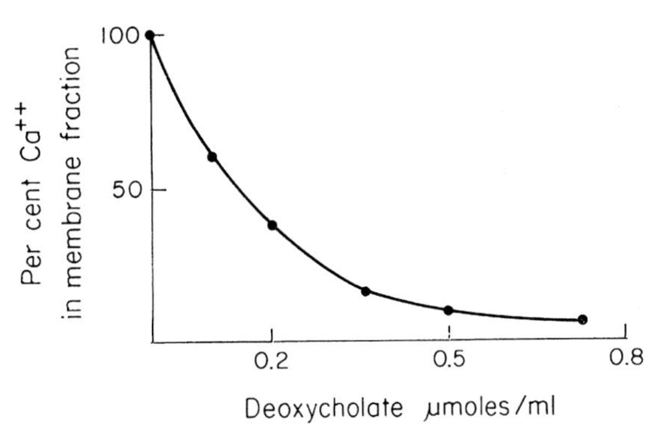

FIGURE 7

Inhibition of calcium ion accumulation by deoxycholate. 1.5 mg per ml of membrane fraction was used. See Methods for further details.

feeder system was omitted. With the feeder system present, however, the amount of calcium ion reached its maximal level during the 15 minutes' spin at 0°C and was maintained at this level with further incubation. Furthermore, addition of ~P-feeder system after decomposition of ATP lifted the calcium ion concentration again to nearly maximum level. Thus, when the ATP level is maintained, the calcium ion remains bound and, when it falls, calcium ion is released reversibly.

The effect of calcium ion concentration on fixation is shown in Fig. 4. The system appears to be saturated near 10^{-4} M; higher concentrations are slightly inhibitory. The level of ATP that would effect a maximal calcium ion binding could be determined, in spite of the ATPase activity in the preparation, by maintaining a steady-state level of ATP with a ~P-feeder system. Fig. 5 shows that the saturation level of ATP is 10^{-5} M and, furthermore, that it is unchanged with a variation of calcium ion concentration. The calcium ion concentration eventually attained in the membrane fraction exceeds the steady-state ATP concentration in the surrounding fluid by a factor of 1000 or more. It is, therefore, a dynamic transfer in or across a membrane, and may be interpreted as an ATP-dependent osmotic concentration of calcium ion. As will be described later, electron micrographs indicate the fraction to contain vesicles of endoplasmic reticulum, some of which appear to be closed, possibly representing resealed tube fragments.

As would be expected for an ATP-linked reaction, the calcium ion binding is dependent upon magnesium, as shown in Fig. 6; this requirement appears most clearly with an EDTA-washed preparation. It was shown in Fig. 2 that the calcium ion concentration function diminishes and eventually disappears on aging. Similarly, deoxycholate inhibits, as seen in Fig. 7; it may be significant that both aging and deoxycholate also inhibit relaxation in contrast to ATPase.

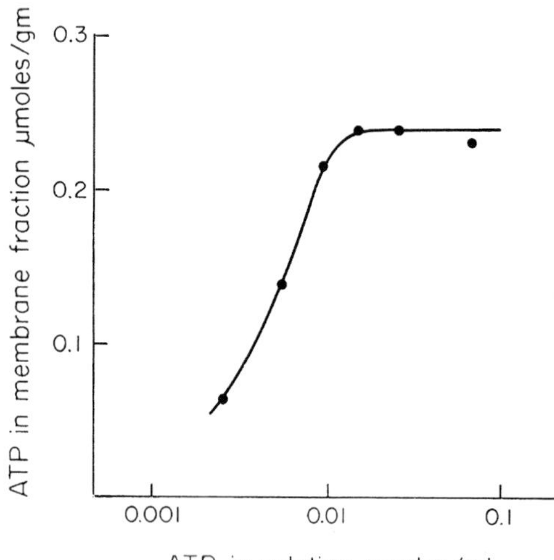

FIGURE 8

ATP binding of membrane fraction as a function of ATP concentration in solution. The solution contained 2.5 mg per ml of membrane fraction. See Methods for further details.

ATP-Binding

If C^{14}-ATP is incubated with the preparation, including a ~P-feeder system, some excess of ATP is found in the membrane fraction (Fig. 8), the concentration of ATP being 20-fold over that in the surrounding medium. As shown in Fig. 8, above 0.01 μmoles per ml the ATP-binding levels off and does not respond to further increase in ATP concentration. Comparison of ATP binding with that of calcium ion shows that about 50 to 70 moles of calcium ion may be held in steady-state per mole of ATP bound possibly to the membrane.

Ebashi had already studied the ATPase in this preparation (1). We find now that, in addition to the ATPase, the preparation also catalyzes a rather rapid exchange between ATP and ADP and that the two effects seem to be closely connected with each other. Fig. 9, for example, shows that the pH optimum of the two reactions overlaps. Both are dependent upon the addition of magnesium and, here again, the effect of mag-

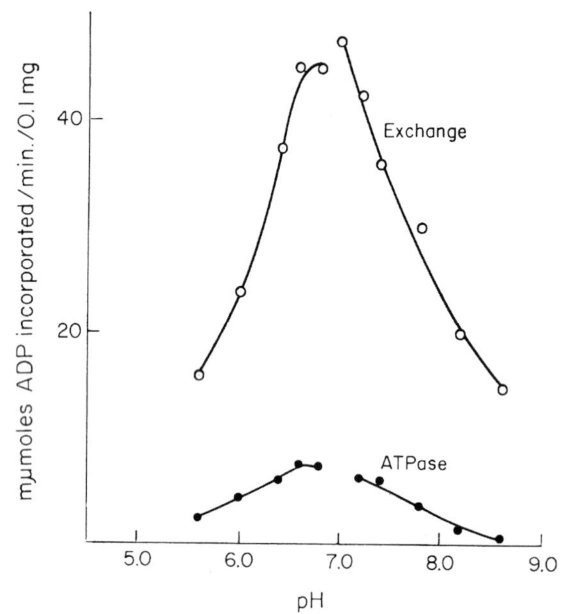

FIGURE 9

ATP-ADP exchange reaction and ATPase of membrane fraction as a function of pH. Each ml of reaction mixture contained 0.1 mg of membrane fraction. Tris-maleate buffer was used below pH 6.8 and tris buffer above pH 7.0. See Methods for further details.

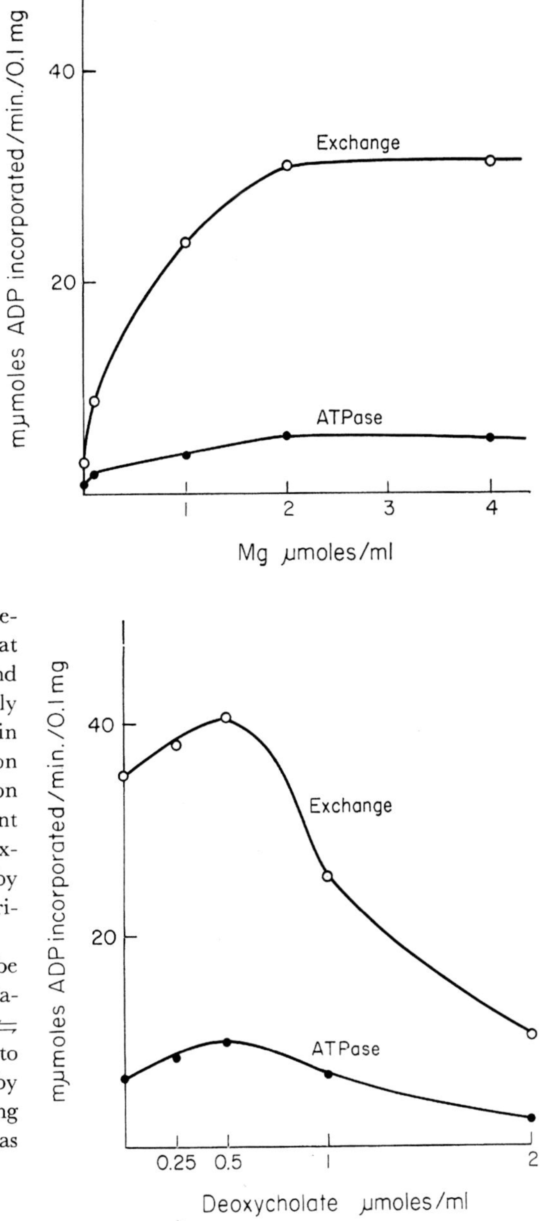

FIGURE 10
Effect of varying concentrations of magnesium ion on the ATP-ADP exchange reaction and ATPase of the particulate fraction. 0.1 mg per ml of membrane fraction was used in each case. See Methods for details.

FIGURE 11
Effect of deoxycholate on the ATP-ADP exchange reaction and ATPase of the membrane fraction. Each reaction mixture contained 0.1 mg of membrane fraction per ml. See Methods for details.

nesium concentration is parallel with both reactions, as shown in Fig. 10. Deoxycholate at lower concentrations causes a stimulation and only at relatively high concentrations eventually inhibits both reactions. This effect is shown in Fig. 11, and it contrasts with the strong inhibition by deoxycholate of the calcium ion concentration and the relaxation effects. It may be significant that calcium ion inhibits both ATPase and exchange and that the inhibition is reversed by deoxycholate. This is illustrated by the experiments presented in Fig. 12(*A* and *B*).

The catalysis of ATP-ADP exchange may be interpreted as indicating a reversible phosphorylation of the membrane: ATP + membrane $\leftrightarrows$ ADP + membrane-P. However, in attempts to prove a transfer of phosphate to the membrane by using terminally labelled ATP^{32}, no convincing phosphate binding by the membrane fraction was observed.

ELECTRON MICROGRAPHS

Methods

Pellets of the particulate fraction obtained as described above were fixed *in situ* (at the bottom of the tube) by overlaying them with a 2 per cent OsO_4 solution in 0.03 M $KHCO_3$. After dehydration in ethanol and embedding in an 80:20 mixture of butyl-methyl methacrylate, the preparations were sectioned parallel to the direction of centrifugation to make possible their systematic survey, from top to bottom, by electron microscopy.

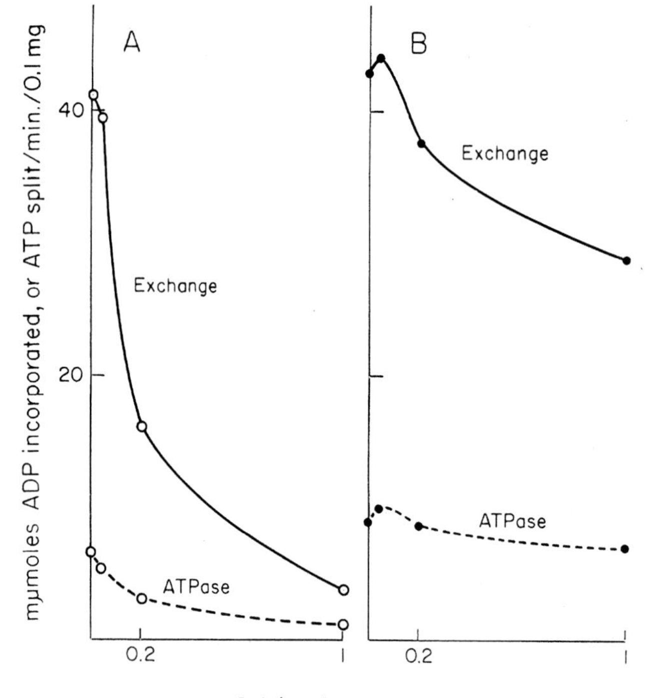

FIGURE 12

Inhibition of ATPase and ATP-ADP exchange by calcium. *A*, effect of increased calcium ion concentration. *B*, antagonism by deoxycholate to calcium ion inhibition.

In both experiments, 0.1 mg per ml of membrane fraction and 0.5 μmoles per ml of deoxycholate were used. See Methods for details.

Findings

The survey showed that the pellets were stratified and heterogeneous, the following components forming three successive and unequal layers: *Top:* (less than 5 per cent of the pellet thickness) dense particles, 10 to 15 mμ in diameter, similar to ribosomes isolated from other sources but more varied in size; *Middle:* (about 10 per cent of the pellet thickness) dense granules ~20 mμ in diameter, reminiscent of the glycogen particles seen in intact muscle fibers; *Bottom:* vesicles (diameter: 60 to 200 mμ) either spherical or flattened (cisternae), and tubules (diameter: 15 to 30 mμ; length: indefinite) mixed with a small amount of 20 mμ particles (Fig. 13).

Most vesicles and tubules appeared as closed structures bound by a continuous, well defined,

FIGURE 13

Representative field in the bottom layer of a particulate fraction pellet. Activity for calcium ion concentration and relaxation, good.

The circular profiles (v_1) represent sectioned vesicles or transversely cut tubules; the elongated profiles (v_2) correspond to normally cut cisternae and longitudinally sectioned tubules. Oblique (v_3) or grazing (v_4) sections through vesicles are seen in many places. Most of these elements are bounded by a well defined continuous membrane. Vesicles with membrane defects or discontinuities (v_5) are rare.

Particles (p) of ~ 20 mμ diameter (glycogen?) represent a minor component of the preparation at this level in the pellet. × 80,000.

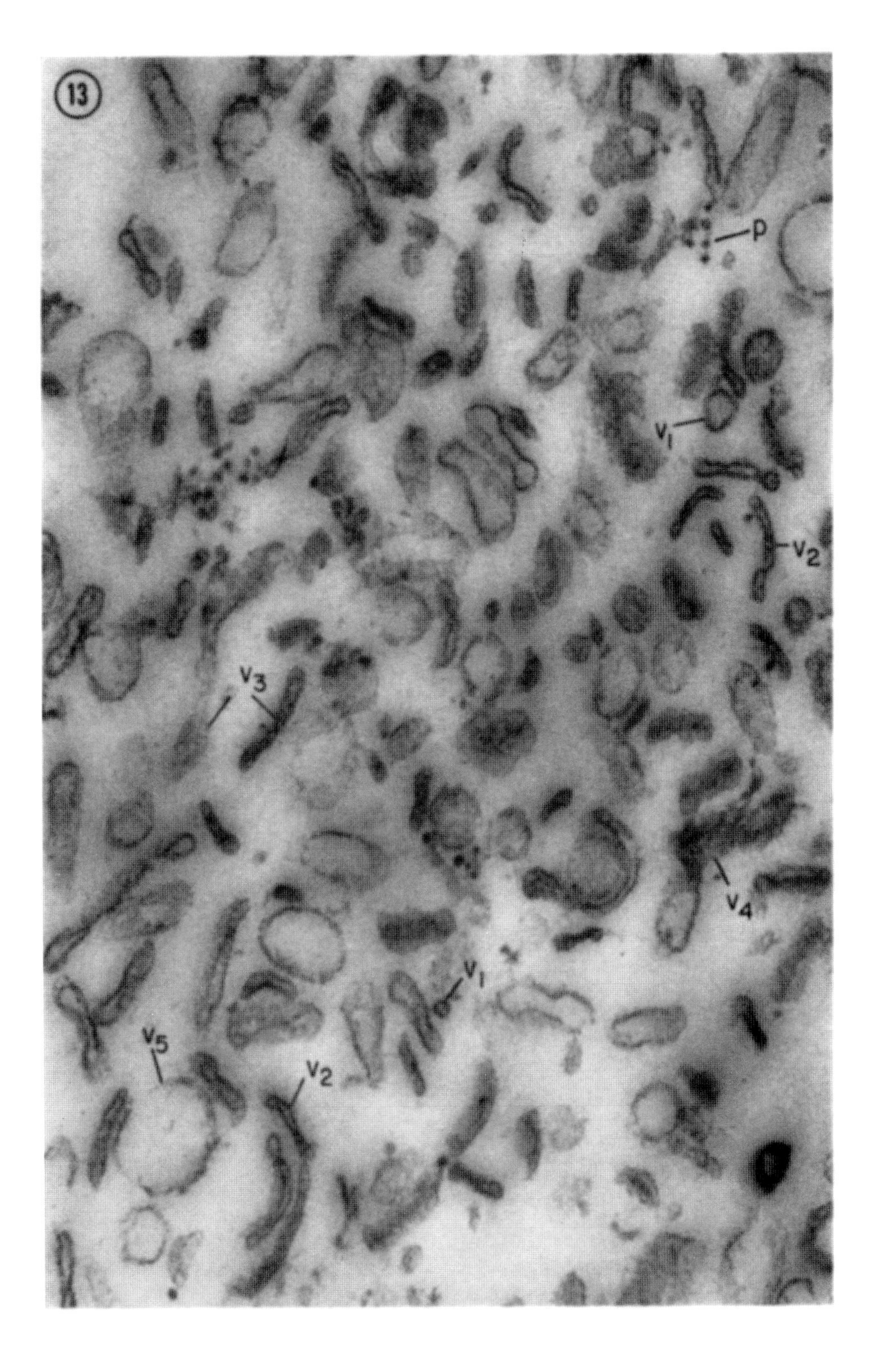
13
P
V_1
V_2
V_3
V_4
V_1
V_5
V_2

smooth surfaced membrane. Broken open elements were rarely encountered.

Since the bottom layer made about 85 per cent of the total thickness of the pellet, the vesicles and tubules mentioned were the dominant structural element of the preparation. Their appearance and dimensions varied to a certain extent with position and degree of packing in the pellet: larger elements were found in the bottom layers of the preparation, and most vesicles appeared flattened in tightly packed sediments. Notwithstanding these variations, the isolated elements were similar in shape and size to the tubules and cisternae of the sarcoplasmic reticulum of the intact muscle fiber. Therefore the preparation appeared to consist mainly of "healed" or "closed"[2] fragments of the sarcoplasmic reticulum, a conclusion strengthened by the occasional encounter of vesicles with the localized membrane thickening (Fig. 14) that characterizes the terminal cisternae *in situ*, or vesicles with the typical grouping shown by the triads of the system in the intact muscle fiber (Fig. 15). The pellets did not contain nuclei, myofilaments, mitochondria, or recognizable mitochondrial fragments.

In preparations aged (Fig. 16) or treated with deoxycholate, which had rather poor activity, the limiting membranes of the vesicular elements were less sharply outlined and frequently discontinuous.

[2] This feature suggests that the fragmentation of the sarcoplasmic reticulum is due to a generalized pinching off process which occurs during tissue homogenization. A similar process has been postulated in the formation of hepatic and pancreatic microsomes from the corresponding endoplasmic reticula (12, 13).

CONCLUSIONS

A particulate fraction from rabbit muscle known to contain ATPase (6, 7) and relaxing factor (1) is found to catalyze an ATP-linked calcium ion concentration effect. This was studied by comparing calcium concentrations in the supernatant and particulate fractions. The reaction is so fast that during 15 minutes' centrifugation in the cooled Spinco it is complete and calcium may be concentrated in the particulate fraction up to 1400-fold. This effect is dependent upon a continuous supply of ATP. A 20-fold concentration of ATP against the surrounding fluid was found in the same fraction; the ratio of calcium ion to ATP is approximately 70 to 1. This seems to leave, as the only interpretation, a dynamic function of ATP in the process of calcium ion accumulation. The reaction appears to be calcium ion–specific since no concentration of monovalent ions was observed with this fraction.

This effect is unstable to aging and to deoxycholate or other detergents in a manner similar to that found for the relaxation effect of the same fraction (1). In contrast, ATPase is either not affected or is rather stimulated by aging or detergents; the same is true for ADP-ATP exchange which was found with this fraction and which appears to run pretty much parallel with the ATPase.

The electron micrographic picture of the fraction consists mainly of membrane-bounded elements, apparently resealed fragments of the endoplasmic reticulum. It is rather attractive to consider the accumulation of calcium ion as being due to an ATP-dependent transport of calcium

FIGURES 14 AND 15

Small fields in the bottom layer of a particulate preparation. Activity for calcium ion concentration and relaxation, good. × 60,000.

In Fig. 14 the arrows point to the local membrane thickening that characterizes terminal cisternae *in situ*.

In Fig. 15 the arrow marks an intermediate vesicle still located between two terminal cisternae, as in the triads of intact muscle fibers.

FIGURE 16

Bottom layer in a particulate preparation aged for 6 days. Activity rather poor.

Profiles of apparently intact vesicles are still present (v_1) but many elements show limited (v_2) or gross (v_3)discontinuity of their limiting membranes. The arrows point to local membrane thickenings of the type encountered in terminal cisternae *in situ*. × 60,000.

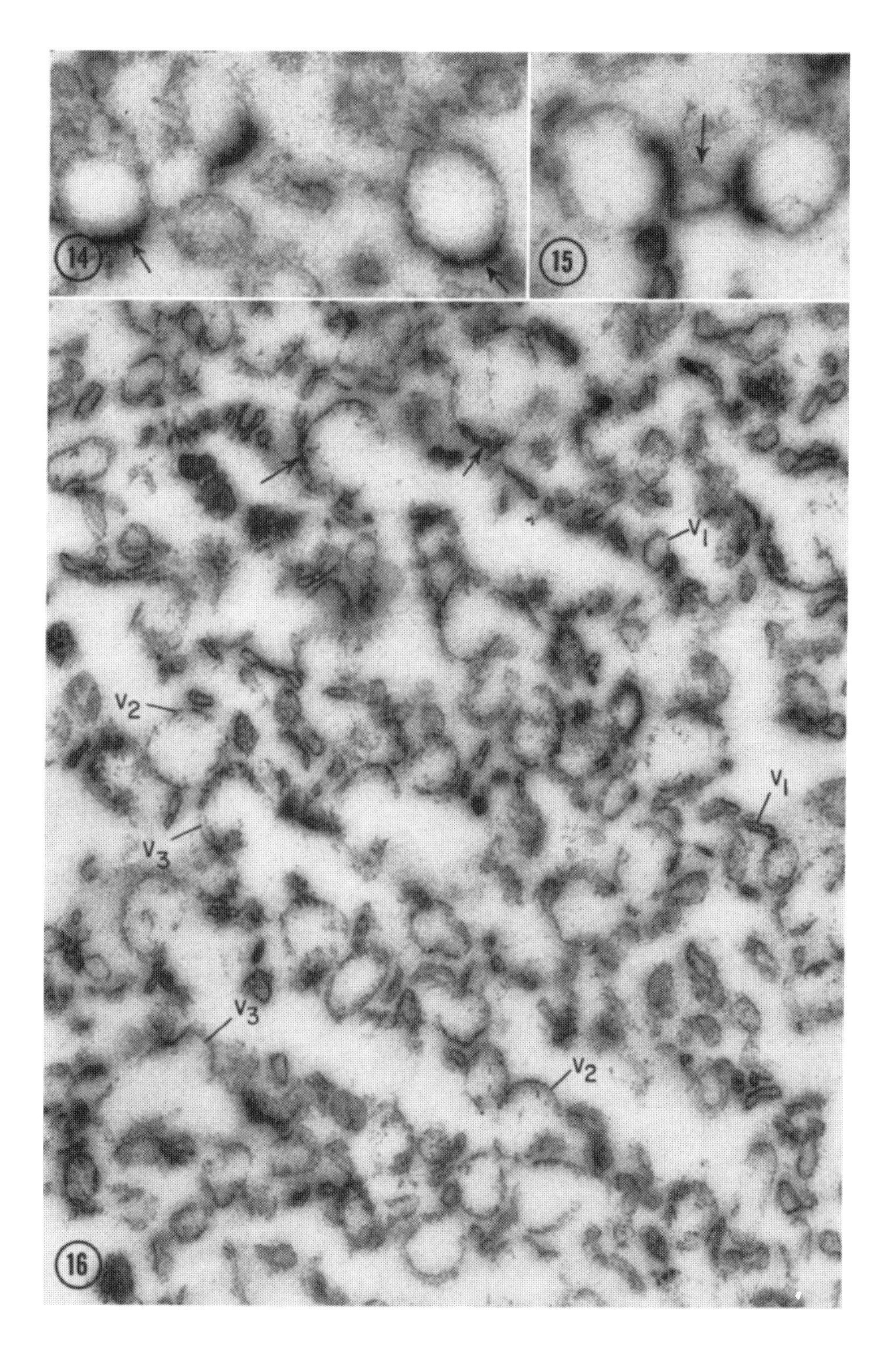
14
15
16
V_1
V_1
V_2
V_2
V_3
V_3

ion across the membrane which holds it reversibly inside the vesicles.

We are greatly indebted to Dr. George Palade for his interest and help, and comments on this paper. We are very grateful for permission to include his electron micrographs, which he prepared and annotated.

This work was supported by grants G-4341 from the National Science Foundation, and C-3159 from the National Cancer Institute, National Institutes of Health, United States Public Health Service. Dr. Ebashi was a 1959–60 Rockefeller Foundation Fellow.

Received for publication, April 11, 1962.

BIBLIOGRAPHY

1. Ebashi, S., *Arch. Biochem. and Biophysics*, 1958, **76**, 410.
2. Ebashi, S., *J. Biochem.*, 1961, **50**, 236.
2a. Ebashi, S., *J. Biochem.*, 1960, **48**, 150.
3. Ebashi, S., Takeda, F., Otsuka, M., and Kumagai, H., *Symposium on Enzyme Chem. (Japan)*, 1956, **11**, 11.
4. Fiske, C. H., and Subbarow, Y., *J. Biol. Chem.*, 1925, **66**, 375.
5. Hasselbach, W., and Makinose, M., *Biochem. Z.*, 1961, **333**, 518.
6. Kielley, W. W., and Meyerhof, O., *J. Biol. Chem.*, 1948, **176**, 591.
7. Kielley, W. W., and Meyerhof, O., *J. Biol. Chem.*, 1950, **183**, 391.
8. Kumagai, H., Ebashi, S., and Takeda, F., *Nature*, 1955, **176**, 166.
9. Markham, R., and Smith, J. D., *Biochem. J.*, 1952, **52**, 552.
10. Nagai, T., Makinose, M., and Hasselbach, W., *Biochim. et Biophvsica Acta*, 1960, **43**, 223.
11. Noda, L., and Kuby, S. A., *J. Biol. Chem.*, 1957, **226**, 541.
12. Palade, G. E., and Siekevitz, P., *J. Biophysic. and Biochem. Cytol.*, 1956, **2**, 171.
13. Palade, G. E., and Siekevitz, P., *J. Biophysic. and Biochem. Cytol.*, 1956, **2**, 671.
14. Porter, K. R., and Palade, G. E., *J. Biophysic. and Biochem. Cytol.*, 1957, **3**, 269.

Inositol 1,4,5-trisphosphate releases Ca^{2+} from intracellular stores

9

The paper by Hokin and Hokin [1] that we have chosen demonstrated for the first time an increase in membrane phospholipid turnover in response to surface membrane receptor occupation. In a paper published two years later, Hokin and Hokin [2] identified phosphoinositide and phosphatidic acid as the main lipids responding with increased turnover. This discovery led, after many years of tribulations involving a large number of groups, to the now well established concept that many surface membrane receptors are coupled, via a G-protein, to phospholipase C acting on phosphatidylinositol 4,5-bisphosphate [PtdIns(4,5)P_2] to generate the two messengers, diacylglycerol and inositol 1,4,5-trisphosphate [Ins(1,4,5)P_3] [3].

Hokin and Hokin's paper [1] provides an interesting example of an important discovery that was initially, because of incomplete information, not entirely understood even by the authors themselves. The Hokins suspected that the stimulant-evoked increase in membrane phospholipid turnover was an essential component of the dynamic events involved in exocytotic secretion [1]. Later, however, they and others found that it could be dissociated from exocytosis in some secretory cells and that it also occurred in non-secretory cells (see Michell et al. [4]).

In an important review article Michell [5] proposed that agonist–receptor interaction would provoke phosphatidylinositol breakdown, which in turn would cause opening of Ca^{2+} gates in the cell membrane, allowing Ca^{2+} inflow. Studies on insect salivary glands provided experimental evidence for this hypothesis [6]. At the same time, electrophysiological studies on pancreatic acinar cells demonstrated that hormone–receptor interaction activated a Ca^{2+}-effector resulting in an increase in intracellular Ca^{2+} [7]. In these experiments two sources of Ca^{2+} were identified: in the initial phase after start of stimulation the Ca^{2+} involved in the stimulus–permeability coupling was derived from an intracellular source, while in the sustained phase of stimulation extracellular Ca^{2+} was needed [7]. Indeed, early studies of Ca^{2+} homeostasis in exocrine gland cells had indicated that the primary result of receptor activation was release of Ca^{2+} stored intracellularly [8]. (For a detailed review of these developments, see [9].)

We now know that the primary receptor-activated event (Figure 9.1) is a breakdown of PtdIns(4,5)P_2 [3] into the dual second messengers diacylglycerol and Ins(1,4,5)P_3. The experiments that finally resolved the question about the molecular link between receptor activation and intracellular Ca^{2+} release were reported in the second paper we have selected [10]. Streb, Irvine, Berridge and Schulz showed in experiments on permeabilized pancreatic acinar cells that Ins(1,4,5)P_3 in micromolar concentrations released Ca^{2+} from a non-mitochondrial store. They provided evidence indicating that Ins(1,4,5)P_3-evoked Ca^{2+} release occurred from the same store as that induced by acetylcholine, and concluded that Ins(1,4,5)P_3 is the second messenger mediating the acetyl-

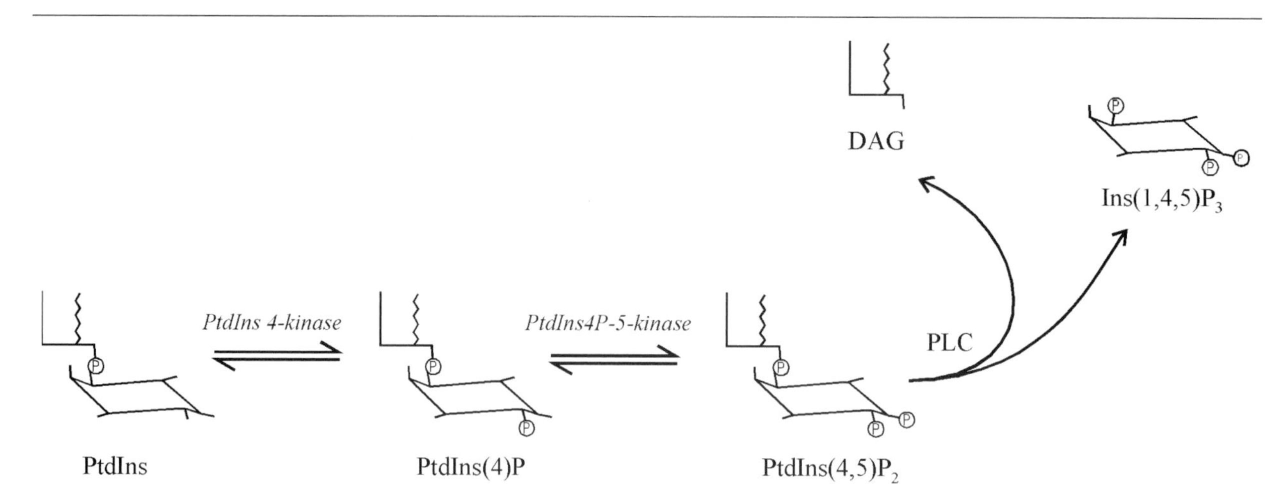

Figure 9.1 Formation of PtdIns(4,5)P_2 and its breakdown following cell stimulation by phospholipase C (PLC) to generate the two messengers, diacylglycerol (DAG) and Ins(1,4,5)P_3.

choline-evoked Ca^{2+} release. Further work on pancreatic acinar cells demonstrated that Ins(1,4,5)P_3 releases Ca^{2+} from endoplasmic reticulum (ER) vesicles [11] and that hormone-evoked Ca^{2+} release is associated with inositol polyphosphate production, an effect that is not mediated by Ca^{2+} [12].

The Ins(1,4,5)P_3 receptor was cloned in 1989 [13–15]. Multiple isoforms are encoded by at least four genes which have considerable similarity to one another [16] The Ins(1,4,5)P_3 receptor is a homotetramer of 310 kDa subunits that make up a relatively non-selective cation channel that has no significant homology with voltage-sensitive Ca^{2+} channels [16]. The regulation of the Ins(1,4,5)P_3 receptor is exceedingly complex. The receptor displays a bell-shaped Ca^{2+}-sensitive response to Ins(1,4,5)P_3, with low levels of Ca^{2+} (below 300 nM) enhancing Ins(1,4,5)P_3-evoked Ca^{2+} release and higher levels (in excess of 300 nM) inhibiting release [17]. Ca^{2+} therefore exhibits both positive and negative feedbacks on intracellular Ca^{2+} release. This may result from differential activation of Ca^{2+}-dependent kinases and phosphatases [18,19]. The positive feedback of Ca^{2+} on Ins(1,4,5)P_3-evoked Ca^{2+} release is responsible for the steep rising phase of cytosolic Ca^{2+} spikes (see Section 10), whereas the negative feedback is important for terminating the release, thereby allowing the cytosolic Ca^{2+} concentration to return to resting levels. These properties of the Ins(1,4,5)P_3 receptor are similar to those of the other major Ca^{2+}-release channel, the ryanodine receptor. Elementary Ca^{2+}-release units caused by clusters of Ins(1,4,5)P_3 or ryanodine receptors have been identified in a number of different cell types as, for example, local Ca^{2+} spikes in the secretory pole of pancreatic acinar cells, Ca^{2+} puffs in oocytes or Ca^{2+} sparks in cardiac myocytes [20].

Although it is generally accepted that Ins(1,4,5)P_3 receptors are found in the ER [16] recent data indicate that mobilizable Ca^{2+} stores can also be found in other organelles. In the nuclear envelope in liver cells, the high-affinity Ins(1,4,5)P_3 receptors are found in the inner nuclear membrane and not in the outer nuclear membrane which has ER properties [21]. This explains the intra-nucleoplasmic Ca^{2+} signals that can be evoked by Ins(1,4,5)P_3-evoked Ca^{2+} release from the nuclear envelope Ca^{2+}

store [22]. The physiologically important local Ca^{2+} spikes in the secretory granule area of pancreatic acinar cells evoked by low agonist concentrations are most likely caused by Ins(1,4,5)P_3-mediated Ca^{2+} release from ER terminals surrounded by zymogen granules followed by Ca^{2+} release from granules [23].

Finally, what is the link between the phosphatidylinositol breakdown and the gating of Ca^{2+} entry already mentioned [5,6]? In all those many cell types where the primary agonist-evoked cytosolic Ca^{2+} signal is caused by Ins(1,4,5)P_3-mediated release of Ca^{2+} from intracellular Ca^{2+} stores, the secondary sustained phase of the Ca^{2+} signal is due to Ca^{2+} entry across the plasma membrane activated by depletion of the intracellular Ca^{2+} store [24]. This so-called capacitative Ca^{2+} entry pathway appears to be controlled by the degree of filling of the Ca^{2+} stores [24], and a number of different mechanisms that could explain how this works have been proposed [25]. The capacitative Ca^{2+} entry appears to be co-localized with the intracellular Ca^{2+} release sites, at least in *Xenopus* oocytes, and recent evidence suggests that there is not a highly diffusible Ca^{2+} influx factor involved [26]. The most likely mechanism linking Ins(1,4,5)P_3-evoked Ca^{2+} release to Ca^{2+} entry appears currently to be protein–protein interaction between the plasma membrane Ca^{2+} channel (probably similar to the *Drosophila* trp protein [27,28]) and the Ins(1,4,5)P_3 receptor on a neighbouring intracellular membrane [25].

References

1. Hokin, M.R. and Hokin, L.E. (1953) *J. Biol. Chem.* **203**, 967–977
2. Hokin, L.E. and Hokin, M.R. (1955) *Biochim. Biophys. Acta* **18**, 102–110
3. Berridge, M.J. and Irvine, R.F. (1984) *Nature (London)* **312**, 315–321
4. Michell, R.H., Kirk, C.J., Jones, L.M., Downes, C.P. and Creba, J.A. (1981) *Philos. Trans. R. Soc. London Ser. B* **296**, 123–137
5. Michell, R.H. (1975) *Biochim. Biophys. Acta* **415**, 81–147
6. Berridge, M.J. (1981) in *Drug Receptors and their Effectors* (Birdsall, N.J., ed.), pp. 75–85, MacMillan, London
7. Petersen, O.H., Nishiyama, A., Laugier, R. and Philpott, H.G. (1981) in *Drug Receptors and their Effectors* (Birdsall, N.J., ed.), pp. 63–73, Macmillan, London
8. Nielsen, S.P. and Petersen, O.H. (1972) *J. Physiol. (London)* **223**, 685–697
9. Petersen, O.H. (1992) *J. Physiol. (London)* **448**, 1–51
10. Streb, H., Irvine, R.F., Berridge, M.J. and Schulz, I. (1983) *Nature (London)* **306**, 67–69
11. Streb, H., Bayerdorffer, E., Haase, W., Irvine, R.F. and Schulz, I. (1984) *J. Membr. Biol.* **81**, 241–253
12. Streb, H., Heslop, J.P., Irvine, R.F., Schulz, I. and Berridge, M.J. (1985) *J. Biol. Chem.* **260**, 7309–7315
13. Furuichi, T., Yoshikawa, S., Miyawaki, A., Wada, K., Maeda, N. and Mikoshiba, K. (1989) *Nature (London)* **342**, 32–38
14. Mignery, G.A., Südhof, T.C., Takei, K. and DeCamilli, P. (1989) *Nature (London)* **342**, 192–195
15. Mignery, G.A., Newton, C.L., Archer, B.T. and Südhof, T.C. (1990) *J. Biol. Chem.* **265**, 12679–12685
16. Clapham, D.E. (1995) *Cell* **80**, 259–268
17. Iino, M. (1990) *J. Gen. Physiol.* **95**, 1103–1122
18. Zhang, B., Zhao, H. and Muallem, S. (1993) *J. Biol. Chem.* **268**, 10997–11001
19. Cameron, A.M., Steiner, J.P., Roskams, A.J., Ali, S.M., Ronnett, G.V. and Snyder, S.H. (1995) *Cell* **83**, 463–472
20. Bootman, M.D. and Berridge, M.J. (1995) *Cell* **83**, 675–678
21. Humbert, J.-P., Matter, N., Artault, J.-C., Köpler, P. and Malviya, A.N. (1996) *J. Biol. Chem.* **271**, 478–485
22. Gerasimenko, O.V., Gerasimenko, J.V., Tepikin, A.V. and Petersen, O.H. (1995) *Cell* **80**, 439–444
23. Mogami, H., Nakano, K., Tepikin, A.V. and Petersen, O.H. (1997) *Cell* **88**, 49–55

24. Putney, J.W. (1990) *Cell Calcium* **11**, 611–624
25. Berridge, M.J. (1995) *Biochem. J.* **312**, 1–11
26. Petersen, C.C.H. and Berridge, M.J. (1996) *Pflügers Arch.* **432**, 286–292
27. Petersen, C.C.H., Berridge, M.J., Borgese, M.F. and Bennett, D.L. (1995) *Biochem. J.* **311**, 41–44
28. Wes, P.D., Chevesich, J., Jeromin, A., Rosenberg, C., Stetten, G. and Montell, C. (1995) *Proc. Natl. Acad. Sci. U.S.A.* **92**, 9652–9656

Hokin & Hokin (1953) J. Biol. Chem. **203**, 967–977

ENZYME SECRETION AND THE INCORPORATION OF P^{32} INTO PHOSPHOLIPIDES OF PANCREAS SLICES

By MABEL R. HOKIN* and LOWELL E. HOKIN†

(*From the Research Institute, the Montreal General Hospital, Montreal, Canada*)

(Received for publication, December 17, 1952)

It has been previously shown that the addition of cholinergic drugs to respiring pancreas slices *in vitro* stimulates the secretion (active extrusion) of amylase; the synthesis of amylase, on the other hand, is not stimulated by cholinergic drugs (1). The present paper is concerned with studies on the uptake of P^{32} into the phospholipides during enzyme secretion in pancreas slices *in vitro*. It has been found that the stimulation of amylase secretion by acetylcholine or carbamylcholine is accompanied by a 5- to 9-fold increase in the specific activity of the phospholipides after a 2 hour incubation.

EXPERIMENTAL

Preparation of Tissue Slices—For studies of enzyme secretion pigeons were fasted for approximately 48 hours prior to killing, unless otherwise stated. The fasting was carried out in order to bring about a non-secreting repleted gland. The functional state of the pancreas could be judged by its appearance. Non-secreting glands from fasted pigeons were small, pale, and friable. Glands from fed pigeons were larger and pink to reddish. When enzyme synthesis was studied, the pigeons were fed; 0.15 mg. of carbamylcholine was administered intramuscularly 1 hour before killing.

The animals were killed by decapitation, and the organs were removed immediately and chilled in iced saline. The slices were prepared by a slight modification of the method described earlier (1–3). As soon as the slices were cut, they were placed in a chilled covered crystallizing dish. After the slicing was completed, the slices were weighed and placed over appropriate numbers in a second chilled crystallizing dish. They were then ready for placing in incubation vessels.

In order to insure uniform results in the determination of the Q_{O_2}, the specific activities of the acid-soluble phosphate esters and the specific activities of the phospholipides at least 100 mg. of tissue were needed. It was usually necessary to use slices from both lobes in order to obtain sufficient tissue for four to six vessels. The proportion of tissue from each lobe was maintained as constant as possible in each vessel to minimize possible differences in the activities of the two lobes.

* Fellow of the National Cancer Institute of Canada.
† Merck Postdoctoral Fellow in the Natural Sciences.

Incubation of Tissues—In the majority of experiments on enzyme secretion the pancreas slices were incubated for 15 minutes at 40° in 30 ml. stoppered conical flasks containing 3 ml. of oxygenated saline with 200 mg. per cent of glucose. After this preliminary incubation the slices were transferred to the experimental vessels. The preliminary incubation was carried out in order to remove from the slices a considerable quantity of amylase which is rapidly discharged into the medium and which is probably mainly derived from damaged cells.

In a few experiments in which measurements of respiration were carried out, the slices were incubated in Medium III of Krebs (4) in Warburg vessels gassed with oxygen. The sodium salts of the organic acids in Medium III were replaced by an equivalent quantity of NaCl. In the majority of experiments the slices were incubated in bicarbonate-saline (5), gassed with 7 per cent CO_2 in O_2, in Warburg vessels or stoppered conical flasks of 30 ml. capacity. In anaerobic experiments slices were incubated in bicarbonate-saline gassed with 7 per cent CO_2 in N_2; yellow phosphorus was placed in the center well or side arm. Alkali and filter paper were placed in the center well when Medium III was used. The slices were shaken in a Warburg bath at 40° for 2 hours.

The following drugs were added to give final concentrations as indicated in Tables I to IV: carbamylcholine, pilocarpine, acetylcholine, eserine, and atropine. Approximately 20 μc. of P^{32} as phosphate were added to each vessel.

Treatment of Tissues after Incubation—Immediately after incubation the tissues were placed in conical 15 ml. centrifuge tubes surrounded by iced water. They were then ground with sand and about 0.2 ml. of water in a mortar surrounded by chipped ice. After the mixture was ground to a pasty consistency, 2.8 ml. of cold water were added, and the mixture was further ground for about 30 seconds. 1.0 ml. of cold 20 per cent trichloroacetic acid was then added. The suspension was stirred and poured back into the centrifuge tube. The mortar was rinsed with 1 ml. of 20 per cent cold trichloroacetic acid, the rinse was added to the centrifuge tube, and the mixture was then centrifuged. The supernatant fluid was poured into another conical centrifuge tube for estimation of the specific activity of the acid-soluble phosphate esters, as described below. The acid-insoluble residue in the first tube was washed twice with 5 ml. portions of cold 10 per cent trichloroacetic acid. All operations were carried out at approximately 0°.

Extraction of Phospholipides—The washed trichloroacetic acid-insoluble residue obtained above was extracted at room temperature with 5 ml. of 95 per cent ethanol. The mixture was centrifuged, and the residue was extracted a second time with 95 per cent ethanol. The residue was then extracted twice for about 5 minutes with 5 ml. portions of a 3:1 ethanol-ether mixture at 60°. The ethanol and ethanol-ether extracts were pooled.

The pooled phospholipide extracts were then freed of contaminating inorganic P^{32} by a method similar to that previously described by Fishler *et al.* (6) and Friedkin and Lehninger (7). The method was carried out as follows: To each of the pooled ethanol-ether extracts was added 1 ml. of M Na_2HPO_4. The extracts were then evaporated to about 1 ml. under an infra-red lamp. The residues were extracted three times with 15 ml. portions of freshly distilled ether. Under these conditions no measurable inorganic phosphate was extracted by the ether. The ether extracts were pooled. To each of the ether extracts was added another 1 ml. of M Na_2HPO_4, and the mixture was again evaporated to 1 ml. The extraction with ether, addition of molar phosphate, and evaporation to 1 ml. were repeated again. This was followed by yet a further extraction with ether as above, addition of 1 ml. of water, and evaporation to about 10 ml. The remaining 10 ml. were then extracted again with three 10 ml. portions of ether. These extracts were evaporated to about 4 ml. Aliquots of the final concentrated ether extract were evaporated on aluminum disks (area 3.8 sq. cm.) with slightly elevated edges and counted at infinite thinness in a helium flow counter. All counts were corrected for background and dead time. A standard plate of the inorganic P^{32} sample added to the medium was always counted along with the phospholipide samples.

Another aliquot of the ether extract of the phospholipides was pipetted into a 30 ml. Kjeldahl flask and dried under an infra-red lamp. Total phosphorus was then estimated by a modification (8) of the method of Fiske and Subbarow (9). All specific activities are expressed as counts per minute per microgram of P corrected to an initial specific activity of 100,000 c.p.m. per γ of P for the inorganic P in the medium. Medium III contained 108 γ of P per ml.; bicarbonate-saline contained 37 γ of P per ml.

Determination of Specific Activity of Acid-Soluble Phosphate Esters—The cold trichloroacetic acid tissue extract was treated with magnesia mixture to remove inorganic phosphate (7). A 0.1 ml. aliquot of the remaining acid-soluble phosphate ester fraction was evaporated on an aluminum disk and counted. Total phosphorus was estimated on a 2 ml. sample.

Amylase Assays—0.2 ml. of the incubation medium was diluted to 10 ml. and assayed for amylase activity by a modification (3) of the method of Smith and Roe (10). Amylase activities are expressed as units of Smith and Roe per mg. of initial wet weight of tissue.

Results

Effects of Cholinergic Drugs and Atropine on Enzyme Secretion and Uptake of P^{32} into Phospholipides—When enzyme secretion was stimulated in pancreas tissue by the addition of carbamylcholine, there was a marked increase in the rate of incorporation of P^{32} into the phospholipide fraction. After 2 hours of incubation the specific activities of the phospholipides of slices stimulated by carbamylcholine were 4.8 to 8.7 times greater (average, 7.0 in eight experiments) than the specific activities of the phospholipides of the control slices. Fig. 1 shows the increase in specific activity of phospholipides of stimulated and unstimulated slices during the 2 hour

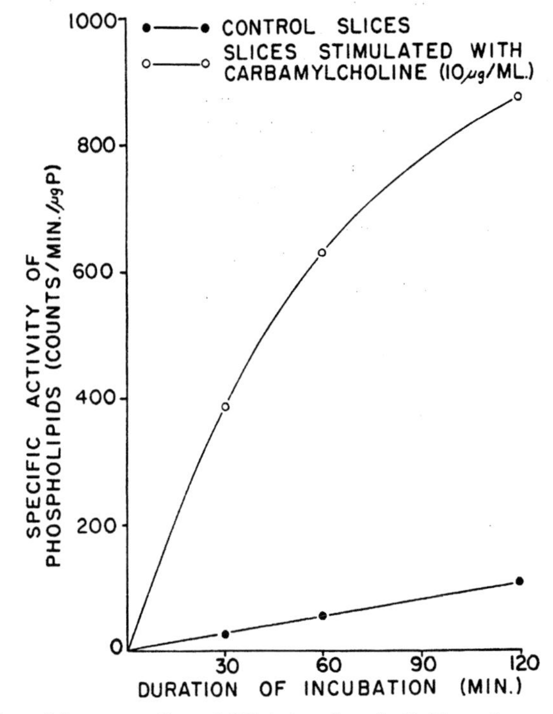

FIG. 1. Rates of incorporation of P^{32} into phospholipides of secreting and non-secreting slices of pigeon pancreas. Medium, bicarbonate-saline. Other conditions of incubation as described in the text. All counts per minute per microgram of P corrected to 100,000 c.p.m. per γ of P for the inorganic P in the medium.

incubation period. Table I gives the quantities of amylase in the medium and the specific activities of the acid-soluble phosphate esters in the same experiments. As described previously, amylase secretion was stimulated by carbamylcholine (1). The major portion of the amylase in the medium of unstimulated slices from glands of fasted pigeons was probably due to passive discharge of the enzyme rather than to active secretion, since it was not reduced under anaerobic conditions. On the other hand, the increment of amylase discharged into the medium in response to cholinergic drugs is abolished under anaerobic conditions (1) and may thus be regarded as true secretion. There was no increase in the specific activities of the acid-soluble phosphate esters in slices in contact with carbamylcholine. This suggests that the increased incorporation of P^{32} into the phospholipides of secreting slices is specific and not merely due to an increased permeability of the cell to phosphate or to an increased rate of phosphorylation. The fact that the specific activities of the acid-soluble phosphate esters were still rising after 2 hours shows that they had not reached equilibrium with inorganic phosphate of the medium.

In connection with the lack of stimulatory effect of carbamylcholine on the rate of incorporation of P^{32} into the acid-soluble phosphate esters, it should be pointed out that cholinergic drugs did not stimulate respiration of pigeon pancreas slices under these conditions. Deutsch and Raper (11) have reported a stimulation of respiration by secretin in slices of cat pancreas and by cholinergic drugs in slices of cat parotid and submaxillary glands. The effect of cholinergic drugs on respiration in cat pancreas was not studied by these workers. In the pancreas, secretin stimulates ion and water secretion and thus differs from vagal stimulation or cholinergic drugs, which stimulate only enzyme secretion (12). The observations of Deutsch and Raper and those of the authors can best be explained by assuming that water and ion secretion, which is stimulated by secretin, requires considerably more energy than does enzyme secretion, which is stimulated by cholinergic drugs.

TABLE I

Amylase Secretion and Specific Activity of Acid-Soluble Organic P in Experiment Illustrated in Fig. 1

Duration of incubation	Amylase in medium, units per mg. fresh weight of tissue		Specific activity of acid-soluble organic P, c.p.m. per γ P	
	Without carbamylcholine	With carbamylcholine	Without carbamylcholine	With carbamylcholine
min.				
30	8.2	14.7	1810	1670
60	9.8	17.0	2520	2300
120	13.9	25.7	5180	4520

No difference between slices of fed and fasted pigeons was observed in the percentage stimulation of P^{32} incorporation into phospholipides by carbamylcholine. However, the actual specific activities of phospholipides and acid-soluble phosphate esters of slices from glands of fed pigeons tended to be higher than those of glands from fasted pigeons. The respiratory rates of slices from fed pigeons also tended to be higher than those from fasted pigeons.

A few experiments on the effect of carbamylcholine on amylase secretion and on the uptake of P^{32} into the phospholipides were performed on slices of duck pancreas. The results were similar to those reported here for slices of pigeon pancreas.

TABLE II

Comparative Effects of Cholinergic Drugs and Atropine on Amylase Secretion and Incorporation of P^{32} into Phospholipides

Experiment No.	Additions	Final concentration, γ per ml.	Amylase in medium, units per mg. fresh weight of tissue	Specific activity of phospholipides, c.p.m. per γ P	Specific activity of acid-soluble organic P, c.p.m. per γ P
1	None		9.4	71	5,520
	Carbamylcholine	100	13.6	626	5,520
	Pilocarpine	100	11.6	164	5,880
	Acetylcholine Eserine	100 100	12.9	772	7,160
2	None		13.0	148	9,700
	Carbamylcholine	10	25.6	1108	
	Pilocarpine	10	16.5	212	10,400
	Acetylcholine Eserine	10 10	23.6	1096	9,800
	Atropine Carbamylcholine	10 10	11.6	136	9,500

Medium, bicarbonate-saline; duration of incubation, 2 hours. All counts per minute per microgram of P corrected to 100,000 c.p.m. per γ of P for the inorganic P in the medium.

TABLE III

Incorporation of P^{32} into Phospholipides of Pigeon Pancreas under Anaerobic Conditions

Experiment No.	Medium	Gas phase	Additions	Final concentration, γ per ml.	Specific activity of phospholipides, c.p.m. per γ P	Specific activity of acid-soluble organic P, c.p.m. per γ P
1	Bicarbonate-saline	93% N_2 + 7% CO_2	None		3	
		Same	Carbamylcholine	10	4	
2	Krebs' Medium III	O_2	None		48	4600
		"	Carbamylcholine	100	390	4800
		N_2	None		2	1510
		"	Carbamylcholine	100	3	1740

Duration of incubation, 2 hours. All counts per minute per microgram of P corrected to 100,000 c.p.m. per γ of P for the inorganic P in the medium.

In Table II are compared the effects of acetylcholine (with eserine), pilocarpine, carbamylcholine, and carbamylcholine with atropine on amylase secretion and the uptake of P^{32} into phospholipides. Acetylcholine (with eserine) stimulated amylase secretion and the incorporation of P^{32} into phospholipides to about the same extent as did carbamylcholine. Pilocarpine was much less effective in stimulating the secretion of amylase, as previously reported (1); the effect of this drug on the uptake of P^{32} into the phospholipides was also much less. The stimulatory effects of carbamylcholine on both amylase secretion and the incorporation of P^{32} into the phospholipides were abolished by atropine. None of the above drugs had any significant effect on the incorporation of P^{32} into the acid-soluble phosphate esters.

Incorporation of P^{32} into Phospholipides of Pancreas Slices under Anaerobic Conditions—The specific activities of phospholipides and acid-soluble phosphate esters observed after incubation of slices of pigeon pancreas under anaerobic conditions for 2 hours are shown in Table III. Although the anaerobic incorporation of P^{32} into the acid-soluble phosphate esters was as much as 34 per cent of the aerobic incorporation, the anaerobic incorporation of P^{32} into the phospholipides was less than 5 per cent of the aerobic incorporation. This suggests that the incorporation of P^{32} into phospholipides is due mainly to phospholipide synthesis, rather than to an enzyme-catalyzed exchange of phosphate in preformed phospholipide. This also follows from a consideration of the structure of phospholipides, in which the phosphate is doubly esterified and does not occupy a terminal position from which it could readily exchange. The proportionately higher anaerobic incorporation of P^{32} into the acid-soluble phosphate esters as compared to phospholipides is probably due to the presence in the acid-soluble fraction of many of the phosphorylated intermediates of glycolysis.

The marked stimulation of P^{32} uptake into phospholipides by cholinergic drugs, observed aerobically, did not occur anaerobically. There may have been some slight increase under anaerobic conditions, but, in view of the very low specific activities being measured under these conditions, it is doubtful whether the difference is significant.

Effects of Cholinergic Drugs on Incorporation of P^{32} into Phospholipides of Tissues Other Than Pancreas—The increased incorporation of P^{32} into phospholipides of secreting pancreas slices suggested that phospholipides play a rôle in enzyme secretion. However, to investigate the possibility that the stimulation of P^{32} uptake into phospholipides by cholinergic drugs might be a general phenomenon, slices of tissue from organs other than pancreas were incubated with and without cholinergic drugs under the same conditions as those used for pancreas slices. The specific activities of the phospholipides, acid-soluble phosphate esters, and, when measured, the respiration of these tissues are shown in Table IV. Cholinergic drugs were found to have little or no effect on the specific activities of phospholipides of slices of pigeon and guinea pig liver, guinea pig kidney cortex, guinea pig heart ventricle, and pigeon gizzard (smooth muscle). A relatively slight stimulation of P^{32} uptake into phospholipides was observed in slices

TABLE IV

Effect of Cholinergic Drugs on Incorporation of P^{32} into Phospholipides of Slices of Various Tissues Other Than Pigeon Pancreas

Species	Tissue	Additions	Final concentration, γ per ml.	Specific activity of phospholipides, c.p.m. per γ P	Specific activity of acid-soluble organic P, c.p.m. per γ P	Q_{O_2}
Pigeon	Liver	None		992	24,400	8.5
		Carbamylcholine	10	1010	32,800	8.3
Guinea pig	"	None		588		
		Acetylcholine	100	594		
		Eserine	100			
" "	Kidney cortex	None		644	13,500	
		Carbamylcholine	100	616	15,800	
" "	Heart ventricle	None		60	9,800	
		Acetylcholine	100	72	11,320	
		Eserine	100			
" "	" "	None		70	13,100	
		Carbamylcholine	100	66	14,700	
Pigeon	Gizzard, smooth muscle	None		1260	24,200	2.8
		Carbamylcholine	10	1364	32,800	2.9
"	Whole brain	None		214	22,400	8.1
		Carbamylcholine	10	352	21,400	8.1
Guinea pig	Brain cortex	None		272	18,300	
		Carbamylcholine	100	356	19,600	

Medium for experiments in which Q_{O_2} is reported, Krebs' Medium III; bicarbonate-saline medium in other experiments. Duration of incubation, 2 hours. All counts per minute per microgram of P corrected to 100,000 c.p.m. per γ of P for the inorganic P in the medium.

of pigeon brain (65 per cent) and guinea pig brain cortex (40 per cent). This possible effect of cholinergic drugs on the incorporation of P^{32} into phospholipides in brain is being investigated further. It is not known whether any of these other tissues studied respond to cholinergic drugs *in vitro*, but the results suggest that the effect of cholinergic drugs on the rate of incorporation of P^{32} into the phospholipides of pancreas is related to physiological function.

Incorporation of P^{32} into Phospholipides during Stimulation of Amylase

Synthesis with Amino Acids—Amylase synthesis by slices of pigeon pancreas can be stimulated by the addition of an appropriate amino acid mixture (1, 3). The stimulation of amylase synthesis is not accompanied by any appreciable stimulation of amylase secretion (1). It was thought of interest to study correlations between amylase synthesis and the uptake of P^{32} into phospholipides. The results of such a study are shown in Table V. In this experiment amylase synthesis was increased almost 3-fold by amino acids, but there was no significant increase in the specific activity of the phospholipides.

TABLE V

Specific Activities of Phospholipides after Stimulation of Amylase Synthesis with Amino Acids

Incubation period	Additions	Total amylase (medium + tissue), units per mg. fresh weight	Amylase synthesized, units per mg. fresh weight	Specific activity of phospholipides, c.p.m. per γ P
min.				
0		15.3, 16.8		
120		21.5, 22.4	5.9	58
120	Amino acid mixture*	32.5, 29.2	16.1	62

Medium, bicarbonate-saline; duration of incubation, 2 hours. All counts per minute per microgram of P corrected to 100,000 c.p.m. per γ of P for the inorganic P in the medium.

* 0.2 ml. of Aminosol (Abbott Laboratories, Chicago, Illinois).

DISCUSSION

The experiments reported here suggest that the process of enzyme secretion is associated with accelerated phospholipide synthesis. It is too early to state with certainty how phospholipide synthesis may be related to enzyme secretion. However, the observations of Ling *et al.* (13) may be of significance in relation to this problem. These workers found that pepsin secretion by the stomach was paralleled by phospholipide secretion. It is, therefore, possible that the accelerated rate of phospholipide synthesis in the secreting pancreas is secondary to the extrusion of phospholipides from the cell during enzyme secretion. However, it should be emphasized that the synthesis of amylase *in vitro* is not accelerated when amylase secretion is stimulated in slices of pigeon pancreas (1). Since the classical work of Heidenhain (14) it has been known that the enzymes are stored in the pancreas in the form of zymogen granules. Claude (15) has found that these granules contain about 20 per cent lipide, mostly phospholipide. There is no clear cut evidence that the zymogen granules are secreted as such. Many cytological studies indicate that the granular material is concentrated into vacuoles during the secretory process and the vacuolar contents are then extruded (16, 17). Irrespective of the cytological form in which the enzymes are secreted, it is reasonable to assume as a working hypothesis that these enzymes are attached to phospholipides in the resting cell and remain attached during their journey into the glandular lumen. According to this view the enzymes would be secreted, not as free enzymes, but as enzyme-phospholipide complexes, or lipoproteins. In this way the destructive action of these enzymes on the cytoplasm during the secretory process would be avoided.

Xeros (18) has recently provided good evidence that the classical Golgi apparatus observed in tissues after treatment with fixatives represents, in the unfixed cell, lipide droplets or lipochondria. These lipochondria lie at the base of the mass of zymogen granules, which occupy the apical regions of the acinar cell. Both the Golgi apparatus of fixed tissues and the lipochondria of unfixed tissues have been observed to undergo cytological changes during enzyme secretion[1] (19). These lipide structures may thus play a rôle in the secretory process; the evidence presented here, linking phospholipides to enzyme secretion, is consistent with this view.

SUMMARY

1. When enzyme secretion was stimulated by carbamylcholine or acetylcholine (with eserine) in slices of pigeon pancreas, the incorporation of P^{32} into the phospholipide fraction of the stimulated slices was, after 2 hours, 4.8 to 8.7 (average, 7.0) times greater than the incorporation of P^{32} into the phospholipides of control slices. Neither respiration nor the incorporation of P^{32} into acid-soluble phosphate esters was increased.
2. Pilocarpine, which on a weight for weight basis was much less effective than carbamylcholine or acetylcholine in stimulating enzyme secretion in pancreas slices, was also much less effective in stimulating the uptake of P^{32} into phospholipides.
3. The stimulatory effects of carbamylcholine on both enzyme secretion and the incorporation of P^{32} into phospholipides were abolished by atropine.
4. The specific activity of the phospholipides from slices incubated anaerobically was less than 5 per cent of that observed aerobically. Anaerobically, carbamylcholine did not stimulate the incorporation of P^{32} into phospholipides to any significant extent. The specific activity of the acid-soluble phosphate esters after anaerobic incubation was 34 per cent of that found aerobically.

[1] Xeros, N., personal communication (1951).

5. Cholinergic drugs had little or no effect on the incorporation of P^{32} into the phospholipides of the following tissue slices: pigeon and guinea pig liver, guinea pig heart ventricle, pigeon gizzard (smooth muscle), and guinea pig kidney cortex. A relatively slight stimulation of P^{32} uptake into phospholipides was observed in slices of pigeon brain (65 per cent) and guinea pig brain cortex (40 per cent).

6. Stimulation of amylase synthesis in slices of pigeon pancreas by the addition of a mixture of amino acids had no effect on the incorporation of P^{32} into phospholipides.

The authors wish to thank Dr. J. H. Quastel for his interest and encouragement.

BIBLIOGRAPHY

1. Hokin, L. E., *Biochem. J.*, **48**, 320 (1951).
2. Hokin, L. E., *Biochim. et biophys. acta*, **8**, 224 (1952).
3. Hokin, L. E., *Biochem. J.*, **50**, 216 (1951).
4. Krebs, H. A., *Biochim. et biophys. acta*, **4**, 249 (1950).
5. Krebs, H. A., and Henseleit, K., *Z. physiol. Chem.*, **210**, 33 (1932).
6. Fishler, M. C., Entenman, C., Montgomery, M. L., and Chaikoff, I. L., *J. Biol. Chem.*, **150**, 47 (1943).
7. Friedkin, M., and Lehninger, A. L., *J. Biol. Chem.*, **177**, 775 (1949).
8. LePage, G. A., in Umbreit, W. W., Burris, R. H., and Stauffer, J. F., Manometric techniques and tissue metabolism, Minneapolis (1949).
9. Fiske, C. H., and Subbarow, Y., *J. Biol. Chem.*, **66**, 375 (1925).
10. Smith, B. W., and Roe, J. H., *J. Biol. Chem.*, **179**, 53 (1949).
11. Deutsch, W., and Raper, H. S., *J. Physiol.*, **87**, 275 (1936).
12. Babkin, B. P., Secretory mechanism of the digestive glands, New York, 2nd edition, 846 (1950).
13. Ling, S. M., Liu, A. C., and Lim, R. K. S., *Chinese J. Physiol.*, **2**, 305 (1928); cited by Babkin (12).
14. Heidenhain, R., *Arch. ges. Physiol.*, **10**, 557 (1875).
15. Claude, A., *Biol. Symposia*, **10**, 111 (1943).
16. Babkin, B. P., Rubashkin, V. J., and Savitch, V. V., *Arch. mikr. Anat.*, **12**, 297 (1909).
17. Covell, W. P., *Anat. Rec.*, **40**, 213 (1928).
18. Xeros, N., *Nature*, **167**, 448 (1950).
19. Nassonov, D. N., *Arch. mikr. Anat.*, **100**, 433 (1924).

Streb et al. (1983) Nature (London) **306**, 67–69

Release of Ca^{2+} from a nonmitochondrial intracellular store in pancreatic acinar cells by inositol-1,4,5-trisphosphate

H. Streb*, R. F. Irvine†, M. J. Berridge‡ & I. Schulz*

* Max-Planck-Institut für Biophysik, Kennedyallee 70, 6000 Frankfurt (Main) 70, FRG
† Department of Biochemistry, ARC Institute of Animal Physiology, Babraham, Cambridge CB2 4AT, UK
‡ ARC Unit of Insect Neurophysiology and Pharmacology, Department of Zoology, University of Cambridge, Downing Street, Cambridge CB2 3EJ, UK

Activation of receptors for a wide variety of hormones and neurotransmitters leads to an increase in the intracellular level of calcium. Much of this calcium is released from intracellular stores but the link between surface receptors and this internal calcium reservoir is unknown. Hydrolysis of the phosphoinositides, which is another characteristic feature of these receptors[1–3], has been implicated in calcium mobilization[1]. The primary lipid substrates for the receptor mechanism seem to be two polyphosphoinositides, phosphatidylinositol 4-phosphate (PtdIns4P) and phosphatidylinositol 4,5-bisphosphate ($PtdIns4,5P_2$), which are rapidly hydrolysed following receptor activation in various cells and tissues[4–10]. The action of phospholipase C on these polyphosphoinositides results in the rapid formation of the water-soluble products inositol 1,4-bisphosphate ($Ins1,4P_2$) and inositol 1,4,5-trisphosphate ($Ins1,4,5P_3$)[9,11,12]. In the insect salivary gland, where changes in $Ins1,4P_2$ and $Ins1,4,5P_3$ have been studied at early time periods, increases in these inositol phosphates are sufficiently rapid to suggest that they might mobilize internal calcium[12]. We report here that micromolar concentrations of $Ins1,4,5P_3$ release Ca^{2+} from a nonmitochondrial intracellular Ca^{2+} store in pancreatic acinar cells. Our results strongly suggest that this is the same Ca^{2+} store that is released by acetylcholine.

The effect of inositol phosphates on intracellular Ca^{2+} stores was studied using isolated rat pancreatic acinar cells with permeable plasma membranes as described previously[13]. The plasma membrane was permeabilized by washing cells with a nominally Ca^{2+}-free solution. Cells were then incubated in a high-potassium solution with respiratory substrates and ATP. Ca^{2+} uptake or release from intracellular Ca^{2+} stores was determined by measuring the changes in free Ca^{2+} concentration of the incubation medium with a Ca^{2+}-specific macroelectrode[14]. In these conditions, intracellular organelles of the leaky cells take up Ca^{2+} from the medium until a steady state is reached at an external free calcium concentration of $\sim 0.4\ \mu mol\ l^{-1}$ (Fig. 1). Subsequent addition of $Ins1,4,5P_3$ (isolated from red blood cell ghosts as described in Table 1 legend) resulted in an increase of medium $[Ca^{2+}]$ followed by re-uptake to the original steady-state concentration (Fig. 1). In the absence of cells, the addition of $Ins1,4,5P_3$ produced only a small increase of medium $[Ca^{2+}]$. This Ca^{2+} contamination of $Ins1,4,5P_3$ accounted for less than 5% of the effect shown in Fig. 1 and was corrected for in the numerical data. If cells were preincubated with the Ca^{2+} ionophore A23187 ($10\ \mu mol\ l^{-1}$) to deplete intracellular Ca^{2+} stores, addition of $Ins1,4,5P_3$ resulted in the same small increase of medium $[Ca^{2+}]$ as observed in the absence of cells. These results show that the increase in medium $[Ca^{2+}]$ following addition of $Ins1,4,5P_3$ in the presence of cells is indeed due to Ca^{2+} release from membrane-bound cellular Ca^{2+} stores.

The possibility that calcium might be derived from the release of secretion granules was excluded by the observation that no amylase was detected following stimulation with either carbachol[13] or $Ins1,4,5P_3$ (this study). To exclude the possibility that calcium release might be due to contaminants introduced during the preparation of $Ins1,4,5P_3$, some of the red blood cell ghosts were taken through the complete preparative procedure (see Table 1 legend) except that they were incubated in calcium-

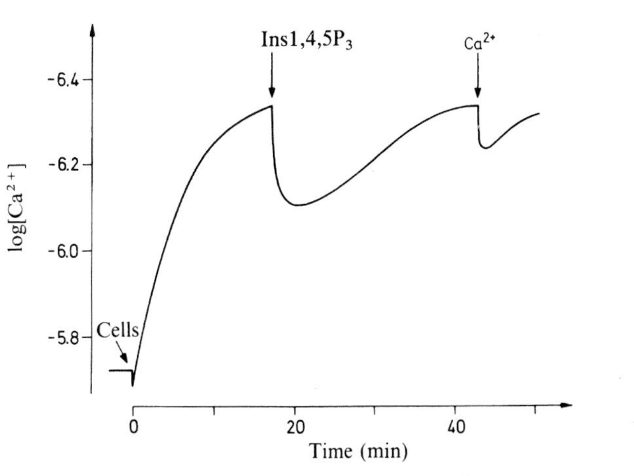

Fig. 1 Effect of $Ins1,4,5P_3$ on intracellular Ca^{2+} stores of leaky acinar cells. Isolated acinar cells from rat pancreas were prepared as described elsewhere[13]. Their plasma membrane was permeabilized by washing with a nominally Ca^{2+}-free solution (Trypan blue uptake about 70%). Cells (6.9 mg protein) were added to 3 ml of the following solution ($mmol\ l^{-1}$): 110 KCl, 6 $MgCl_2$, 5 K-succinate, 5 K-pyruvate, 25 HEPES, 5 K_2ATP, 10 creatine phosphate, $10\ U\ ml^{-1}$ creatine kinase, *p*H 7.4, at 25 °C. The medium $[Ca^{2+}]$ was recorded continually with a Ca^{2+}-specific macroelectrode (neutral carrier ETH 1001). Electrodes were made and calibrated as described before[13,14]. $Ins1,4,5P_3$ (see Table 1 legend) to a final concentration of $2.5\ \mu mol\ l^{-1}$ and $CaCl_2$ (10 nmol) were added where indicated. This trace is representative of a typical experiment that was repeated at least 20 times.

free conditions to prevent the formation of inositol phosphates. Control samples prepared in this way failed to release calcium from pancreatic acinar cells.

The $Ins1,4,5P_3$-induced release of calcium showed a steep dose–response relationship with a clear effect at concentrations as low as 0.2 μM and a near-maximum release at 5 μM (4.3 ± s.e. 0.5 nmol per mg protein, $n = 16$; Fig. 2). A double reciprocal plot yielded an apparent K_m of 1.1 μM (Fig. 2, inset). Weiss *et al.*[7] have shown that the breakdown of $PtdIns4,5P_2$ in stimulated parotid cells is of the order of 0.3 nmol per mg protein per min. This would give an intracellular concentration of $Ins1,4,5P_3$ of ~30 μM which is in excess of the doses required to release calcium in these permeabilized cells.

The specificity of the calcium-releasing response was studied by testing the effect of $Ins1,4P_2$, inositol 1-phosphate (Ins1P), inositol 1,2-cyclic phosphate (cyclic IMP) and *myo*-inositol; none of these substances was able to release Ca^{2+} from intracellular stores (Table 1). When tested at the higher concentration of $5\ \mu mol\ l^{-1}$, $Ins1,4P_2$ induced release of ~10% as much calcium as did the control ($Ins1,4,5P_3$), which can be fully accounted for by $Ins1,4,5P_3$ contamination of $Ins1,4P_2$.

If the $Ins1,4,5P_3$ sample was hydrolysed at 100 °C for 30 min in 5 M HCl (conditions which randomize the phosphates by bond migration[11,15]), its potency to release Ca^{2+} at a concentration of 1 μM decreased to <50% of the control (100 °C, 30 min, no acid). Thus, although quantitative conclusions cannot be drawn from these observations, they clearly demonstrate that the calcium-releasing activity depends on the distribution of phosphates on the inositol ring.

In a previous study on leaky acinar cells, it was shown that Ca^{2+} was taken up both by mitochondria and by at least one nonmitochondrial ATP-dependent Ca^{2+} store[13]. Mitochondrial Ca^{2+} uptake could be abolished by a combination of the mitochondrial inhibitors antimycin A ($10\ \mu mol\ l^{-1}$) and oligomycin ($5\ \mu mol\ l^{-1}$), whereas nonmitochondrial Ca^{2+} uptake was completely inhibited by high concentrations of the ATPase inhibitor vanadate ($2\ mmol\ l^{-1}$)[13]. Two separate calcium stores have also been identified in rat[16] and guinea pig[17] pancreas after permeabilization with saponin or digitonin. To identify which intracellular calcium store was sensitive to $Ins1,4,5P_3$, we compared the effect of these inhibitors on calcium release in our system. Incubation of cells in the presence of

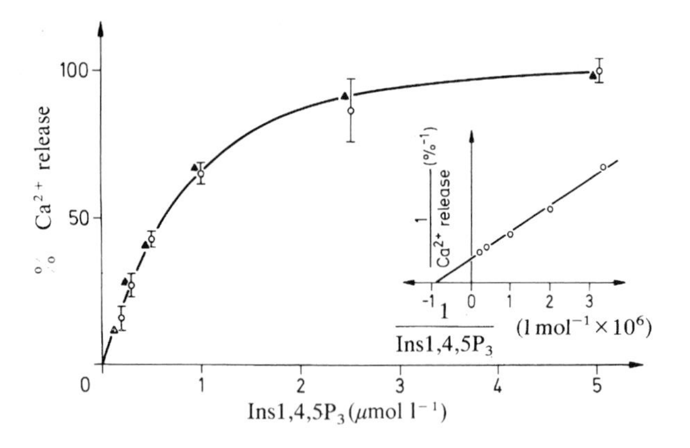

Fig. 2 Dose–response relationship of Ins1,4,5P_3-induced Ca^{2+} release. Ins1,4,5P_3-induced Ca^{2+} release was determined by comparison to known Ca^{2+} additions as described in Fig. 1 legend in either the standard incubation medium (○; mean ±s.e. of three different cell preparations) or the standard incubation medium supplemented with antimycin (10 μmol l^{-1}) and oligomycin (5 μmol l^{-1}) (▲; one experiment). Inset: double-reciprocal plot of the data without inhibitors.

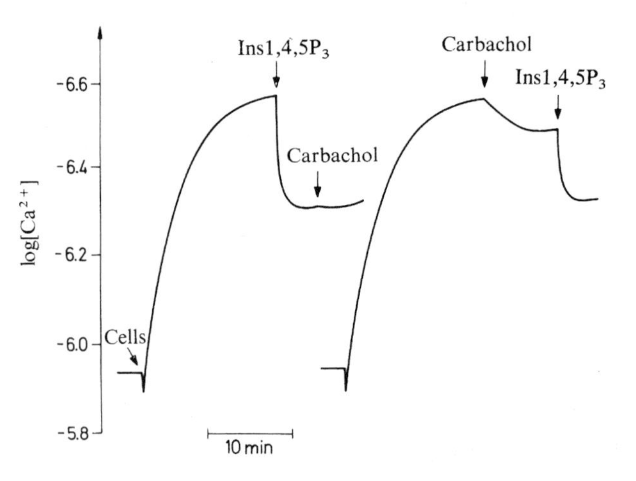

Fig. 3 Effect of sequential additions of Ins1,4,5P_3 and carbamylcholine. Isolated cells (2.5 mg protein ml^{-1}) were prepared and incubated as described in Fig. 1 legend. Where indicated carbachol (final concentration 40 μmol l^{-1}) or Ins1,4,5P_3 (final concentration 5 μmol l^{-1}) were added. Typical for four determinations.

mitochondrial inhibitors did not impair the ability of Ins1,4,5P_3 to release Ca^{2+} (Table 1). The dose–response curve to Ins1,4,5P_3 was similar both in the presence and in the absence of antimycin and oligomycin (Fig. 2). This strongly suggests that Ins1,4,5P_3 does not release Ca^{2+} from mitochondria. In contrast, a clear, though not complete, inhibition was observed with the ATPase inhibitor vanadate, which inhibits Ca^{2+} uptake and thus reduces the amount of calcium in the nonmitochondrial pool (Table 1). Which of the nonmitochondrial pools is sensitive to inositol phosphates is not yet clear however.

Leaky cells retain the ability to react to carbamylcholine (carbachol) with Ca^{2+} release from intracellular stores[13]. It was possible, therefore, to test the interaction of the release mechanisms sensitive to carbachol and Ins1,4,5P_3. Addition of carbachol, like Ins1,4,5P_3, results in Ca^{2+} release from leaky cells followed by re-uptake to the prestimulation value[13]. If carbachol (40 μmol l^{-1}) was added after Ins1,4,5P_3 (5 μmol l^{-1}), the carbachol-induced Ca^{2+} release was completely abolished (Fig. 3). If, on the other hand, Ins1,4,5P_3 was added after carbachol, the Ins1,4,5P_3-induced release of calcium was decreased by about one-third (Fig. 3, Table 1). The sum of the Ca^{2+} released by carbachol and Ins1,4,5P_3 was constant. Thus, in the presence of saturating concentrations of exogenous Ins1,4,5P_3, carbachol can no longer release Ca^{2+}, suggesting that both agents are acting on the same pool of releasable calcium. This in turn indicates that the carbachol-induced release of Ca^{2+} from the intracellular stores ('trigger Ca^{2+} pool') of these permeabilized pancreatic cells may be mediated by Ins1,4,5P_3.

A characteristic feature of calcium-mobilizing receptors is that they induce a rapid decline in the polyphosphoinositides[5–10] with a corresponding increase in the levels of Ins1,4,5P_3 and Ins1,4P_2 (refs 9, 11, 12). Despite the probable relationship between this phosphoinositide effect and calcium signalling, the link between these two events has been missing. Our observations reported here may provide this missing link by suggesting that Ins1,4,5P_3 functions as second messenger to mobilize internal calcium. Whether phosphoinositides also regulate the permeability of the plasma membrane remains to be determined.

Table 1 Calcium release by inositol phosphates and the effects of carbachol and metabolic inhibitors

Additions	Ca^{2+} release (% of Ins1,4,5P_3 control)	
Ins1,4,5P_3 (5 μM)	100	(n = 12)
Ins1,4P_2 (1 μM)	2 ± 1	(n = 4)
Ins1P (5 μM)	3	(n = 2)
Cyclic IMP (5 μM)	3	(n = 2)
myo-Inositol (100 μM)	0	(n = 2)
Ins1,4,5P_3 (5 μM) + antimycin A (10 μM) + oligomycin (5 μM)	103 ± 11	(n = 4)
Ins1,4,5P_3 (5 μM) + vanadate (2 mM)	42 ± 10	(n = 5)
Ins1,4,5P_3 (5 μM) + carbachol (40 μM)	57 ± 4	(n = 4)

Calcium release was determined as described in Fig. 1 legend. The amount of calcium released is expressed as a percentage of that released by Ins1,4,5P_3 using the same cell preparation. The inhibitors antimycin A, oligomycin and vanadate were present in the incubation medium before addition of cells (about 20 min before addition of Ins1,4,5P_3). In the carbachol experiment, cells were stimulated with carbachol about 10 min before addition of Ins1,4,5P_3 as shown in Fig. 3. Results are expressed as mean ±s.e. Ins1,4,5P_3 and Ins1,4P_2 were prepared by incubating human red blood cell ghosts with $CaCl_2$ followed by a Dowex-formate column separation, and desalted by elution from a Dowex-Cl column with 1M LiCl, followed by removal of the LiCl with ethanol[18]. Ins1P and cyclic IMP were prepared by ionophoretic separation[19] of the products resulting from enzymatic hydrolysis of yeast PtdIns[19] by a rat brain[20] or celery[21] soluble supernatant.

We thank A. J. Letcher for help with the inositol phosphate preparations; Professor K. J. Ullrich and Drs R. M. C. Dawson and C. P. Downes for helpful discussions; and the Cambridge Regional Blood Transfusion Centre for supply of blood. H. S. was supported in part by the Deutsche Gesellschaft zur Bekämpfung der Mucoviscidose and by the Deutsche Forschüngsgemeinschaft (grant No. Schu 429/2-1).

Received 28 June; accepted 25 August 1983.

1. Michell, R. H. *Biochim. biophys. Acta* **415,** 81–147 (1975).
2. Berridge, M. J. *Molec. cell. Endocr.* **24,** 115–140 (1981).
3. Hokin, M. R. & Hokin, L. E. *J. biol. Chem.* **203,** 967–977 (1953).
4. Abdel-Latif, A. A., Akhtar, R. A. & Hawthorne, J. N. *Biochem. J.* **162,** 61–73 (1977).
5. Kirk, C. J., Creba, J. A., Downes, P. & Michell, R. H. *Biochem. Soc. Trans.* **9,** 377–379 (1981).
6. Thomas, A. P., Marks, J. S., Coll, K. E. & Williamson, J. R. *J. biol. Chem.* **258,** 5716–5725 (1983).
7. Weiss, S. J., McKinney, J. S. & Putney, J. W. *Biochem. J.* **206,** 555–560 (1982).
8. Billah, M. M. & Lapetina, E. G. *J. biol. Chem.* **257,** 12705–12708 (1982).
9. Agranoff, B. W., Murthy, P. & Seguin, E. B. *J. biol. Chem.* **258,** 2076–2078 (1983).
10. Putney, J. W., Burgess, G. M., Halenda, S. P., McKinney, J. S. & Rubin, R. P. *Biochem. J.* **212,** 483–488 (1983).
11. Berridge, M. J., Dawson, R. M. C., Downes, C. P., Heslop, J. P. & Irvine, R. F. *Biochem. J.* **212,** 473–482 (1983).
12. Berridge, M. J. *Biochem. J.* **212,** 849–858 (1983).
13. Streb, H. & Schulz, I. *Am. J. Physiol.* (in the press).
14. Affolter, H. & Sigel, E. *Analyt. Biochem.* **97,** 315–319 (1979).
15. Tomlinson, R. V. & Ballou, C. E. *J. biol. Chem.* **236,** 1902–1906 (1961).
16. Wakasugi, H., Kimura, T., Haase, W., Kribben, A., Kaufmann, R. & Schulz, I. *J. Membrane Biol.* **65,** 205–220 (1982).
17. Lucas, M., Galvan, A., Solano, P. & Goberna, R. *Biochim. biophys. Acta* **731,** 129–136 (1982).
18. Downes, C. P., Mussat, M. C. & Michell, R. H. *Biochem. J.* **203,** 169–177 (1982).
19. Irvine, R. F., Hemington, N. & Dawson, R. M. C. *Biochem. J.* **176,** 475–484 (1978).
20. Irvine, R. F., Letcher, A. J. & Dawson, R. M. C. *Biochem. J.* **178,** 497–500 (1979).
21. Irvine, R. F., Letcher, A. J. & Dawson, R. M. C. *Biochem. J.* **192,** 279–283 (1980).

Cytosolic Ca^{2+} spiking

10

Cellular regulation by oscillations in the cytosolic free calcium ion concentration ($[Ca^{2+}]_i$) is a phenomenon known at the undergraduate textbook level. The classic example is the heart. The membrane potential is the primary controller regulating the opening of voltage-sensitive Ca^{2+} channels, and the repetitive depolarizations therefore elicit repetitive cytosolic Ca^{2+} transients that cause repetitive contractions. Many other cell types have Ca^{2+} oscillations of this kind. For example, in the insulin-secreting pancreatic β-cells, glucose-evoked oscillations in membrane potential control the periodic opening of voltage-sensitive Ca^{2+} channels causing repetitive cytosolic Ca^{2+} spikes, that give rise to pulsatile insulin secretion [1]. The Ca^{2+} oscillations (spiking) to be discussed here are of a different kind, where repetitive release of Ca^{2+} from intracellular stores gives rise to repetitive $[Ca^{2+}]_i$ spikes.

During the 1980s many investigators discovered that neurotransmitters or hormones could evoke $[Ca^{2+}]_i$ spikes in exocrine glands [2,3], fibroblasts [4], endothelial cells [5], HeLa cells [6], macrophages [7] and oocytes [8]. Our selected paper by Woods, Cuthbertson and Cobbold [9] demonstrated repetitive $[Ca^{2+}]_i$ spikes in hepatocytes evoked by the α-adrenergic agonist phenylephrine and the hormone vasopressin. This paper is generally regarded as the starting point for discussions of cytosolic Ca^{2+} oscillations [10]. Our reasons for signposting this paper as the key contribution, when several other investigators had discovered the same phenomenon in other cells at the same time [3,4] or even earlier [2], are based on the following: (i) the clarity of the records shown; (ii) the direct nature of the experiments (measurement of $[Ca^{2+}]_i$ using a Ca^{2+}-sensitive photo-protein injected into single isolated hepatocytes); and (iii) the demonstration that the frequency of Ca^{2+} spiking increased with increasing hormone concentration (although the latter had already been shown indirectly in *Calliphora* salivary glands [2]).

The repetitive cytosolic Ca^{2+} spikes evoked by hormones or neurotransmitters have important physiological effects as they can cause repetitive pulses of exocytotic secretion, as revealed by capacitance measurements [11]. In addition, they can activate Ca^{2+} pumps in the plasma membrane, evoking repetitive spikes of Ca^{2+} extrusion [12], and induce repetitive Ca^{2+} spikes in the mitochondria, thereby activating Ca^{2+}-sensitive mitochondrial dehydrogenases [13]. The decoding of cytosolic Ca^{2+} oscillations in the mitochondria has been studied in some detail. Each mitochondrial Ca^{2+} spike is sufficient to cause a maximal transient activation of the Ca^{2+}-sensitive mitochondrial dehydrogenases, and Ca^{2+} oscillations at frequencies above 0.5 per minute cause sustained activation of mitochondrial metabolism. In contrast, a sustained rise in $[Ca^{2+}]_i$ only evokes a transient mitochondrial dehydrogenase activation [13]. It seems likely that, as a general rule, the $[Ca^{2+}]_i$ rise, evoked by physiological hormone concentrations in their various target cells, consists of repetitive spikes rather than being sustained and that a continued elevation of $[Ca^{2+}]_i$ is harmful, for example by activating Ca^{2+}-sensitive proteases and particularly endonucleases [14]. It is known that abusive stimulation of cells leading to substantial and sustained $[Ca^{2+}]_i$ elevations can trigger apoptosis [14] and such toxic Ca^{2+} signals may trigger other disease processes, for example, acute pancreatitis evoked by excessive hormone stimulation [15].

It was believed initially that the hormone-evoked $[Ca^{2+}]_i$ oscillations occurred as relatively uniform rises in Ca^{2+} concentration throughout the cytosol. In hepatocytes, however, each Ca^{2+} transient is organized as a wave that originates at a specific subplasmalemmal region [16]. In addition, low (physiological) agonist concentrations evoke local cytosolic Ca^{2+} spikes in the secretory pole (secretory granule area) of single pancreatic acinar cells [17,18]. These subcellular spikes can occur alone or repetitively or can precede longer lasting Ca^{2+} signals that spread throughout the cells [17,18]. The short-lasting repetitive local Ca^{2+} spikes in the secretory pole, that can also be evoked by intracellular infusion of the Ca^{2+}-releasing messenger inositol 1,4,5-trisphosphate [Ins(1,4,5)P_3] [17], provide an economical signalling mechanism and are of physiological significance since they activate exocytosis and Ca^{2+}-dependent ionic currents important for fluid secretion in the exocrine glands [17].

The mechanisms underlying cytosolic Ca^{2+} spiking have been clarified to some extent. In general, spikes are initiated by release of Ca^{2+} from intracellular stores mediated via Ins(1,4,5)P_3 activation of Ca^{2+}-release channels [19]. Ins(1,4,5)P_3 is generated by hormone stimulation of phospholipase C [20]. It is known that a constant level of Ins(1,4,5)P_3 can evoke repetitive Ca^{2+} spikes [17] and a number of different theories have been advanced to account for this finding [19].

Ca^{2+} has many different target molecules, namely the Ca^{2+}-binding proteins (CaBPs). Co-ordinated activation of different CaBPs is essential for integrated cellular function which is achieved by the spatio-temporal organization of the Ca^{2+} signal and the heterogeneous distribution, affinity, kinetics and function of CaBPs. Ca^{2+} spikes have at least three advantages over a graded amplitude regulation of $[Ca^{2+}]_i$. First, Ca^{2+} spikes are more resistant to noise than graded rises in $[Ca^{2+}]_i$. Secondly, certain frequencies of Ca^{2+} spikes could selectively activate CaBPs with particular association and dissociation rate constants for Ca^{2+}. Thirdly, local short-lasting Ca^{2+} spikes may have advantages over a global rise from at least two points of view: i.e. the local spikes are both economical (the small amounts of Ca^{2+} released reduce the need for ATP-dependent Ca^{2+} re-uptake and extrusion) and prevent undesirable Ca^{2+}-dependent activation processes elsewhere in the cell. Since it is very difficult to envisage a sustained local rise in Ca^{2+} concentration, the latter point may by itself have forced the evolution of Ca^{2+} spiking mechanisms.

The very convincing records of repetitive transient rises in $[Ca^{2+}]_i$ provided in the paper by Woods, Cuthbertson and Cobbold [9] have had an enormous impact on the Ca^{2+} signalling field, and most investigators would now agree with the authors' assertion that: "It is an attractive possibility that the frequency of the transients determines the intensity of a cell's response to hormone concentrations within the physiological range".

References

1. Bokvist, K., Eliasson, L., Ammala, C., Renstrom, E. and Rorsman, P. (1995) *EMBO J.* **14**, 50–57
2. Rapp, P.E. and Berridge, M.J. (1981) *J. Exp. Biol.* **93**, 119–132
3. Evans, M.G. and Marty, A. (1986) *Proc. Natl. Acad. Sci. U.S.A.* **83**, 4099–4103
4. Ueda, S., Oiki, S. and Okada, Y. (1986) *J. Membr. Biol.* **91**, 65–72
5. Jacob, R., Merritt, J.E., Hallam, T.S. and Rink, T.J. (1988) *Nature (London)* **335**, 40–45
6. Sauve, R., Simoneau, C., Parent, L., Monette, R. and Roy, G. (1987) *J. Membr. Biol.* **96**, 199–208
7. Kruskal, B.A. and Maxfield, F.R. (1987) *J. Cell Biol.* **105**, 2685–2693
8. Cuthbertson, K.S.R. and Cobbold, P.H. (1985) *Nature (London)* **316**, 541–542
9. Woods, N.M., Cuthbertson, K.S.R. and Cobbold, P.H. (1986) *Nature (London)* **319**, 600–602
10. Berridge, M.J. (1995) in *Calcium Waves, Gradients and Oscillations*, pp. 1–3, CIBA Foundation Symposium 188, John Wiley and Sons, Chichester
11. Tse, A., Tse, F.W., Hille, B. and Almers, W. (1993) *Science* **260**, 82–84
12. Tepikin, A.V., Voronina, S.G., Gallacher, D.V. and Petersen, O.H. (1992) *J. Biol. Chem.* **267**, 14073–14076
13. Hajnoczky, G., Robb-Gaspers, L.D., Seitz, M.B. and Thomas, A.P. (1995) *Cell* **82**, 415–424
14. Nicotera, P., Bellomo, G. and Orrenius, S. (1992) *Annu. Rev. Pharmacol. Toxicol.* **32**, 449–470
15. Ward, J.B., Petersen, O.H., Jenkins, S.A. and Sutton, R. (1995) *Lancet* **346**, 1016–1019
16. Rooney, T.A., Sass, E. and Thomas, A.P. (1990) *J. Biol. Chem.* **265**, 10792–10796
17. Thorn, P., Lawrie, A.M., Smith, P.M., Gallacher, D.V. and Petersen, O.H. (1993) *Cell* **74**, 661–668
18. Kasai, H., Li, Y.X. and Miyashita, Y. (1993) *Cell* **74**, 669–677
19. Petersen, O.H., Petersen, C.C.H. and Kasai, H. (1994) *Annu. Rev. Physiol.* **56**, 297–319
20. Exton, J.H. (1994) *Annu. Rev. Physiol.* **56**, 349–369

Woods et al. (1986) Nature (London) 319, 600–602

Repetitive transient rises in cytoplasmic free calcium in hormone-stimulated hepatocytes

Niall M. Woods, K. S. Roy Cuthbertson & Peter H. Cobbold

Department of Zoology, University of Liverpool, PO Box 147, Liverpool L69 3BX, UK

In the stressed animal, the vasoactive hormones vasopressin and angiotensin-II and the neurotransmitter noradrenaline induce liver cells to release glucose from glycogen. The intracellular signal that links the cell-surface receptors for noradrenaline (α_1) and vasoactive peptides to activation of glycogenolysis is known to be a rise in the cytoplasmic concentration of free calcium ions (free Ca)[1–3]. The receptors for these agonists induce the hydrolysis of phosphatidylinositol 4,5-bisphosphate, a minor plasmalemma lipid, to produce inositol trisphosphate and diacylglycerol[3–5]. Inositol trisphosphate has been shown to mobilize intracellular calcium in hepatocytes[3,6–8]. We show here, by means of aequorin measurements in single, isolated rat hepatocytes, that the free Ca response to these agonists consists of a series of transients. Each transient rose within 3 s to a peak free Ca of at least 600 nM and had a duration of approximately 7 s. The transients were repeated at intervals of 0.3–4 min, depending on agonist concentration. Between transients, free Ca returned to the resting level of ~200 nM. Clearly, the mechanisms controlling free Ca in hepatocytes are more complex than hitherto suspected.

Hepatocytes were isolated from fed, mature male rats and were kept at 37 °C in a complete culture medium for use within 10 h. Cells of about 25 μm diameter were injected with aequorin (see Fig. 1 legend). Stable resting signals could be recorded from individual hepatocytes for several hours. The mean resting signal for 55 cells was 1.2 photon counts s^{-1} above background (0.8 or 1.6 c.p.s.) and mean total counts per cell 2.6×10^5. These measurements give an upper estimate for mean resting free Ca of 211 ± 9 ($\pm$s.e.m.) nM ($n = 55$), assuming 1 mM free Mg and uniform calcium distribution. Arsenazo(III) and quin2 measurements also give values of ~200 nM (refs 1, 2).

The α_1-preferring adrenergic agonist phenylephrine induced repetitive transient rises in free Ca which usually peaked at above 600 nM and were blocked reversibly by phentolamine, an α_1-antagonist (Fig. 1*a*). Propranolol, a β-adrenergic antagonist (2 μM), had no effect on transients induced by 2 μM phenylephrine, while yohimbine, an α_2-preferring antagonist (2 μM), reduced the frequency of transients induced by 2 μM phenylephrine (the period rose from 40 s to 60 s; data not shown). Only a proportion of the hepatocytes responded to phenylephrine (34/42 = 81%), a rather higher percentage than respond to α_1-adrenergic stimulation by entering into DNA synthesis (~30%)[9]. The frequency of the transients was dependent on the concentration of phenylephrine (Fig. 1*b*). Low concentrations (0.6 μM), which activate phosphorylase by ~10% of maximum[1], induced transients at ~100-s intervals, while 2 μM phenylephrine produced transients every 20-30 s (Fig. 1*b*). At the higher concentrations (2–10 μM), the initial few transients were usually of higher magnitude and more frequent than those later in the series, when the period stabilized at ~30 s. At very high phenylephrine concentrations (100–1,000 μM), free Ca briefly reached 1,100 nM and fell slowly over ~2 min before repetitive transients appeared with a period of ~20 s (Fig. 1*c*). Adrenaline induced similar response patterns but was more potent (0.3 μM adrenaline was comparable to 2 μM phenylephrine; data not shown).

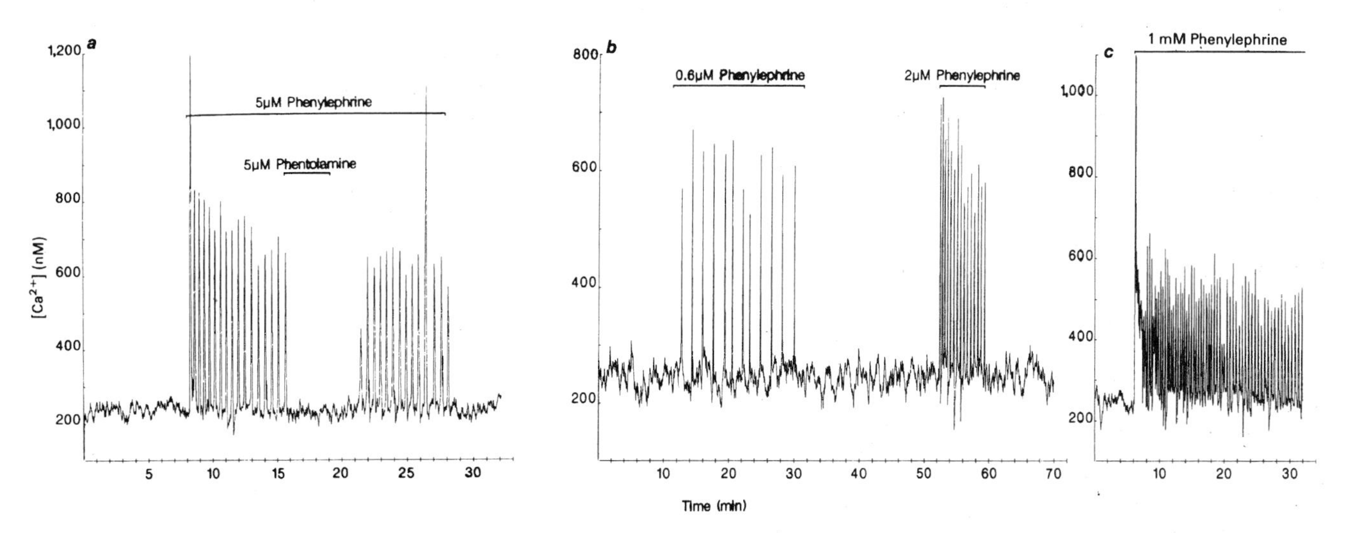

Fig. 1 Aequorin signals, transformed by microcomputer into free Ca values (nM), from individual rat hepatocytes exposed to phenylephrine. *a*, Phentolamine (5 μM) reversibly blocked free Ca transients induced by 5 μM phenylephrine. *b*, The frequency of the transients was dependent on the concentration (0.6 and 2 μM) of phenylephrine. *c*, Very high concentrations of phenylephrine (1 mM) generated a slow-falling rise, briefly reaching ~1,200 nM free Ca (peak not shown), followed by transients. (Time constants were 15 s for resting levels and 1s for the transients.)
Methods. Hepatocytes were prepared by perfusion of collagenase[24] (Boehringer) through livers from fed, male rats (150–250 g) and kept at 37 °C in CO_2-bicarbonate-buffered Williams' medium E (Flow Labs) containing 1.8 mM Ca, 0.05% bovine serum albumin but lacking insulin or serum. The aequorin technique was modified from refs 25 and 26 as follows. Single hepatocytes of ~25 μm diameter, selected under a ×180 stereomicroscope for their healthy appearance (no blebs, no vacuolation, distinct nuclei), were transferred to a 0.1-mm microslide (Camlab) containing a pool of 1% agarose (Sigma VII) and held at 37 °C under a layer of liquid paraffin. The cell was positioned within 100 μm of the agarose-oil interface before gelling the agarose at 4 °C for 2 min. Cells were injected on the stage of an inverted microscope (×600, Nomarski) at ~35 °C with aequorin solution from a micropipette whose tip had been filled with a 50-μm-long column of aequorin. (The tips of freshly pulled pipettes when immersed in liquid paraffin for a few seconds will subsequently allow a short column of aequorin solution to enter spontaneously.) Injected cells that had retained the distinct concave appearance of their nuclei were transferred in the microslide to a perfusable cup held at ~37 °C positioned under a low-noise photomultiplier[25]. Photon counts were sampled every 50 ms by a Sirius microcomputer. Data were plotted using exponential smoothing with time constants as indicated and calibration as before[25], except that 1 mM free Mg^{2+} was used.

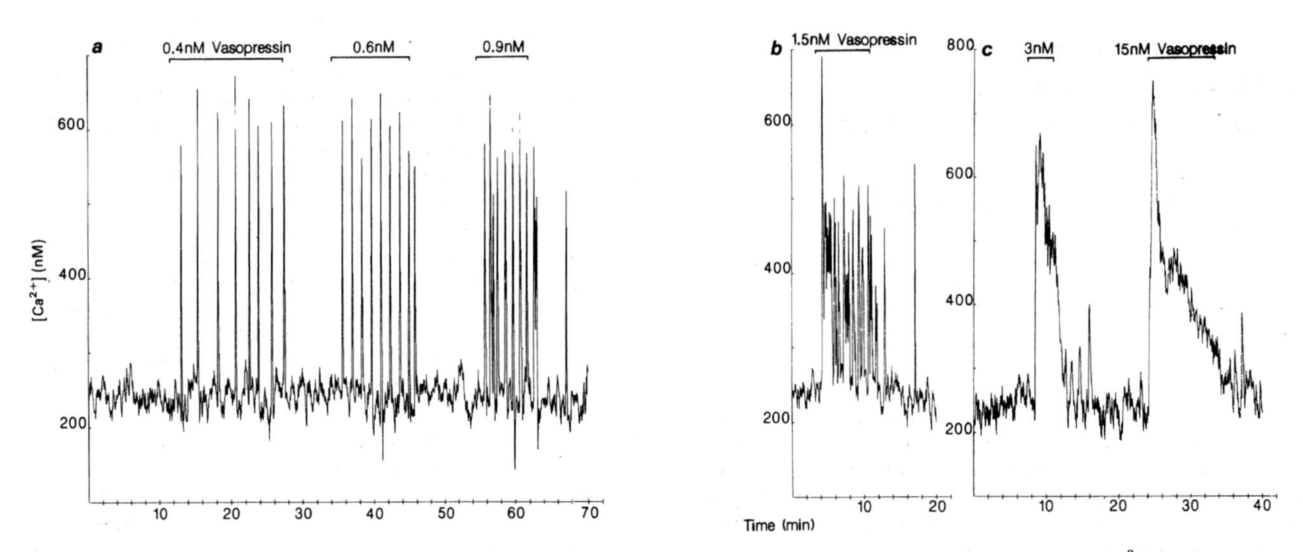

Fig. 2 Aequorin signals plotted as free Ca (nM) from a single hepatocyte exposed to various concentrations of [Arg^8]vasopressin (time constants as Fig. 1). *a*, Physiological concentrations of vasopressin (0.4, 0.6, 0.9 nM) induced free Ca transients at increasing frequencies (periods were 120, 80 and 60 s, respectively). Before the 0.4 nM recording, the cell had failed to generate transients when exposed to 0.2 nM vasopressin for 8.5 min. *b*, *c*, High concentrations of vasopressin (1.5, 3 and 15 nM) induced a more sustained initial free Ca rise. Repetitive transients occurred in 1.5 nM, but not at the higher concentrations until the hormone was removed (time constants in *c* were 10 and 3s).

Vasopressin and angiotensin-II at concentrations of ~1 nM, comparable to the K_m for activation of glycogen phosphorylase[10,11], induced a series of free Ca transients whose frequency was again dependent on hormone concentration (Fig. 2*a*). The onset of low-frequency transients (period 1-4 min) was preceded by a lag comparable to the period. Lags have been reported previously in free Ca measurements on populations of hepatocytes[2,3,12]. Very high concentrations of peptide hormones (≥10 nM), which would induce measurable degradation of phosphatidylinositol 4,5-bisphosphate and receptor occupancy[13,14], induced a more sustained, slowly decaying rise in free Ca which peaked at 600-800 nM (Fig. 2*c*). The time course of these sustained free Ca responses is similar to that of changes in phosphatidylinositol levels and phosphorylase activation described at high hormone doses[13,14]. At intermediate vasopressin concentrations (1.5 nM; Fig. 2*b*), an initial sustained rise in free Ca was followed by transients, similar to the response to very high doses of phenylephrine (Fig. 1*c*). Most cells responded to vasopressin (32/34), including 5 cells that had failed to respond to phenylephrine. Angiotensin induced a response in 21/27 cells (78%).

Figure 3 shows that the rate of rise of free Ca in a transient was rapid, taking ~2-3 s to rise from resting levels (200 nM) to peak values (800-1,200 nM). The peak concentration was sustained for ~2 s, while the fall to resting levels took a further 3-5 s. Overall duration was ~7 s. No appreciable change in these parameters was detected at different hormone concentrations. Although quin2 measurements on populations of hepatocytes[2,3] have not revealed transient behaviour, perhaps because of buffering of free Ca or poor cell synchrony, oscillations have been detected in the extracellular calcium concentration in the perfusate from intact liver exposed to 10 μM phenylephrine (H. Sies, personal communication).

It is an attractive possibility that the frequency of the transients determines the intensity of a cell's response to hormone concentrations within the physiological range. Calcium concentrations within the range we observe can activate phosphorylase kinase[15,16], which contains covalently bound calmodulin. This kinase in turn activates glycogen phosphorylase, whose activity will be a function of the rates of phosphorylation and dephosphorylation. Since the frequency of the transients is the only parameter of free Ca which varies with hormone concentration, it is possible that frequency determines the rate of glycogenolysis in an individual cell. Rasmussen and Barrett have postulated that a transient rise in free Ca to micromolar levels followed by a fall to ~500 nM would initiate and sustain calmodulin activity without promoting mitochondrial calcium overload[17]. In hepatocytes, the free Ca response is more complex than this, as free Ca falls to resting levels between transients, but our data offer a plausible explanation for the lack of mitochondrial calcium overload in chronically stimulated cells. Campbell has proposed that calcium acts as an intracellular regulator for 'threshold phenomena' that are either 'switched on' or 'switched off'[18]. Each free Ca transient conforms to this concept, while the occurrence of frequency-modulated series of transients may allow the amplitude of the glycogenolytic response in a single cell to be varied according to hormone concentration. There are indications from blowfly salivary glands[19], pancreatic islet β-cells[20] and oocytes during fertilization[21-23] that the phenomenon

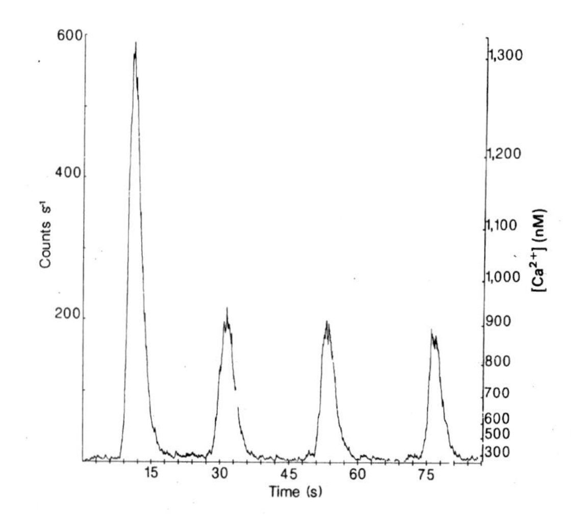

Fig. 3 The aequorin signal, expressed as photoelectron counts s^{-1} and as free Ca (nM), from a single hepatocyte responding to 5 μM phenylephrine. Time constant, 0.5 s. The signal rose from resting to peak values within 2-3 s. Owing to dead space in the perfusion chamber, the exact time of arrival of phenylephrine at the cell was unknown. The background was 1.6 counts s^{-1} and the resting level 2.4 counts s^{-1} above background, corresponding to 200 nM free Ca. The cell contained a total of 4.2×10^5 photoelectron counts.

of repetitive transient rises in free Ca may be widespread in cells that use free Ca as a messenger for extracellular ligands.

We thank the Wellcome Trust for funding this work.

Received 31 October 1985; accepted 6 January 1986.

1. Murphy, E., Coll, K., Rich, T. L. & Williamson, J. R. *J. biol. Chem.* **255**, 6600-6608 (1980).
2. Charest, R., Blackmore, P. F., Berthon, B. & Exton, J. H. *J. biol. Chem.* **258**, 8769-8773 (1983).
3. Thomas, A. P., Alexander, J. & Williamson, J. R. *J. biol. Chem.* **259**, 5574-5584 (1984).
4. Creba, J. *et al. Biochem. J.* **212**, 733-747 (1983).
5. Williamson, J. R., Cooper, R. H., Joseph, S. K. & Thomas, A. P. *Am. J. Physiol.* **248**, C203-C216 (1985).
6. Berridge, M. J. & Irvine, R. F. *Nature* **312**, 315-321 (1984).
7. Burgess, G. M., Irvine, R. F., Berridge, M. J., McKinney, J. S. & Putney, J. W. *Biochem. J.* **224**, 741-746 (1984).
8. Joseph, S. K., Thomas, A. P., Williams, R. J., Irvine, R. F. & Williamson, J. R. *J. biol. Chem.* **259**, 3077-3081 (1984).
9. Cruise, J. L., Houck, K. A. & Michalopoulos, G. K. *Science* **227**, 749-751 (1985).
10. Kirk, C. J., Rodrigues, L. M. & Hems, D. A. *Biochem. J.* **178**, 493-496 (1979).
11. Campanile, C. P., Crane, J. K., Peach, M. J. & Garrison, J. C. *J. biol. Chem.* **257**, 4951-4958 (1982).
12. Cooper, R. H., Coll, K. E. & Williamson, J. R. *J. biol. Chem.* **260**, 3281-3288 (1985).
13. Kirk, C. J., Creba, J. A., Downes, C. P. & Michell, R. H. *Biochem. Soc. Trans.* **9**, 377-379 (1981).
14. Kirk, C. J. *Cell Calcium* **3**, 399-411 (1982).
15. Chrisman, T. D., Jordan, J. F. & Exton, J. H. *J. biol. Chem.* **257**, 10798-10804 (1982).
16. Erdödi, F., Gergely, P. & Bot, G. *Int. J. Biochem.* **16**, 1391-1394 (1984).
17. Rasmussen, H. & Barrett, P. Q. *Physiol. Rev.* **64**, 938-984 (1984).
18. Campbell, A. K. *Intracellular Calcium, Its Universal Role as Regulator* (Wiley, Chichester, 1983).
19. Berridge, M. J. & Prince, W. T. *J. exp. Biol.* **56**, 139-153 (1972).
20. Matthews, E. K. & O'Connor, M. D. L. *J. exp. Biol.* **81**, 75-91 (1979).
21. Cuthbertson, K. S. R., Whittingham, D. G. & Cobbold, P. H. *Nature* **294**, 754-757 (1981).
22. Poeinie, M., Alderton, J., Tsien, R. Y. & Steinhardt, R. A. *Nature* **315**, 147-149 (1985).
23. Cuthbertson, K. S. R. & Cobbold, P. H. *Nature* **316**, 541-542 (1985).
24. Burgess, G. M., Claret, M. & Jenkinson, D. H. *J. Physiol., Lond.* **317**, 67-90 (1981).
25. Cobbold, P. H., Cuthbertson, K. S. R., Goyns, M. M. & Rice, V. R. *J. Cell Sci.* **61**, 123-136 (1983).
26. Cobbold, P. H. & Bourne, P. K. *Nature* **312**, 444-446 (1984).

Protein kinase C

The occupancy by agonists of many different cell surface receptors stimulates the activity of phospholipase C and the rapid breakdown of plasma membrane phosphatidylinositol 4,5-bisphosphate [PtdIns(4,5)P_2] leading to the generation of two products, diacylglycerol and inositol 1,4,5-trisphosphate [Ins(1,4,5)P_3]. These molecules then act as signalling messengers in a bifurcating pathway [1]. Ins(1,4,5)P_3 mobilizes Ca^{2+} from intracellular stores to elevate cytoplasmic Ca^{2+} concentration, and diacylglycerol activates protein kinase C (PKC), which in turn phosphorylates many cellular protein substrates and is known to regulate a multitude of cellular functions. Phosphoinositide turnover in response to cell stimulation had been known since the 1950s (Section 9 and [2]) but the significance of this for cell regulation was not clear at the time. The identification of PKC as a diacylglycerol-sensitive protein kinase provided the first evidence for a signalling pathway activated following PtdIns(4,5)P_2 breakdown (see our first selected paper [3]).

PKC was originally identified as a proenzyme that could give rise to a constitutively activate protein kinase (protein kinase M) after Ca^{2+}-dependent proteolysis [4,5]. The physiological mode of activation was initially unclear and, since the Ca^{2+}-dependent proteolysis required very high Ca^{2+} levels and was obviously not reversible, it did not seem to be a likely regulatory mechanism involved in normal cell signalling. A further advance was the finding that the intact PKC could be activated by micromolar levels of Ca^{2+} in the presence of an appropriate phospholipid, most particularly phosphatidylserine [6]. The key observation, however, was that PKC activation could be achieved at much lower Ca^{2+} levels in the presence of phospholipid if an unsaturated diacylglycerol such as diolein was included in the reaction mixture [3]. The significance of this observation was that it demonstrated how the enzyme could be activated at physiological Ca^{2+} concentrations, and it also led to the correct suggestion of a second messenger role for diacylglycerol in the activation of PKC. The data showed that diacylglycerol could activate PKC at resting Ca^{2+} levels with further activation if the cytosolic Ca^{2+} concentration was also increased.

Subsequent work resulted in the identification of a wide range of proteins that can be phosphorylated *in vitro* and *in vivo* by PKC. Study of the cellular role of PKC activation has, however, benefited considerably from the fact that PKC enzymes bind tightly and are activated by the tumour-promoting phorbol esters. The phorbol ester 12-*O*-tetradecanoylphorbol-13-acetate (TPA), otherwise known as phorbol 12-myristate 13-acetate (PMA), was found to be able to replace diacylglycerol and potently activate PKC [7]. In the presence of TPA, the affinity of PKC for Ca^{2+} and phospholipid is increased sufficiently that TPA activates PKC in intact, unstimulated cells. PKC enzymes are major targets for phorbol ester action. The phorbol esters have, therefore, been crucial tools for the manipulation of PKC in intact cells, and have allowed the range of cellular processes regulated by PKC to be determined [8]. It has been realized more recently, however, that

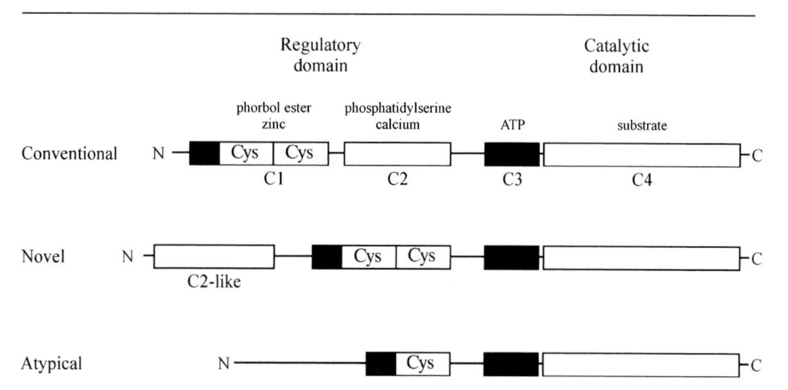

Figure 11.1. Conserved and variable domains in PKC enzymes. The proteins can be seen to contain two major domains. The regulatory domain contains a repeated cysteine-rich sequence in C1, and the kinase domain contains ATP-binding and substrate-binding sites. The classic PKCs similar to those first sequenced [9] also have a C2 motif in the regulatory domain that is required for Ca^{2+} and phospholipid binding. The novel PKCs do not possess the full C2 motif and have Ca^{2+}-independent activity. The C2 domain is absent from atypical PKCs. Modified from [13] with permission.

phorbol esters can bind to other cellular proteins and potentially affect cell function via these additional targets.

More detailed molecular understanding of PKC came from the molecular cloning and sequencing of a bovine brain PKC (PKCα) described in the second of the highlighted papers [9]. Following its purification, tryptic fragments of PKC were generated and sequenced. These allowed the design of oligonucleotide probes for the identification of cDNA clones containing the full coding sequence. This PKC was revealed to have a C-terminal catalytic domain related in sequence to other protein kinases (Figure 11.1). The N-terminal half of the protein appears to be a regulatory domain, containing a duplicated cysteine-rich sequence (C1), which is now known to be a zinc-binding domain that binds diacylglycerol and phorbol esters, and a further domain, designated C2. C2 domains are required for Ca^{2+} and phospholipid binding and have also been found in other Ca^{2+}- and phospholipid-binding proteins.

Extensive molecular cloning studies have resulted in the identification of multiple PKCs related in sequence to that originally identified [10–13]. It has been found that distinct subfamilies of PKC exist, which show differential sensitivity to activators, thus allowing increased diversity of PKC regulation and function. In addition, the PKC isoforms have different tissue distributions and distinct subcellular localizations within the cell. The originally characterized and cloned PKCs were Ca^{2+}-dependent, but a subfamily of PKC forms has been identified that lack the full C2 domain and instead have a C2-like domain unable to bind Ca^{2+}. These novel PKCs are activated by diacylglycerol in a Ca^{2+}-independent manner. A third family of 'atypical' PKC species is known that also lack the C2 domain, do not require either Ca^{2+} or diacylglycerol for their activation and do not bind phorbol ester (Table 11.1). The duplicated cysteine-rich sequences in the C1 domain each co-ordinate a Zn^{2+} ion and are involved in phorbol ester binding. The atypical PKCs contain only one of these repeats in their C1 domain and so both are essential for phorbol ester binding [12]. The lack of constitutive activity of PKC is believed to be due to the presence of a pseudo-substrate inhibitory region in the regulatory domain that normally prevents activity by occupying the substrate-binding site of the enzyme. Activation of the kinase requires a conformational change to remove this

Table 11.1. Mammalian PKC isoforms

	Subspecies	Amino acid residues	Ca^{2+} and lipid activators	Phorbol ester binding	Tissue expression
cPKC	α	672	Ca^{2+}, DAG, PS, FFAs, lysoPC	Yes	Universal
	βI	671	Ca^{2+}, DAG, PS, FFAs, lysoPC	Yes	Some tissues
	βII	671	Ca^{2+}, DAG, PS, FFAs, lysoPC	Yes	Many tissues
	γ	697	Ca^{2+}, DAG, PS, FFAs, lysoPC	Yes	Brain only
nPKC	δ	673	DAG, PS	Yes	Universal
	ε	737	DAG, PS, FFA, Ins(3,4,5)P_3	Yes	Brain and others
	η	683	DAG, PS, Ins(3,4,5)P_3, cholesterol sulphate	Yes	Skin, lung, heart
	θ	707	DAG, PS	Yes	Muscle, T cell, etc.
	μ	912	DAG, PS	Yes	NRK cells
aPKC	ζ	592	PS, FFA, Ins(3,4,5)P_3?	No	Universal
	λ	587	?	No	Many tissues

Abbreviations used: DAG, diacylglycerol; FFA, free fatty acid; lysoPC, lysophosphatidylcholine; PS, phosphatidylserine; PtdIns(3,4,5)P_3, phosphatidylinositol (3,4,5)-trisphosphate.

inhibition, or, as in the original discovery of PKC, limited proteolysis to remove the regulatory domain. Recent work has begun to define specific sequences in certain PKC isoforms that allow their interaction with, for example, the actin cytoskeleton [14]. Analysis of the sequences of the PKC family of enzymes has thus generated important structural insights into the function and regulation of these key enzymes.

References

1. Berridge, M.J. and Irvine, R.F. (1989) *Nature (London)* **341**, 197–205
2. Hokin, M.R. and Hokin, L.E. (1953) *J. Biol. Chem.* **205**, 967–977
3. Takai, Y., Kishimoto, A., Kikkawa, U., Mori, T. and Nishizuka, Y. (1979) *Biochem. Biophys. Res. Commun.* **91**, 1218–1224
4. Takai, Y., Kishimoto, A., Inoue, M. and Nishizuka, Y. (1977) *J. Biol. Chem.* **252**, 7603–7609
5. Inoue, M., Kishimoto, A., Takai, Y., and Nishizuka, Y. (1977) *J. Biol. Chem.* **252**, 7610–7616
6. Takai, Y., Kishimoto, A., Iwasa, Y., Kawahara, Y., Mori, T. and Nishizuka, Y. (1979) *J. Biol. Chem.* **254**, 3692–3695
7. Castagna, M., Takai, Y., Kaibuchi, K., Sano, K., Kikkawa, U. and Nishizuka, Y. (1982) *J. Biol. Chem.* **257**, 1847–1851
8. Nishizuka, Y. (1984) *Nature (London)* **308**, 693–698
9. Parker, P.J., Coussens, L., Totty, N., Rhee, L., Young, S., Chen, E., Stabel, S., Waterfield, M.D. and Ullrich, A. (1986) *Science* **233**, 853–859
10. Nishizuka, Y. (1988) *Nature (London)* **334**, 661–665
11. Nishizuka, Y. (1995) *FASEB J.* **9**, 484–496
12. Dekker, L.V., Palmer, R.H. and Parker, P.J. (1995) *Curr. Opin. Struct. Biol.* **5**, 396–402
13. Newton, A.C. (1995) *J. Biol. Chem.* **270**, 28495–28498
14. Prekeris, R., Mayhew, M.W., Cooper, J.B. and Terrian, D.M. (1996) *J. Cell. Biol.* **132**, 77–90

Takai et al. (1979) Biochem. Biophys. Res. Commun. **91**, 1218–1224

UNSATURATED DIACYLGLYCEROL AS A POSSIBLE MESSENGER FOR THE ACTIVATION OF CALCIUM-ACTIVATED, PHOSPHOLIPID-DEPENDENT PROTEIN KINASE SYSTEM*

Yoshimi Takai, Akira Kishimoto, Ushio Kikkawa, Terutoshi Mori†
and Yasutomi Nishizuka

Department of Biochemistry
Kobe University School of Medicine, Kobe 650, Japan

Received October 29, 1979

SUMMARY: A small quantity of unsaturated diacylglycerol (DG) sharply decreased the Ca^{2+} and phospholipid concentrations needed for full activation of a Ca^{2+}-activated, phospholipid-dependent multifunctional protein kinase described earlier (Takai, Y., Kishimoto, A., Iwasa, Y., Kawahara, Y., Mori, T. and Nishizuka, Y. (1979) J. Biol. Chem. 254. 3692-3695). In the presence of unsaturated DG and micromolar order of Ca^{2+}, phosphatidylserine (PS) was most relevant with the capacity to activate the enzyme, whereas phosphatidylethanolamine and phosphatidylinositol (PI) were far less effective. Phosphatidylcholine was practically inactive. It is possible, therefore, that unsaturated DG, which may be derived from PI turnover provoked by various extracellular stimulators, acts as a messenger for activating the enzyme, and that Ca^{2+} and various phospholipids such as PI and PS seem to play a role cooperatively in this unique receptor mechanism.

Hokin and Hokin (1) first presented evidence that PI[1/] turns over very rapidly in response to acetylcholine. Early work on such PI turnover was carried out with various types of secretory tissues such as pancreas (1,2), salivary gland (3) and salt-secreting gland (4). Subsequent studies developed by many investigators (for reviews see Refs. 5,6) have shown that the PI response can be provoked in a variety of tissues which are activated by various extracellular stimulators including α-adrenergic and muscarinic cholinergic neurotrans-

---

* This investigation has been supported in part by research grants from the Scientific Research Fund of Ministry of Education, Science and Culture, Japan (1979), the Intractable Diseases Division, Public Health Bureau, the Ministry of Health and Welfare, Japan (1979), the Yamanouchi Foundation for Research on Metabolic Disorders (1979) and from the Foundation for the Promotion of Research on Medical Resources, Japan (1977-1979).
† On leave from the Faculty of Nutrition, Kobe Gakuin University, Kobe 673.
1/ Abbreviations used are: PI, phosphatidylinositol; PS, phosphatidylserine; PE, phosphatidylethanolamine; and PC, phosphatidylcholine.

mitters as well as some peptide hormones. Nevertheless, all attempts to clarify the physiological significance of such PI turnover have been thus far uniformly unsuccessful. In preceding reports from this laboratory (7,8) a new species of multifunctional protein kinase has been identified in mammalian tissues which may be selectively activated by the simultaneous presence of Ca^{2+} and phospholipid. This communication will present evidence suggesting that PI turnover may be coupled with the activation of this protein kinase, and possible roles of various phospholipids in this receptor mechanism will be proposed. In order to relate to our previous papers (7,8) the Ca^{2+}-activated, phospholipid-dependent protein kinase will be referred to as protein kinase C.

EXPERIMENTAL PROCEDURES

Protein kinase C was purified partially from rat brain cytosol as described previously (9), and the preparation used was essentially free of endogenous phosphate acceptor proteins and interfering enzymes. The enzyme was assayed with H1 histone as phosphate acceptor in the presence of Ca^{2+}, phospholipid and neutral lipid. The detailed conditions are given in each experiment. PI (pig liver) was purchased from Serdary Research Laboratories, and was purified by thin layer chromatography on a Silica Gel H (E. Merck) plate as described previously (8). PS (bovine brain), PE and PC (human erythrocyte) were generous gifts of Dr. T. Fujii and Dr. A. Tamura, Kyoto College of Pharmacy. All samples employed were chromatographically pure. Mono-, di- and triacylglycerols employed were synthetic products which were obtained from commercial sources. Unless otherwise specified each sample of diacylglycerol was a mixture of 1,2- and 1,3-diacyl derivatives as judged by thin layer chromatography. 1-Stearoyl-2-oleoyl diglyceride and 1-stearoyl-2-linoleoyl diglyceride were products of Serdary Research Laboratories. Samples of monoacylglycerols were also mixtures of 1- and 2-acyl derivatives. H1 histone was prepared from calf thymus as described earlier (10). [γ-^{32}P]ATP was prepared by the method of Glynn and Chappell (11). All materials and reagents employed for the present studies were taken up in water which was prepared by a double distillation apparatus followed by passing through a Chelex-100 column to remove Ca^{2+} as much as possible as specified by Teo and Wang (12). Protein was determined by the method of Lowry *et al.* (13) with bovine serum albumin as a standard protein.

RESULTS AND DISCUSSION

Protein kinase C normally present as an inactive form in the soluble fraction of mammalian tissues was activated by reversible association with membranes in the presence of Ca^{2+} (7). The active

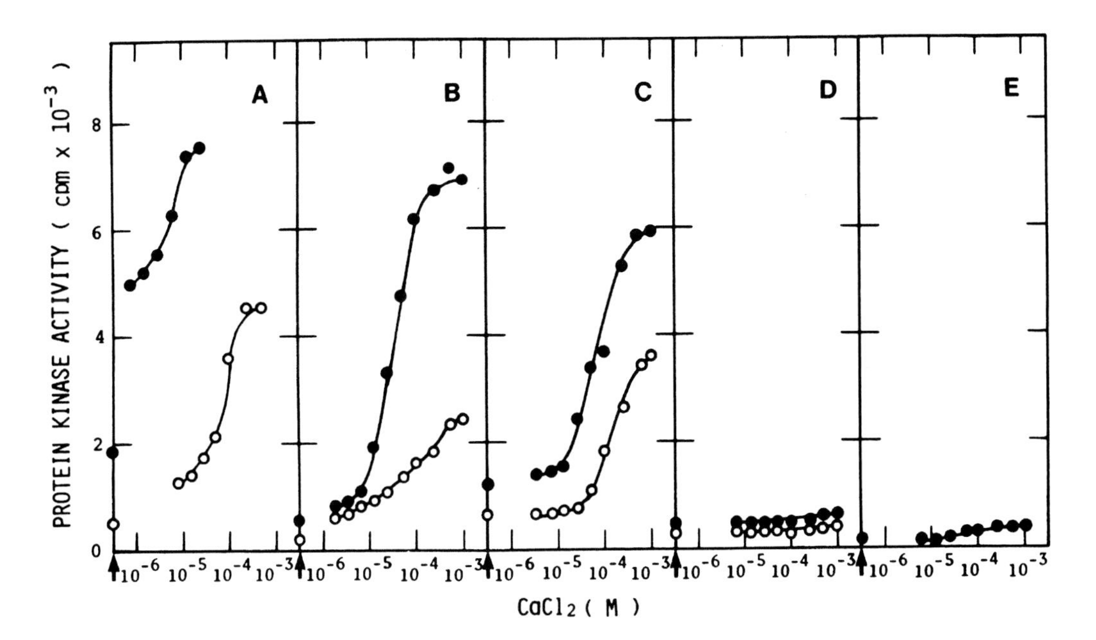

Fig. 1. Effect of diolein on reaction velocity of protein kinase C at various concentrations of $CaCl_2$. The complete reaction mixture (0.25 ml) contained 5 µmol of Tris/HCl at pH 7.5, 1.25 µmol of magnesium nitrate, 50 µg of H1 histone, 2.5 nmol of [γ-^{32}P]ATP (5 x 10^4 cpm/nmol), 0.4 µg of protein kinase C, 2 µg each of phospholipid indicated, and various concentrations of $CaCl_2$ as indicated. Where indicated diolein (0.2 µg/tube) was added. Each phospholipid was first mixed with diolein in a small volume of chloroform. After chloroform was removed in vacuo, the residue was suspended in 20 mM Tris/HCl at pH 7.5 by sonication with a Kontes sonifier K881440 for 5 min at 0°C, and employed for the assay. The incubation was carried out for 3 min at 30°C. The reactions were stopped by the addition of 25% trichloroacetic acid, and acid-precipitable materials were collected on a Toyo-Roshi membrane filter (pore size, 0.45 µm). The radioactivity was determined as described (14). Abscissa indicates the final concentration of $CaCl_2$ added. Where indicated with an arrow, ethylene glycol bis(β-aminoethyl ether) N,N,N',N'-tetraacetic acid (0.5 mM at final concentration) was added instead of $CaCl_2$. A, with PS; B, with PE; C, with PI; D, with PC; and E, with diolein alone. (●——●), assayed in the presence of diolein; and (○——○), assayed in the absence of diolein.

factor in membranes was identified as phospholipid; particularly PI and PS were most effective to support enzymatic activity (8). Subsequent analysis on the mechanism of action of phospholipid revealed that coexistence of a very small quantity of diacylglycerol possessing unsaturated fatty acid particularly at the position 2 greatly enhanced the phospholipid-dependent activation of enzyme especially at lower concentrations of Ca^{2+}. A typical result of such experiments is shown in Fig. 1. In this figure the reaction velocities in the presence and absence of diolein were plotted against Ca^{2+} concentrations in a logarithmic scale. The enhancement of reaction by diolein was most remarkable when PS was employed (Fig. 1A). Namely, supplement of a small quantity of diolein to PS greatly enhanced the reaction velocity with the concomitant decrease in Ca^{2+} concentrations giving rise to full activation of the enzyme. If, however, PE or PI was employed instead of PS, only reaction velocity was accelerated by the addition of diolein and relatively higher concentrations of Ca^{2+} were needed for activation of the enzyme (Fig. 1,B and C). PC was practically ineffective to support enzymatic activity irrespective of the presence and absence of diolein (Fig. 1D). Diolein alone showed a very little or no effect over a wide range of Ca^{2+} concentrations (Fig. 1E).

The enhancement of reaction by diolein in the presence of PS did not appear to be attributed simply to the increase in reaction velocity but was accompanied by the decrease in Ca^{2+} concentration which was needed for full activation of the enzyme as described above. Kinetic analysis indicated that the addition of diolein greatly increased an apparent affinity of the enzyme for PS as well as for Ca^{2+}. In the experiments shown in Fig. 2, Ka value for Ca^{2+}, the concentration needed for half maximum activation, was plotted against mono-, di- or triolein which was added together with either PS or PI. The results showed that, when PS was employed, Ka value for Ca^{2+} was decreased from about 5×10^{-5} $\underline{M}$ sharply to the micromolar order, and that diolein in an amount of less than 10% of that of PS showed remarkable effect (Fig. 2A). Again, diolein showed a very little effect when PS was replaced by PI (Fig. 2B). It may be noted that monoolein and triolein did not enhance reaction velocity nor decreased Ka value for Ca^{2+} under comparable conditions. Such a unique effect of neutral lipid was specific for diacylglycerol possessing unsaturated fatty acid, and essentially similar results were obtained for dilinolein,

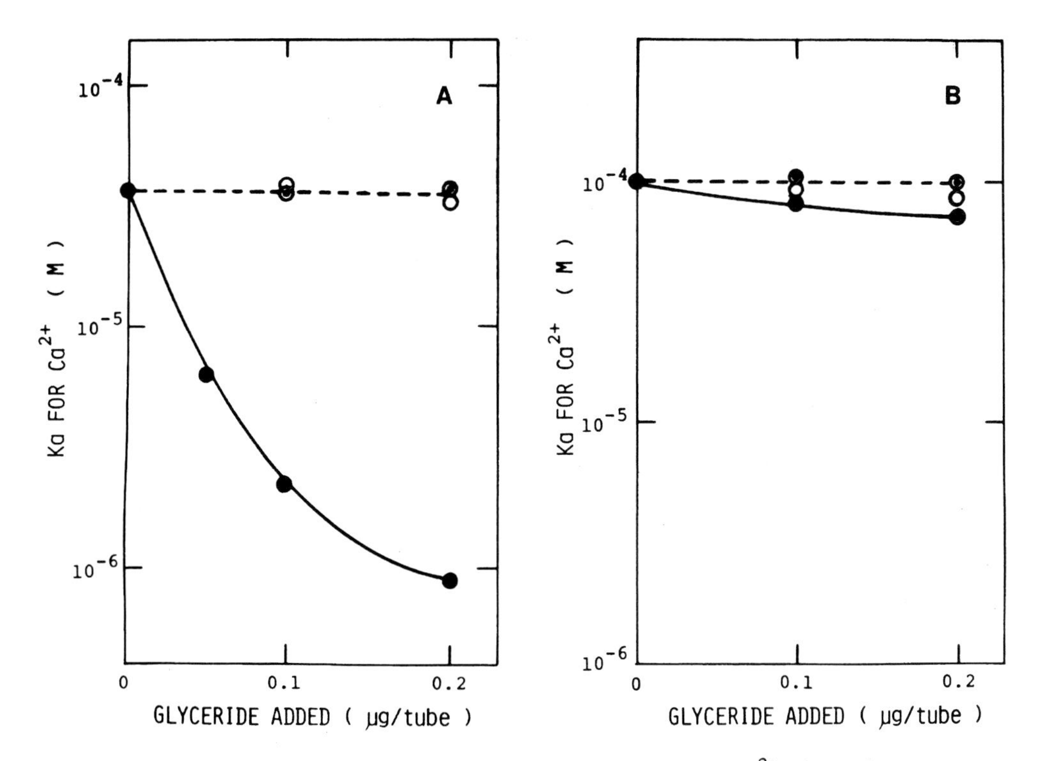

Fig. 2. Effects of mono-, di- and triolein on Ka value for Ca^{2+} of protein kinase C. The reaction mixture contained a fixed amount (2 μg each) of either PS or PI, and various amounts of glyceride as indicated. Other assay conditions were the same to those described in Fig. 1, and Ka value for Ca^{2+} was estimated. A, with PS; and B, with PI. (●——●), assayed in the presence of diolein; (⊙--⊙), assayed in the presence of monoolein; and (○--○), assayed in the presence of triolein.

diarachidonin, 1-stearoyl-2-oleoyl diglyceride and also for 1-stearoyl-2-linoleoyl diglyceride as shown in Table I. Both dipalmitin and distearin were less effective. Monoacylglycerol and triacylglycerol tested thus far were totally ineffective irrespective of the fatty acyl moieties. Neither cholesterol nor glycolipid could substitute for unsaturated diacylglycerols mentioned above.

It appears to be established that PI turnover which is provoked by various extracellular stimulators is initiated by hydrolysis of the phosphodiester linkage in a manner of phospholipase C (15-20). Thus, the primary product of this reaction is expected to be diacylglycerol which is very effective to potentiate the Ca^{2+} and phospholipid-dependent activation of protein kinase C, since PI of most

Table I

Effects of various diacylglycerols on Ka value for Ca^{2+} and reaction velocity of protein kinase C

Diacylglycerol added	Ka for Ca^{2+}	Protein kinase activity
	(μM)	(cpm)
None	50	970
Diolein	2	6,830
Dilinolein	3	5,040
Diarachidonin	6	4,270
1-Stearoyl-2-oleoyl diglyceride	4	4,890
1-Stearoyl-2-linoleoyl diglyceride	5	6,630
Dipalmitin	20	2,340
Distearin	50	950

The reaction mixture contained PS (2 μg) and diacylglycerol indicated (0.1 μg each). Other conditions were the same to those described in Fig. 1, and Ka value for Ca^{2+} was estimated. The protein kinase activity at 6.4×10^{-6} $\underline{M}$ $CaCl_2$ is given.

mammalian origins is well known to be composed of unsaturated fatty acid such as arachidonic or oleic acid particularly at the position 2 (21). Therefore, PI turnover may be directly related to the activation of this unique protein kinase in such a way that signals of extracellular stimulators induce the activation of a phospholipase C-type enzyme which is presumably specific for PI. This activation of phospholipase C may initiate PI turnover on one hand and, on the other hand, the resulting unsaturated diacylglycerol may serve as a messenger which in turn activates protein kinase C in the presence of Ca^{2+} and phospholipid. At lower concentrations of Ca^{2+} the highest enzymatic activity was obtained with the combination of PS and unsaturated diacylglycerol as described above. Presumably, some lipid bilayer structure is necessary for rendering the enzyme more active, and better physiological picture will be clarified by further investigations. Nevertheless, Ca^{2+} and various phospholipids such as PI and PS seem to play a role cooperatively in this unique receptor mechanism.

It may also be emphasized that in this mechanism protein kinase C can be activated without net increase in Ca^{2+} concentrations within the cell, since the unsaturated diacylglycerol markedly increases the affinity of this protein kinase system for this divalent cation.

Acknowledgements——Authors are grateful to Dr. T. Fujii and Dr. A. Tamura for valuable discussion and advice for isolation of phospholipid. Skilful secretarial assistance of Mrs. S. Nishiyama and Miss K. Yamasaki are also greatly acknowledged.

REFERENCES

1. Hokin, L.E. and Hokin, M.R. (1955) Biochim. Biophys. Acta 18, 102-110
2. Hokin, L.E. and Hokin, M.R. (1958) J. Biol. Chem. 233, 805-810
3. Eggman, L.D. and Hokin, L.E. (1960) J. Biol. Chem. 235, 2569-2571
4. Hokin, L.E. and Hokin, M.R. (1960) J. Gen. Physiol. 44, 61-85
5. Hawthorne, J.N. and White, D.A. (1975) Vitam. Horm. (New York) 33, 529-573
6. Michell, R.H. (1975) Biochim. Biophys. Acta 415, 81-147
7. Takai, Y., Kishimoto, A., Iwasa, Y., Kawahara, Y., Mori, T. Nishizuka, Y., Tamura, A. and Fujii, T. (1979) J. Biochem. 86, 575-578
8. Takai, Y., Kishimoto, A., Iwasa, Y., Kawahara, Y., Mori, T. and Nishizuka, Y. (1979) J. Biol. Chem. 254, 3692-3695
9. Inoue, M., Kishimoto, A., Takai, Y. and Nishizuka, Y. (1977) J. Biol. Chem. 252, 7610-7616
10. Hashimoto, E., Takeda, M., Nishizuka, Y., Hamana, K. and Iwai, K. (1976) J. Biol. Chem. 251, 6287-6293
11. Glynn, I.M. and Chappell, J.B. (1964) Biochem. J. 90, 147-149
12. Teo, T.S. and Wang, J.H. (1973) J. Biol. Chem. 248, 5950-5955
13. Lowry, O.H., Rosebrough, N.J., Farr, A.L. and Randall, R.J. (1951) J. Biol. Chem. 193, 265-275
14. Takai, Y., Kishimoto, A., Inoue, M. and Nishizuka, Y. (1977) J. Biol. Chem. 252, 7603-7609
15. Friedel, R.O., Brown, J.D. and Durell, J. (1969) J. Neurochem. 16, 371-378
16. Durell, J., Garland, J.T. and Friedel, R.O. (1969) Science, 165, 862-866
17. Dawson, R.M.C., Freinkel, N., Jungalwala, F.B. and Clarke, N. (1971) Biochem. J., 122, 605-607
18. Lapetina, E.G. and Michell, R.H. (1973) Biochem. J. 131, 433-442
19. Irvine, R.F. and Dawson, R.M.C. (1978) J. Neurochem. 31, 1427-1434
20. Rittenhouse-Simmons, S. (1979) J. Clin. Invest. 63, 580-587
21. Holub, B.J., Kuksis, A. and Thompson, W. (1970) J. Lipid Res. 11, 558-564

Parker et al. (1986) Science **233**, 853–859

The Complete Primary Structure of Protein Kinase C—the Major Phorbol Ester Receptor

PETER J. PARKER, LISA COUSSENS, NICK TOTTY, LUCY RHEE, SUSAN YOUNG, ELLSON CHEN, SILVIA STABEL, MICHAEL D. WATERFIELD, AXEL ULLRICH

Protein kinase C, the major phorbol ester receptor, was purified from bovine brain and through the use of oligonucleotide probes based on partial amino acid sequence, complementary DNA clones were derived from bovine brain complementary DNA libraries. Thus, the complete amino acid sequence of bovine protein kinase C was determined, revealing a domain structure. At the amino terminal is a cysteine-rich domain with an internal duplication; a putative calcium-binding domain follows, and there is at the carboxyl terminal a domain that shows substantial homology, but not identity, to sequences of other protein kinase.

ANALYSIS OF GROWTH FACTORS AND THEIR ACTION HAS provided important insights into the mechanisms used to subvert the control of normal cell proliferation. Thus there is evidence that certain genes capable of transforming cells encode growth factors (*1, 2*) or abnormal growth factor receptors (*3–5*); the expression of these genes allows cells to divide in a constitutive manner. In elucidating the responses of cells to growth factors, it has become evident that postreceptor events may also be in some way involved in cellular transformation. Therefore, a detailed molecular description of the intracellular pathways responsible for cell division induced by growth factors is necessary if we are to understand the normal mechanisms involved in growth factor action and in so doing to identify critical links open to subversion.

The phosphorylation of proteins plays a key role in regulating cellular functions (*6–8*). The kinases and phosphatases responsible for governing such phosphorylations are themselves targets for the action of growth factors, hormones, and other extrinsic agents participating in the control of cellular events (*6–8*). One of the major signal transduction pathways defined recently involves the enzyme protein kinase C (*9–11*), a multifunctional kinase that appears to play a central regulatory role akin to that of cyclic nucleotide-dependent and calcium-calmodulin–dependent enzymes.

Protein kinase C is a serine- and threonine-specific protein kinase that is dependent upon calcium and phospholipid for activity (*12*). However, at physiological calcium concentrations diacylglycerol is required for activity (*13*). Thus diacylglycerol has been defined as a second messenger responsible for the activation of protein kinase C in vivo (*9–11*). Agonist-induced generation of diacylglycerol has been widely described and forms part of a bifurcating signal pathway (*14*). It is thought that agonist-induced receptor-mediated activation of phospholipase C acts to generate two important second messengers from inositol phospholipids; the first, inositol 1,4,5-triphosphate, appears to be responsible for the release of calcium from intracellular stores (*15*) and the second, diacylglycerol, leads to protein kinase C activation (*13*). There is as yet only circumstantial evidence for the functioning of such a pathway in vivo (*16*).

From studies on protein kinase C in vitro, it has become apparent that those phorbol esters capable of tumor promotion can mimic the effect of diacylglycerol in enzyme activation (*17*). More recently, other structurally related and unrelated tumor promoters have also been shown to activate protein kinase C in vitro (*18–21*). The implication is that these tumor promoters elicit responses through protein kinase C and that activation of this enzyme is at least in part responsible for the activity of these hyperplasiogenic tumor-promot-

P. J. Parker, N. Totty, S. Young, S. Stabel, and M. D. Waterfield are at the Ludwig Institute for Cancer Research, Imperial Cancer Research Fund, Lincoln's Inn Fields, London WC2A 3PX, United Kingdom. L. Coussens, L. Rhee, E. Chen, and A. Ullrich are in the Department of Developmental Biology, Genentech, Inc., 460 Point San Bruno Boulevard, South San Francisco, CA 94080.

ing agents. Indeed, there is evidence that phorbol esters do act via high-affinity binding sites to provoke cellular responses (*22*), and there is both indirect (*23–25*) and direct evidence (*26–28*) that identifies this high-affinity phorbol ester receptor as protein kinase C itself.

To probe the role of protein kinase C in growth factor action and also to define the way the function of this signal transducing protein is undermined by the action of certain tumor promoters, we have undertaken a structural analysis of the protein. We now report the complete amino acid sequence of protein kinase C obtained by sequence analysis and recombinant DNA techniques. The predicted domain structure of the enzyme has been probed with antibodies to specific regions of the polypeptide, and this provides a basis for a detailed understanding of the structure and function of this protein kinase.

Protein kinase C complementary DNA clones. Protein kinase C was purified from bovine brain as described (*27*). The yield of the enzyme was optimized by taking broad cuts at each purification step. The resulting material was approximately 70 percent pure as judged by sodium dodecyl sulfate (SDS)–polyacrylamide gel electrophoresis (PAGE) and specific activity (3500 units per milligram). Minor contaminating polypeptides were completely resolved from protein kinase C by gel-permeating high-performance liquid chromatography (HPLC) in SDS-phosphate (Fig. 1A). The pooled fractions from this column were dialyzed and digested with trypsin, and the peptides were purified by HPLC (Fig. 1, B to D). The high sensitivity gas phase sequence analysis of two of these tryptic peptides is shown in Fig. 1, E and F. In all, the sequence of more than 100 amino acid residues was established from analysis of eight tryptic peptides.

A complementary DNA (cDNA) library of 4×10^7 independent clones was prepared from total poly(A)$^+$-containing calf brain RNA with the use of the λgt10 vector system (*29*) and previously described procedures (*4*). Recombinant phage (2×10^6) were initially screened with radioactively labeled synthetic oligonucleotide probes that had been designed on the basis of partial amino acid sequences of peptides 3, 4, and 5 (Fig. 2A). The sequences of the double-stranded oligonucleotide probes 3, 4, and 5 were chosen on the basis of codon frequency analyses and were 54, 42, and 54 bases in length, respectively (Fig. 2A). Initial plaque hybridization with pooled probes 3, 4, and 5 resulted in three strongly hybridizing recombinant phage with cDNA inserts of similar length. Southern hybridization analysis of Eco RI–digested phage DNA with the three probes on separate blots revealed that all three hybridized with probe 5 only. Subsequent nucleotide sequence analysis, however, revealed that clone λbPKC21 contained an open reading frame that included sequences of peptide 5, as expected, and also of peptide 4. Only four codons of the corresponding probe 4 were predicted correctly (Fig. 3), resulting in 11 mismatches (26 percent) and only

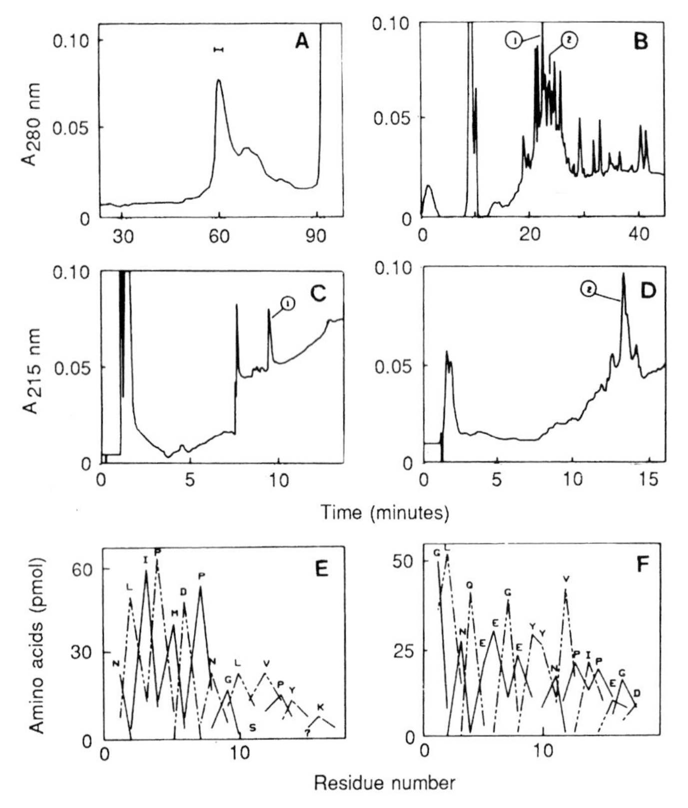

Fig. 1. Purification, trypsinolysis, and sequence analysis of protein kinase C. Protein kinase C was purified from bovine brain as described (*27*). In order to obtain optimum yields, broad cuts were taken at each step, and the final preparation was contaminated with polypeptides ≤60,000 daltons. In order to resolve these contaminants, samples (0.2 mg) of the preparation were subjected to gel permeation on a TSK 3000-GW column equilibrated in 0.1*M* sodium phosphate, *p*H 7.0, containing 0.1 percent SDS (A). In order to monitor recoveries during processing, samples of protein kinase C (1 to 2 μg) were autophosphorylated in the presence of ovalbumin as carrier (0.1 mg) and used to spike the bulk of the unlabeled material. The purified protein, which showed a single species on SDS-PAGE, was pooled from three runs as indicated (4.5 ml) and dialyzed twice against 5 liters of 20 percent methanol, 0.1*M* ammonium bicarbonate, and finally with two portions (1 liter each) of 0.05*M* ammonium bicarbonate. The protein was then digested with trypsin (1:50 by weight). The resulting tryptic per· ·des were resolved on a Vydak C_{18} reversed-phase column equilibrated in .08 percent trifluoroacetic acid and developed with an acetonitrile gradient (0 to 60 percent) (B). Individual absorbing peaks from this column were further purified on a Pharmacia C_8 reversed-phase column (C and D). Absorbing material eluted from this second column was subjected to gas phase microsequencing. Examples of amino yields for two sequencer runs are shown in (E) and (F).

A

```
             *       *   *       *           *   *
Probe 3  Ile His Thr Tyr Gly Glu Pro Val Phe Val Asp His Leu Gly Tyr Leu Leu Tyr
         ATC CAC TTC TAC GGC GAG CCC GTG TTC GTG GAC CAC CTC GGC TAC CTC CTC TAC
         TAG GTG AAG ATG CCG CTC GGG CAC AAG CAC CTG GTG GAG CCG ATG GAG GAG ATG

                                                 *
         Asn Leu Ile Pro Met Asp Pro Asn Gly Leu Ser Val Pro Tyr
Probe 4  AAC CTC ATC CCC ATG GAC CCC AAC GGC CTC TCC GTC CCC TAC
         TTG GAG TAG GGG TAC CTG GGG TTG CCG GAG AGG CAG GGG ATG

          *
         Gly Leu Asn Gln Glu Glu Gly Gly Tyr Tyr Asn Val Pro Ile Pro Glu Gly Asp
Probe 5  GGC CTC AAC CAG GAG GAG GGC GAG TAC TAC AAC GTG CCC ATC CCC GAG GGC GAC
         CCG GAG TTG GTC CTC CTC CCG CTC ATG ATG TTG CAC GGG TAG GGG CTC CCG CTG
```

B

ATG
TGA
bPKC
BH
H
200 bp
λbPKC21
P
BH
H
λbPKC306

Fig. 2. Protein kinase C cDNA clones. (A) Synthetic oligonucleotide probes used for screening of cDNA libraries. Oligonucleotide sequences were predicted from codon usage frequency analyses (*37*). The bases underlined delineate differences between predicted and determined sequence (Fig. 3). The extent of mismatch in probes 3 (19 of 54) and 4 (11 of 42) resulted in neither probe hybridizing to λbPKC21 or λbPKC306 under the conditions employed. (B) Bovine brain protein kinase C cDNA clones. Recombinant λgt10 clone λbPKC21 and λbPKC306 cDNA inserts are shown as black bars. Locations of restriction endonuclease sites for Pst I (P), Bam HI (B) and Hind III (H) are indicated. The structure of bovine protein kinase C mRNA is schematically shown above with the coding region (shaded box) flanked by initiation (ATG) and termination (TGA) signals and untranslated sequences (heavy line).

```
                                                             1                         10                            20
                                                             MetAlaAspValPheProAlaAlaGluProAlaAlaProGlnAspValAlaAsnArgPheAlaArgLys
   1 CCCTCTCGGCCGCCGCCCCGCGCCCCCCGCGGCAGGAGGCGGCGAGGGACCATGGCTGACGTCTTCCCGGCCGCCGAGCCGGCGGCGCCGCAGGACGTGGCCAACCGCTTCGCCCGCAAA

                     30                            40                            50                            60  [1]
     GlyAlaLeuArgGlnLysAsnValHisGluValLysAsnHisArgPheIleAlaArgPhePheLysGlnProThrPheCysSerHisCysThrAspPheIleTrpGlyPheGlyLysGln
 120 GGGGCGCTGAGGCAGAAGAACGTGCACGAGGTGAAGAACCACCGCTTCATCGCGCGCTTCTTCAAGCAGCCCACCTTCTGCAGCCACTGCACCGACTTCATCTGGGGGTTTGGGAAACAA

                     70                       [2]  80                            90                            100
     GlyPheGlnCysGlnValCysCysPheValValHisLysArgCysHisGluPheValThrPheSerCysProGlyAlaAspLysGlyProAspThrAspAspProArgSerLysHisLys
 240 GGCTTCCAGTGCCAAGTTTGCTGTTTTGTGGTTCACAAGAGGTGCCATGAATTTGTTACTTTTTCTTGTCCGGGGGCGGATAAAGGACCCGACACAGATGACCCGAGGAGCAAGCACAAG

        [3]          110                           120                           130                           140
     PheLysIleHisThrTyrGlySerProThrPheCysAspHisCysGlySerLeuLeuTyrGlyLeuIleHisGlnGlyMetLysCysAspThrCysAspMetAsnValHisLysGlnCys
 360 TTCAAGATCCACACGTATGGCAGCCCCACCTTCTGTGATCACTGCGGCTCCCTGCTCTACGGACTCATCCACCAGGGGATGAAATGTGACACCTGTGATATGAACGTGCACAAGCAGTGC

                     150                           160                           170                           180 [4]
     ValIleAsnValProSerLeuCysGlyMetAspHisThrGluLysArgGlyArgIleTyrLeuLysAlaGluValThrAspGluLysLeuHisValThrValArgAspAlaLysAsnLeu
 480 GTGATCAACGTGCCCAGCCTCTGCGGGATGGACCACACGGAGAAGAGGGGCCGCATCTACCTGAAGGCCGAGGTCACGGATGAAAAGCTGCACGTCACAGTACGAGACGCGAAAAACCTA

                     190                           200                           210                           220
     IleProMetAspProAsnGlyLeuSerAspProTyrValLysLeuLysLeuIleProAspProLysAsnGluSerLysGlnLysThrLysThrIleArgSerThrLeuAsnProArgTrp
 600 ATCCCTATGGATCCAAATGGGCTTTCAGATCCTTACGTGAAGCTGAAGCTTATTCCTGACCCCAAGAACGAGAGCAAACAGAAAACCAAGACCATCCGCTCGACGCTGAACCCCCGGTGG

                     230                           240                           250                           260
     AspGluSerPheThrPheLysLeuLysProSerAspLysAspArgArgLeuSerGluGluIleTrpAspTrpAspArgThrThrArgAsnAspPheMetGlySerLeuSerPheGlyVal
 720 GACGAGTCCTTCACGTTCAAATTAAAACCTTCTGATAAAGACCGGCGACTGTCCGAGGAAATCTGGGACTGGGATCGAACCACACGGAACGACTTCATGGGGTCCCTTTCCTTTGGGGTC

                     270                      [5]  280                           290
     SerGluLeuMetLysMetProAlaSerGlyTrpTyrLysLeuLeuAsnGlnGluGluGlyGluTyrTyrAsnValProIleProGluGlyAspGluGluGlyAsnValGluLeuArgGln
 840 TCGGAGCTGATGAAGATGCCGGCCAGCGGATGGTACAAGCTGCTGAACCAAGAGGAGGGCGAGTACTACAACGTGCCGATCCCCGAAGGCGACGAGGAAGGCAATGTGGAGCTCAGGCAG

                     310                           320                           330                           340
     LysPheGluLysAlaLysLeuGlyProAlaGlyAsnLysValIleSerProSerGluAspArgArgGlnProSerAsnAsnLeuAspArgValLysLeuThrAspPheAsnPheLeuMet
 960 AAATTCGAGAAAGCCAAGCTTGGCCCTGCCGGCAACAAAGTCATCAGTCCCTCCGAGGACAGGAGACAGCCTTCCAACAACCTGGACAGAGTGAAGCTCACGGACTTCAACTTCCTCATG

            *      *     350 *                         360                    ↓      370    [6]                    380
     ValLeuGlyLysGlySerPheGlyLysValMetLeuAlaAspArgLysGlyThrGluGluLeuTyrAlaIleLysIleLeuLysLysAspValValIleGlnAspAspValGluCys
1080 GTGCTGGGCAAAGGCAGCTTTGGGAAGGTGATGCTGGCCGACCGGAAGGGGACAGAGGAGCTGTACGCCATCAAGATCCTGAAGAAGGACGTGGTCATCCAGGACGACGTGGAGTGC

                     390                           400                           410                           420
     ThrMetValGluLysArgValLeuAlaLeuLeuAspLysProProPheLeuThrGlnLeuHisSerCysPheGlnThrValAspArgLeuTyrPheValMetGluTyrValAsnGlyGly
1200 ACCATGGTGGAGAAGCGGGTCCTGGCGCTGCTCGACAAGCCGCCGTTCCTGACGCAGCTGCACTCCTGCTTCCAGACGGTGGACCGGCTGTACTTCGTCATGGAGTACGTCAACGGCGGG

                     430                           440                           450                           460
     AspLeuMetTyrHisIleGlnGlnValGlyLysPheLysGluProGlnAlaValPheTyrAlaAlaGluIleSerIleGlyLeuPhePheLeuHisLysArgGlyIleIleTyrArgAsp
1320 GACCTCATGTACCACATCCAGCAGGTCGGGAAGTTCAAGGAGCCGCAAGCAGTGTTCTATGCAGCAGAGATTTCCATCGGGCTGTTCTTTCTTCATAAAAGAGGAATCATTTATCGGGAC

                     470                           480                           490           [7]             500
     LeuLysLeuAspAsnValMetLeuAspSerGluGlyHisIleLysIleAlaAspPheGlyMetCysLysGluHisMetMetAspGlyValThrThrArgThrPheCysGlyThrProAsp
1440 CTGAAGTTAGACAACGTCATGCTGGACTCGGAAGGACACATTAAGATCGCGGACTTCGGGATGTGCAAGGAGCACATGATGGACGGCGTCACGACCAGGACCTTCTGCGGGACCCCCGAC

                     510                           520                           530                           540
     TyrIleAlaProGluIleIleAlaTyrGlnProTyrGlyLysSerValAspTrpTrpAlaTyrGlyValLeuLeuTyrGluMetLeuAlaGlyGlnProProPheAspGlyGluAspGlu
1560 TACATCGCCCCAGAGATAATCGCCTATCAGCCGTACGGGAAGTCCGTGGACTGGTGGGCCTACGGCGTCCTGTTGTACGAGATGTTGGCCGGGCAGCCTCCGTTCGACGGCGAGGACGAG

                     550                           560                           570                           580
     AspGluLeuPheGlnSerIleMetGluHisAsnValSerTyrProLysSerLeuSerLysGluAlaValSerIleCysLysGlyLeuMetThrLysHisProGlyLysArgLeuGlyCys
1680 GACGAGCTGTTCCAGTCCATCATGGAGCACAACGTCTCGTACCCCAAGTCCTTGTCCAAGGAGGCCGTGTCCATCTGCAAAGGGCTGATGACCAAGCACCCCGGGAAGCGGCTGGGCTGC

                     590    [8]                    600                           610                           620
     GlyProGluGlyGluArgAspValArgGluHisAlaPhePheArgArgIleAspTrpGluLysLeuGluAsnArgGluIleGlnProProPheLysProLysValCysGlyLysGlyAla
1800 GGGCCCGAGGGCGAGCGCGACGTGCGGGAGCATGCCTTCTTCCGGAGGATCGACTGGGAGAAGCTGGAGAACCGTGAGATCCAGCCACCCTTCAAGCCCAAAGTGTGCGGCAAAGGAGCA

                     630                           640                           650                           660
     GluAsnPheAspLysPhePheThrArgGlyGlnProValLeuThrProProAspGlnLeuValIleAlaAsnIleAspGlnSerAspPheGluGlyPheSerTyrValAsnProGlnPhe
1920 GAGAACTTTGACAAGTTCTTCACGCGAGGGCAGCCTGTCTTGACGCCCGCCGACCAGCTGGTCATCGCTAACATCGACCAGTCTGATTTTGAAGGCTTCTCCTACGTCAACCCCCAGTTC

                     670
     ValHisProIleLeuGlnSerAlaValEnd
2040 GTGCACCCCATCCTGCAGAGCGCGGTATGAGACGCCTCGCGGAAGCCTGGTCCGCGCCCCCGCCCCCGCCTCCGCCCCCGCCGTGGGAAGCGACCCCCACCCTAGGGTTTGCCGGCCTCG

2160 GCCCTCCCTGTTCCAGGTGGAGGCCTGAAAACTGTAGGGTGGTTGTCCCCGCGTGCTCGGCTGCGTCATCTCAGCGGAAGATGACGTCACGTCGGCATCTGCTTGACGTAGAGGTGACAT

2280 CTGGCGGGGGATTGACCCTTTCTGGAAAGCAAACAGACTCTGGCC
```

Fig. 3. Complete nucleotide sequence and deduced amino acid sequence of bovine brain protein kinase C. The nucleotide sequence was obtained from clones λbPKC21 and λbPKC306 by M13-based chain termination analysis (*38*) and is shown in 5′ to 3′ orientation. The amino acid sequence of the longest open reading frame beginning with an initiation codon is translated and contains sequences coding for eight protein kinase C peptides (overlined and sequentially numbered). Sequences constituting the proposed calcium-binding site are emphasized (cross-hatched bar), and residues predicted to be involved in formation of an ATP-binding site are marked by asterisks (Gly X Gly XX Gly) and an arrow (Lys[368]).

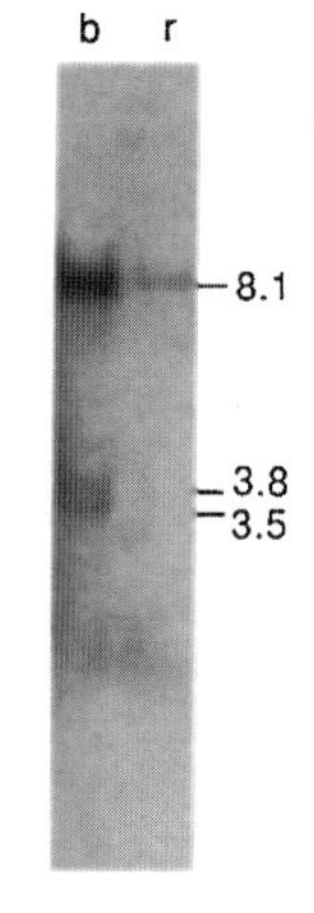

Fig. 4. Protein kinase C mRNA analysis. Polyadenylated RNA's from 4 μg of rat whole brain (r) and calf cerebellum (b) were analyzed by Northern blot hybridization (*39*); the entire λbPKC306 cDNA as a ^{32}P-labeled (10^8 cpm/μg) hybridization probe. RNA sizes are given in kilobases (kb) and were determined by comparison with RNA size standards of 9.5, 7.4, 4.4, 2.4, 1.4 and 0.3 kb (Bethesda Research Labs).

short stretches of matching sequence (Fig. 2A). Under the conditions used, this configuration was not sufficient to yield a hybridization signal.

Subsequent screening of another fraction (2×10^6 clones) of our bovine brain cDNA library with the radioactively labeled cDNA insert of λbPKC21 yielded clone λbPKC306 with a 2323-nucleotide cDNA insertion. Complete sequence analysis revealed the presence of an open reading frame (672 amino acids), which starts with a potential initiation codon. This ATG is flanked by sequences that fulfill the Kozak criteria for initiation codons (*30*) and lies within a 100-nucleotide GC-rich sequence (83 percent), which results in the absence of an in-frame stop codon upstream. Four additional cDNA clones were subsequently isolated, and all terminated within this GC-rich 5′ terminal sequence. The open reading frame initiated by this ATG codes for a 76.8-kD polypeptide, a molecular size close to that observed for the purified bovine brain enzyme (*27*). All experimentally determined protein kinase C peptide sequences are found within this predicted amino acid sequence, which provides further proof regarding the identity of clone λbPKC306. The 2016-nucleotide protein kinase C open reading frame is flanked by untranslated sequences of 50 (5′) and 258 (3′) nucleotides. Neither a poly(A) tail nor a polyadenylation signal (AATAAA) is contained in the λbPKC306 cDNA insert, suggesting that this clone was generated by nonspecific priming and that protein kinase C messenger RNA (mRNA) extends further 3′.

Protein kinase C mRNA's and their translation products. To examine protein kinase C mRNA size and complexity, we carried out Northern blot analysis with ^{32}P-labeled cDNA as a probe. There were one major and two minor mRNA hybridization signals of 8.1 kb, 3.8 kb, and 3.5 kb, respectively, in bovine brain poly(A)$^+$, and there was only one hybridization signal of ~8.6 kb in rat brain RNA (Fig. 4). Multiple mRNA's can be the product of alternative initiation, termination, or splicing events during transcription and posttranscriptional processing of a primary transcript. Alternatively, these hybridizing mRNA's may represent products of closely related but distinct genes [Coussens *et al.* (*31*) provide further information regarding this aspect]. All bovine protein kinase C RNA's are larger than our λbPKC306 cDNA and hybridize with coding as well as 3′ untranslated sequence probes, suggesting that extensions on 5′ or 3′ ends (or both) account for the observed mRNA size heterogeneity.

Since none of the protein kinase C cDNA clones described above have stop codons at the 5′ end of the open reading frame, a comparison was made between the primary protein kinase C translation products of poly(A)$^+$ RNA purified from bovine brain and in vitro transcripts from a cDNA construct covering the entire predicted amino acid sequence. The full-length cDNA in pSP65 was generated from λbPKC21 by taking the Xho I fragment covering the entire coding region and ligating into the Sal I site of the plasmid yielding pSP65-PKC1. The orientation was confirmed by restriction mapping. RNA from pSP65-PKC1 was generated in vitro with the use of SP6 polymerase, and this uncapped RNA was used as a template for translation in rabbit reticulocyte lysates.

Translation of poly(A)$^+$ RNA from bovine brain yielded a single major polypeptide that was specifically recognized by an antiserum to bovine protein kinase C. A single band at 76 kD is selectively extinguished by the addition of the protein kinase C–based peptide antigen to the reaction (Fig. 5A). Immunoprecipitation of the in

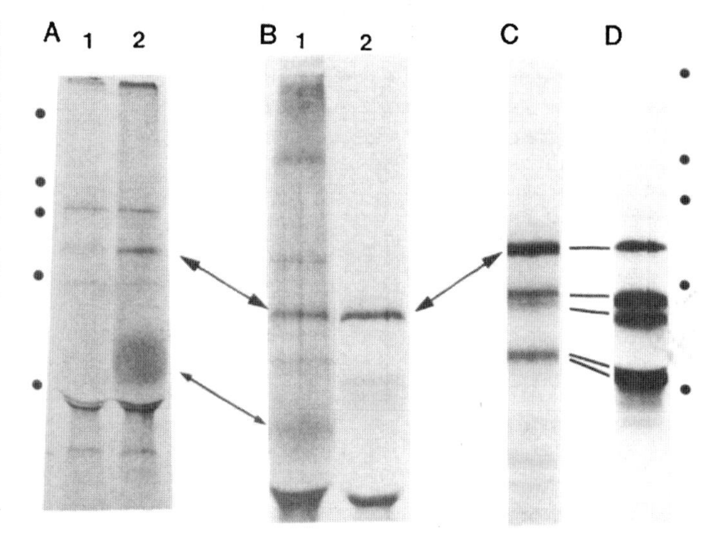

Fig. 5. In vitro translation products. An Xho I fragment covering the complete open reading frame of protein kinase C was excised from clone λbPKC306 and ligated into the Sal I site of the polylinker of plasmid pSP65 (Promega Biotec). The correct orientation of the insert was confirmed by restriction mapping. After linearization with Aha III the plasmid was used for in vitro transcription with SP6 polymerase (BioLabs) according to the manufacturer's protocols. The uncapped RNA produced was translated in vitro with a rabbit reticulocyte lysate (Amersham). In order to compare the translation product (or products) with authentic protein kinase C synthesized in vitro, total RNA was isolated from bovine brain and selected on oligo(dT)–cellulose. Approximately 0.5 μg of poly(A)$^+$ RNA from bovine brain was translated in vitro as described above. The translation products were boiled in SDS-PAGE sample buffer and diluted (1 to 10) with 1 percent deoxycholate, and 1 percent Triton X100 in phosphate-buffered saline and analyzed either directly by SDS-PAGE or immunoprecipitated. Samples were cleared with nonimmune rabbit serum and precipitated with antiserum 0442 in the presence and absence of competing peptide. Immune complexes were precipitated with protein A–Sepharose, washed with 0.5*M* NaCl in phosphate-buffered saline, and separated on 10 percent SDS-polyacrylamide gels. (A) In vitro translation of poly(A)$^+$ RNA from bovine brain yielded a single 76-kD species (large arrow) that was precipitated by protein kinase C antiserum 0442 (lane 2). Precipitation of the 76-kD protein and cross-reactive tulin (small arrow) was specifically blocked by inclusion of the peptide antigen (lane 1). Molecular size markers are indicated by dots and are in descending order: 200, 116, 93, 68, and 45 kD. (B) In vitro translation of poly(A)$^+$ RNA from bovine brain (lane 1) and manufactured putative full-length coding RNA transcribed in vitro from pSP65-PKC1 (lane 2). Antiserum 0442 was used to immunoprecipitate protein kinase C–related polypeptides. (C) In vitro translation of uncapped RNA transcribed in vitro from pSP65-PKC1 yielded a series of polypeptides ranging from the full-length translation product to several smaller species. These appear to be derived from initiation at internal methionines. Shorter translation products show apparent molecular weights of 60, 57, 54, 47 and 46 kD, and appear to be derived from initiation of translation at each of the downstream methionine residues (residues 130 to 137, 153, 186, 256, and 266 to 268). Some smaller fragments (<40 kD) are also observed; these are likely to be generated from initiation of translation further downstream. These fragments are also precipitated by antibody 0442, which is visible on longer exposure of (B) lane 2. (D) Immunoprecipitation by polyclonal antiserum (BG 36) (*32a*) of in vitro translation products shown in (C).

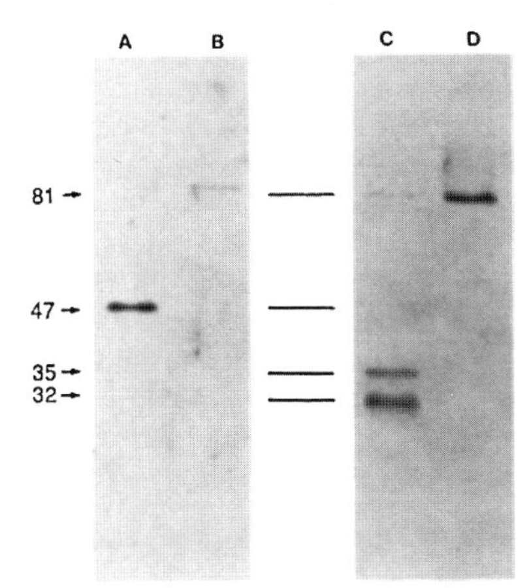

Fig. 6. Definition of the kinase domain. Antibodies to synthetic peptides based on protein kinase C residues 280 to 292 (serum 0442) and 317 to 328 (serum PP2) were obtained as described (*40*). Protein kinase C was purified to apparent homogeneity and subjected to SDS–polyacrylamide electrophoresis either without trypsin treatment (lanes B and D) or after digestion with trypsin at a final concentration of 20 μg/ml for 30 seconds at 30°C in the presence of 50 percent glycerol, 20 m*M* tris-HCl, *p*H 7.5, 2 m*M* EDTA, and 50 m*M* β-mercaptoethanol (lanes A and C). Trypsinization was terminated by addition of a tenfold molar excess of trypsin inhibitor and subsequent boiling. After SDS gel electrophoresis and transfer to nitrocellulose, protein kinase C and its fragments were subjected to immunoblot analysis with antibody 0442 (lanes C and D) or PP2 (lanes A and B). Arrows indicate molecular sizes of the immunoreactive polypeptides.

vitro translation products (Fig. 5B) from either bovine brain poly(A)$^+$ RNA (lane 1) or the SP6 polymerase transcript of the putative full-length open reading frame (Fig. 5B, lane 2, and D) yields a polypeptide of 76 kD in each case. This empirical value is close to that predicted from the sequence (76.8 kD). That these in vitro translation products are indistinguishable by size, provides some direct evidence that the initiator methionine shown in Figs. 2 and 3 is indeed the amino terminal residue of the primary translation product. Further circumstantial evidence for this is provided by the observation that the shorter in vitro translation products from these synthetic RNA's appear to be derived from each of the methionines downstream from residue 1 (Fig. 3); this pattern would not be consistent if methionine 130 was the major site of translational initiation (Fig. 5C).

The kinase domain. Protein kinase C can undergo proteolysis to generate a catalytically active fragment that is no longer dependent upon Ca^{2+} and phospholipid (*32*). This catalytic moiety can be readily generated in vitro by limited trypsinolysis (*32*); the fragment migrates as a ~50-kD species on gel filtration (*32*) and as a major band of ~47 kD on SDS-polyacrylamide gels (minor fragments of 45 and 43 kD are also present). We have used antisera to peptide fragments to define the limits of this catalytic domain. A Western blot of trypsinized protein kinase C shows that antiserum 0442 recognizes a species of 81 kD (intact protein kinase C) and a doublet at 33 to 35 kD (Fig. 6); in contrast, antiserum PP2 recognizes the intact polypeptide (81 kD) and a species at 47 kD. The polypeptide fragment at 47 kD comigrates with activity on gel filtration and therefore we can conclude that PP2 reacts with an epitope in the kinase domain, while 0442 reacts with the regulatory domain. These two antisera were made against amino acid sequences 280 to 292 (serum 0442) and 317 to 328 (serum PP2). Thus the limit on the catalytic domain is 43 kD (residues 292 to 717), this correlates with an empirical molecular size of 43 to 47 kD.

The reaction of antibody PP2 with only the uppermost of the 47- to 43-kD triplet suggests that conversion to smaller 45- and 43-kD species is associated with further cleavage at the amino terminus of the 47-kD fragment with subsequent loss of the PP2 epitope. It has not yet been possible to subfractionate this triplet in order to determine whether all these species are catalytically active.

Structural complications for protein kinase C function. Our results were designed to describe elucidation of the complete amino acid sequence of protein kinase C, the major receptor for phorbol esters. This was obtained through the isolation of cDNA clones coding for protein kinase C. The positive identification of these cDNA's has come from the presence of predicted tryptic peptides throughout the protein, whose sequences were in part predetermined from isolated bovine brain protein kinase C. In addition, translation of RNA that was derived from λbPKC-306 cDNA with the SP6 vector system (see below) yielded a 76-kD translation product that was recognized by a polyclonal antiserum to human protein kinase C (*32a*). Furthermore, antisera to a determined (antibody 0442) or predicted peptide sequence [antibody PP2, specific for alpha type protein kinase C; see legend to Fig. 5 and (*31*)] both recognize the purified bovine enzyme (see Fig. 5D).

The predicted amino acid sequence for protein kinase C shows a number of interesting features. At the amino terminus there is a tandem repeat that contains a series of six cysteine residues; the spacing of the cysteines between these two repeats is precisely conserved. Within these repeats is the sequence $C\text{-}X_2\text{-}C\text{-}X_{13}\text{-}C\text{-}X_2\text{-}C$; this type of pattern has been observed for various metalloproteins and also for certain DNA-binding proteins (*33*). While protein kinase C is known to bind calcium ions, this particular structure has not previously been shown to confer calcium binding and indeed other potential calcium-binding sites are present in the predicted sequence (as discussed below). The presence of this repeated pattern is of great interest; and it is now important to establish whether protein kinase C is itself a metalloprotein and whether this structure plays a role either in interaction with phospholipid (by analogy with the hydrophobic and ionic interactions of these structures with DNA) or indeed with DNA itself.

After the cysteine repeats there is a stretch of approximately 200 amino acids preceding the catalytic domain. Within this sequence lies the only potential calcium-binding site that comes close to the "E-F hand" structure characteristic of calmodulin and related calcium-binding proteins (*34*). This sequence (residues 292 to 303) is a predicted coil lying amino terminal to a predicted α-helix, with a glycine residue at the crucial turn point (Gly^{297}). This residue is surrounded by alternating amino acids capable of coordinating the calcium ion through their side chains (Glu^{292}, Asp^{294}, Glu^{296}, Asn^{298}, Glu^{300}, Gln^{303}). While there would appear to be an F helix, the presence of proline residues on the amino side of this putative binding site would not provide an equivalent E helix. There is no direct evidence that this is indeed the calcium-binding site, or that such a site in protein kinase C should fit the E-F hand model, indeed the presence of glutamic residues in such a structure would be unusual. Nevertheless, it is well documented that protein kinase C, like calmodulin, is inhibited by the phenothiazines and it might be surmised that some structural homology would exist between the calcium-binding sites of these proteins. Recent reports have suggested the possible existence of two Ca^{2+}-binding domains, one associated with catalytic activation, the other with phospholipid (*35*, *36*). The primary sequence does contain other regions that may function as Ca^{2+}-binding sites, and there is now an opportunity to directly assess the structure-function relations of these individual regions.

The catalytic domain shares substantial homology with other serine, threonine, and tyrosine protein kinases (Fig. 7). Thus by analogy with these protein kinases, one would predict that lysine-368 would lie at the ATP (adenosine triphosphate)-binding site. This residue is 17 amino acids carboxyl terminal to the last of the three conserved glycines that are also associated with the predicted nucleotide-binding site. The remainder of this domain shows further homology with these kinases; this reflects the conservation of particular sequences between all the members of this family and one would predict a functional role for these conserved residues.

A definitive proof of the location of the amino terminus has not been provided. However there are a number of considerations that

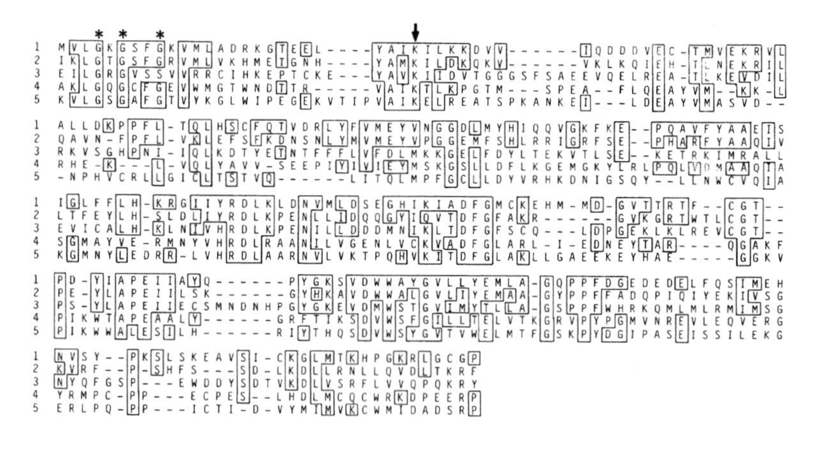

Fig. 7. Kinase domain homology to other protein kinases. The sequences of the putative catalytic (kinase) domains of a series of protein kinases are compared. Homologies to protein kinase C are shown in boxes. The asterisks indicate residues thought to be involved in the nucleotide binding site. (1) Protein kinase C; (2) cyclic AMP–dependent protein kinase (*41*); (3) γ-subunit of phosphorylase kinase (*42*); (4) v-*src* (*43*); (5) EGF-receptor (*4*).

are consistent with the present assignment: (i) The predicted size 77 kD is in reasonable agreement with previous determinations of the apparent molecular size of the purified protein; (ii) a related protein (*31*) is colinear at the amino terminus and retains the same amino-terminal Met-Ala-Asp triplet within a stretch that is otherwise divergent; (iii) the proteins synthesized in vitro from purified mRNA and from RNA transcribed from a putative full-length coding sequence display an identical apparent molecular size on SDS-polyacrylamide electrophoresis. Direct determination of the amino-terminal sequence has not been possible. The lack of success in obtaining this amino-terminal sequence suggests that the polypeptide might be blocked; the nature of this blocking group is not known at present.

There is evidence that protein kinase C in vivo may under certain circumstances be cleaved into a catalytically active fragment that displays an apparent molecular size of 50 kD on gel filtration. The sequence for protein kinase C would suggest that the region of this proteolytic cleavage is between the putative calcium-binding domain and the region encoding the kinase domain; within this sequence are a number of basic residues that may act as calpain or tryptic cleavage sites. Through the use of antibodies to specific regions of the polypeptide, it has been possible to identify the region that is cleaved by trypsin in vitro. Thus the catalytic domain (which retains activity) carries antigenic epitopes contained in residues 317 to 328, while not retaining those in residues 280 to 292. This puts an upper limit on the size of the fragment carrying catalytic activity and this is entirely consistent with the region of homology shared with other protein kinases (Fig. 7).

The proteolytic activation of protein kinase C generates a fragment with kinase activity that is not dependent on calcium and phospholipid (*32*). It would appear then that the regulatory domain acts to maintain the catalytic activity, and removal of this domain by proteolysis (or perhaps partial denaturation) leads to activation. Presumably the binding of calcium and phospholipid provokes a conformational change with consequent activation. It is thus possible to form a model for the activation of protein kinase C based upon these considerations (Fig. 8). It is proposed that the "hinge" region between the regulatory and catalytic domains is the exposed site of cleavage (residues 292 to 317). This type of structure has functional homology to the cyclic GMP–dependent protein kinase (GMP, guanosine monophosphate) where within a single polypeptide, a regulatory domain maintains a catalytic domain in an inactive state, until ligand is bound. This similarity suggests that perhaps as predicted for cyclic GMP–dependent protein kinase, protein kinase C was derived from the fusion of two genes and that these define the basic domain structure of the polypeptide.

When we used the isolated cDNA clones to probe the size of the mRNA for protein kinase C, we observed three major size species. We have some initial data indicating that the 8.1-kb and 3.8- or 3.5-kb messages all direct the synthesis of a 76-kD polypeptide that is recognized by antibodies to protein kinase C. This would suggest that these mRNA's are derived by alternative splicing or alternative processing outside the coding region (that is, in the 5′ or 3′ untranslated region) or that splicing variants of the coding region are indistinguishable by our criteria.

The ability of phorbol esters to bind to and activate protein kinase C is shared by a number of other tumor promoters including mezerin, aplysiatoxin, debromoaplysiatoxin, and teleocidin (*18–21*). Elucidation of the protein kinase C–binding site for these diverse structures should provide insight into the chemical features that appear to be in part responsible for tumor-promoting activity. The definition of the primary structure of protein kinase C allows us now to probe this structure, delineate the binding site and perhaps design antagonists.

Protein kinase C evidently does not show homology to any oncogenes that have been described to date. Given the important role that protein kinase C can play in the control of growth and differentiation, it might be anticipated that the expression of a mutated form of this protein, in particular cell lineages, may induce constitutive proliferation. However, it should be noted that, physiologically, protein kinase C activation by diacylglycerol is probably coupled to mobilization of Ca^{2+} and that the cellular responses are therefore a consequence of both events (*14*). Thus it may transpire that protein kinase C lies too far down a controlling hierarchy for

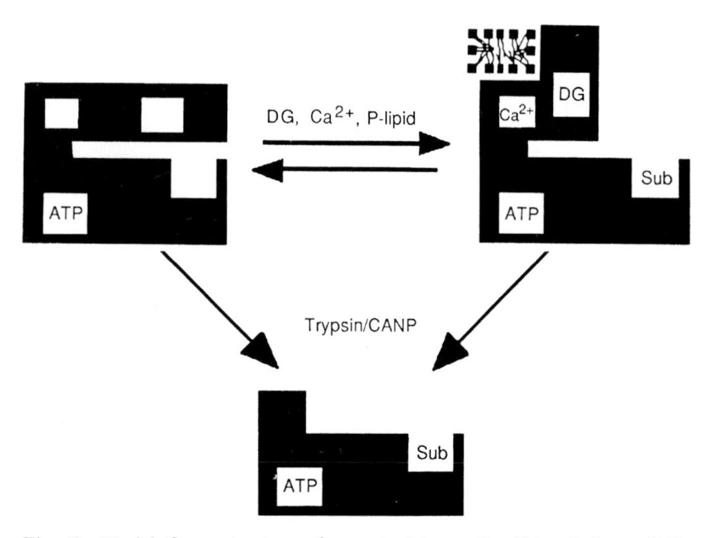

Fig. 8. Model for activation of protein kinase C. Abbreviations: DG, diacylglycerol; P-lipid, phospholipid; CANP, calpain; Sub, substrate-binding site.

aberrant activation to generate such a constitutive proliferation response (at least in the absence of other events). The generation of suitable protein kinase C cDNA constructs should make it possible to determine whether the expression of this kinase is abnormal in any human neoplasias.

REFERENCES AND NOTES

1. M. D. Waterfield *et al.*, *Nature (London)* **304**, 35 (1983).
2. R. F. Doolittle *et al.*, *Science* **221**, 275 (1983).
3. J. Downward *et al.*, *Nature (London)* **307**, 521 (1984).
4. A. Ullrich *et al.*, *ibid.* **309**, 418 (1984).
5. C. J. Sherr *et al.*, *Cell* **41**, 665 (1985).
6. P. Cohen, *Bioessays* **2**, 63 (1985).
7. ______, *Curr. Top. Cell. Reg.* **27**, 23 (1985).
8. T. Hunter and J. A. Cooper, *Annu. Rev. Biochem.* **54**, 897 (1985).
9. Y. Nishizuka, *Science* **225**, 1365 (1984).
10. ______, *Nature (London)* **308**, 693 (1984).
11. ______, *J. Natl. Cancer Inst.* **76**, 363 (1985).
12. Y. Takai *et al.*, *J. Biol. Chem.* **254**, 3692 (1979).
13. A. Kishimoto, Y. Takai, T. Mori, U. Kikkawa, Y. Nishizuka, *ibid.* **255**, 2273 (1980).
14. M. J. Berridge and R. F. Irvine, *Nature (London)* **312**, 315 (1984).
15. H. Streb, R. F. Irvine, M. J. Berridge, I. Schulz, *ibid.* **306**, 67 (1983).
16. Y. Kawahara, Y. Takai, R. Minakuchi, K. Sano, Y. Nishizuka, *Biochem. Biophys. Res. Commun.* **97**, 309 (1980).
17. M. Castagna *et al.*, *J. Biol. Chem.* **257**, 7847 (1982).
18. H. Fujiki *et al.*, *Biochem. Biophys. Res. Commun.* **120**, 339 (1984).
19. R. Miyake *et al.*, *ibid.* **121**, 649 (1984).
20. A. Couturier, S. Bazgar, M. Castagna, *ibid.*, p. 448.
21. J. P. Arcoleo and I. B. Weinstein, *Carcinogene* **6**, 213 (1985).
22. M. K. L. Collins and E. Rozengurt, *J. Cell. Physiol.* **122**, 42 (1982).
23. J. E. Niedel, L. J. Kuhn, G. R. Vandenbark, *Proc. Natl. Acad. Sci. U.S.A.* **80**, 36 (1983).
24. C. L. Ashendel, J. M. Staller, R. K. Boutwell, *Cancer Res.* **43**, 4333 (1983).
25. K. L. Leach, M. L. James, P. M. Blumberg, *Proc. Natl. Acad. Sci. U.S.A.* **80**, 4208 (1983).
26. U. Kikkawa, Y. Takai, Y. Tanaka, R. Miyake, Y. Nishizuka, *J. Biol. Chem.* **258**, 11442 (1983).
27. P. J. Parker, S. Stabel, M. D. Waterfield, *EMBO J.* **3**, 953 (1984).
28. T. Uchida and C. R. Filburn, *J. Biol. Chem.* **259**, 12311 (1984).
29. T. Huynh, R. Young, R. Davis, in *Practical Approaches in Biochemistry*, D. Grover, Ed. (IRL Press, Oxford, 1984).
30. M. Kozak, *Microbiol. Rev.* **47**, 1 (1983).
31. L. Coussens *et al.*, *Science* **233**, 859 (1986).
32. M. Inoue, A. Kishimoto, Y. Takai, Y. Nishizuka, *J. Biol. Chem.* **252**, 7610 (1977).
32a. M. J. Fry, A. Grebhardt, P. J. Parker, J. G. Foulkes, *EMBO J.* **4**, 3173 (1985).
33. J. M. Berg, *Science* **232**, 485 (1986).
34. L. J. van Eldik, J. G. Zendoqui, D. R. Marshak, D. M. Wattenson, *Int. Rev. Cytol.* 77, 1 (1982).
35. M. Wolf, H. Levine, W. S. May, P. Cuatrecasas, N. Sahyoun, *Nature (London)* **317**, 546 (1985).
36. W. S. May, N. Sahyoun, M. Wolf, P. Cuatrecasas, *ibid.*, p. 549.
37. R. Grantham, C. Gautier, M. Govy, M. Jacobzone, R. Mercier, *Nucleic Acids Res.* **9**, 43 (1981).
38. J. Messing, R. Crea, P. H. Seeburg, *ibid.*, p. 309.
39. H. Lehrach, D. Diamond, J. M. Wozney, H. Boedtker, *Biochemistry* **16**, 4743 (1977).
40. P. J. Parker *et al.*, in preparation.
41. S. Shoji, L. H. Ericsson, K. A. Walsh, E. H. Fischer, K. Titani, *Biochemistry* **22**, 3702 (1983).
42. E. M. Reimann *et al.*, *ibid.* **23**, 4185 (1984).
43. A. P. Czernilofsky *et al.*, *Nature (London)* **287**, 198 (1980).
44. We thank A. Pereira for expert preparation of the manuscript.

17 July 1986; accepted 25 July 1986

Calmodulin

12

Ca^{2+} regulates many aspects of cell function. With the exception of muscle contraction, however, the means by which changes in cytosolic Ca^{2+} concentration affect events in the cell are not well understood. In skeletal muscle contraction, the key Ca^{2+} receptor is the small Ca^{2+}-binding protein, troponin C. In non-muscle cells this protein is not expressed, but instead many Ca^{2+}-dependent events are regulated by a closely related, ubiquitous and abundant Ca^{2+}-binding protein known as calmodulin. Calmodulin regulates multiple intracellular pathways, in response to changes in Ca^{2+} concentration, but, interestingly, its discovery (see selected paper [1]) arose from studies aimed at investigating the metabolism of another second messenger, cyclic AMP (see Section 1). In addition, the initial discovery of calmodulin as an activator of cyclic 3′,5′-nucleotide phosphodiesterase, the enzyme responsible for the breakdown of cyclic AMP, was purely fortuitous and came from the analysis of apparently anomalous results in the phosphodiesterase assay.

Cheung had been attempting to purify the phosphodiesterase activity to homogeneity in order to characterize the enzyme more fully, and found that, after a purification step involving ion-exchange chromatography on DEAE-cellulose, there was a dramatic loss of enzyme activity [2]. Such a loss of activity can often occur during protein purification due to instability of the protein. In the case of phosphodiesterase, however, Cheung noticed an unusual result that was dependent on exactly how the assay was performed. The enzyme preparation was usually incubated with cyclic AMP, and then, to assay formation of its product (5′-adenosine monophosphate), treated in a second step with snake venom as a source of 5′-nucleotidase to release the easily assayable inorganic phosphate from the reaction product. Surprisingly, Cheung found that if the snake venom was present during the first step, while cyclic AMP was being hydrolysed, the enzyme activity of the purified phosphodiesterase was markedly increased. The activity of the crude enzyme, in contrast, was unaffected. The correct interpretation arrived at was that snake venom contained an activator of the purified phosphodiesterase. The presence of an activator separated from the phosphodiesterase during purification was confirmed by adding back the more acidic protein fractions removed during ion-exchange chromatography. These fractions were shown to be at least as effective as snake venom in activation of phosphodiesterase activity [1]. The activator was found to be protein in nature and to be remarkably stable to heat, a property which was subsequently exploited in its purification. The results were interpreted as suggesting a stoichiometric interaction between the activator and the enzyme. It is now known that calmodulin does indeed bind to the phosphodiesterase and activate it in a stoichiometric complex.

The activation of phosphodiesterase by the cytosolic activator in a Ca^{2+}-dependent manner was described by Kakiuchi and co-workers [3], and within a short time it was demonstrated that the activator, later to be named calmodulin, was a Ca^{2+}-binding protein [4]. Subsequent work showed that calmodulin is expressed

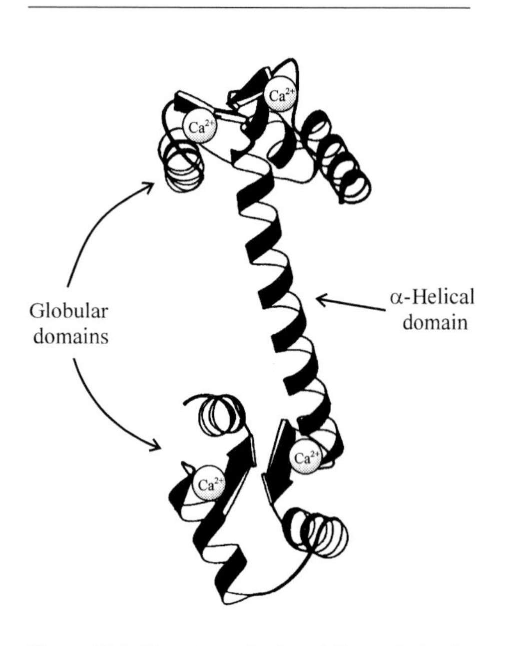

Figure 12.1 Structure of calmodulin as derived from X-ray crystallography. The figure is based on the structure determined in [13]. The positions of bound Ca^{2+} are shown (not to scale) in the two globular domains which are linked by a flexible α-helical domain.

in all tissues examined, that it is found in plants as well as animals [5,6], and that it is a small 16700 Da highly acidic protein containing 148 amino acids [7]. Analysis of the amino acid sequence of calmodulin [7] revealed that it shares around 50% identical amino acids with muscle troponin C, and that both proteins possess four putative Ca^{2+}-binding sites of a type that have been designated E-F hands [8]. The presence of four Ca^{2+}-binding sites was subsequently confirmed experimentally.

Studies on calmodulin in many laboratories have established this protein as a regulator of diverse cellular processes and enzymic activities. Among the many enzymes, in addition to phosphodiesterase, known to be directly regulated by Ca^{2+}-dependent interactions with calmodulin are adenyl cyclase, the plasma membrane Ca^{2+}-ATPase, myosin light chain kinase, NAD kinase, guanylate cyclase, phospholipase A_2, nitric oxide synthase (Section 14), calmodulin-dependent protein kinases I and II, and phosphorylase kinase [5,6,9]. In the case of phosphorylase kinase, which is a key enzyme in the control of glycogen breakdown, calmodulin not only binds reversibly to the kinase but is also one of the stable enzyme subunits [10]. The activation of calmodulin-dependent protein kinases, which phosphorylate numerous substrate proteins [11], further broadens the regulatory roles of calmodulin, explaining how it can be involved in the control of so many cellular processes [12].

As suggested by Cheung [1], calmodulin activates its target proteins by direct binding to form an activated complex, and work during the 1970s and early 1980s led to a model in which Ca^{2+} binding to calmodulin resulted in a conformational change in the protein to expose hydrophobic sites which could interact with binding partners. More recent work has been aimed at understanding this conformational change and binding interaction in molecular detail. The three-dimensional structure of calmodulin (Figure 12.1) was determined by X-ray crystallography, revealing it to be a dumbbell-shaped molecule with two globular domains, each containing two Ca^{2+}-binding sites joined by an α-helical domain [13]. The crystal structure of calmodulin complexed to synthetic peptides based on binding proteins such as myosin light chain kinase has also been described [14]. More recently, new information has come from the NMR determination of calmodulin structure in Ca^{2+}-free and Ca^{2+}-bound forms in solution, which has allowed examination of the conformational change that occurs on Ca^{2+} binding [15], and is giving new insights into the way in which this Ca^{2+}-binding protein acts as a key transducer of Ca^{2+} signals.

References

1. Cheung, W.Y. (1970) *Biochem. Biophys. Res. Commun.* **38**, 533–538
2. Cheung, W.Y. (1969) *Biochim. Biophys. Acta.* **191**, 303–309
3. Kakiuchi, S., Yamazaki, R. and Nakajima, H. (1970) *Proc. Jpn. Acad.* **46**, 587–592
4. Teo, T.S. and Wang, J.H. (1973) *J. Biol. Chem.* **248**, 5950–5955
5. Klee, C.B., Crouch, T.H. and Richman, P.G. (1980) *Annu. Rev. Biochem.* **49**, 489–515
6. Means, A.R. and Dedman, J.R. (1980) *Nature (London)* **285**, 73–77
7. Watterson, D.M., Sharief, F. and Vanaman, T.C. (1980) *J. Biol. Chem.* **255**, 962–975
8. Kretsinger, R.H. and Barry, C.D. (1975) *Biochim. Biophys. Acta.* **405**, 40–52
9. Cheung, W.Y. (1980) *Science* **207**, 19–27
10. Cohen, P., Burchell, A., Foulkes, J.G., Cohen, P.T.W., Vanaman, T.C. and Nairn, A.C. (1978) *FEBS Lett.* **92**, 287–293

11. Schulman, H. and Greengard, P. (1978) *Proc. Natl. Acad. Sci. U.S.A.* **75**, 5432–5436
12. Means, A.R., Tash, J.S. and Chafouleas, J.G. (1982) *Physiol. Rev.* **62**, 1–39
13. Babu, Y.S., Sack, J.S., Greenhough, T.J., Bugg, C.E., Means, A.R. and Cook, W.J. (1985) *Nature (London)* **315**, 37–40
14. Meador, W.E., Means, A.R. and Quiocho, F.A. (1992) *Science* **257**, 1251–1255
15. Chazin, W.J. (1995) *Curr. Opin. Struct. Biol.* **2**, 707–710

Cheung (1970) Biochem. Biophys. Res. Commun. **38**, 533–538

CYCLIC 3',5'-NUCLEOTIDE PHOSPHODIESTERASE

Demonstration of an Activator

Wai Yiu Cheung

Department of Biochemistry, St. Jude Children's
Research Hospital and University of Tennessee
Medical Units, Memphis, Tennessee 38101

Received December 16, 1969

SUMMARY. The activity of a mixture of a purified and a crude cyclic 3',5'-nucleotide phosphodiesterase was greater than the summed activities of the individual preparations. The crude enzyme contained an activator, which was removed from the purified enzyme during purification. The activator, isolated free of phosphodiesterase activity, effectively reconstituted the activity of the purified enzyme. The relative inactivity of purified phosphodiesterase was due to removal of the activator from the enzyme.

INTRODUCTION

Cyclic AMP is believed to be a mediator of a variety of different hormones (1). Cyclic 3',5'-nucleotide phosphodiesterase catalyzes the hydrolysis of this nucleotide to 5'-AMP. Thus the activity of phosphodiesterase is critically related to the tissue levels of cyclic AMP, and therefore to the extent and duration of the related hormonal action.

In an attempt to elucidate mechanisms regulating tissue levels of cyclic AMP, we have studied factors which affect phosphodiesterase activity (2,3). We found that while the crude phosphodiesterase of bovine brain cerebra was fully active, the purified enzyme was relatively inactive (4). The purified enzyme depended on a stimulatory factor(s) in snake venoms for maximal activity. We now show that an activator of phosphodiesterase is present in a crude homogenate of bovine brain cerebra. The activator, which has been prepared free of phosphodiesterase, markedly stimulates the purified enzyme.

METHODS

Phosphodiesterase was prepared and assayed as described elsewhere (4). Briefly, the enzyme was extracted with water from bovine brain cerebra, and was purified by pH and ammonium sulfate fractionation, calcium phosphate gel and DEAE-cellulose column chromatography. The purified enzyme was relatively inactive unless supplemented with its activator. To demonstrate activation, enzyme activity was measured using a two-stage procedure, modified slightly after that of Butcher and Sutherland (5). At the end of the first stage of incubation, enzymic activity was arrested by boiling. After thermal equilibration, snake venom (*Crotalus atrox*) was added to all assayed tubes for a second incubation. Venom was used as a source of 5'-nucleotidase, which converted 5'-AMP into adenosine and inorganic phosphate. The latter was measured colorimetrically. Unless otherwise indicated, the venom was not present in the first incubation. It should be pointed out that the venom was inactive towards cyclic AMP and that all data were corrected for a control containing no phosphodiesterase.

The activator was isolated free of phosphodiesterase at the last stage of our purification procedure. It was eluted after the enzyme on the DEAE-cellulose column. Tubes following the peak activity of phosphodiesterase were pooled. The combined eluate was lyophilized and then dialyzed against 20 m*M* Tris-Cl, pH 7.5. The activator was retained inside the dialysis tubing. The dialyzed sample usually contained a small amount of denatured protein, which was removed by centrifugation.

Proteins were determined by the biuret reagent containing deoxycholate or by the spectrophotometric method of Warburg and Christian (6).

RESULTS AND DISCUSSION

Table I compares the activity of a crude and a purified phosphodiesterase assayed individually to that of a mixture of these two enzymes assayed together. The activity of the mixture was more than twice the summed activities of the

TABLE I

COMPARISON OF THE ACTIVITY OF A CRUDE AND A PURIFIED PHOSPHODIESTERASE WITH THAT OF A MIXTURE OF THESE TWO.

	Fraction	Activity (mμ moles)
[a]	Crude phosphodiesterase	66
[b]	Purified phosphodiesterase	40
[c]	Crude + purified phosphodiesterase	260

Phosphodiesterase was assayed using a two-stage procedure as summarized under Methods. A homogenate of bovine brain cerebra was dialyzed extensively against 20 m$\underline{M}$ Tris-Cl, pH 7.5 and was then centrifuged at 40,000 g for 20 minutes. An aliquot of the supernatant fluid was used as a source of the crude phosphodiesterase. The purified phosphodiesterase was prepared from a DEAE-cellulose column and was relatively inactive prior to activation (4). Protein concentration in μg per 0.5 ml of the reaction mixture: crude phosphodiesterase, 110; purified phosphodiesterase, 50. Activity is expressed as mu moles of inorganic phosphate formed per 10 minutes.

individual preparations. This experiment demonstrated the synergistic effect of the mixture but did not reveal which one of the two components was the activating agent.

Our studies on the stimulation of purified phosphodiesterase by snake venom is illuminating in this respect. Table II shows the effect of the venom on the activity of a crude and a purified phosphodiesterase. The venom activated the purified but not the crude enzyme (compare [d] with [c] and [b] with [a]). Also, the activity of the mixture was markedly greater than the summed activities of the individual preparations, as was shown in Table I. Of particular interest is the fact that the activity of the mixture was equivalent to the summed activities of the crude and the purified enzyme subsequent to its activation by the venom. This experiment suggested two things. First, since the purified,

TABLE II

EFFECT OF SNAKE VENOM ON THE ACTIVITY OF CRUDE AND PURIFIED PHOSPHODIESTERASE.

Fractions	Activity (mμ moles)
[a] Crude phosphodiesterase	262
[b] Crude phosphodiesterase + venom	279
[c] Purified phosphodiesterase	80
[d] Purified phosphodiesterase + venom	454
[e] Crude + purified phosphodiesterase	658

Phosphodiesterase was assayed using a two-stage procedure. When indicated, venom was present in the first stage of incubation. Protein concentration in μg per one ml of the reaction mixture: crude phosphodiesterase, 245; purified phosphodiesterase, 100; snake venom (Crotalus atrox), 100. Activity is expressed as mu moles of inorganic phosphate formed per 10 minutes.

---

but not the crude enzyme, was stimulated by the venom, the purified enzyme was the component whose activity had been stimulated. In other words, the activating agent (or activator) was probably associated with the crude enzyme. Second, this activating agent was as effective as the venom in stimulating the purified enzyme.

Phosphodiesterase is a stable enzyme in our hands. Other experiments have shown that the stimulation of the purified enzyme by the venom is not due to protection of some unstable phosphodiesterase activity afforded by proteins in the venom (4). This would suggest that the increased activity of the purified enzyme in the presence of the homogenate was not due to protection of some unstable activity, but rather to a stimulation of the enzyme.

TABLE III

EFFECT OF ACTIVATOR ON THE ACTIVITY OF PURIFIED PHOSPHODIESTERASE.

Fractions	Activity (mμ moles)
[a] Purified phosphodiesterase	56
[b] Activator	0
[c] Purified phosphodiesterase + activator	315
[d] Purified phosphodiesterase + venom	306

Phosphodiesterase was assayed using a two-stage procedure. When indicated, venom was present in the first stage of incubation. The activator was a preparation which had been stored at -20° for six months prior to use. Protein concentration in μg per one ml of the reaction mixture: purified phosphodiesterase, 100; activator, 50; snake venom (Crotalus atrox), 100. Activity is expresses as mu moles of inorganic phosphate formed per 10 minutes.

From such considerations, we reasoned that the relative inactivity of the purified enzyme was due to removal of an activator during the course of enzyme purification. Indeed, we found that the activator was present originally in the crude homogenate and that it was removed from phosphodiesterase as purification proceeded. Fractions obtained early in the purification procedure were rich in the activator while those obtained at a later stage were deficient. To illustrate this point, the preparation of the activator from the DEAE-cellulose column as described under Methods may be cited. The enzyme prior to this step was fully active (4). When the activator was dissociated from the enzyme on the column, the enzyme became relatively inactive. Table III shows the effect of the activator on the activity of a purified phosphodiesterase. In this experiment, the activator, itself inactive [b], increased the activity of the purified enzyme more than five-fold (compare [c] with [a]). The venom, as usual, caused a pronounced stimulation [d]. It is noted that the extent of stimulation by the activator and by the venom was comparable.

The activator was inactivated by trypsin, and not by DNase or RNase, indicating that it was a protein. Gel filtration using a calibrated Sephadex G-100 column gave a molecular weight of about 40,000. The activator was remarkably stable to heat, to acidic pH's and to 8*M* urea. No proteolytic activity could be demonstrated. Stimulation of the purified enzyme by the activator appeared specific, as several proteins of various molecular weights were unable to mimic its stimulatory effect.

Stimulation of the enzyme was directly proportional to the concentration of the activator, indicating a stoichiometric interaction between the two. Kinetic analysis showed that the increase of Vmax was coupled to a decrease of Km for cyclic AMP. Although these results do not reveal the mechanisms of stimulation, it is conceivable that the activator may be important in regulating phosphodiesterase *in vivo*.

Acknowledgement: This work was supported by U.S. Public Health Service Grants NB08059 and CA08480. It was also supported by ALSAC.

REFERENCES

1. Robinson, G.A., Butcher, R.W., and Sutherland, E.W., Ann. Rev. Biochem., 37,149(1968).
2. Cheung, W.Y., Biochemistry, 6,1079(1967).
3. Cheung, W.Y., and Jenkins, A., Federation Proc., 28,473(1969).
4. Cheung, W.Y., Biochim. Biophys. Acta, 191,303(1969).
5. Butcher, R.W., and Sutherland, E.W., J. Biol. Chem., 237,1244(1962).
6. Warburg, O., and Christian, W., Biochem. Z., 310,384(1941).

The site of action of Ca^{2+} in exocytosis in neurons and neuroendocrine cells

13

The work of Bernard Katz and his co-workers in the 1950s established that the release of neurotransmitter at the frog neuromuscular junction was quantal in nature. This led to the idea that the release of these quanta or packets of neurotransmitter resulted from the fusion of synaptic vesicles (containing fixed amounts of neurotransmitter) with the presynaptic membrane to allow release of their contents by exocytosis. The paper by Katz and Miledi that we have chosen [1] was one of a series [1–4] published by these authors between 1965 and 1967 that further characterized neurotransmission at the frog neuromuscular junction, with the aim of determining the site of action of Ca^{2+}. This work established that Ca^{2+} acts on the neurotransmitter-release process, and the chosen paper showed that external Ca^{2+} was required during the depolarization phase and thus immediately preceding release. This led to the suggestion by Katz and Miledi that it was Ca^{2+} entering the nerve terminal that stimulated the release process. A direct demonstration of the activation of a vesicular-release process by Ca^{2+}, and determination of the internal Ca^{2+} concentration required, was first achieved by Baker and Knight (see the second selected paper [5]) using adrenal chromaffin cells whose plasma membranes had been made leaky and thus permeable to Ca^{2+} and other small molecules. These studies led to the general acceptance of the idea that a rise in cytosolic Ca^{2+} concentration is the trigger for the release of neurotransmitters, hormones and other extracellularly acting molecules. More recent work has characterized the Ca^{2+}-dependent steps in exocytosis, the Ca^{2+} concentration required to activate them and many of the proteins that are involved.

The requirement for external Ca^{2+} for neuromuscular transmission was first demonstrated by Locke in 1894 [6] and the importance of calcium for neurotransmission was generally accepted. It had been suggested that Ca^{2+} played a part in the mechanism leading to quantal (vesicular) release, but an alternative idea was that external Ca^{2+} was required for propagation of the nerve impulse [7]. Using a variety of experimental manipulations, Katz and Miledi were able to show quite clearly that Ca^{2+} was not required for impulse propagation but for the release process itself.

The approaches used by Katz and Miledi were as follows. First, the use of localized application of Ca^{2+} from a micropipette to preparations in which transmission was blocked by high Mg^{2+} showed that Ca^{2+} acted rapidly on the nerve terminal, and that its action was not due to any change in the action potential [2]. Secondly, blockade of the voltage-sensitive Na^{+} channels of the nerves using tetrodotoxin to prevent impulse conduction showed that neurotransmission still occurred following local depolarization of the nerve terminal in a Ca^{2+}-dependent manner [3]. Thirdly, lengthening the depolarization pulse led to an increase in the latency period before neurotransmission occurred, suggesting that the first step towards release is the entry into the nerve terminal of a positively charged substance, Ca^{2+} or a Ca^{2+} compound, CaR^{+} [4]. The conclusion at this stage of the investigation was that: "The subsequent steps could be pictured in various ways, e.g. according to the vesicular hypothesis as a reaction on the inside of the membrane by which Ca or CaR causes momentary fusion of vesicular and axon envelopes at their point of collision....". Fourthly, the final paper in this series examined the time at which Ca^{2+} is again required during neurotransmission, using local ionophoretic application of Ca^{2+} to nerve terminals. These studies revealed that Ca^{2+} is required to be present before or during the depolarization period and is ineffective if added after the depolarization of the nerve ending but before neurotransmitter release is expected to occur [1]. The correct interpretation provided by Katz and Miledi was that "the most likely picture on the present evidence is this: (i) depolarization of the axon terminal opens a 'gate' to calcium; (ii) calcium moves to the inside of the axon membrane and (iii) becomes involved in a reaction which causes the rate of transmitter release to increase and which contributes a large part of the synaptic delay".

Later work based on microinjection of Ca^{2+} into the presynaptic nerve terminals of the squid giant synapse [8], or the use of Ca^{2+} ionophores to enable Ca^{2+} entry [9], confirmed that the site of action of Ca^{2+} on exocytosis is indeed intracellular. However, further progress in the study of the factors involved was hampered by the difficulty of gaining access to the intracellular sites of exocytosis in synapses or in other secretory cell types. A new approach was opened up by the work of Baker and Knight [5] who used high-voltage discharges to permeabilize the plasma membrane of adrenal chromaffin cells to low-molecular-mass compounds, and then followed the exocytotic release of catecholamines. This allowed the determination of the concentration of internal free Ca^{2+} required to activate exocytosis (half-maximal at 1 μM), as well as the demonstration of a requirement for MgATP. Since the work of Baker and Knight, a whole range of cell-permeabilization methods has been developed to allow the characterization of, not only exocytosis, but all aspects of the secretory pathway in cells, and to make possible the examination of the proteins involved. In addition, the introduction of whole-cell patch-clamp recording has allowed both manipulation of the intracellular environment and high-resolution recording of the kinetics of exocytosis by monitoring of plasma membrane capacitance [10].

Baker and Knight suggested that their permeabilized cell preparation might lead to the discovery of the role of MgATP in exocytosis. Despite considerable work on many cell types it is still not clear where and how MgATP acts in exocytosis, and the idea that it plays an essential role has been revised. Baker and Knight showed that if the chromaffin cells were permeabilized and left for 30 min in the absence of MgATP, then they no longer responded to a challenge with a Ca^{2+} trigger. The initial interpretation was that MgATP is essential for Ca^{2+} to be able to trigger exocytosis. Work by the laboratory of Holz [11], however, has produced a different picture. Using digitonin-permeabilized chromaffin cells it was shown that there exists a labile MgATP-primed state that is rapidly lost if MgATP is removed, but that Ca^{2+}-triggered exocytosis can occur in the absence of MgATP if the cells are already primed by prior exposure to MgATP.

The general picture that we now have of exocytosis is a multi-stage phenomenon in which early steps require MgATP and ATP hydrolysis, and later steps closer to the membrane fusion events are MgATP-independent. In addition, it appears that two or more steps are regulated by Ca^{2+} [12]. What has also become clear more recently is that the concentration of Ca^{2+} required to activate exocytosis in adrenal chromaffin and other neuroendocrine cells is much lower than that in synaptic terminals. This information has come from studies using rapid release of Ca^{2+} from caged Ca^{2+} compounds within cells, and determination of the kinetics of exocytosis using patch-clamp capacitance measurement. Both early and late steps in exocytosis in neuroendocrine cells require Ca^{2+}, and the final step leading to fusion is half-maximally activated at around 15–30 μM free Ca^{2+} [13,14]. In contrast, in the one neuronal cell type that has been examined in this way, the retinal bipolar neuron of the goldfish [15], exocytosis involves a lower Ca^{2+} affinity process which is half-maximal at 190 μM free Ca^{2+}. The difference in Ca^{2+} dependencies relates to the fact that secretory vesicles in neuroendocrine cells are some distance from the plasma membrane and must respond to lower Ca^{2+} concentrations as Ca^{2+} diffuses through and is buffered by the cytoplasm. In contrast, synaptic vesicles in nerve terminals are docked in close physical contact with Ca^{2+} channels in the presynaptic membrane. They must respond very rapidly when Ca^{2+} is raised to high levels at the mouth of the Ca^{2+} channels, but must be switched off just as rapidly as the depolarization ends and Ca^{2+} is buffered back to resting levels.

Progress has been made in recent years in the identification of some of the proteins that are involved in neurotransmitter and hormone release, and it is clear that many proteins must interact to control the multiple steps leading to membrane fusion [16,17] and that considerably more work will be needed to elucidate this complex process. The identity of the various proteins that act as Ca^{2+} receptors in the activation of exocytosis in neurons and neuroendocrine cells is still a matter of debate [12], and it is thought that multiple Ca^{2+}-binding proteins may control membrane fusion. One protein that appears to act as at least one of the neuronal Ca^{2+} receptors is synaptotagmin I [18]. Evidence for the role of this protein has come from electrophysiological measurements of neurotransmission between hippocampal neurons in culture using methods not dramatically dissimilar to those used by Katz and Miledi. The difference now is the added power of molecular biology to generate transgenic mice in which the synaptotagmin I gene has been specifically disrupted to allow the assay of transmission in neurons lacking this protein. It will be the continued application of electrophysiological, biochemical and molecular biological approaches in concert that will eventually lead to a full description of how intracellular Ca^{2+} activates the fusion of vesicular and plasma membranes, leading to neurotransmission.

References

1. Katz, B. and Miledi, R. (1967) *J. Physiol. (London)* **189**, 535–544
2. Katz, B. and Miledi, R. (1965) *Proc. R. Soc. London Ser. B* **161**, 496–503
3. Katz, B. and Miledi, R. (1967) *Proc. R. Soc. London Ser. B* **167**, 8–22
4. Katz, B. and Miledi, R. (1967) *Proc. R. Soc. London Ser. B* **167**, 23–38
5. Baker, P.F. and Knight, D.E. (1978) *Nature (London)* **276**, 620–622
6. Locke, F.S. (1894) *Zb. Physiol.* **8**, 166–167
7. Franke, G.B. (1963) *J. Pharmacol.* **139**, 261–268
8. Miledi, R. (1973) *Proc. R. Soc. London Ser. B.* **183**, 421–425

9. Foreman, J.C., Mongar, J.L. and Gomperts, B.D. (1973) *Nature (London)* **245**, 249–251
10. Neher, E. and Marty, A. (1982) *Proc. Natl. Acad. Sci. U.S.A.* **79**, 6712–6719
11. Holz, R.W., Bittner, M.A., Peppers, S.C., Lenter, R.A. and Eberhard, D.A. (1989) *J. Biol. Chem.* **264**, 5412–5419
12. Burgoyne, R.D. and Morgan, A. (1995) *Trends Neurosci.* **18**, 191–196
13. Heinemann, C., Chow, R.A., Neher, E. and Zucker, R.S. (1994) *Biophys. J.* **67**, 2546–2557
14. Thomas, P., Wong, J.G., Lee, A.K. and Almers, W. (1993) *Neuron* **11**, 93–104
15. Heidelberger, R., Heinemann, C., Neher, E. and Matthews, G. (1994) *Nature (London)* **371**, 513–515
16. Burgoyne, R.D. and Morgan, A. (1993) *Biochem. J.* **293**, 305–316
17. Morgan, A. (1995) *Essays Biochem.* **30**, 77–95
18. Geppert, M., Geppert, M., Goda, Y., Hammer, R.E., Li, C., Rosahl, T.W., Stevens, C.F. and Sudhof, T.C. (1994) *Cell* **79**, 717–727

Katz & Miledi (1967) J. Physiol. (London) **189**, 535–544

THE TIMING OF CALCIUM ACTION DURING NEUROMUSCULAR TRANSMISSION

By B. KATZ and R. MILEDI

From the Department of Biophysics, University College London

(*Received* 7 *November* 1966)

SUMMARY

1. When a nerve–muscle preparation is paralysed by tetrodotoxin, brief depolarizing pulses applied to a motor nerve ending cause packets of acetylcholine to be released and evoke end-plate potentials (e.p.p.s), provided calcium ions are present in the extracellular fluid.

2. By ionophoretic discharge from a 1 M-$CaCl_2$ pipette, it is possible to produce a sudden increase in the local calcium concentration at the myoneural junction, at varying times before or after the depolarizing pulse.

3. A brief application of calcium facilitates transmitter release if it occurs immediately before the depolarizing pulse. If the calcium pulse is applied a little later, during the period of the synaptic delay, it is ineffective.

4. It is concluded that the utilization of external calcium ions at the neuromuscular junction is restricted to a brief period which barely outlasts the depolarization of the nerve ending, and which precedes the transmitter release itself.

5. The suppressing effect of magnesium on transmitter release was studied by a similar method, with ionophoretic discharges from a 1 M-$MgCl_2$-filled pipette. The results, though not quite as clear as with calcium, indicate that Mg pulses also are only effective if they precede the depolarizing pulses.

INTRODUCTION

The presence of extracellular calcium ions is known to be essential for neuromuscular transmission. The principal point of calcium action is the process by which the nerve impulse releases acetylcholine from the motor nerve endings (Katz & Miledi, 1965*c*). This process can be studied even when the nerve impulse and its accompanying sodium current have been eliminated by tetrodotoxin; under these conditions a brief depolarizing pulse locally applied to the nerve ending causes 'packets' of acetylcholine to be released, provided calcium ions are present in the extracellular medium (Katz & Miledi, 1967*a*). The time course of the transmitter release could be determined with great accuracy, by measuring the statistically varying intervals between pulse and quantal e.p.p.s in a long series of observations. After a brief pulse (0·5–1 msec), there is a short delay, about 1–2 msec at 5° C, during which the probability of release does not perceptibly exceed the low background level. This is followed by a rapid rise to a peak and a gradual decline of the probability of release which may extend over 10 msec or more.

It was of interest to find out at what stage during this sequence of events the external calcium ions come into play. By close ionophoretic application, it is possible to produce a sudden increase in the local calcium concentration, and to time it fairly accurately in relation to the depolarizing pulse. The question which the present experiments are meant to answer is whether the extracellular calcium is utilized during the period preceding the transmitter release (i.e. during depolarization plus initial latency), or whether it becomes effective during the transmitter release itself.

METHODS

The procedure follows that described in previous papers (Katz & Miledi, 1965*a*, *d*; 1967*a*, *b*). The muscle (sartorius of *Rana pipiens* or *temporaria*) was placed in a Ringer solution of low calcium (< 0·1 mM) and added magnesium (about 1 mM) content, paralysed by tetrodotoxin (about 10^{-6} g/ml.) and kept at low temperature (about 4° C). The main difference from previous work is that a twin-pipette was employed, containing 1 M solutions of NaCl and $CaCl_2$ respectively in the two barrels. The sodium channel was used to apply depolarizing and various electrical control pulses to the surface of the nerve terminal, while the other channel was used to raise the local calcium concentration at desired moments. Once a calcium effect was seen, the strength and duration of the pulse was reduced and the pipette carefully re-positioned until an optimum effect was obtained. E.p.p.s were recorded with an intracellular electrode. Prostigmine (10^{-6} g/ml.) was used in most experiments. In other experiments, twin- or triple-pipettes were used one of whose barrels contained 1 M-$MgCl_2$ replacing, or in addition to, the calcium pipette.

The use of multiple-barrel pipettes made special precautions necessary. A strong current pulse through one of the barrels was apt to cause a transient resistance change in an adjacent barrel, possibly by dislodging charged particles from the tip. This was checked by monitoring the current intensities, and it was verified that, usually, barrel interactions of this kind were not large enough to vitiate the results.

When applying a strong positive-going pulse to the calcium pipette, the possibility of electrical (as distinct from ionophoretic) effects had to be considered (see Katz & Miledi, 1967 *b*, Methods). Certain control experiments have already been reported (Katz & Miledi, 1967*b*): depolarizing pulses were followed after intervals of 20–140 μsec by similar hyperpolarizing (positive-going) pulses from the same sodium pipette. It was shown that there is no significant interference by the positive pulse, if it was applied more than 100 μsec after the end of the depolarization, and only a small reduction of the transmitter release when applied within 20–50 μsec. These earlier findings are also relevant to the observations described below.

RESULTS

Figures 1 and 2 illustrate results of two experiments. A series of depolarizing pulses (*P*) was applied via the sodium barrel, at intervals of several seconds. The pulses given by themselves failed to evoke more than very infrequent unit e.p.p.s. If P was preceded by a brief ionophoretic discharge of calcium from the other barrel (positive-going pulse Ca), the failure rate was greatly reduced, and unit responses occurred at varying times as shown in Figs. 1–4. (Italicized symbols (Ca, Mg) refer to ionophoretically applied doses of these substances.)

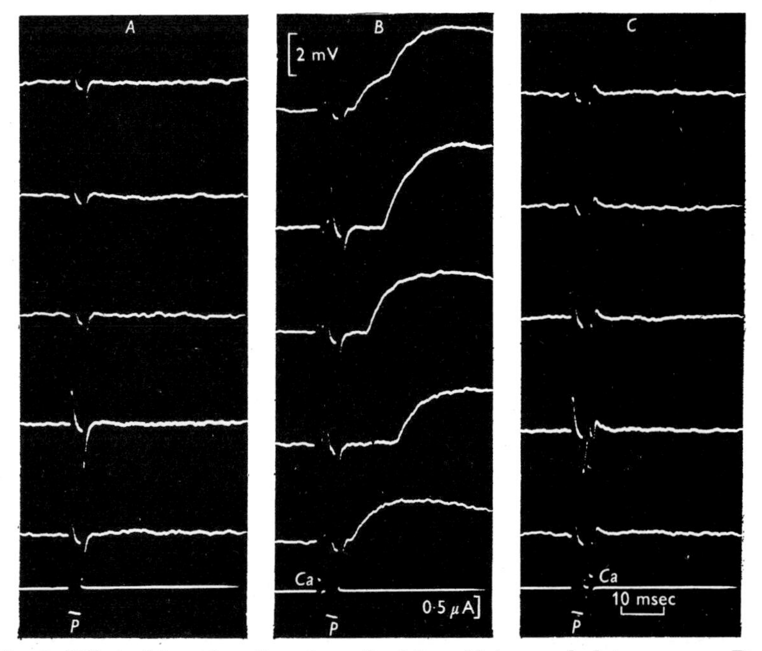

Fig. 1. Effect of ionophoretic pulses of calcium (Ca) on end-plate response. Depolarizing pulses (P), and calcium were applied from a twin-barrel micropipette to a small part of the nerve–muscle junction. Intracellular recording from the end-plate region of a muscle fibre. Bottom traces show current pulses through the pipette. Column *A*. Depolarizing pulse alone. *B*. Calcium pulse precedes depolarizing pulse. *C*. Depolarization precedes calcium pulse. Temperature 4° C.

To obtain a maximum effect, the interval between Ca and P pulses had to be adjusted so as to allow the calcium concentration to reach a peak at the critical site and time. In the case of Fig. 5, the optimum Ca–P interval was about 10–20 msec. But a significant effect could be obtained with much shorter intervals, and this was important for the present study. With careful placing of the pipette, a facilitating action of calcium could be observed when a Ca pulse as brief as 1 msec was applied, separated from the start of the depolarizing pulse by as little as 50–100 μsec.

In the experiments illustrated in Figs. 1 and 2, pairs of $Ca+P$ pulses alternated with pairs in which the time sequence of Ca and P was reversed. The result was unequivocal: Ca pulses given *after* the depolarization were ineffective. They failed to raise the level of response even if they were applied immediately after *P*, i.e. during the minimum latent period of the recorded response. It should be noted that it was not possible to synchronize the *Ca* and *P* pulses, because simultaneous application of the two pulses of opposite sign cancelled some of the depolarization produced by *P* alone.

One may conclude from these experiments that the utilization of external calcium ions occurs during, and possibly immediately after, the depolarization, that is during a period preceding that of increased probability of transmitter release. It is true that the time resolution of the ionophoretic method is limited because of inevitable diffusion delay. Nevertheless, the difference in timing between effective *Ca* pulses, immediately before *P*,

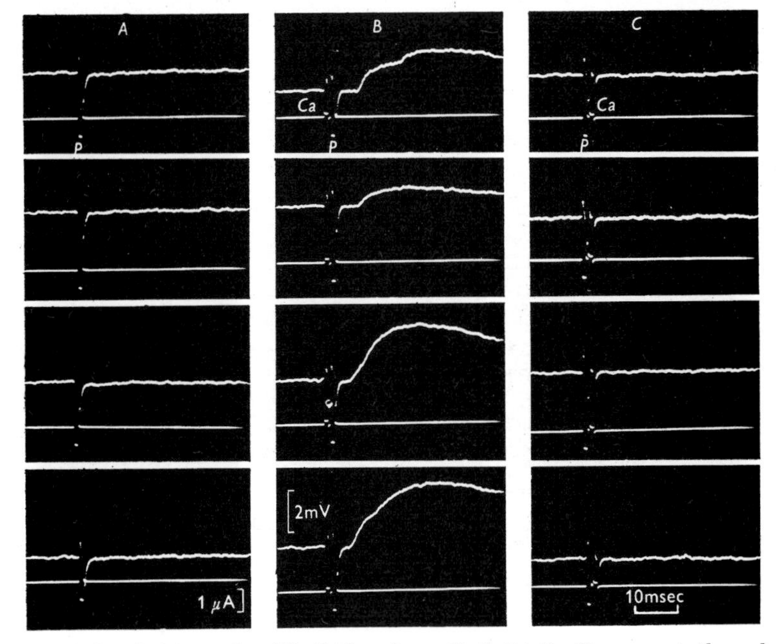

Fig. 2. *A*, depolarizing pulses (*P*); *B*, *Ca* pulses + *P*; *C*, *P* + *Ca*. The currents through the twin-pipette are shown in the bottom trace of each record. Temperature 3° C.

and ineffective *Ca* pulses, after *P*, was so small that our interpretation cannot be in doubt. If one were to argue that the responses obtained with *Ca* + *P* are due to external calcium ions reaching the critical membrane sites only after the initial latent period, then there would be no explanation for the fact that calcium ions discharged ionophoretically during the latency fail to promote the occurrence of later units of response (Figs. 3, 4).

A number of controls were made to check on any electric interaction between the two barrels. In addition to the observations mentioned in Methods, a test was made to see whether the negative *P* pulse from the sodium barrel interfered in any way with the discharge of calcium by an immediately following pulse from the other barrel. This point was examined by giving a triple sequence of pulses $P_1 + Ca + P_2$, and comparing the effect with the usual $Ca + P$ action (relating, in each case, the response to that obtained without the Ca pulse). The result is shown in Table 1. The calculated values of 'm' are not at all accurate, but there was clearly no evidence for any substantial reduction of the calcium effect by the immediately preceding negative pulse P.

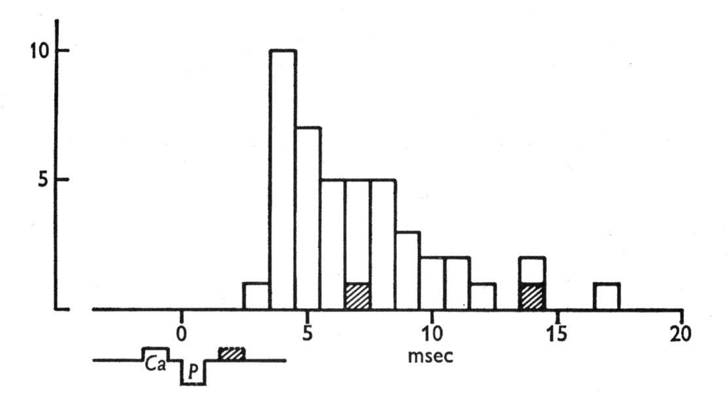

Fig. 3. Histogram of 'synaptic delays'. Abscissa: time interval between start of depolarizing pulse (P) and beginning of a unit end-plate potential. The time relations between calcium and depolarizing pulse are indicated below. Ordinate: number of observed unit potentials. The main histogram shows responses evoked by forty-nine ($Ca + P$) pulses. Shaded blocks: responses evoked by 48 ($P + Ca$) pulses. (These were as infrequent as responses due to P alone.) Temperature 3° C.

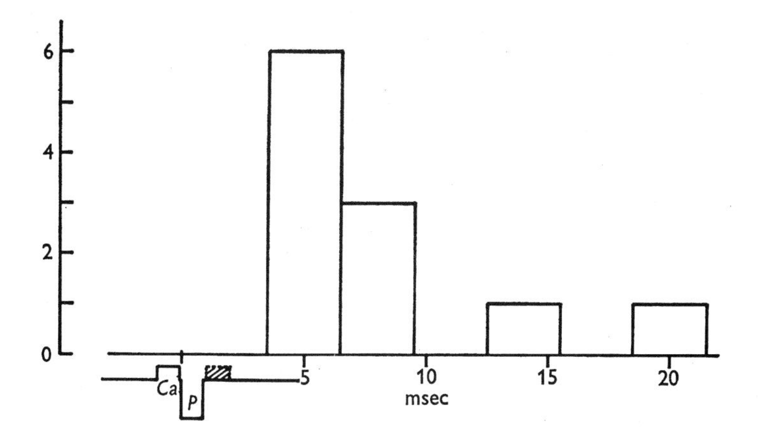

Fig. 4. Another histogram, obtained with very brief separation between P and Ca pulses. $Ca + P$ gave eleven unit responses (as shown) to twenty-nine pulses. $P + Ca$ (see shaded Ca block below) produced no response in fourteen trials. P alone also failed to evoke any response. Temperature 4·5° C.

Another test was to verify that a *Ca* pulse following *P* did not *reduce* the response. This was done simply by raising the steady Ca-efflux from the pipette (see Katz & Miledi, 1965*a*, *c*) and repeating the experiment with an increased initial rate of *P* responses.

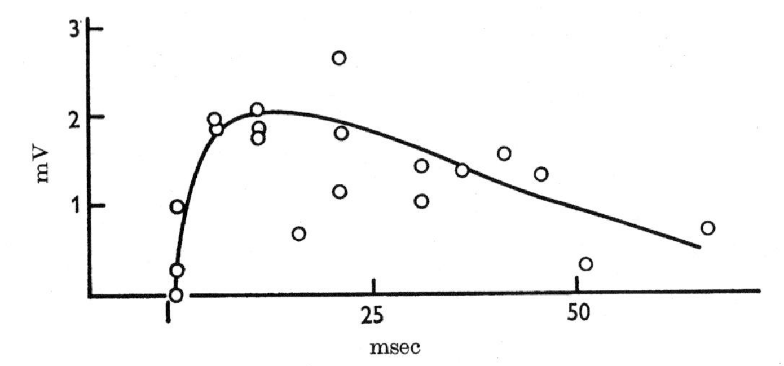

Fig. 5. Effect of time interval between *Ca* and *P* on end-plate response. Abscissa: Interval between start of *Ca* pulse (1 msec duration) and 2·7 msec depolarizing pulse *P*. Ordinates are amplitudes of individual e.p.p.s evoked by *P*. Temperature 3° C.

TABLE 1. Control experiment

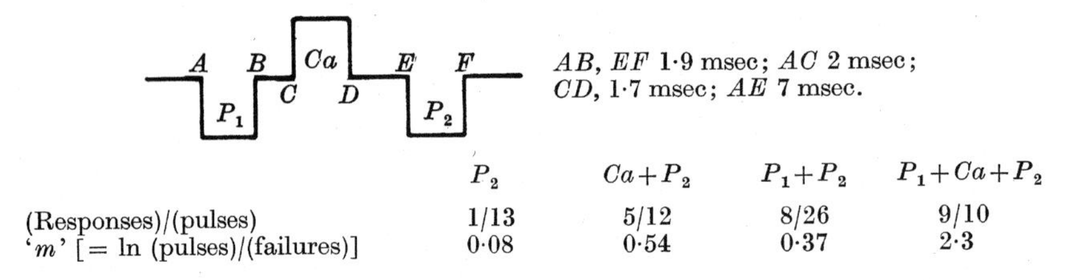

AB, *EF* 1·9 msec; *AC* 2 msec; *CD*, 1·7 msec; *AE* 7 msec.

	P_2	$Ca+P_2$	P_1+P_2	P_1+Ca+P_2
(Responses)/(pulses)	1/13	5/12	8/26	9/10
'*m*' [= ln (pulses)/(failures)]	0·08	0·54	0·37	2·3

In a few experiments, the inhibitory influence of magnesium on transmitter release was studied by ionophoretic application (Figs. 6–8). Magnesium is less potent in equimolar concentration than its antagonist calcium, and as a consequence it was difficult to produce a sufficiently intense suppression by brief ionophoretic discharges from the *Mg*-barrel. Indeed, the positive pulses had to be made so strong that local membrane break-down was a serious risk (see Katz & Miledi, 1967*b*). It was possible, nevertheless, to obtain clear inhibitory effects with *Mg* pulses of 5 msec duration provided they were applied *before P*. Figs. 7 and 8 illustrate an experiment with a triple pipette (the three barrels containing respectively, Ca, Na and Mg). A *Ca* pulse was given throughout, preceding *P* by about 15 msec, in order to elicit a background rate of response sufficiently high for testing the suppressing action of *Mg*. Figure 8 combines the results of two complete series in which somewhat different interval settings for *P* and *Mg* pulses were chosen. The bottom part shows the responses to eighty-three *P* pulses in the absence of magnesium. When *Mg* pulses preceded *P*, the average value of m (i.e. ln (number of pulses)/(number of failures)) was reduced from 0·96 to 0·22, and from 0·4 to 0·03 in the two series, respectively. The residual responses (to sixty-one pulses) are represented in the histogram at the top (Fig. 8*A*). When the *Mg* pulses *followed* *P* (middle histogram, Fig. 8*B*, sixty-five pulses), no suppression was observed.

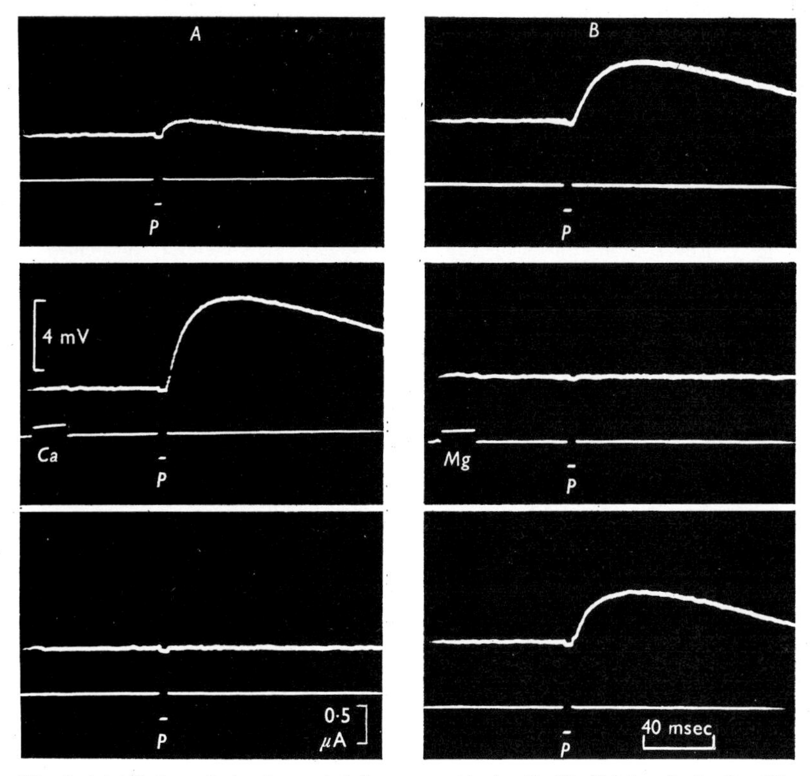

Fig. 6. A triple-barrel pipette containing, respectively, Ca, Na (for depolarization *P*), and Mg, was used. Column *A* shows the facilitating action of a calcium pulse; column *B* shows the inhibiting influence of a magnesium pulse. To demonstrate the inhibitory effect, the bias on the calcium barrel was adjusted between *A* and *B*, so as to increase the response to *P* alone. Temperature 5° C.

Several other experiments of a similar kind were made; in some the result differed from that shown in Fig. 8, in that the *Mg* pulse reduced the response slightly even when it was applied immediately after *P*. It is not certain whether this was attributable to the magnesium ion, or to the small suppressing effect which a hyperpolarizing pulse had occasionally been found to produce (Katz & Miledi, 1967*b*). Apart from this small and inconsistent effect, the results were in line with those of Fig. 8.

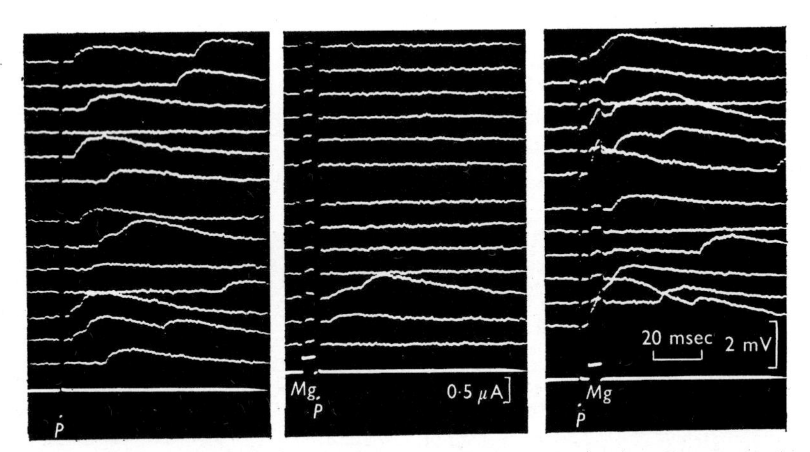

Fig. 7. Effect of *Mg*-pulses on end-plate response. Left column: Responses to depolarizing pulses *P* alone. Middle column: *Mg* pulse precedes *P*. Right column: *Mg* pulse follows *P*. Note: in the right column, the positive-going *Mg* pulse overlapped in time with the rising phase of some of the responses, and the associated focal hyperpolarization caused a transient increase in the amplitude of the e.p.p. (see Katz & Miledi, 1965*a*). Temperature 4° C. In this figure and Fig. 8, a *Ca*-pulse was applied in all trials, preceding *P* by about 15 msec, to raise the level of response sufficiently for inhibition to be demonstrable.

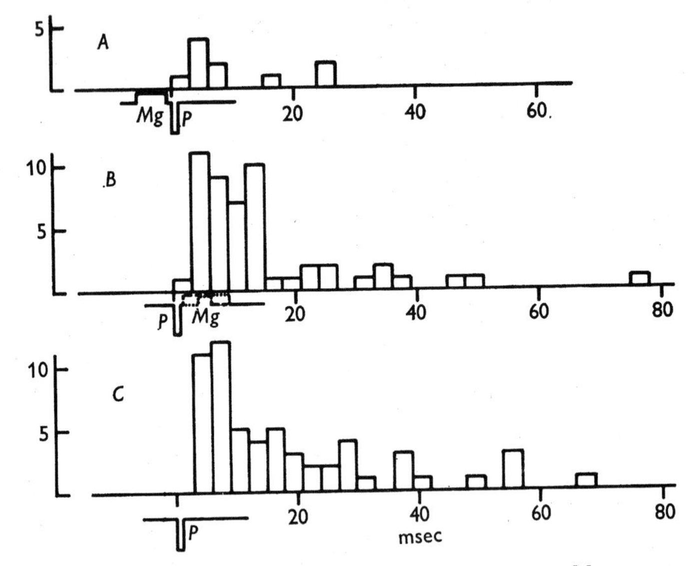

Fig. 8. Histograms of synaptic latencies. Abscissa: time interval between start of depolarizing pulse and start of unit e.p.p. Ordinate: number of observed unit potentials. Top histogram *A* shows distribution of responses to 61 (*Mg*+*P*) pulses. *Middle* (*B*): Responses to 65 (*P*+*Mg*) pulses. Two series with different *P*–*Mg* intervals as indicated by the interrupted and dotted outlines of the *Mg* pulses. Bottom histogram *C*: responses to eighty-three depolarizing pulses alone. Temperature 3° C.

DISCUSSION

It has been suggested that the first step by which depolarization of nerve endings leads to quantal release of acetylcholine is influx of calcium through the axon membrane (Hodgkin & Keynes, 1957; Katz & Miledi, 1967*b*). Recent observations have shown that there is a considerable delay between a brief depolarizing pulse and the release. The release may not even commence until sometime after the end of the depolarization (Katz & Miledi, 1965*b*, 1967*b*). The question, therefore, arises what happens during this latent period. Does the inward movement of calcium not begin until sometime after the end of the depolarizing pulse? Or is the delay largely due to subsequent reaction steps? In a recent paper (Katz & Miledi, 1967*b*), we suggested that calcium enters the membrane carrying net positive charge, either as Ca^{2+} or in the form of a compound CaR^{+}. This view was put forward to explain the peculiar 'latency shift', that is an increase in the minimum latency when the depolarizing pulse was lengthened.

The present findings suggest that the postulated entry of external calcium ions into the axon membrane is halted very soon after the end of the depolarizing pulse. Calcium ions which are made available on the outside after that brief initial period, have no influence on the process which has been set in motion by the depolarization. It may be that the opening of the external membrane 'gates' to Ca^{2+} or CaR^{+} is a transient event much briefer than the subsequent rise and fall of the probability of release. Or some calcium 'carrier' or 'receptor' only appears for a brief initial interval of time on the external surface of the axon membrane.

It will have been noted that the present technique is capable of resolving the time course of the calcium action with only limited accuracy. One may conclude that 'acceptance' of external calcium is terminated before the actual transmitter release commences. We cannot, however, distinguish between the period of depolarization and the latent period which immediately follows it. To do this, a method would have to be devised which enables one to reduce diffusion time even further than the present ionophoretic pulse technique.

In summary, the most likely picture on the present evidence is this: (i) depolarization of the axon terminal opens a 'gate' to calcium; (ii) calcium moves to the inside of the axon membrane and (iii) becomes involved in a reaction which causes the rate of transmitter release to increase and which contributes a large part of the synaptic delay.

REFERENCES

Hodgkin, A. L. & Keynes, R. D. (1957). Movements of labelled calcium in squid giant axons. *J. Physiol.* **138**, 253–281.

Katz, B. & Miledi, R. (1965*a*). Propagation of electric activity in motor nerve terminals. *Proc. R. Soc.* B **161**, 453–482.

Katz, B. & Miledi, R. (1965*b*). The measurement of synaptic delay, and the time course of acetylcholine release at the neuromuscular junction. *Proc. R. Soc.* B **161**, 483–495.

Katz, B. & Miledi, R. (1965*c*). The effect of calcium on acetylcholine release from motor nerve terminals. *Proc. R. Soc.* B **161**, 496–503.

Katz, B. & Miledi, R. (1965*d*). The effect of temperature on the synaptic delay at the neuromuscular junction. *J. Physiol.* **181**, 656–670.

Katz, B. & Miledi, R. (1967*a*). Tetrodotoxin and neuromuscular transmission. *Proc. R. Soc.* B **167**, 8–22.

Katz, B. & Miledi, R. (1967*b*). The release of acetylcholine from nerve endings by graded electric pulses. *Proc. R. Soc.* B **167**, 23–38.

Baker & Knight (1978) Nature (London) **276**, 620–622

Calcium-dependent exocytosis in bovine adrenal medullary cells with leaky plasma membranes

SECRETION from nerve terminals and many other cell types occurs by exocytosis, a process in which the contents of small intracellular vesicles are released after fusion of the vesicle membrane with the plasma membrane of the cell. Exocytosis generally requires calcium ions[1,2], but little is known about the site at which these Ca ions act. The electrophysiological experiments of Katz and Miledi[3] strongly point to an intracellular site, a conclusion that is supported both by the observation that transmitter release is increased after microinjection of Ca into the presynaptic terminal of the squid giant synapse[4] and the finding that the Ca ionophore A23187 promotes secretion from a variety of cells[5]. Further progress has been severely hampered by lack of preparations in which this presumed intracellular site is freely accessible to experimental investigation. We now present evidence that the plasma membrane of medullary cells from the bovine adrenal gland can be rendered permeable to small molecular weight substances without blocking Ca-dependent exocytosis. This preparation has been used to examine the dependence of exocytosis on Ca and Mg ions.

Cells were obtained from thin slices of bovine adrenal medulla by protease digestion[6]. Catecholamine can be released in a Ca-dependent manner by stimulation with carbamylcholine, veratridine or potassium chloride. The Ca-dependence of catecholamine release in response to these secretagogues is shown in Fig. 1. In all cases secretion is half-maximal at Ca concentrations between 0.1 and 1 mM, and is associated with the release of the vesicular enzyme dopamine-β-hydroxylase but note cytosolic enzyme lactate dehydrogenase. Secretion is inhibited by the Ca channel blocker D600 (10^{-4} M).

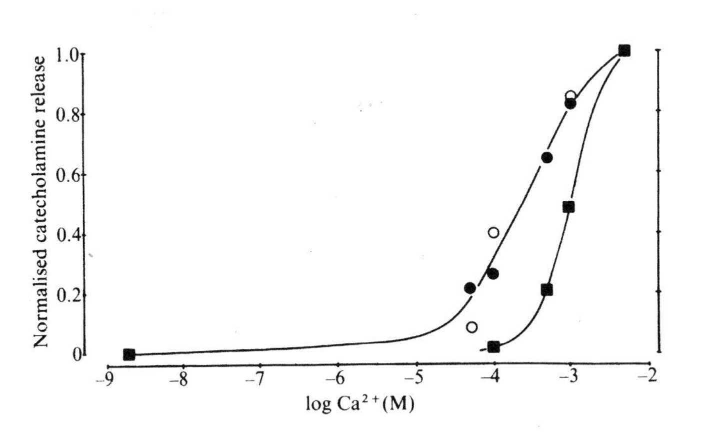

Fig. 1 Ca-dependence of catecholamine release from intact bovine adrenal medullary cells in response to 5×10^{-4} M carbamyl choline (●), 5×10^{-4} M veratridine (■) and to a potassium challenge (○). Ordinate: catecholamine release normalised to that in 5 mM Ca. Catecholamine was assayed fluorimetrically by the trihydroxyindole method[7]. Incubation was for 15 min at 37 °C after which time the cells were removed by centrifugation and samples of the supernatant taken for analysis. The medium contained (mM) NaCl, 136; KCl, 2.7; glucose, 5; Mg, 2; HEPES, 16 *p*H 7.2 and for the potassium challenge NaCl, 68; KCl, 70.7; glucose, 5; Mg, 2; HEPES, 16 *p*H 7.2. The media containing 2×10^{-9} M Ca^{2+} (calculated) contained in addition 5 mM EGTA.

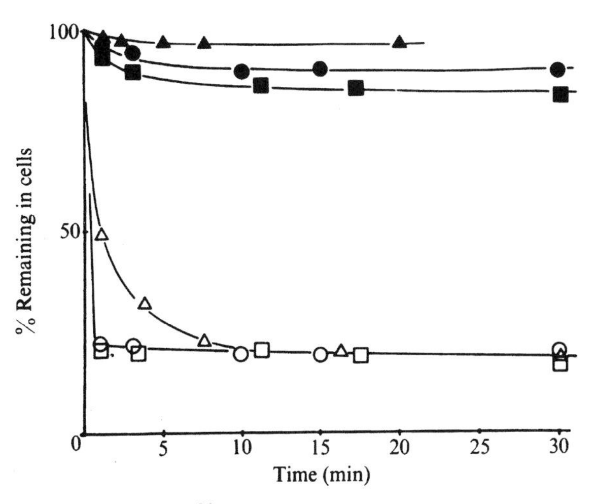

Fig. 2 Release of ^{86}Rb and catecholamine after exposure to high-voltage discharges. 1 ml of a suspension of medullary cells (10^6 per ml) preloaded with ^{86}Rb was placed between a pair of stainless steel electrodes (0.9 cm apart and each 1.0 cm^2) in a square perspex container and subjected to the discharge of a 2 μF capacitor. Cells exposed at zero time to 1 (▲, △), 5 (●, ○) and 10 (■, □) discharges, τ approx 200 μs. Field 2 kV cm^{-1}. Medium (mM): potassium glutamate, 138.7; glucose, 5; Mg, 2; PIPES, 20 *p*H 6.6; Ca-EGTA buffer, 5. Calculated free Ca^{2+} 10^{-5}M. Temperature 37 °C. Closed symbols represent catecholamine release and open symbols ^{86}Rb release. Similar results were obtained with cells preloaded with the non-metabolisable sugar 3-*O*-[^{14}C]methyl-D-glucose showing that the increase in permeability is not restricted to charged molecules.

To provide ready access to the cell interior, suspensions of cells were exposed to a small number of high-voltage discharges[8]. This technique was used previously to alter the intracellular environment of erythrocytes[9,10]. It presumably works by causing localised dielectric breakdown generating holes in the plasma membrane. This increase in permeability is most readily demonstrated in cells preloaded with ^{86}Rb where a suitable shocking regime can release 80% of the cellular ^{86}Rb within a few minutes (Fig. 2). As we wished to gain ready access to the cell interior for Ca-EGTA buffers, we also examined the uptake of ^{51}Cr-EDTA. In conditions similar to those used in Fig. 2, the ^{51}Cr-EDTA space increased within 5 min to a new steady level that varied between 50 and 75% of the cell volume. This suggests that the cells should have also been rendered freely permeable to Ca-EGTA which is only a little larger than Cr-EDTA. Although erythrocytes reseal after exposure to high voltage discharges, this does not seem to happen in adrenal cells where measurements of the ^{51}Cr-EDTA space show that the cells remain freely permeable for at least 20 min at 37 °C.

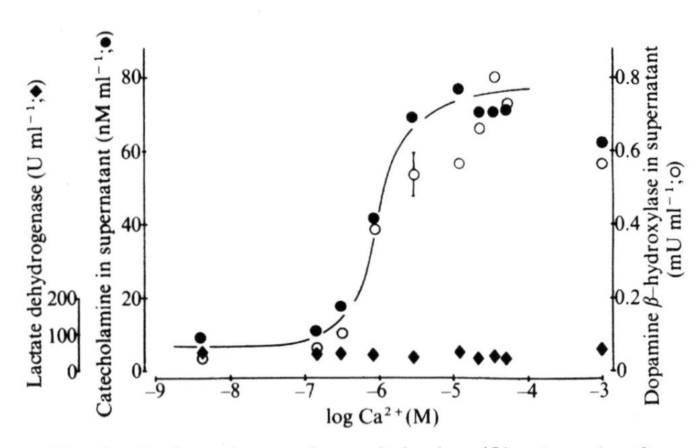

Fig. 3 Ca-dependence of catecholamine (●); dopamine-β-hydroxylase (○), and lactate dehydrogenase (◆) release in response to five discharges (2 kV cm^{-1}, $\tau \sim 200$ μs) over about 30 s. The samples were centrifuged and assayed 15 min after the discharges. Medium (mM): potassium glutamate, 138.7; glucose, 5; Mg^{2+} 2; PIPES, 30; Ca-EGTA buffer, 5; *p*H 6.6. Temperature 37 °C. Catecholamine was assayed fluorimetrically[7]. Dopamine-β-hydroxylase was assayed spectrophotometrically by measuring conversion of tyramine to octopamine[11]. Lactate dehydrogenase was assayed spectrophotometrically[12].

When cells are exposed to high-voltage discharges in media containing Ca, catecholamine is released (Fig. 2). This release is Ca-dependent, but is not mediated by depolarisation nor does it involve Ca channels. Release is not brought about by the temperature jump generated by the high-voltage discharge. Less than 1% of the total intracellular catecholamine is normally released after exposure to high-voltage discharges in nominally Ca-free media containing EGTA though the cells are rendered freely permeable to Cr-EDTA. With calcium present more than 10% of the total catecholamine is released and this response persists unchanged in cells fully depolarised by potassium and in the presence of a Ca channel blocker D600 (10^{-4} M). This Ca-dependent release of catecholamine seems to reflect exocytosis because it is associated with the selective release of the vesicular enzyme dopamine-β-hydroxylase but not lactate dehydrogenase which is present in the cytosol and is very temperature-dependent, being reduced by over 90% when the temperature is lowered from 37 °C to 5 °C. The catecholamine release is unlikely to result from a direct effect on the vesicles because exposure of isolated chromaffin granules to the same shock regime gives no release of catecholamine or dopamine-β-hydroxylase.

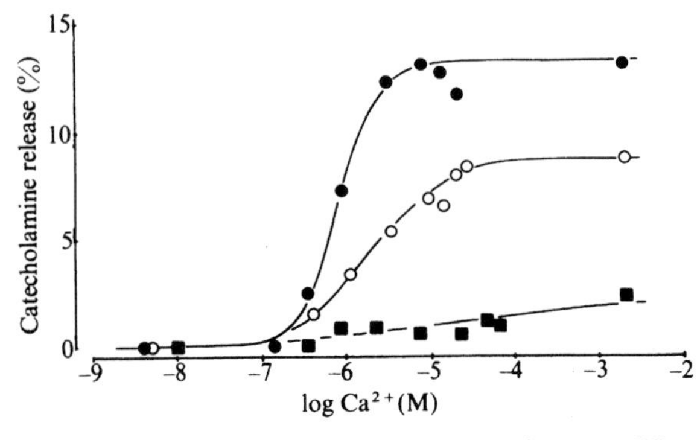

Fig. 4 Ca-dependence of catecholamine release at different magnesium levels in response to five discharges of 2 kV cm^{-1}. The samples were centrifuged and assayed 15 min after the discharges. Medium (mM): potassium glutamate, 138.7; glucose, 5; PIPES 20; Ca-EGTA buffer, 10.4, *p*H 6.6; $MgCl_2$ 2, (●); 10, (○); 50, (■). Temperature 37 °C. The time constant of the discharge is dependent on the conductivity of the medium and would have been close to 200 μs for media containing 2 mM Mg Cl_2 and 140 μs for media containing 50 mM Mg Cl_2.

Working with cells rendered permeable by exposure to high-voltage discharges, we have investigated the dependence of exocytosis on the level of ionised Ca and Mg in the medium in which the shocks are delivered. Using two Ca-buffers, EGTA (Fig. 3) and EDTA, the Ca concentration giving half-maximal activation of the release of both catecholamine and dopamine-β-hydroxylase is close to 1 μM in the presence of 2 mM Mg. This is more than two orders of magnitude lower than the Ca concentration required for half-maximal activation of secretion from intact cells (Fig. 1), providing further evidence that exposure to high voltage discharges renders the interior of the cell freely accessible to externally applied Ca-buffers. Figure 4 shows that raising the Mg concentration both reduces the apparent affinity for Ca and also leads to a decrease in the total amount of catecholamine released.

Another interesting feature of Figs 3 and 4 is that only about 10% of the total cellular catecholamine and dopamine-β-hydroxylase is released even though the cells are freely permeable to Cr-EDTA. At first sight this is very surprising as one might expect that if Ca has free access to an intracellular site controlling exocytosis all the vesicular catecholamine and dopamine-β-hydroxylase should ultimately be released. There may, however, be only a limited number of release sites and the conditions used in our experiments may not permit repriming these sites for more than one round of exocytosis.

In all these experiments cells have been suspended in the test medium and then exposed to a series of high-voltage discharges.

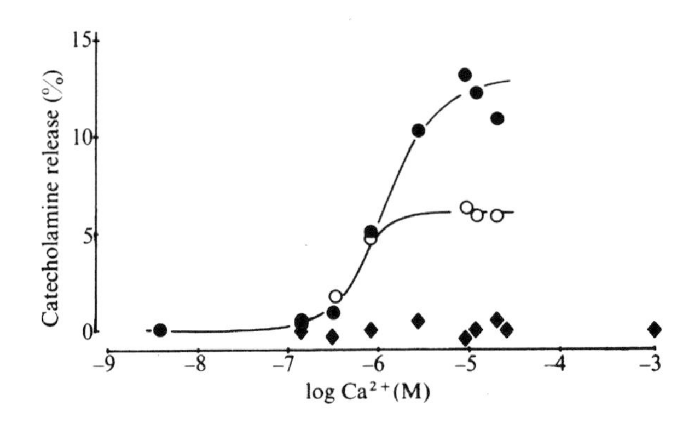

Fig. 5 Influence of ATP on Ca-dependent catecholamine release. Ten discharges 2 kV cm^{-1}, $\tau \sim 200$ μs were applied to cells in a medium of (mM): potassium glutamate, 138.7; glucose, 5; PIPES, 20; EGTA, 0.4; ATP, 0 (◆), 1 (○), 5 (●); *p*H 6.6. The cells were incubated for 30 min at 37 °C and then Ca-EGTA buffer at *p*H 6.6 was added to give a final buffer concentration of 10.4 mM. After a further 15 min at 37 °C the samples were centrifuged and assayed.

This means that the site at which Ca acts is still exposed to a fairly normal intracellular environment. In order to probe further the ionic and metabolic requirements for exocytosis it is necessary to render cells freely permeable in the absence of Ca and only expose them to Ca after adequate time has been allowed for small molecular weight materials to equilibrate. Figure 5 shows that this is technically possible. Preliminary results indicate that after exposure at 37 °C for longer than 5 min to Ca-free solutions lacking ATP, addition of Ca fails to promote exocytosis whereas Ca remains fully effective for at least 30 min provided ATP is present throughout the period of exposure to nominally Ca-free media. These results suggest that ATP or something derived from it is necessary for Ca-dependent exocytosis in adrenal medulla cells. A similar conclusion had been reached from experiments on intact cells[13,14], but the present preparation seems to offer an excellent opportunity to discover the part played by ATP in exocytosis. Recently there have been a number of claims that Ca-dependent exocytosis can occur in cell-free systems[15–17]. A feature common to these systems and the present preparation is that all are activated by micromolar concentrations of Ca which seems well within the physiologically possible range[18]. It is also within the range in which troponin and related Ca-binding proteins are activated.

This work was supported by the MRC and the Wellcome Trust.

P. F. BAKER
D. E. KNIGHT

Department of Physiology,
King's College,
The Strand, London WC2, UK

Received 18 August; accepted 6 October 1978.

1. Douglas, W. W. *Br. J. Pharmac. Chemother.* **34,** 451 (1968).
2. Baker, P. F. in *Recent Advances in Physiology* (ed. Linden, R. J.) 51 (Churchill Livingstone, London, 1974).
3. Katz, B. & Miledi, R. *J. Physiol., Lond.* **192,** 407 (1967).
4. Miledi, R. *Proc. R. Soc.* B **183,** 421 (1973).
5. Foreman, J. C., Mongar, J. L. & Gomperts, B. D. *Nature* **245,** 249 (1973).
6. Knight, D. E. & Whitaker, M. J. *J. Physiol., Lond.* **281,** 18P (1978).
7. von Euler, U. S. & Floding, I. *Acta physiol. scand.* **33,** Suppl. 118, 45 (1955).
8. Baker, P. F. & Knight, D. E. *J. Physiol., Lond.* (in the press).
9. Zimmermann, U., Reimann, F. & Pilwot, G. *Biochim. biophys. Acta* **394,** 449 (1975).
10. Kinosita, I. & Tsong, T. Y. *Nature* **272,** 258 (1978).
11. Aunis, D., Serck-Hanssen, G. & Helle, K. B. *Gen. Pharmac.* **9,** 37 (1978).
12. Bergmeyer, H. U., Bernt, E. & Hess, B. in *Methods of Enzymatic Analysis* (ed. Bergmeyer, H. U.) 737 (Academic, New York, 1965).
13. Kirshner, N. & Smith, W. J. *Life Sci.* Part 1 **8,** 799 (1969).
14. Rubin, R. P. *J. Physiol., Lond.* **206,** 181 (1970).
15. Davis, B. & Lazarus, N. R. *J. Physiol., Lond.* **256,** 709 (1976).
16. Gratzl, M., Dahl, G., Russell, J. T. & Thorn, N. A. *Biochim. biophys. Acta* **470,** 45 (1977).
17. Baker, P. F. & Whitaker, M. J. *J. Physiol., Lond.* (in the press).
18. Baker, P. F. *Prog. Biophys. molec. Biol.* **24,** 177 (1972).

Endothelium-derived relaxing factor is nitric oxide

14

Endothelial cells separate the intravascular compartment from the surrounding smooth muscle cells. They receive many signals from the lumen of the blood vessels in the form of physical forces, hormones, cytokines, coagulation products and platelets. The endothelial cells respond to these stimuli by the synthesis and release of messengers that diffuse to the vascular smooth muscle cells, inducing them to contract, relax or proliferate [1,2] (Figure 14.1a). The first messenger of this type to be described was prostacyclin. Weksler, Ley and Jaffe [3] showed that prostacyclin can be released from endothelial cells by thrombin, trypsin and the

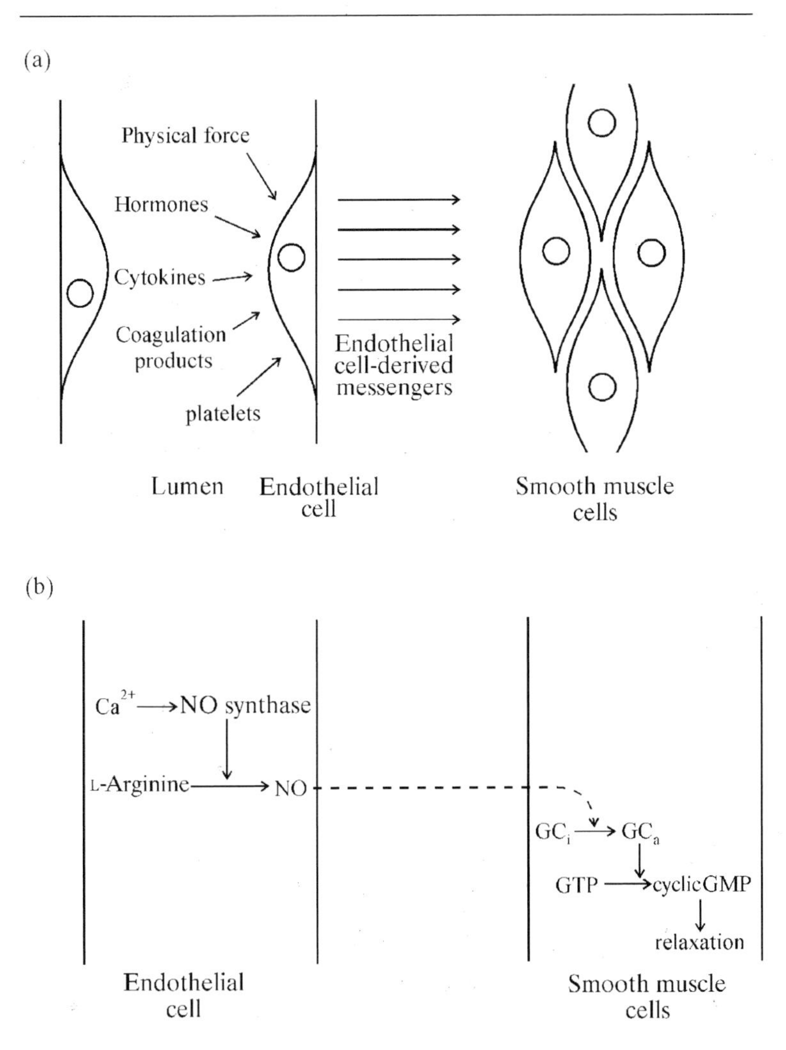

Figure 14.1. Signalling pathways from endothelial to smooth muscle cells. (a) Diagram showing how various physical and chemical influences on endothelial cells can affect the surrounding smooth muscle cells via endothelial cell-derived diffusible messengers. (b) Diagram illustrating Ca^{2+} stimulation of NO synthase with subsequent NO formation from L-arginine in an endothelial cell. NO diffuses into an adjacent smooth muscle cell where it makes an inactive guanyl cyclase (GC_i) active (GC_a), thereby allowing the formation of cyclic GMP from GTP. The increased level of cyclic GMP evokes relaxation of smooth muscle cells (see text for further details).

calcium ionophore A23187. Further work has indicated that this messenger contributes very little to the vasorelaxation response, but is efficient at inhibiting platelet aggregation.

The breakthrough in this field came in 1980 when Furchgott and Zawadzki [4] demonstrated the obligatory role of endothelial cells in the relaxation of arterial smooth muscle by acetylcholine (ACh). ACh (and also bradykinin) evoked relaxation of intact vascular muscle strips, but, if the endothelium was removed, the relaxation response was absent. The response could be re-initiated when the vascular strip from which the endothelium had been removed was placed in contact with another strip with intact endothelium. These experiments indicated that a messenger passed from the endothelial cells to smooth muscle cells. This so-called endothelium-derived relaxing factor (EDRF) [4] has a short half-life and is released by, for example, shear stress, bradykinin, histamine, ACh, thrombin, and products from platelets like ATP and ADP [2]. Nitrovasodilators, which are likely to act by releasing nitric oxide (NO), mimic the actions of EDRF. Thus, Furchgott [5] proposed that EDRF may actually be NO.

In the paper we have selected by Palmer et al. [6], Furchgott's proposal was examined by critical experiments. The release of EDRF and NO from endothelial cells in culture was assessed both chemically and by bioassay. The authors found that the relaxation of the bioassay tissues evoked by EDRF was identical to that induced by NO. These two substances were equally unstable, and bradykinin evoked a dose-dependent liberation of NO from cells in sufficient amounts to explain the biological action of EDRF. Palmer et al. [6] also found that the relaxations evoked by NO and EDRF were both inhibited by haemoglobin and enhanced by superoxide dismutase to the same degree. The clear and correct conclusion from this study was that, since the NO released from the endothelial cells is indistinguishable from EDRF with regard to stability and susceptibility to an inhibitor as well as to a potentiator, and since the two messengers have the exact same biological activity, EDRF and NO are identical [6].

There is a Ca^{2+}-calmodulin-NADPH-dependent synthase that catalyses oxidation of the *N*-guanidine terminal of L-arginine to produce citrulline and diffusible NO [2]. NO can then diffuse into the smooth muscle cell layer where it will activate guanyl cyclase, which in turn will catalyse the formation of cyclic GMP. Cyclic GMP causes relaxation of the smooth muscle cell via a mechanism that is not fully understood (Figure 14.1). The relaxation is dependent on a decrease in the free cytosolic Ca^{2+} concentration $[Ca^{2+}]_i$ that could be brought about by an increased level of Ca^{2+} pumping into stores and across the plasma membrane, and/or by a decrease in the opening of Ca^{2+}-release channels in stores as well as Ca^{2+} channels in the plasma membrane.

The trigger event for NO production and release is a rise in $[Ca^{2+}]_i$ in the endothelial cells. This Ca^{2+} signal is produced by both release of Ca^{2+} from internal stores as well as Ca^{2+} inflow from the extracellular solution [2]. The initiation sites for intracellular Ca^{2+} release have recently been studied by Missiaen et al. [7]. The group looked at the spatial organization of the cytosolic Ca^{2+} signal in single endothelial cells stimulated with ATP, and found that the long, thin processes of the endothelial cells had a higher agonist sensitivity and showed a faster rise in $[Ca^{2+}]_i$ and wave propagation than the cell body at the same agonist concentration. The Ca^{2+} waves therefore originated preferentially in one of these processes and then invaded the cell body.

Since the discovery of NO as a messenger released from endothelial cells mediating the relaxation of surrounding smooth muscle cells [4–6], it has become clear that this very unusual messenger also has a role in the central nervous system. Garthwaite et al. [8] showed that, in cerebellar cultures, NO was the unstable intercellular factor that mediated the increase in cyclic GMP levels in glial cells following activation of *N*-methyl-D-aspartate glutamate receptors on cerebellar granule cells in the cultures. NO synthase is only present in neurons [9], but, because of its extreme diffusibility (it is the most diffusible messenger currently known [10]), it can spread out from its site of production to reach not only other neurons but also glial and vascular cells [9].

References

1. Daniel, T.O. and Ives, H.E. (1989) *News Physiol. Sci.* **4**, 139–142
2. Nilius, B. and Casteels, R. (1996) in *Comprehensive Human Physiology* (Greger, R. and Windhorst, U., eds.), pp. 1981–1993, Springer, Heidelberg
3. Weksler, B.B., Ley, C.W. and Jaffe, E.A. (1978) *J. Clin. Invest.* **62**, 923–930
4. Furchgott, R.F. and Zawadzki, J.V. (1980) *Nature (London)* **288**, 373–376
5. Furchgott, R.F. (1988) in *Vasodilation: Vascular Smooth Muscle, Peptides, Autonomic Nerves and Endothelium* (Vanhoutte, P.M., ed.), pp. 401–414, Raven Press, New York
6. Palmer, R.M.J., Ferrige, A.G. and Moncada, S. (1987) *Nature (London)* **327**, 524–526
7. Missiaen, L., Lemaire, F.X., Parys, J.B., De Smedt, H., Sienaert, I. and Casteels, R. (1996) *Pflügers Arch.* **431**, 318–324
8. Garthwaite, J., Charles, S.L. and Chess-Williams, R. (1988) *Nature (London)* **336**, 385–388
9. Garthwaite, J. and Boulton, C.L. (1995) *Annu. Rev. Physiol.* **57**, 683–706
10. Kasai, H. and Petersen, O.H. (1994) *Trends Neurosci.* **17**, 95–101

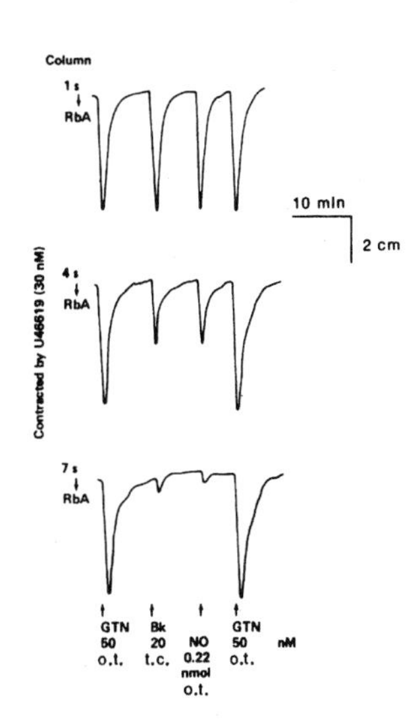

Fig. 1 Relaxation of rabbit aortae by EDRF and NO. A column packed with endothelial cells cultured on microcarriers was perfused with Krebs' buffer (5 ml min^{-1}). The effluent was used to superfuse three spiral strips of rabbit aorta (RbA), denuded of endothelium, in a cascade. The tissues were contracted submaximally by a continuous infusion of 9,11-dideoxy-9α, 11α-methano epoxy-prostaglandin $F_{2\alpha}$ (U46619; 30 nM) and were separated from the cells by delays of 1, 4 and 7 s, respectively. The sensitivity of the bioassay tissues was standardized by administration of glyceryl trinitrate (GTN; 50 nM) over the tissues (o.t.). EDRF was released from the cells by a 1 min infusion through the column (t.c.) of bradykinin (Bk, 20 nM). NO (0.22 nmol) was dissolved in He-deoxygenated H_2O and administered as a 1 min infusion.

Nitric oxide release accounts for the biological activity of endothelium-derived relaxing factor

R. M. J. Palmer, A. G. Ferrige & S. Moncada*

Wellcome Research Laboratories, Langley Court, Beckenham, Kent BR3 3BS, UK

Endothelium-derived relaxing factor (EDRF) is a labile humoral agent which mediates the action of some vasodilators. Nitrovasodilators, which may act by releasing nitric oxide (NO), mimic the effect of EDRF and it has recently been suggested by Furchgott[1] that EDRF may be NO. We have examined this suggestion by studying the release of EDRF and NO from endothelial cells in culture. NO was determined as the chemiluminescent product of its reaction with ozone[2]. The biological activity of EDRF and of NO was measured by bioassay[3]. The relaxation of the bioassay tissues induced by EDRF was indistinguishable from that induced by NO. Both substances were equally unstable. Bradykinin caused concentration-dependent release of NO from the cells in amounts sufficient to account for the biological activity of EDRF. The relaxations induced by EDRF and NO were inhibited by haemoglobin and enhanced by superoxide dismutase to a similar degree. Thus NO released from endothelial cells is indistinguishable from EDRF in terms of biological activity, stability, and susceptibility to an inhibitor and to a potentiator. We suggest that EDRF and NO are identical.

* To whom correspondence should be addressed.

EDRF was released from porcine aortic endothelial cells (2-5×10^7 cells) cultured on microcarriers and was detected by bioassay on spiral strips of rabbit aorta superfused in a cascade as described previously[3]. In some experiments the endothelial cells were replaced by porcine aortic smooth muscle cells or 3T3 cells (1.5-1.9 and 2.4-3.9×10^7 cells respectively). A gas bulb was filled with NO (British Oxygen) from a cylinder and sealed with silicone rubber injection septa. An appropriate volume (10-1,000 μl) was removed with a syringe and injected into another gas bulb filled with H_2O, which had been deoxygenated by gassing with He for 1 h, to give stock solutions of NO of 0.01-1.0% (v/v). These solutions were stable for ~10 h. NO was determined by chemiluminescence[2].

The bioassay tissues were relaxed by glyceryl trinitrate (GTN, 50 nM), by EDRF released from the cells by bradykinin (20 nM) and by authentic NO (0.22 nmol; Fig. 1; $n=30$). The relaxations induced by EDRF and NO declined during passage down the cascade at rates which were not significantly different from each other, and were 50% of that of the uppermost bioassay tissue (the 'half-life') after 3.6 ± 0.1 and 4.1 ± 0.2 s (mean ± s.e.m., $n=4$) respectively. In contrast, the relaxations caused by GTN did not decline during passage down the cascade at any of the concentrations used (1-50 nM).

The stability of EDRF or of authentic NO in Krebs' buffer was determined using bioassay by interposing lengths of tubing, between either the generator cells or the point of infusion of NO (0.22 nmol) and the uppermost detector tissue, so as to vary the transit time. The half-life in transit of EDRF was not significantly different from that of NO (30.8 ± 1.9 and 30.4 ± 2.2 s; $n=3$ respectively). The fact that both EDRF and NO disappear more rapidly during passage down the cascade than they do in Krebs' buffer alone may indicate that the tissues contribute in some way to their decay, either by generating superoxide anions (O_2^-) or by some unknown mechanism. Alternatively, the inter-

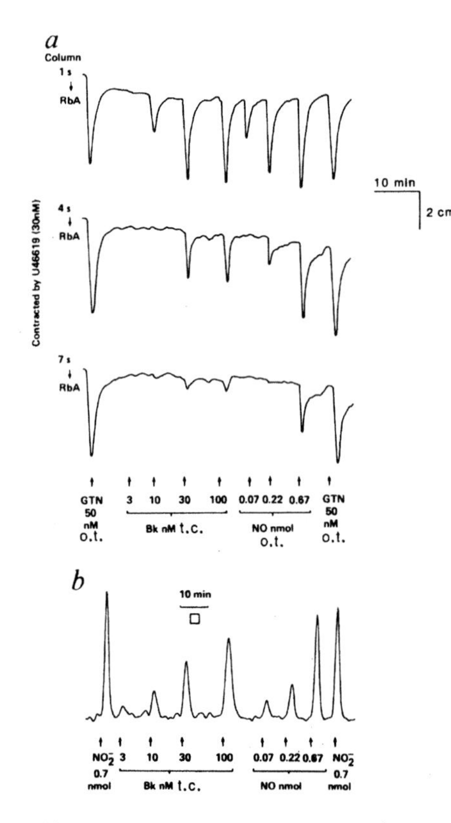

Fig. 2 *a*, Bioassay. Relaxation of rabbit aortae by EDRF and NO. The bioassay tissues were relaxed in a concentration-dependent manner by EDRF released from the cells by bradykinin (Bk; 3-100 nM t.c.) and by NO (0.07-0.67 nmol, o.t.) as in Fig. 1. *b*, Chemiluminescence. Release of NO by bradykinin (Bk) from a replicate column of the cells used in the bioassay. The amounts of NO (administered as 1 min infusion into the column effluent) which relaxed the bioassay tissues were also detectable by chemiluminescence. Effluent from the column, or Krebs' buffer into which authentic NO was injected, was passed continuously (5 ml min^{-1}) into a reaction vessel containing 75 ml 1.0% sodium iodide in glacial acetic acid under reflux. NO was removed from the refluxing mixture under reduced pressure in a stream of N_2, mixed with ozone and the chemiluminescent product measured with a photomultiplier. The amounts of NO detected were quantified after correcting for baseline drift using a polynomial fit and reducing electrical noise, by Fourier transformation and application of a Gaussian function. The areas under the peaks were converted to nmol of NO by reference to a NO_2^- standard curve. Similar results were obtained in two other experiments. □, Area equivalent to 0.22 nmol NO.

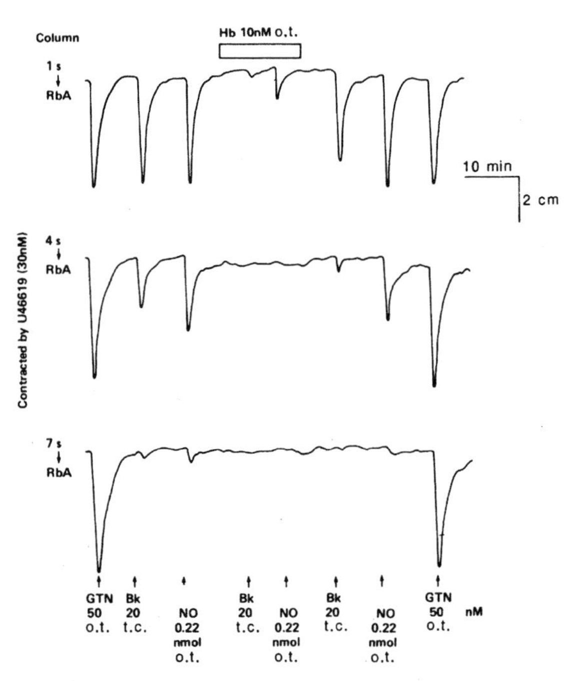

Fig. 3 The effect of purified human haemoglobin (Hb; <3% Met haemoglobin) on the relaxation of rabbit aortae induced by EDRF and NO. Effluent of a column containing endothelial cells was used to superfuse three spiral strips of rabbit aorta in a cascade (as in Fig. 1). The bioassay tissues were relaxed by NO (0.22 nmol o.t.) or by EDRF released by bradykinin (Bk, 20 nM t.c.) as in Fig. 1. Infusion of Hb (10 nM o.t.) inhibited the relaxations induced by both EDRF and NO to a similar extent. Larger concentrations (>100 nM) of Hb were needed to inhibit the relaxations induced by GTN (data not shown) confirming a previous report[7]. Similar results were obtained in two other experiments.

action of NO and O_2 in the delay tubes may differ from that during passage down the cascade.

Bradykinin released EDRF from all the endothelial cell cultures tested ($n > 30$). At concentrations ranging from 3 nM to 100 nM the release of EDRF was concentration-dependent and caused relaxations of the bioassay tissues similar to those induced by NO (0.07-0.67 nmol; Fig. 2*a*). Bradykinin (3-100 nM) also caused concentration-dependent release of amounts of NO sufficient to account for the relaxations of the bioassay tissues induced by EDRF (Fig. 2*b*). The release induced by 3, 10, 30 and 100 nM bradykinin was 0.09 ± 0.02, 0.19 ± 0.01, 0.44 ± 0.02, and 0.68 ± 0.02 nmol respectively ($n = 3$). Furthermore, there was a correlation ($r = 0.94$, $n = 12$) between the amount of NO released and the relaxation of the bioassay tissues induced by EDRF. Release of EDRF or NO by bradykinin (100 nM) could not be detected (<0.05 nmol) from porcine aortic smooth muscle cells (2 experiments) or 3T3 cells (2 experiments).

NO reacts readily with O_2 to produce NO_2, which then forms NO_2^- and NO_3^- in neutral aqueous solution according to the following reactions:

$$2NO + O_2 \rightarrow 2NO_2 \quad (1)$$

$$2NO_2 + H_2O \rightarrow NO_2^- + NO_3^- + 2H^+ \quad (2)$$

Neither NO_2^- nor NO_3^- (as their sodium salts) cause relaxation of the vascular strips at concentrations below 10 μM (data not shown). These reactions therefore represent an inactivation mechanism for NO in terms of the biological activity we measure and thus the decay of NO can be studied by bioassay.

There was a delay of 1 s between the endothelial cells, or the point of infusion of authentic NO, and the reflux vessel from which NO was recovered before measurement. Because the half-life of NO and EDRF determined by bioassay is ~30 s, little decay of NO to NO_2^- would occur in this time. Therefore we detect NO directly rather than NO derived from the reduction of NO_2^- in the reflux vessel. This is supported by the finding that the chemiluminescence detected from 0.7 nmol NO_2^- and from 0.67 nmol authentic NO are similar (Fig. 2*b*). If NO decayed completely then a maximum of 50% of the NO would be detected, for NO_3^- is not reduced to NO in the reflux vessel. Thus the chemiluminescence method cannot be used to determine the decay of NO when longer delays are used.

NO_2^- is not converted to NO when refluxing glacial acetic acid (without sodium iodide) is diluted >50% with Krebs' buffer. Under these conditions, NO released from the endothelial cells could be detected. Furthermore, the decay of NO could also be determined although larger amounts were required (2.25 nmol). The decay was found to approximate to second-order kinetics (data not shown), consistent with the reaction (1) above. Second-order decay kinetics may help to explain the variations in half-life reported for EDRF[4,5].

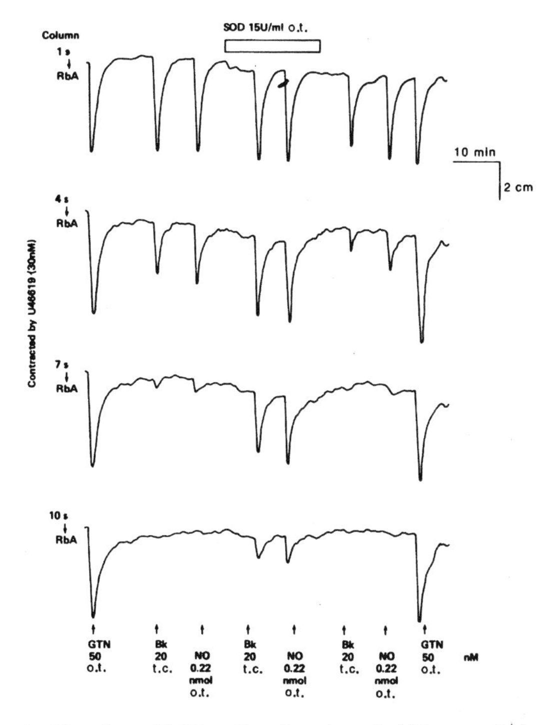

Fig. 4 The effect of SOD on the relaxation of rabbit aortae induced by EDRF and NO. Effluent of a column of endothelial cells was used to superfuse four strips of rabbit aorta in a cascade (as in Fig. 1). The bioassay tissues were relaxed by NO (0.22 nmol, o.t.) or by EDRF released by bradykinin (20 nM Bk t.c.). Infusion of SOD (15 U ml^{-1} o.t.) increased the stability of EDRF and NO to a similar extent, so that the fourth tissue was relaxed by both compounds. Similar results were obtained in two other experiments.

Haemoglobin, which has been reported to bind EDRF[6,7] and NO[8], inhibited the relaxations of the bioassay tissues induced by EDRF and NO (Fig. 3), with an IC_{50} of 3.6 ± 0.6 and 8.1 ± 1.4 nM ($n = 3$) respectively. Superoxide dismutase (SOD) reduces the inactivation of EDRF[9,10]. Infusions of SOD (15 U ml^{-1}) over the tissues reduced the decay of both EDRF and authentic NO to a similar extent (Fig. 4), indicating that O_2^- contributes to the inactivation of both compounds.

NO is therefore released from vascular endothelial cells, but not from smooth muscle cells or 3T3 cells, in amounts sufficient to account for the relaxations attributed to EDRF. Furthermore, it has the same biological activity and chemical stability as EDRF and its action is as susceptible to inhibition by haemoglobin and to potentiation by SOD. Thus, the biological activity of EDRF is accounted for by NO when considered in terms of the classical criteria proposed by Dale[11] for the identification of a mediator. The proposal that EDRF is NO is consistent with our suggestion that EDRF is inactivated by O_2^- generated in the oxygenated Krebs' buffer[9] or by the action of some inhibitors of EDRF[12], for O_2^- reacts with NO to form NO_3^- ultimately[13]. Whether NO is released from vascular endothelial cells as NO itself, or as a precursor which gives rise to NO following release, deserves further investigation.

We thank Dr C. L. Walters for advice and assistance with the chemiluminescence method, N. A. Foxwell, C. J. Saxby, B. C. Sweatman, Ms L. J. Wallis and M. J. Ashton for technical assistance, and our colleagues at the Wellcome Research Laboratories for helpful discussions.

Received 28 January; accepted 10 April 1987.

1. Furchgott, R. F. in *Mechanisms of Vasodilatation* Vol. IV (ed. Vanhoutte, P. M.) (Raven, New York, in the press).
2. Downes, M. J., Edwards, M. W., Elsey, T. S. & Walters, C. L. *Analyst* **101**, 742–748 (1976).
3. Gryglewski, R. J., Moncada, S. & Palmer, R. M. J. *Br. J. Pharmac* **87**, 685–694 (1986).
4. Griffith, T. M., Edwards, D. H., Lewis, M. J., Newby, A. C. & Henderson, A. H. *Nature* **308**, 645–647 (1984).
5. Forstermann, U., Trogisch, G. & Busse, R. *Eur. J. Pharmac.* **106**, 639–643 (1985).
6. Cocks, T. M. & Angus, J. A. in *Vascular Neuroeffector Mechanisms* (eds Bevan, J. A., Godfraind, T., Maxwell, R. A., Stoclet, J. C. & Worcel, M.) 131–136 (Elsevier, Amsterdam, 1985).
7. Martin, W., Villani, G. M., Jothianandan, D. & Furchgott, R. F. *J. Pharmac. exp. Ther.* **232**, 708–716 (1985).
8. Gibson, Q. H. & Roughton, F. J. W. *J. Physiol., Lond.* **136**, 507–526 (1957).
9. Gryglewski, R. J., Palmer, R. M. J. & Moncada, S. *Nature* **320**, 454–456 (1986).
10. Rubanyi, G. M. & Vanhoutte, P. M. *Am. J. Physiol.* **250**, H222–H227 (1986).
11. Dale, H. H. *Bull. Johns Hopkins Hosp.* **53**, 297–347 (1933).
12. Moncada, S., Palmer, R. M. J. & Gryglewski, R. J. *Proc. natn. Acad. Sci. U.S.A.* **83**, 9164–9168 (1986).
13. Blough, N. V. & Zafiriou, O. C. *Inorg. Chem.* **24**, 3502–3504 (1985).

Src: tyrosine phosphorylation and src-homology domains in signal transduction pathways

15

The study of the transforming v-*src* gene of Rous sarcoma virus has led to important insights into the mechanisms of cellular transformation by oncogenes and into general features of protein–protein interactions in signal transduction pathways that control cell growth and proliferation. The v-*src* gene encodes a phosphoprotein, pp60$^{v\text{-}src}$, which is required for the induction of cancerous growth by the virus. The papers that we have chosen [1,2] initiated two important areas of current research. First, the demonstration that pp60$^{v\text{-}src}$ possesses the ability to phosphorylate substrate proteins on tyrosine [1] led to a rapid appreciation of the widespread importance of protein tyrosine phosphorylation. Second, the discovery of domains in pp60$^{v\text{-}src}$ that are conserved in a range of protein tyrosine kinases [2] led to the realization that these are widespread protein modules that are required for direct protein–protein interactions by binding to specific sequences on the binding partner.

The Rous sarcoma virus can produce tumours as a consequence of the expression of pp60$^{v\text{-}src}$. This viral protein has a cellular counterpart that is believed to play a role in the normal control of cell growth. An important advance was the demonstration that the cellular phosphoprotein, pp60$^{c\text{-}src}$, had an associated protein kinase activity [3]. In fact, both pp60$^{v\text{-}src}$ and pp60$^{c\text{-}src}$ are protein kinases, and detailed analysis of the phosphoamino acids in phosphorylated immunoglobulin-heavy-chain immunoprecipitates of pp60$^{v\text{-}src}$ showed that tyrosine was the amino acid phosphorylated [1]. In addition, cells transformed by Rous sarcoma virus contained substantially more phosphorylated tyrosine in protein than control cells. It is now known that the tyrosine kinase activity of pp60$^{v\text{-}src}$ is crucial to its transforming activity. Until this study [1] was carried out, it had been believed that cellular protein kinases only phosphorylated either serine or threonine residues in substrate proteins. This view was as a consequence of cells containing relatively few phosphotyrosine residues.

The observation of tyrosine phosphorylation by pp60$^{v\text{-}src}$ was quickly followed by the demonstration that the plasma membrane receptor for epidermal growth factor (EGF) has an associated protein kinase activity which also phosphorylates tyrosine [4]. It was subsequently shown that the EGF receptor is itself the tyrosine kinase [5]. Many other tyrosine-specific protein kinases have now been identified, and the current view is that cellular protein tyrosine kinases fall into two classes: the cytoplasmic tyrosine kinases such as cellular pp60$^{c\text{-}src}$, and the plasma membrane receptors with intrinsic tyrosine kinase activity, such as the EGF and insulin receptors.

Our second chosen paper [2] concerns the observation that all of the cytoplasmic protein tyrosine kinases known at that time possessed a catalytic domain similar to that of pp60$^{v\text{-}src}$ (the src-homology domain 1; SH1) and a nearby domain designated the src-homology domain 2 (SH2), which consists of around 100 amino acids. This paper assessed the functional role of the conserved SH2 domain in the p130$^{gag-fps}$ tyrosine kinase of the Fujinami sarcoma virus. It was shown that the SH2 domain was not essential for the intrinsic protein tyrosine kinase activity of the isolated protein. Mutation or deletion of this domain, however, abolished the ability of the p130$^{gag-fps}$ kinase to transform cells, and, within cells, the mutant proteins showed reduced protein tyrosine phosphorylation. These data indicated that the SH2 domain in some way regulates the protein tyrosine phosphorylation and plays an essential role in the signal transduction pathway leading to cell transformation to a neoplastic phenotype.

Later work revealed the presence of SH2 domains in a variety of proteins (Figure 15.1), including those that are not protein tyrosine kinases, and also a third conserved domain in pp60^{v-src} consisting of around 60 amino acids, called the SH3 domain [6–9]. The SH2 and SH3 domains function to bind proteins to one another to form protein complexes required for signal transduction cascades. The SH2 domains bind to sequences containing a phosphorylated tyrosine, and the domains in various SH2-containing proteins show differing sequence specificities providing for specific interaction with target proteins after the activation of protein tyrosine kinases. Since the cytoplasmic and membrane receptor protein tyrosine kinases phosphorylate themselves, they are able to recruit SH2-containing proteins to form an activated complex. The structures of several SH2 domains bound to a phosphorylated peptide have been solved. Characteristically, the first of these to be published was for the pp60^{v-src} SH2 domain complexed with peptides based on a binding sequence in the platelet-derived growth factor or EGF receptor

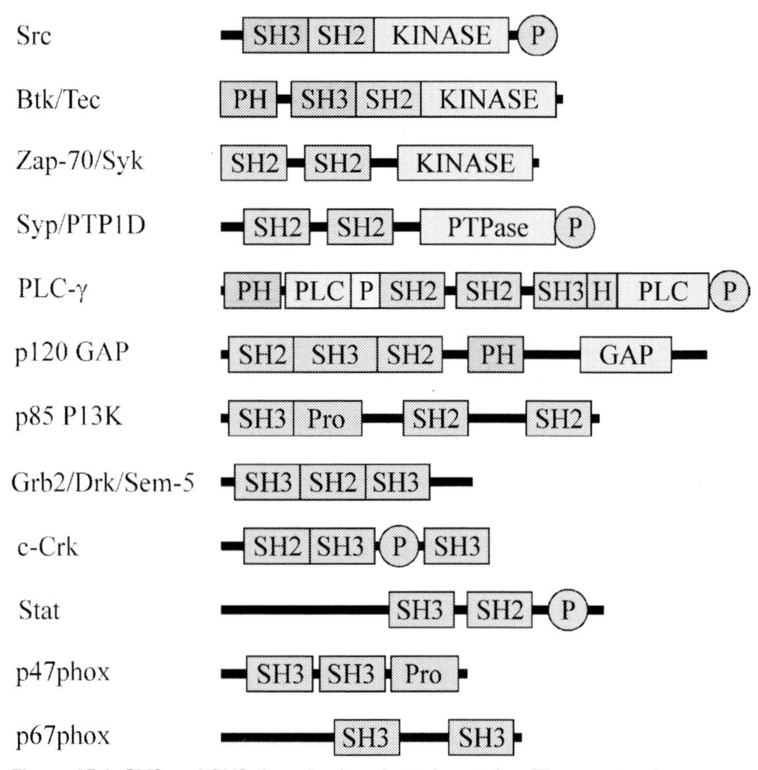

Figure 15.1 SH2 and SH3 domains in selected proteins. The proteins shown include protein kinases with a kinase domain, a protein tyrosine phosphatase (Syp/PTP1D) and a range of other proteins involved in signal transduction pathways (see [8]). Abbreviations used: Pro, proline-rich SH3-binding site; P, tyrosine phosphorylation site; PH, pleckstrin-homology domain.

[10], and subsequently with peptides that show higher-affinity binding [11]. The SH3 domains recognize and bind to specific, short proline-rich sequences with a Pro-Xaa-Xaa-Pro core in their partner proteins, and are found not only in proteins involved in signal transduction but also in cytoskeletal proteins [7].

A series of proteins have been discovered that consist essentially of SH2 and SH3 domains, and must function, therefore, as adaptor or linking proteins in protein complex assembly [7,8]. Such proteins have key roles in a signalling pathway (the Ras/MAP kinase pathway; see Section 17) that is activated by the occupancy of the receptor protein tyrosine kinases of, for example, the EGF and insulin receptor class [9,11].

The study of the pp60$^{v\text{-}src}$ kinase has been crucial in both revealing the existence of tyrosine phosphorylation and the conserved SH2 and SH3 domains, and for continuing analysis of the function of these domains. The discoveries leading from pp60$^{v\text{-}src}$ have been crucially important for our understanding of the activation of protein–protein interactions in a multitude of signal transduction cascades. An account of the discovery and significance of tyrosine phosphorylation has been published recently [12].

References

1. Hunter, T. and Sefton, B.M. (1980) *Proc. Natl. Acad. Sci. U.S.A.* **77**, 1311–1315
2. Sadowski, I., Stone, J.C. and Pawson, T. (1986) *Mol. Cell. Biol.* **6**, 4396–4408
3. Collet, M.S. and Erikson, R.L. (1978) *Proc. Natl. Acad. Sci. U.S.A.* **75**, 2021–2024
4. Ushiro, H. and Cohen, S. (1980) *J. Biol. Chem.* **255**, 8363–8365
5. Buhrow, S.A., Cohen, S. and Staros, J.V. (1982) *J. Biol. Chem.* **257**, 4019–4022
6. Koch, C.A., Anderson, D., Moran, M.F. and Pawson, T. (1991) *Science* **252**, 668–674
7. Mayer, B.J. and Baltimore, D. (1993) *Trends Cell Biol.* **3**, 8–13
8. Feller, S.M., Ren, R., Hanafusa, H. and Baltimore, D. (1994) *Trends Biochem. Sci.* **19**, 453–458
9. Pawson, T. (1995) *Nature (London)* **373**, 573–580
10. Waksman, G., Kominas, D., Robertson, S.C., Pant, N., Baltimore, D., Birge, R.B., Cowburn, D., Hanfusa, H., Mayer, B.J., Overduin, M., Resh, M.D., Rios, C.B., Silverman, L. and Kuriyan, J. (1992) *Nature (London)* **358**, 646–653
11. Waksman, G., Shoelson, S.E., Pant, N., Cowburn, D. and Kuriyan, J. (1993) *Cell* **72**, 779–790
12. Hunter, T. (1996) *Biochem. Soc. Trans.* **24**, 307–327

Hunter & Sefton (1980) Proc. Natl. Acad. Sci. U.S.A. **77**, 1311–1315

Transforming gene product of Rous sarcoma virus phosphorylates tyrosine

(phosphotyrosine/protein kinase/*src* gene/phosphoproteins)

TONY HUNTER AND BARTHOLOMEW M. SEFTON

Tumor Virology Laboratory, The Salk Institute, P. O. Box 85800, San Diego, California 92138

Communicated by Robert W. Holley, December 3, 1979

ABSTRACT **The protein kinase activity associated with pp60** ***src*****, the transforming protein of Rous sarcoma virus, was found to phosphorylate tyrosine when assayed in an immunoprecipitate. Despite the fact that a protein kinase with this activity has not been described before, several observations suggest that pp60** ***src*** **also phosphorylates tyrosine** ***in vivo*****. First, chicken cells transformed by Rous sarcoma virus contain as much as 8-fold more phosphotyrosine than do uninfected cells. Second, phosphotyrosine is present in pp60** ***src*** **itself, at one of the two sites of phosphorylation. Third, phosphotyrosine is present in the 50,000-dalton phosphoprotein that coprecipitates with pp60** ***src*** **extracted from transformed chicken cells. We infer from these observations that pp60** ***src*** **is a novel protein kinase and that the modification of proteins via the phosphorylation of tyrosine is essential to the malignant transformation of cells by Rous sarcoma virus. pp60** ***sarc*****, the closely related cellular homologue of viral pp60** ***src*****, is present in all vertebrate cells. This normal cellular protein, obtained from both chicken and human cells, also phosphorylated tyrosine when assayed in an immunoprecipitate. This is additional evidence of the functional similarity of these structurally related proteins and demonstrates that all uninfected vertebrate cells contain at least one protein kinase that phosphorylates tyrosine.**

The product of the *src* gene of Rous sarcoma virus (RSV) is a 60,000-dalton phosphoprotein, pp60*src* (1–4), which is necessary for the transformation of cells in culture and for the formation of sarcomas in birds (5). Immunoprecipitates containing pp60*src* invariably have a protein kinase activity that is capable of phosphorylating the immunoglobulin heavy chain (2, 4, 6, 7). It has been proposed that the protein kinase activity associated with pp60*src* is critical in its transforming function (6). Substantial support for this idea comes from the observation that temperature-sensitive mutations that render the virus unable to transform also decrease the protein kinase activity induced at the nonpermissive temperature and cause this activity to be extremely labile after lysis of infected cells (2, 4, 6–9).

All normal vertebrate cells possess one or a few genes (*sarc*) that are homologous to the viral *src* gene (10, 11). It has been shown that these sequences are expressed, at a low level, as a 60,000-dalton phosphoprotein, pp60*sarc* (12–16). Like viral pp60*src*, this protein possesses a protein kinase activity capable of phosphorylating the immunoglobulin heavy chain (13–16). Hanafusa and colleagues (17, 18) and Vogt and colleagues (19) have shown that viruses with a partial deletion in the *src* gene can recover a fully functional *src* gene during passage through a chicken. Partial analysis of the sequence of the *src* gene of the recovered viruses suggests that much of the gene is derived, by some unknown mechanism, from the cellular *sarc* gene (18). This in turn suggests that RSV transforms by producing abnormally high levels of a protein very similar to a normal cellular protein kinase.

Immunoprecipitates prepared from polyoma virus-infected cells and antipolyoma tumor antiserum contain a novel activity that is capable of phosphorylating a tyrosine in the 60,000-dalton large tumor antigen of polyoma virus present in the precipitate (20). This observation stimulated us to examine whether pp60*src* also had such an activity. We have found that viral pp60*src* and the endogenous pp60*sarc* of birds and of mammals catalyze the phosphorylation of a tyrosine in the immunoglobulin heavy chain when their protein kinase activity is assayed in an immunoprecipitate. It appears that viral pp60*src* and cellular pp60*sarc* are protein kinases with unique specificities.

MATERIALS AND METHODS

Cells and Viruses. The preparation and growth of normal and infected cultures of chicken cells were as described (3, 21). The origin and the cultivation of the normal human mammary cell line HBL-100 is described elsewhere (21).

Radioactive Labeling, Cell Lysis, and Immunoprecipitation. Chicken cells were labeled with [^{32}P]orthophosphate (carrier-free, 0.3–1.0 mCi/ml; 1 Ci = 3.7×10^{10} becquerels; ICN) for 15–18 hr (3). Cell lysis was as before (3) except that the modified RIPA buffer was supplemented with 2 mM EDTA. Immunoprecipitation was as described (3, 8); the antitumor antisera were prepared by the protocol of Brugge and Erikson (1).

Protein Kinase Assay and Gel Electrophoresis. Incubation of immunoprecipitates with [γ-^{32}P]ATP (2000–3000 Ci/mmol; New England Nuclear) and isolation of the phosphorylated heavy chain were as described (7).

Extraction of Gel-Purified Proteins for Acid Hydrolysis and Tryptic Peptide Mapping. Proteins were recovered from gels exactly as described (22) up to and including the organic solvent washes of the trichloroacetic acid precipitates. For acid hydrolysis the precipitates were dissolved directly in 6 M HCl by heating at 100°C for 1 min and then were incubated for 2 hr at 110°C under N_2. The HCl was removed under reduced pressure and the hydrolysates were dissolved in a marker mixture containing phosphoserine, phosphothreonine, and O^4-phosphotyrosine [Tyr(P)], each at 1 mg/ml. For peptide mapping the protein precipitates were treated with performic acid and digested with trypsin (22).

Electrophoresis and Chromatography. Acid hydrolysates were analyzed on cellulose thin-layer plates (100 μm) by electrophoresis at pH 1.9 for 60 min at 1.5 kV in glacial acetic acid/formic acid (88% by vol)/H_2O, 78:25:897 (vol/vol), and at pH 3.5 for 45 min at 1 kV in glacial acetic acid/pyridine/H_2O, 50:5:945 (vol/vol). Ascending chromatography in the second dimension was performed with isobutyric acid/0.5 M NH_4OH, 5:3 (vol/vol). The markers were detected by staining with ninhydrin.

The publication costs of this article were defrayed in part by page charge payment. This article must therefore be hereby marked "*advertisement*" in accordance with 18 U. S. C. §1734 solely to indicate this fact.

Abbreviations; RSV, Rous sarcoma virus; Tyr(P), O^4-phosphotyrosine.

Tryptic digests were resolved by electrophoresis at pH 8.9 for 27 min at 1 kV in buffer containing 1% ammonium carbonate. Ascending chromatography in the second dimension was performed with *n*-butanol/pyridine/glacial acetic acid/H_2O, 75:50:15:60 (vol/vol).

Extraction of ^{32}P-Labeled Proteins from Cell Lysates. ^{32}P-Labeled cells were lysed in RIPA buffer, and insoluble nuclear and cytoskeletal elements were removed by centrifugation. Phosphoproteins were then extracted by treatment with an equal volume of phenol saturated with 0.1 M NaCl/0.05 M Tris·HCl, pH 7.5/5 mM EDTA. The first aqueous phase was reextracted with an equal volume of phenol. The combined phenol phases including the interfaces in both cases were extracted three times with 2 vol of buffer, the aqueous phase being discarded in each case. The final phenol phase with the interface was diluted 1:40 with water and the proteins were precipitated by addition of trichloroacetic acid to a concentration of 15%. The precipitate was recovered by centrifugation and extracted twice with a large volume of chloroform/methanol, 2:1 (vol/vol). The resulting proteins were dissolved in 6 M HCl and hydrolyzed as outlined above.

Materials. Phosphoserine and phosphothreonine were obtained from Sigma. Tyr(P) was synthesized in two different ways: by reaction of tyrosine with $POCl_3$ as reported elsewhere (20), and by condensation of P_2O_5 with tyrosine (23). Both preparations of Tyr(P) had identical mobilities on four different chromatographic and electrophoretic separation systems. Treatment of the preparations of Tyr(P) with alkaline phosphatase led to release of orthophosphate and generation of free tyrosine.

RESULTS

Amino Acid Substrate Specificity of Viral pp60^{src}. The protein kinase activity associated with pp60^{src} phosphorylates the heavy chain of immunoglobulin when immunoprecipitates containing pp60^{src} are incubated with Mg^{2+} and ATP (2, 4, 6, 7). The linkage of the phosphate incorporated into the heavy chain was completely stable to treatment with 1 M HCl for 2 hr at 55°C. This ruled out the possibility that the phosphate was attached to the protein via histidine or as an acyl phosphate (24). The phosphate linkage was much more stable to alkali (60% resistance to 1 M KOH for 2 hr at 55°C) than expected for either phosphoserine or phosphothreonine (24). This suggested that the phosphate acceptor in the heavy chain was not threonine as reported (6) but rather some other amino acid. Hydrolysis of the isolated heavy chain with 6 M HCl for 2 hr at 110°C and two-dimensional separation of the products revealed that about 25% of the radioactivity was released as Tyr(P) which, in turn, comprised >95% of the radioactivity in phosphoamino acids (Fig. 1). Identification of the phosphorylated amino acid as Tyr(P) was corroborated by the comigration of the phosphate label with authentic Tyr(P) during electrophoresis at pH 3.5 (see Fig. 2) and also on chromatography in saturated ammonium sulfate/1 M sodium acetate/isopropanol, 40:9:1 (23) (data not shown). It seems unlikely that the occurrence of Tyr(P) in the heavy chain is an artifact of acid hydrolysis because Tyr(P) was released from the phosphorylated heavy chain by exhaustive digestion with Pronase (data not shown).

Strong support for the idea that the protein kinase activity associated with pp60^{src} isolated from transformed cells is due to pp60^{src} itself and not to a contaminating protein kinase is the observation that pp60^{src} isolated from a reticulocyte lysate after synthesis by *in vitro* translation has protein kinase activity (7, 25). We therefore examined which amino acid became phosphorylated when the protein kinase activity associated with pp60^{src} produced by *in vitro* translation of RSV virion RNA was assayed in the immunoprecipitate. Tyr(P) was again the principal phosphorylated amino acid (Fig. 2).

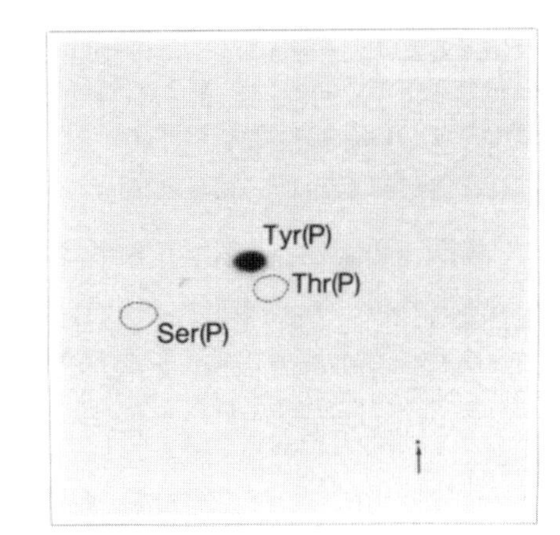

FIG. 1. Identification of phosphoamino acids in phosphorylated immunoglobulin heavy chain. Immunoglobulin heavy chain phosphorylated in an immunoprecipitate conzaining Schmidt–Ruppin RSV-D pp60^{src} was isolated and subjected to partial acid hydrolysis as described in *Materials and Methods*. A sample (2500 cpm) of the products was subjected to electrophoresis toward the anode at pH 1.9 and chromatography in isobutyric acid/ammonia. The origin is indicated with an arrow. Here and in subsequent figures, the positions of the stained marker phosphoamino acids are indicated by broken lines. The orthophosphate was run off the plate during electrophoresis. The minor spots are probably incompletely hydrolyzed phosphopeptides. Autoradiography was performed with a fluorescent screen for 12 hr. Thr(P), phosphothreonine; Ser(P), phosphoserine.

Amino Acid Substrate Specificity of Endogenous pp60^{sarc}. Tryptic peptide analysis has shown that although

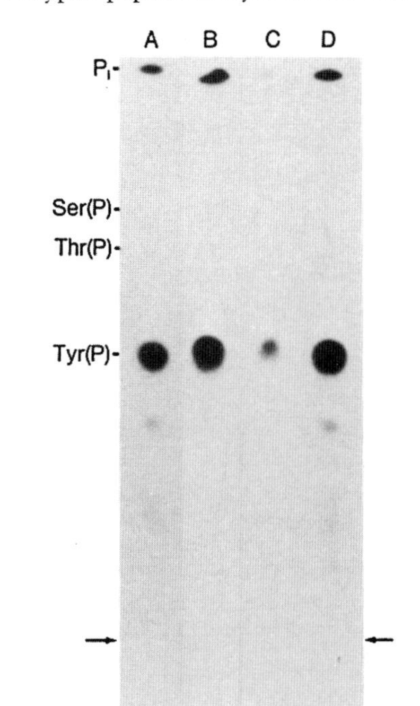

FIG. 2. Identification of amino acids phosphorylated by viral pp60^{src} and endogenous pp60^{sarc}. Immunoglobulin heavy chains phosphorylated by pp60^{src} from SR-RSV-D infected cells, pp60^{src} of SR-RSV-D synthesized *in vitro* (7), and pp60^{sarc} from chicken and human cells were isolated and subjected to partial acid hydrolysis. The hydrolysates were analyzed on a single plate by electrophoresis toward the anode at pH 3.5. The origin is indicated by arrows. Lanes, the exposure times (with a fluorescent screen), and the protein that phosphorylated the heavy chain were: A, pp60^{src} from infected cells, 25,000 cpm applied, 1 hr; B, pp60^{src} synthesized *in vitro*, 1400 cpm; 2 days; C, human pp60^{sarc}, 800 cpm, 10 hr; D, chicken pp60^{sarc}, 4000 cpm, 10 hr.

the pp60^{sarc} and pp60^{src} polypeptides are closely related they are not identical (21). We examined whether the protein kinase activity associated with chicken and human pp60^{sarc}s shared with that associated with viral pp60^{src} the ability to phosphorylate tyrosine when assayed in the immunoprecipitate. In both cases the heavy chain of the phosphorylated immunoglobulin was found to contain predominantly Tyr(P) (Fig. 2).

Phosphoamino Acids in pp60 ^{src}. pp60^{src} contains two sites of phosphorylation: one in the NH_2-terminal half of the molecule (which has been shown clearly to be a phosphoserine), and one in the COOH-terminal half (26). The phosphoamino acid in the COOH-terminal half was identified as phosphothreonine on the basis of its electrophoretic mobility at pH 1.9 (26). Stimulated by indications that pp60^{src} could undergo autophosphorylation and knowing that Tyr(P) and phosphothreonine are difficult to resolve by electrophoresis at pH 1.9, we decided to reexamine the question of what phosphorylated amino acids were contained in pp60^{src}.

In addition to pp60^{src}, two other polypeptides, of 80,000 and 50,000 daltons, are found in immunoprecipitates made from ^{32}P-labeled transformed chicken cells with antitumor antiserum (Fig. 3) (3). Neither of these is related to pp60^{src} in amino acid sequence (ref. 3; see below). Both appear to be cellular proteins that coprecipitate with pp60^{src} because they are in physical association with some fraction of the pp60^{src} molecules in a transformed chicken cell (our unpublished observations; J. Brugge, personal communication). We analyzed the phosphotryptic peptides and the phosphorylated amino acids present in all three phosphoproteins.

pp60^{src} contained two phosphorylated tryptic peptides (Fig. 4), designated here as α and β (9), as was first observed by Collett *et al.* (26). The 50,000-dalton protein contained two major and one minor phosphotryptic peptides; the 80,000-dalton protein contained an indeterminate number of phosphotryptic peptides that were poorly resolved. The phosphotryptic peptides of the three polypeptides were obviously unrelated. Hydrolysis of each protein and two-dimensional separation of the phosphorylated amino acids revealed that both pp60^{src} and the 50,000-dalton protein contained both Tyr(P) and phosphoserine and that the 80,000-dalton protein contained only phosphoserine (Fig. 4).

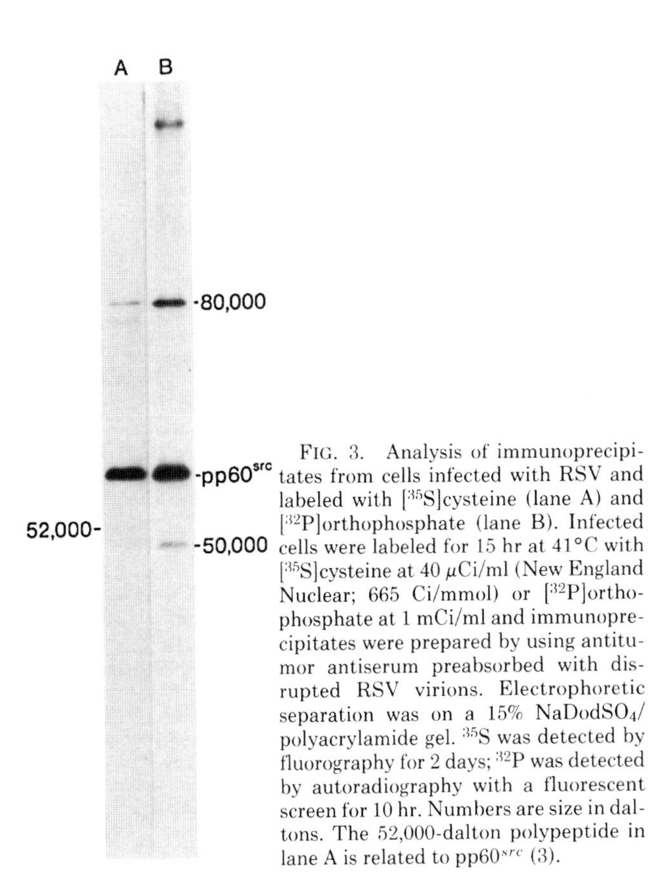

FIG. 3. Analysis of immunoprecipitates from cells infected with RSV and labeled with [^{35}S]cysteine (lane A) and [^{32}P]orthophosphate (lane B). Infected cells were labeled for 15 hr at 41°C with [^{35}S]cysteine at 40 μCi/ml (New England Nuclear; 665 Ci/mmol) or [^{32}P]orthophosphate at 1 mCi/ml and immunoprecipitates were prepared by using antitumor antiserum preabsorbed with disrupted RSV virions. Electrophoretic separation was on a 15% $NaDodSO_4$/polyacrylamide gel. ^{35}S was detected by fluorography for 2 days; ^{32}P was detected by autoradiography with a fluorescent screen for 10 hr. Numbers are size in daltons. The 52,000-dalton polypeptide in lane A is related to pp60^{src} (3).

We had previously determined (9) that pp60^{src} phosphopeptide α is derived from the 34,000-dalton NH_2-terminal fragment generated from pp60^{src} by partial proteolysis with *Staphylococcus aureus* V8 protease, whereas peptide β is derived from the COOH-terminal 24,000-dalton fragment (9). Partial acid hydrolysis of the purified peptides revealed that peptide α, as expected, contained phosphoserine and β contained Tyr(P). In the 50,000-dalton protein, peptide I contained Tyr(P) and peptide II contained phosphoserine (data not shown).

Tyr(P) in Normal and Transformed Cells. Because Tyr(P) had not been detected in normal cells (28, 29) and because we now knew that at least two proteins that contained Tyr(P) were present in RSV-transformed cells, a comparison of the amounts of Tyr(P) in uninfected and RSV-transformed chicken cells seemed warranted. We isolated proteins from cells that had been labeled for 18 hr with [^{32}P]orthophosphate and carried out partial acid hydrolysis. RSV-transformed chicken cells contained 7- to 8-fold more Tyr(P) than did uninfected cells or cells infected with RSV carrying a deletion in the *src* gene (Fig. 5; Table 1).

The hydrolysis conditions chosen here provide a compromise between the efficient release of phosphoamino acids and the competing hydrolysis of the phosphomonoester linkage. Under our conditions, approximately 50% of the ^{32}P was released as orthophosphate and 20–25% as phosphoamino acids. Because of this, we have expressed the radioactivity in the individual phosphoamino acids as a percentage of the sum of the radioactivity recovered as phosphoamino acids. These ratios will deviate from those in whole cells if the three phosphoamino acids are released or destroyed by acid at different rates. Although free Tyr(P) has been reported to be somewhat more acid labile than phosphoserine or phosphothreonine (27), this may not be true for Tyr(P) in a polypeptide. We have not found a major difference between the recovery of phosphoserine and Tyr(P) from pp60^{src} and the 50,000-dalton protein. Therefore, we think that we are not underestimating seriously the relative abundance of Tyr(P). Both the rate of release of any particular amino acid residue and the stability of the phosphomonoester bond will be affected by its polypeptide environment. However, there is no reason to assume *a priori* that the rate of release of Tyr(P) from proteins in transformed cells is significantly greater than the rate of release from proteins in normal cells. We believe therefore that we detect significantly more Tyr(P) in RSV-transformed cells because there is increased phosphorylation of tyrosine in these cells and not because the yield of Tyr(P) is greater from these cells.

DISCUSSION

The protein kinase activity associated with pp60^{src}, the transforming protein of RSV, phosphorylates tyrosine when assayed in an immunoprecipitate. This observation is surprising because protein modification by way of phosphorylation of tyrosine is unprecedented (28, 29). It is nonetheless real. We have found that chicken cells (Table 1) and mouse, rat, and hamster cells (data not shown) all contain readily detectable amounts of Tyr(P). This modified amino acid appears to have escaped detection before because it is rare (phosphoserine and phos-

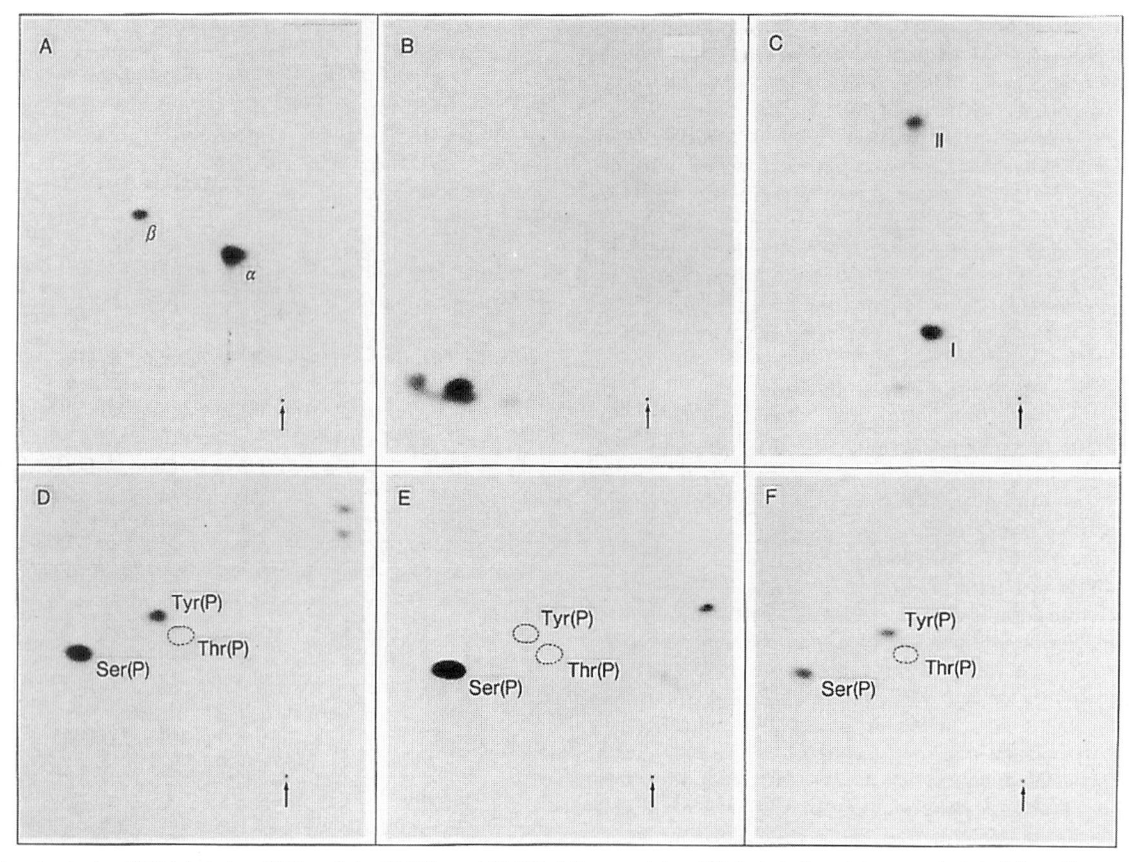

FIG. 4. Analysis of ^{32}P-labeled pp60^{src} and the 80,000- and 50,000-dalton proteins. ^{32}P-Labeled proteins were isolated from the preparation in Fig. 3 and part of each sample was subjected to tryptic digestion and analyzed. The remainder was subjected to partial acid hydrolysis and analyzed as in Fig. 1. The origins were all on the right and are indicated with arrows. All exposures were with a fluorescent screen. (*A*) pp60^{src}; tryptic digest; 6400 cpm applied; 5 hr exposure. (*B*) 80,000 protein; tryptic digest; 2000 cpm; 20 hr. (*C*) 50,000 protein; tryptic digest; 1500 cpm; 20 hr. (*D*) pp60^{src}; acid hydrolysate; 1300 cpm; 2 days. (*E*) 80,000 protein; acid hydrolysate; 900 cpm; 4 days. (*F*) 50,000 protein; acid hydrolysate; 500 cpm; 4 days.

phothreonine together being about 3000 times more abundant) and because it and phosphothreonine are difficult to separate by traditional electrophoretic procedures. Because there is a 7-fold increase in the abundance of Tyr(P) in proteins in cells transformed by RSV and because pp60^{src} itself contains Tyr(P), it seems likely that pp60^{src} phosphorylates tyrosine *in vivo* as well as *in vitro*. We suggest that pp60^{src} is a protein kinase and

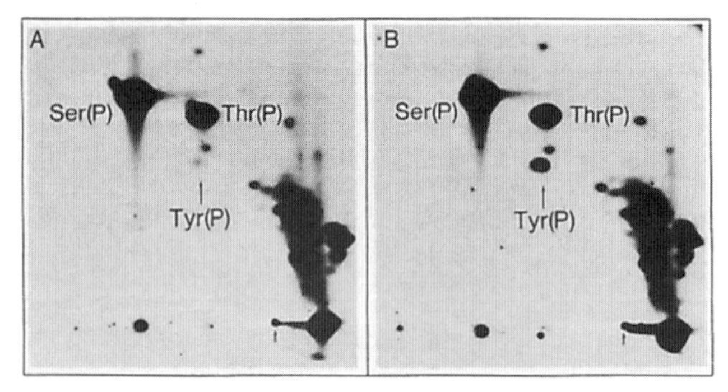

FIG. 5. Analysis of phosphoamino acids of whole cells. ^{32}P-Labeled proteins were extracted and subjected to partial acid hydrolysis; 1.5×10^6 cpm of each hydrolysate was resolved in two dimensions by electrophoresis toward the anode at pH 1.9 followed by electrophoresis toward the anode at pH 3.5. Autoradiography with a fluorescent screen was for 10 hr. The radioactive spots corresponding to the markers are indicated. The origins are designated by arrows. The material migrating toward the cathode at pH 1.9 corresponds to incompletely hydrolyzed phosphopeptides. (*A*) Uninfected chicken cells. (*B*) Chicken cells transformed with Schmidt–Ruppin RSV-A.

that the modification of proteins by phosphorylation of tyrosine is essential to the malignant transformation of cells by RSV.

The viral *src* gene, which encodes pp60^{src}, appears to be descended from the homologous *sarc* gene of the chicken (10–13). The peptide composition of the polypeptide product of the *sarc* gene of the chicken, pp60^{sarc}, is closely related to that of all viral pp60^{src}s but identical to none (21). A funda-

Table 1. Abundance of phosphoamino acids in cells

Cells*	Temp., °C	% of total cpm		
		Ser(P)	Thr(P)	Tyr(P)
Uninfected	36	92.19	7.77	0.03
Uninfected	41	91.42	8.53	0.04
SR-RSV-A	36	92.48	7.24	0.28
SR-RSV-A	41	92.73	6.96	0.31
Uninfected	38	92.40	7.55	0.04
td SR-RSV-D	38	92.22	7.75	0.04
SR-RSV-D	38	92.02	7.66	0.32

Chicken cells were labeled with [^{32}P]orthophosphate. The protein was extracted, and hydrolyzed with acid. Approximately 2×10^6 cpm of each hydrolysate was analyzed by two-dimensional electrophoresis at pH 1.9 in the first dimension and pH 3.5 in the second dimension. The thin-layer plate was autoradiographed with the aid of a fluorescent screen for 10 hr, and the spots were scraped off the plate, eluted, and assayed for radioactivity in an aqueous scintillator. The radioactivity ifseach phosphoamino acid is expressed as a percentage of the total radioactivity recovered in phosphxserine [Ser(P)], phosphothreonine [Thr(P)], and Tyr(P), which amounted to approximately 20% of the applied radioactivity in every case.
* SR, Schmidt–Ruppin; td, transformation defective.

mental question is whether RSV transforms a cell because of the overproduction of a polypeptide functionally equivalent to the normal cellular pp60^{sarc} or through the production of an altered form of the cellular protein. It is not yet possible to answer this question definitively, but it does appear that the two proteins are functionally similar: both are protein kinases with the unusual property of phosphorylating tyrosine. Because both chicken and human pp60^{sarc}s have the ability to phosphorylate tyrosine it is possible that the modification of proteins through the phosphorylation of tyrosine is an indispensable cellular function that has been conserved during evolution.

It has been shown recently that a tyrosine in both the 60,000-dalton tumor antigen of polyoma virus (20) and p120 of Abelson virus (30) can undergo phosphorylation in an immunoprecipitate. The significance of these reactions is not yet clear because neither protein contained Tyr(P) when isolated from infected cells. It has been suggested that these phosphorylated tyrosines are observed only *in vitro* because they are normally short-lived reaction intermediates that are trapped *in vitro* due to the absence of the appropriate acceptor (20, 30).

We favor the idea that the phosphorylation of tyrosine carried out by pp60^{src} is an end-state protein modification that is analogous to the phosphorylation of serine or threonine. Several observations suggest this. One is that Tyr(P) is detectable, albeit at a low level, in a wide variety of cells that have been labeled for 18 hr with [^{32}P]orthophosphate. Another is that the reaction that occurs in precipitates that contain pp60^{src} is clearly a transphosphorylation reaction which appears to involve the pp60^{src}-catalyzed transfer of phosphate from ATP to the heavy chain of immunoglobulin. The third is that pp60^{src} and the 50,000-dalton phosphoprotein that coprecipitates with pp60^{src} both contain a Tyr(P) when isolated from chicken cells. Although we cannot quantify precisely the amount of Tyr(P) in pp60^{src} or the 50,000-dalton phosphoprotein, it appears to be at least half the amount of phosphoserine in the two molecules. This is much more than we would expect were the phosphorylated tyrosine only a reaction intermediate.

Phosphorylated tyrosines are rare in uninfected cells and significantly more abundant in cells transformed by RSV. It is conceivable that all of this increase is due to the presence in these cells of pp60^{src} and its substrates. We suspect that it may be possible to identify cellular substrates of pp60^{src} simply on the basis of their containing a Tyr(P). We have already found one such protein, the 50,000-dalton phosphoprotein that coprecipitates with pp60^{src}. It is possible that this protein is one of the normal substrates of pp60^{src} and that it coprecipitates because of an association with pp60^{src}.

We thank David Shannahoff for help with the synthesis of Tyr(P), Walter Eckhart for encouragement, Tilo Patschinsky for preparing the [^{35}S]cysteine-labeled immunoprecipitate, Jack Rose for suggesting the use of electrophoresis at pH 3.5, and Claudie Berdot for help with all the experiments. This work was supported by Grants CA 14195, CA 17096, and CA 17289 from the U.S. Public Health Service.

1. Brugge, J. S. & Erikson, R. L. (1977) *Nature (London)* **269,** 346–348.
2. Levinson, A. D., Oppermann, H., Levintow, L., Varmus, H. E. & Bishop, J. M. (1978) *Cell* **15,** 561–572.
3. Sefton, B. M., Beemon, K. & Hunter, T. (1978) *J. Virol.* **28,** 957–971.
4. Rübsamen, H., Friis, R. R. & Bauer, H. (1979) *Proc. Natl. Acad. Sci. USA* **76,** 967–971.
5. Hanafusa, H. (1977) in *Comprehensive Virology*, eds. Fraenkel-Conrat, H. & Wagner, R. R. (Plenum, New York), Vol. 10, pp. 401–483.
6. Collett, M. S. & Erikson, R. L. (1978) *Proc. Natl. Acad. Sci. USA* **75,** 2021–2024.
7. Sefton, B. M., Hunter, T. & Beemon, K. (1979) *J. Virol.* **30,** 311–318.
8. Sefton, B. M., Hunter, T. & Beemon, K. (1980) *J. Virol.* **33,** 220–229.
9. Hunter, T., Sefton, B. M. & Beemon, K. (1979) *Cold Spring Harbor Symp. Quant. Biol.* **44,** in press.
10. Stehelin, D., Varmus, H. E., Bishop, J. M. & Vogt, P. K. (1976) *Nature (London)* **260,** 170–173.
11. Spector, D., Varmus, H. E. & Bishop, J. M. (1978) *Proc. Natl. Acad. Sci. USA* **75,** 4102–4106.
12. Collett, M. S., Brugge, J. S. & Erikson, R. L. (1978) *Cell* **15,** 1363–1370.
13. Oppermann, H., Levinson, A. D., Varmus, H. E., Levintow, L. & Bishop, J. M. (1979) *Proc. Natl. Acad. Sci. USA* **76,** 1804–1808.
14. Karess, R. E., Hayward, W. S. & Hanafusa, H. (1979) *Proc. Natl. Acad. Sci. USA* **76,** 3154–3158.
15. Rohrschneider, L. R., Eisenman, R. N. & Leitch, C. R. (1979) *Proc. Natl. Acad. Sci. USA* **76,** 4479–4483.
16. Collett, M. S., Erikson, E., Purchio, A. F., Brugge, J. S. & Erikson, R. L. (1979) *Proc. Natl. Acad. Sci. USA* **76,** 3159–3163.
17. Hanafusa, H., Halpern, C. C., Buchhagen, D. L. & Kawai, S. (1977) *J. Exp. Med.* **146,** 1735–1747.
18. Wang, L.-H., Halpern, C. C., Nadel, M. & Hanafusa, H. (1978) *Proc. Natl. Acad. Sci. USA* **75,** 5812–5816.
19. Vigne, R., Breitman, M. L., Moscovici, C. & Vogt, P. K. (1979) *Virology* **93,** 413–426.
20. Eckhart, W., Hutchinson, M. A. & Hunter, T. (1979) *Cell* **18,** 925–933.
21. Sefton, B. M., Hunter, T. & Beemon, K. (1980) *Proc. Natl. Acad. Sci. USA*, in press.
22. Beemon, K. & Hunter, T. (1978) *J. Virol.* **28,** 551–566.
23. Rothenberg, P. G., Harris, T. J. R., Nomoto, A. & Wimmer, E. (1978) *Proc. Natl. Acad. Sci. USA* **75,** 4868–4872.
24. Bitte, L. & Kabat, D. (1974) *Methods Enzymol.* **30,** 563–590.
25. Erikson, E., Collett, M. S. & Erikson, R. L. (1978) *Nature (London)* **274,** 919–921.
26. Collett, M. S., Erikson, E. & Erikson, R. L. (1979) *J. Virol.* **29,** 770–781.
27. Plimmer, R. H. A. (1941) *Biochem. J.* **35,** 461–469.
28. Taborsky, G. (1974) *Adv. Prot. Chem.* **28,** 1–187.
29. Uy, R. & Wold, F. (1977) *Science* **198,** 890–896.
30. Witte, N. O., Dasgupta, A. & Baltimore, D. (1980) *Nature (London)*, in press.

A Noncatalytic Domain Conserved among Cytoplasmic Protein-Tyrosine Kinases Modifies the Kinase Function and Transforming Activity of Fujinami Sarcoma Virus P130$^{gag\text{-}fps}$

IVAN SADOWSKI,[1] JAMES C. STONE,[2] AND TONY PAWSON[1]*

Division of Molecular and Developmental Biology, Mount Sinai Hospital Research Institute, and Department of Medical Genetics, University of Toronto, Toronto, Ontario M5G 1X5, Canada; and The Jackson Laboratory, Bar Harbor, Maine 04609[2]

Received 25 February 1986/Accepted 4 September 1986

Proteins encoded by oncogenes such as v-*fps*/*fes*, v-*src*, v-*yes*, v-*abl*, and v-*fgr* are cytoplasmic protein tyrosine kinases which, unlike transmembrane receptors, are localized to the inside of the cell. These proteins possess two contiguous regions of sequence identity: a C-terminal catalytic domain of 260 residues with homology to other tyrosine-specific and serine-threonine-specific protein kinases, and a unique domain of approximately 100 residues which is located N terminal to the kinase region and is absent from kinases that span the plasma membrane. In-frame linker insertion mutations in Fujinami avian sarcoma virus which introduced dipeptide insertions into the most stringently conserved segment of this N-terminal domain in P130$^{gag\text{-}fps}$ impaired the ability of Fujinami avian sarcoma virus to transform rat-2 cells. The P130$^{gag\text{-}fps}$ proteins encoded by these transformation-defective mutants were deficient in protein-tyrosine kinase activity in rat cells. However v-*fps* polypeptides derived from the mutant Fujinami avian sarcoma virus genomes and expressed in *Escherichia coli* as *trpE*-v-*fps* fusion proteins displayed essentially wild-type enzymatic activity, even though they contained the mutated sites. Deletion of the N-terminal domain from wild-type and mutant v-*fps* bacterial proteins had little effect on autophosphorylating activity. The conserved N-terminal domain of P130$^{gag\text{-}fps}$ is therefore not required for catalytic activity, but can profoundly influence the adjacent kinase region. The presence of this noncatalytic domain in all known cytoplasmic tyrosine kinases of higher and lower eucaryotes argues for an important biological function. The relative inactivity of the mutant proteins in rat-2 cells compared with bacteria suggests that the noncatalytic domain may direct specific interactions of the enzymatic region with cellular components that regulate or mediate tyrosine kinase function.

Many tyrosine kinases were first identified as the products of activated retroviral or cellular oncogenes in which manifestation they induce neoplastic cell growth (19). In normal vertebrate cells protein-tyrosine kinases are apparently involved in the interpretation of mitogenic and developmental signals (19). Both normal protein-tyrosine kinases and their oncogenic counterparts can be divided into two groups based on their subcellular locations. One group contains transmembrane growth factor receptors such as those for epidermal growth factor, platelet-derived growth factor, insulin, and colony-stimulating factor 1 in which a cytoplasmic kinase domain is linked by a transmembrane segment to an extracellular ligand-binding domain (2, 11, 20, 43, 46, 59, 60). In general, association of a hormone with its receptor stimulates the activity of the intracellular kinase domain toward cellular protein substrates (43). The products of the v-*erbB*, v-*fms*, v-*ros*, c-*neu*, and *trk* oncogenes are apparently aberrant forms of such receptors (10, 18, 30, 33, 45). In contrast tyrosine kinases encoded by genes such as *src*, *fps*/*fes*, *abl*, *fgr*, and *yes* have no obvious transmembrane sequence and are entirely cytoplasmic, although they are generally found in association with the internal face of the plasma membrane or with the cytoskeleton (14, 15, 28, 31, 39–42).

The normal functions of proteins in this latter group are less clear than those of receptors for which specific ligands can be identified. p60$^{c\text{-}src}$ is expressed to high levels in postmitotic neuronal cells and platelets (8, 17, 49), whereas synthesis of the c-*fps*/*fes* product is largely confined to cells capable of myeloid differentiation (12, 27). Such observations have suggested that the cytoplasmic tyrosine kinases may have functions in development or differentiation.

All protein-tyrosine kinases share a region of sequence identity spanning approximately 260 amino acids which apparently corresponds to the catalytic domain. This region is released from p60$^{v\text{-}src}$ and P130$^{gag\text{-}fps}$ by limited proteolysis as a C-terminal 29-kilodalton (kDa) fragment that retains full enzymatic activity (4, 25, 62). One border of this 29-kDa kinase domain is predicted to lie immediately N terminal to an ATP-binding site containing elements conserved between many nucleotide-binding proteins (1, 2, 63a). We have previously employed in-frame linker insertion mutagenesis as a tool to investigate the structural and functional domains of a cytoplasmic protein-tyrosine kinase, the P130$^{gag\text{-}fps}$ transforming protein of Fujinami sarcoma virus (FSV) (53). Dipeptide insertions in P130$^{gag\text{-}fps}$ apparently disrupt local protein conformation in the affected structural domain without perturbing the structure of separately folding domains. Thus insertions within 250 N-terminal v-*fps* residues distort a *fps*-specific domain and impair FSV transforming function without obvious effect on the activity of the C-terminal kinase domain (54). A central putative hinge region accommodates insertion without effect, whereas insertions in the conserved kinase domain entirely abolish transforming ability. One FSV mutant (RX15m) with an insertion approximately 80 residues N terminal to the P130$^{gag\text{-}fps}$ ATP-binding site was not easily categorized since it was unable to elicit focus formation in rat-2 cells, yet was unaltered in either the

* Corresponding author.

N-terminal *fps*-specific domain or the C-terminal kinase domain (53).

We have therefore examined in detail the region immediately N terminal to the predicted kinase domain of P130$^{gag\text{-}fps}$. In all known cytoplasmic protein-tyrosine kinases a sequence of approximately 100 amino acids N terminal to the catalytic domain is highly conserved. Here we describe FSV mutants and bacterial expression vectors designed to test the functional importance of this conserved noncatalytic domain.

MATERIALS AND METHODS

Construction of in-frame linker insertion mutations and mammalian cell expression vector. The procedure for construction of AX (*Alu*I→*Xho*I) and RX (*Rsa*I→*Xho*I) linker insertion mutations in FSV with the plasmid pJ2 was reported previously (53). The RX15m mutant contains an *Xho*I linker (CTCGAG) in the *Rsa*I site at FSV nucleotide position 2839 (numbered according to Shibuya and Hanafusa [47]), whereas the AX9m mutant contains one copy of an *Xho*I linker in the *Alu*I site at position 2872 of the FSV sequence. Sequencing of the mutation sites was by the dideoxynucleotide chain termination method of Sanger et al. (44). For increased expression of mutant proteins in mammalian cells we used a plasmid vector (pIV2; see Fig. 3) in which FSV sequences are driven from the simian virus 40 early promoter and the Neo^r gene from an FSV long terminal repeat. To construct pIV2 a 4.8-kilobase (kb) *Sst*I fragment representing the entire circularly permuted FSV genome was subcloned into pUC18 at the *Sst*I site. The *Hin*dIII-*Eco*RI insert from this plasmid was ligated to the *Hin*dIII-*Eco*RI sites of pSV2neo (pIV1) (50). A 2.5-kb *Hin*dIII-*Bam*HI fragment from pSV2neo containing the Neo^r gene and simian virus 40 early polyadenylation site was inserted after end filling and attachment of *Eco*RI linkers into the *Eco*RI site of pIV1.

Cell culture and DNA transfection. Rat-2 cells (58) were grown in Dulbecco modified Eagle medium supplemented with 10% fetal bovine serum. Transfection of pJ2 or its RX15m or AX9m derivatives was as described previously (53). Plates transfected with plasmids containing RX15m or AX9m mutant FSV genomes were maintained for 8 weeks for analysis of transforming activity. To isolate cell lines expressing equivalent amounts of wild-type (wt), RX15m, or AX9m P130$^{gag\text{-}fps}$ protein, 5×10^6 rat-2 cells were transfected with 500 ng of pIV2 plasmid containing wt or mutant *gag-fps* genes plus 5 μg of rat-2 carrier DNA by using the calcium phosphate coprecipitation technique. At 48 h posttransfection the cells were trypsinized and replated at a 1:20 dilution in Dulbecco modified Eagle medium containing 10% fetal bovine serum and 400 μg of G418 (GIBCO Laboratories) per ml. The medium was changed every 4 to 5 days. After 4 weeks G418-resistant colonies were subcultured to 35-mm wells and subsequently analyzed for expression of P130$^{gag\text{-}fps}$ by metabolic labeling with [^{35}S]methionine and immunoprecipitation with anti-p19^{gag} monoclonal antibody. Cell lines were maintained in 200 μg of G418 per ml.

Antisera. The anti-p19^{gag} mouse monoclonal antibody R254E has been described previously (21). Polyclonal rabbit antisera were raised to the 37-kDa *trpE* product of the bacterial pATH expression vector and to the bacterial pTF822 *trpE*–v-*fps* protein (see Fig. 8 and 9). For this purpose 100-ml cultures of *Escherichia coli* RR1 induced for expression of the pATH or pTF822 proteins (see below) were lysed and electrophoresed on 10% sodium dodecyl sulfate (SDS)-polyacrylamide preparative gels. The positions of *trpE* or *trpE*-v-*fps* proteins were identified by staining of gel strips with Coomassie blue. Proteins were electroeluted and concentrated as described previously (24). In each case 250 μg of gel-purified protein was emulsified in complete Freund adjuvant and injected at various sites intramuscularly and subcutaneously into a 6-month-old New Zealand white rabbit. Rabbits were boosted after 5 and 10 weeks with 100 μg of protein in incomplete Freund adjuvant and were bled for antiserum 7 days after the second boost.

Metabolic labeling and immunoprecipitation of mammalian cell proteins. Rat-2 cells were metabolically labeled for 4 h with [^{35}S]methionine or with $^{32}P_i$ as described previously (62). For pulse-chase experiments, cells were labeled for 2 h with [^{35}S]methionine followed by chases for increasing lengths of time with fresh medium containing 20 mM unlabeled methionine.

Cell lysis, immunoprecipitation of P130$^{gag\text{-}fps}$ with R254E anti-p19^{gag} mouse monoclonal antibody or with rabbit antiserum to bacterial v-*fps* protein, and analysis of immunoprecipitates by SDS-polyacrylamide gel electrophoresis and autoradiography were as described previously (21, 62), except that for the rabbit antiserum 10 μl of serum was added to each immunoprecipitation reaction and collected with 100 μl of 10% (vol/vol) heat-inactivated *Staphylococcus aureus*. To quantitate ^{35}S or ^{32}P radioactivity in metabolically labeled P130$^{gag\text{-}fps}$, the relevant bands were excised from the gel and counted by liquid scintillation spectrophotometry.

In vitro kinase reactions. *gag-fps* or *trpE*-v-*fps* proteins were immunoprecipitated from lysates of rat-2 cells or bacteria with anti-*gag* or anti-*trpE* antibodies for analysis of in vitro kinase activity. Rat-2 cells were lysed in kinase lysis buffer (20 mM Tris hydrochloride [pH 7.5], 150 mM NaCl, 1 mM EDTA, 1% [vol/vol] Nonidet P-40 and 0.5% [wt/vol] sodium deoxycholate) and incubated with anti-p19^{gag} antibody, and the immune complex was precipitated with *S. aureus* cells precoated with rabbit anti-mouse immunoglobulin (21). For preparation of bacterial *trpE*–v-*fps* proteins *E. coli* cells containing the pTd1359 plasmids were grown and induced as described below. The pellet from a 45-ml induced culture was washed two times in 50 mM Tris hydrochloride (pH 7.5)–100 mM NaCl–10 mM $MgCl_2$ and suspended in 800 μl of *E. coli* lysis buffer (50 mM Tris hydrochloride [pH 7.5], 0.3 M NaCl, 1% Nonidet P-40, 20 mM $MgCl_2$). A 200-μl amount of the same buffer and 2 mg of lysozyme per ml were added, and the suspension was incubated on ice for 30 min. A 200-μl amount of *E. coli* lysis buffer and 1 mg of DNase I per ml were then added, and the suspension was held on ice for a further 60 min. The crude lysate was clarified by centrifugation at 26,000 × *g* for 30 min. The supernatant was then incubated with 10 μl of rabbit anti-*trpE* antiserum for 60 min at 4°C and for a further 30 min with 100 μl of 10% (vol/vol) *S. aureus*. Immune complexes were washed in kinase lysis buffer and then in 10 mM $MnCl_2$–50 mM HEPES (*N*-2-hydroxyethylpiperazine-*N'*-2-ethanesulfonic acid), pH 7.5. Kinase activity was assayed in 50 μl immune complex reaction containing 10 mM $MnCl_2$, 50 mM HEPES (pH 7.5), 2.5 μCi of [γ-^{32}P]ATP (3,000 Ci/mmol), and 5 μg of acid-denatured enolase (7, 63, 64). The reaction was allowed to proceed for 15 min at 30°C and was stopped by the addition of 2× SDS gel sample buffer (20% [vol/vol] glycerol, 10% [vol/vol] 2-mercaptoethanol, 4.6% [wt/vol] SDS, 0.125 M Tris [pH 6.8]) for a further 10 min. Bacteria were removed by centrifugation, and the kinase reaction products were heated at 100°C for 3 min and then analyzed by electrophoresis on a 7.5% SDS–polyacrylamide gel.

To measure phosphorylation of the synthetic poly (Glu,Tyr) (80:20) polymer (Sigma Chemical Co.) by bacterial extracts, 20 μl of the crude bacterial lysate was added to 20 μl of 2× kinase reaction buffer (40 mM $MnCl_2$, 200 mM HEPES [pH 7.5]). A 1-μl amount of poly(Glu,Tyr) polymer (10 mg/ml) and 2 μCi of [γ-^{32}P]ATP were added, and the reactions were incubated at 37°C for 15 min. The reactions were stopped by the addition of 1 μl of 100 mM cold ATP, 2 μl of 500 mM EDTA, and 40 μl of 2× sample buffer (4% [vol/vol] Nonidet P-40, 10% [vol/vol] glycerol, 125 mM Tris hydrochloride [pH 6.8], 0.002% bromophenol blue). The poly(Glu,Tyr) was separated from the reaction mixture by electrophoresis on 5% polyacrylamide gels containing 8 M urea, 2% Nonidet P-40, and 0.325 M Tris hydrochloride (pH 8.8) without SDS (37). The gels were fixed and washed extensively in 10% (vol/vol) acetic acid–50 mM sodium pyrophosphate. Incorporation of ^{32}P into poly(Glu,Tyr) was quantitated by excision of the appropriate band from the gel and scintillation counting. Control reactions used for subtraction of background radioactivity contained extracts of bacteria expressing the parental pATH plasmid.

Phosphoamino acid analysis and tryptic phosphopeptide mapping. Cell lines expressing wt or mutant FSV P130$^{gag\text{-}fps}$ were metabolically labeled for 18 h with $^{32}P_i$, lysed, and immunoprecipitated with anti-p19^{gag} antibody. Bacterial proteins were labeled as described below. Labeled proteins were purified by electrophoresis through 7.5% SDS–polyacrylamide gels. Analysis of phosphoamino acid content and two-dimensional tryptic phosphopeptide mapping of gel-purified proteins were undertaken as described previously (62, 64).

Construction of *trpE*-v-*fps* fusion vectors. wt FSV DNA (47) was digested with *Sma*I, which cuts at nucleotide positions 520 and 3942. This 3.4-kb fragment, which encodes the entire *gag-fps* protein except for the first 48 amino acids of *gag*, was cloned into the *Sma*I site of pUC18. The fragment was then excised by digestion with *Eco*RI-*Hin*dIII and subcloned into the bacterial expression plasmid pATH-1, which contains a multiple cloning site to allow joining of foreign sequences to a partial *trpE* coding sequence (51, 57; T. J. Koerner and A. Tzalgoloff, personal communications). The resulting construct, pTF49, encodes a 165-kDa *trpE*–*gag*–v-*fps* fusion protein which was very unstable. pTd1359 was constructed by digesting pTF49 with *Nco*I-*Eco*RV, treating with S1 nuclease, and religating. The plasmid encodes a *trpE*–v-*fps* protein of 100 kDa consisting of 42 amino acids of *trpE* fused to the transformation-competent p91 v-*fps* of F36 virus (13). The RX15m and AX9m mutations were subsequently subcloned into pTd1359 by using the *Bst*XI and *Apa*I restriction sites within the v-*fps* gene (nucleotide positions 2619 and 3423, respectively).

To construct pTF729, pTF729-RX15m and pTF729-AX9m, wt pJ2, or mutant RX15m and AX9m plasmid DNAs were cut with *Pvu*II and *Sma*I to excise a 2.4-kb fragment containing the coding sequence of FSV P130$^{gag\text{-}fps}$ from amino acid 729 to the C terminus. The blunt-ended fragments were cloned into the *Sma*I site of pUC8, and the resulting clones were screened for orientation. *Eco*RI-*Hin*dIII fragments (1.4 kb) containing the v-*fps* inserts were cut out of these constructs and inserted into the *Eco*RI-*Hin*dIII sites of pATH-1. pTF822 and pTF833 were constructed by digesting RX15m and AX9m plasmid DNAs with *Xho*I (which cuts at the linker insertion site in RX15m and AX9m DNA) and *Hin*dIII to generate fragments which would code for the carboxy terminus of P130$^{gag\text{-}fps}$ from amino acids 822 and 833, respectively. The 3.0-kb *Xho*I-*Hin*dIII fragments were

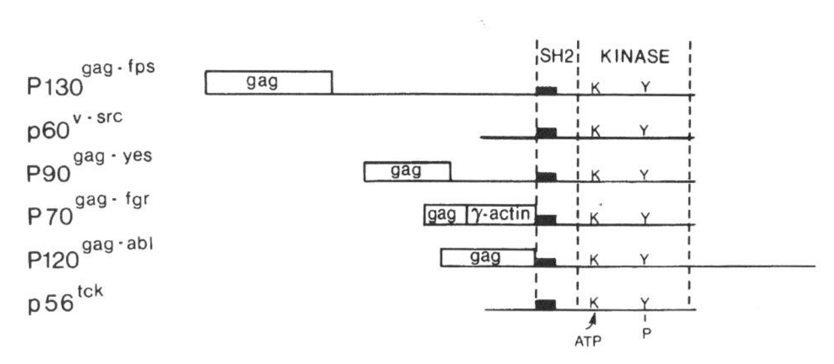

FIG. 1. Retroviral and cellular cytoplasmic protein-tyrosine kinases aligned to show the catalytic (kinase) and noncatalytic SH2 domains. The most highly conserved segment of SH2 is indicated by a raised box.

cloned into the *Sal*I-*Hin*dIII sites of pUC8. The *Eco*RI-*Hin*dIII fragments from these plasmids were cloned into pATH-11 (RX15m DNA) to give pTF822 or into pATH-3 (AX9m DNA) to give pTF833. The pTF893 plasmid was constructed by digesting wt FSV DNA with *Bsp*MI (which cuts at FSV nucleotide 3055) and *Sma*I (nucleotide position 3942). The 5′ protruding overhang resulting from the *Bsp*MI digestion was filled in with the Klenow fragment of DNA polymerase I, and the 887-base-pair fragment was blunt end ligated into the *Sma*I site of pUC18. The fragment was subcloned into pATH-1.

In all cases, when cloning into pATH vectors the bacterial colonies were screened by analyzing whole cell protein on polyacrylamide gels and staining with Coomassie blue. Clones encoding the correctly sized protein were found to have the proper construct by restriction analysis and partial DNA sequencing.

Growth and metabolic labeling of *E. coli* expressing *trpE*–v-*fps* fusion protein. To grow bacteria for analysis of protein expression, 1 ml of M9 minimal medium supplemented with 0.5% (wt/vol) Casamino Acids (Difco Laboratories), 10 μg of thiamine B_1 per ml, 20 μg of tryptophan per ml, and 50 μg of ampicillin per ml was inoculated for growth overnight at 37°C. A 0.1-ml sample of the overnight culture was reinoculated into 1 ml of M9 plus Casamino Acids, thiamine, and ampicillin. The cultures were grown for 1 h at 37°C with strong aeration, and then 5 μl of 1-mg/ml indole acrylic acid in ethanol was added to induce the tryptophan operon. The cultures were grown a further 2 h, and the bacteria were pelleted by centrifugation. The pellets were suspended in 25 μl of SDS-urea buffer (10 mM sodium phosphate [pH 7.2], 1% [vol/vol] 2-mercaptoethanol, 1% [wt/vol]SDS, 6 M urea) and incubated at 37°C for 30 min. A 25-μl amount of 2× SDS buffer was then added, the samples were heated at 100°C for 3 min, and 15-μl portions were analyzed by electrophoresis on 7.5% polyacrylamide gels.

Growth of the bacterial cells for metabolic labeling was exactly as described above. For [^{35}S]methionine labeling, [^{35}S]methionine was added at the time of induction to 10 μCi/ml. For ^{32}P labeling, the cells were pelleted after 2 h of induction with indole acrylic acid and washed twice in 50 mM Tris hydrochloride (pH 7.5)–100 mM NaCl–10 mM $MgCl_2$. The cells were suspended in 50 μl of the same buffer. To 15 μl of the cell suspension 60 μCi of $^{32}P_i$ was added, and the cells were incubated at room temperature for 20 min. A 500-μl amount of TEN buffer (50 mM Tris [pH 7.5], 0.5 mM EDTA, 0.3 M NaCl) was added, and the cells were pelleted and washed three times in TEN. The pellets were then suspended in 20 μl of SDS-urea buffer, incubated at 37°C for 30 min, and analyzed by polyacrylamide gel electrophoresis and autoradiography.

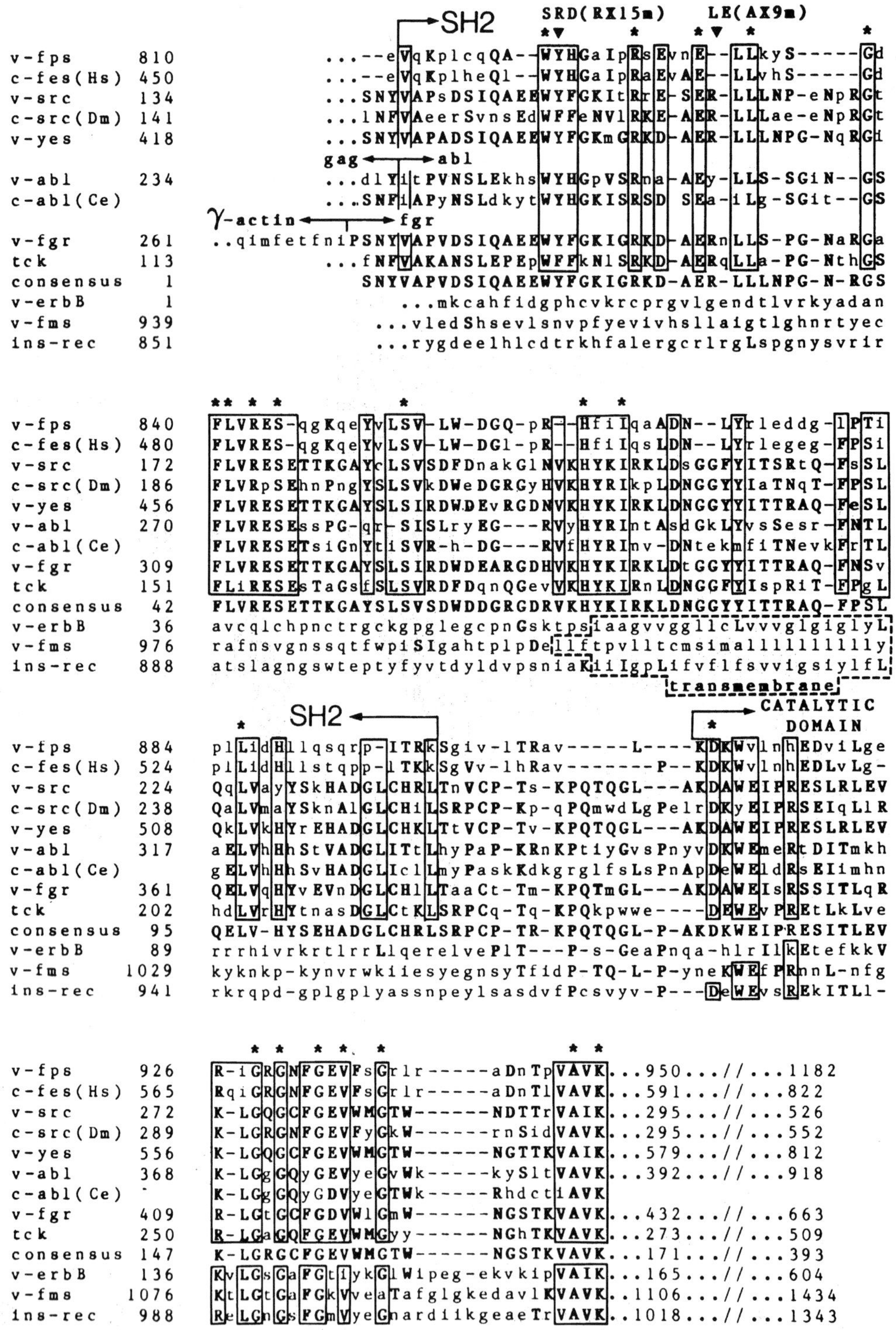

```
                                     ->SH2         SRD(RX15m)      LE(AX9m)
                                                  *▼       *     * ▼   *         *
v-fps       810          ...--eVqKplcqQA--WYHGaIpRsEvnE--LLkyS-----Gd
c-fes(Hs)   450          ...--eVqKplheQl--WYHGaIpRaEvAE--LLvhS-----Gd
v-src       134          ...SNYVAPsDSIQAEEWYFGKItRrE-SER-LLLNP-eNpRGt
c-src(Dm)   141          ...lNFVAeerSvnsEdWFFeNVlRKE-AER-LLLae-eNpRGt
v-yes       418          ...SNYVAPADSIQAEEWYFGKmGRKD-AER-LLLNPG-NqRGi
                         gag<-->abl
v-abl       234          ...dlYitPVNSLEkhsWYHGpVSRna-AEy-LLS-SGiN--GS
c-abl(Ce)                ...SNFiAPyNSLdkytWYHGKISRSD SEa-iLg-SGit--GS
                 γ-actin<-->fgr
v-fgr       261  ..qimfetfniPSNYVAPVDSIQAEEWYFGKIGRKD-AERnLLS-PG-NaRGa
tck         113          ...fNFVAKANSLEPEpWFFkNlSRKD-AERqLLa-PG-NthGS
consensus     1             SNYVAPVDSIQAEEWYFGKIGRKD-AER-LLLNPG-N-RGS
v-erbB        1                ...mkcahfidgphcvkrcprgvlgendtlvrkyadan
v-fms       939              ...vledShsevlsnvpfyevivhsllaigtlghnrtyec
ins-rec     851              ...rygdeelhlcdtrkhfalergcrlrgLspgnysvrir

            ** * *          *                *   *
v-fps       840  FLVRES-qgKqeYvLSV-LW-DGQ-pR--HfiIqaADN--LYrleddg-lPTi
c-fes(Hs)   480  FLVRES-qgKqeYvLSV-LW-DGl-pR--HfiIqsLDN--LYrlegeg-FPSi
v-src       172  FLVRESETTKGAYcLSVSDFDnakGlNVKHYKIRKLDsGGFYITSRtQ-FsSL
c-src(Dm)   186  FLVRpSEhnPngYSLSVkDWeDGRGyHVKHYRIkpLDNGGYYIaTNqT-FPSL
v-yes       456  FLVRESETTKGAYSLSIRDWDEvRGDNVKHYKIRKLDNGGYYITTRAQ-FeSL
v-abl       270  FLVRESEssPG-qr-SISLryEG---RVyHYRIntAsdGkLYvsSesr-FNTL
c-abl(Ce)        FLVRESETsiGnYtiSVR-h-DG---RVfHYRInv-DNtekmfiTNevkFrTL
v-fgr       309  FLVRESETTKGAYSLSIRDWDEARGDHVKHYKIRKLDtGGYYITTRAQ-FNSv
tck         151  FLiRESEsTaGsfSLSVRDFDqnQGevVKHYKIRnLDNGGFYIspRiT-FPgL
consensus    42  FLVRESETTKGAYSLSVSDWDDGRGDRVKHYKIRKLDNGGYYITTRAQ-FPSL
v-erbB       36  avcqlchpnctrgckgpglegcpnGsktpsiaagvvggllcLvvvglgiglyL
v-fms       976  rafnsvgnssqtfwpiSIgahtplpDellftpvlltcmsimallllllllly
ins-rec     888  atslagngswteptyfyvtdyldvpsniaKiiIgpLifvflfsvvigsiylfL
                                                 transmembrane
                                                      -> CATALYTIC
              *      SH2<-                          *      DOMAIN
v-fps       884  plLidHllqsqrp-ITRkSgiv-lTRav-----L----KDKWvlnhEDviLge
c-fes(Hs)   524  plLidHllstqpp-lTKkSgVv-lhRav-------P--KDKWvlnhEDLvLg-
v-src       224  QqLVayYSkHADGLCHRLTnVCP-Ts-KPQTQGL---AKDAWEIPRESLRLEV
c-src(Dm)   238  QaLVmaYSknAlGLCHiLSRPCP-Kp-qPQmwdLgPelrDKyEIPRSEIqLlR
v-yes       508  QkLVkHYrEHADGLCHKLTtVCP-Tv-KPQTQGL---AKDAWEIPRESLRLEV
v-abl       317  aELVhHhStVADGLITtLhyPaP-KRnKPtiyGvsPnyvDKWEmeRtDITmkh
c-abl(Ce)        gELVhHhSvHADGLIclLmyPaskKdkgrglfsLsPnApDeWEldRsEIimhn
v-fgr       361  QELVqHYvEVnDGLCHlLTaaCt-Tm-KPQTmGL---AKDAWEIsRSSITLqR
tck         202  hdLVrHYtnasDGLCtKLSRPCq-Tq-KPQkpwwe----DEWEvPREtLkLve
consensus    95  QELV-HYSEHADGLCHRLSRPCP-TR-KPQTQGL-P-AKDKWEIPRESITLEV
v-erbB       89  rrrhivrkrtlrrLlqerelvePlT---P-s-GeaPnqa-hlrIlkEtefkkV
v-fms      1029  kyknkp-kynvrwkiiesyegnsyTfidP-TQ-L-P-yneKWEfPRnnL-nfg
ins-rec     941  rkrqpd-gplgplyassnpeylsasdvfPcsvyv-P---DeWEvsREkITLl-

                    * *  * *  *             * *
v-fps       926  R-iGRGNFGEVFsGrlr-----aDnTpVAVK...950...//...1182
c-fes(Hs)   565  RqiGRGNFGEVFsGrlr-----aDnTlVAVK...591...//...822
v-src       272  K-LGQGCFGEVWMGTW------NDTTrVAIK...295...//...526
c-src(Dm)   289  K-LGRGNFGEVFyGkW------rnSidVAVK...295...//...552
v-yes       556  K-LGQGCFGEVWMGTW------NGTTKVAIK...579...//...812
v-abl       368  K-LGgGQyGEVyeGvWk-----kySltVAVK...392...//...918
c-abl(Ce)        K-LGgGQyGDVyeGTWk-----RhdctiAVK
v-fgr       409  R-LGtGCFGDVWlGmW------NGSTKVAVK...432...//...663
tck         250  R-LGaGQFGEVWMGyy------NGhTKVAVK...273...//...509
consensus   147  K-LGRGCFGEVWMGTW------NGSTKVAVK...171...//...393
v-erbB      136  KvLGsGaFGtiykGlWipeg-ekvkipVAIK...165...//...604
v-fms      1076  KtLGtGaFGkVveaTafglgkedavlKVAVK..1106...//...1434
ins-rec     988  ReLGnGsFGmVyeGnardiikgeaeTrVAVK..1018...//...1343
```

FIG. 2. Region of sequence identity among cytoplasmic protein-tyrosine kinases located amino terminal to the proposed catalytic domain. The products of the following genes are aligned to display a region of sequence identity spanning approximately 100 residues N terminal to the kinase domain (designated SH2): FSV v-*fps*/*fes* (47), human (Hs) c-*fps*/*fes* (38a), RSV v-*src* (55), *Drosophila melanogaster* (Dm) c-*src* (48), Y73 v-*yes* (22), Ab-MLV *gag-abl* (38), *C. elegans* (Ce) *abl* (16), Gardner-Rasheed feline sarcoma virus γ-actin–v-*fgr* (32), and mouse *tck*/*lsk*^{T} (29, 61). The sequence comparison starts at the *gag-abl*/*actin-fgr* borders and extends to include approximately 37 residues in the catalytic domain up to a conserved lysine within the ATP-binding site (63a). Residues in common between two or more different cytoplasmic tyrosine kinases are depicted in bold uppercase type, and sites showing particularly strong conservation are boxed. Invariant residues are indicated with an asterisk (*). A consensus sequence is derived from this comparison of cytoplasmic tyrosine kinases, which is in turn compared with the transmembrane tyrosine kinases gp74$^{v\text{-}erbB}$ (10), gp140$^{v\text{-}fms}$ (18), and the β-subunit of the human insulin receptor (59). For the transmembrane kinases, residues are in uppercase type where they coincide with the cytoplasmic kinase consensus sequence. Their hydrophobic transmembrane domains are boxed with a broken line. The proposed start of the catalytic kinase domain is indicated. Amino acid insertions introduced into P130$^{gag\text{-}fps}$ by site-directed mutagenesis of the FSV genome are indicated above the v-*fps* polypeptide sequence. Sequences were aligned by eye.

RESULTS

An amino-terminal region of sequence identity in cytoplasmic protein-tyrosine kinases. Figure 1 shows a diagrammatic comparison of the proteins encoded by FSV ($P130^{gag\text{-}fps}$), Rous sarcoma virus (RSV; $p60^{v\text{-}src}$), Y73 ($P90^{gag\text{-}yes}$), Abelson murine leukemia virus ($P120^{gag\text{-}abl}$), and Gardner-Rasheed feline sarcoma virus ($P70^{gag\text{-}actin\text{-}fgr}$), and the mouse lsk^T/*tck* gene. These transforming proteins all have a cytoplasmic location and possess no obvious hydrophobic sequences capable of traversing the plasma membrane (15, 22, 28, 32, 38, 39–42, 47, 55, 61). The 260 residues shared with transmembrane tyrosine kinases are depicted as the kinase domain (*src* homology 1), and the locations of two invariant residues are given: the principal site of tyrosine autophosphorylation (Tyr-1073 of $P130^{gag\text{-}fps}$, Tyr-416 of $p60^{v\text{-}src}$) (19, 64) and a lysine in the ATP-binding site thought to participate in cleaving the γ phosphate of bound ATP (Lys-950 of $P130^{gag\text{-}fps}$, Lys-295 of $p60^{v\text{-}src}$) (1, 63a). N terminal to the kinase domain lies an additional region of sequence identity comprising approximately 100 residues, which we have termed SH2 for *src* homology 2. It is of interest to note that in two of these kinases, $P120^{gag\text{-}abl}$ and $P70^{gag\text{-}actin\text{-}fgr}$, a foreign polypeptide is fused directly N terminal to SH2. In both cases the junction of the heterologous (*gag* or γ-actin) sequence with *abl* or *fgr* occurs within a few residues in the consensus tyrosine kinase sequence, defining the start of the SH2 region (Fig. 2). A detailed sequence analysis of the SH2 domain is given in Fig. 2, starting at the actin-*fgr*/*gag-abl* junctions and extending to the conserved lysine in the ATP-binding site. The comparison has been expanded to include a variety of cellular proteins related to the retroviral cytoplasmic tyrosine kinases, including the products of the human *fps*/*fes*, mouse lsk^T/*tck*, *Drosophila melanogaster src*, and *Caenorhabditis elegans abl* genes. A consensus sequence was deduced for these viral and cytoplasmic tyrosine kinases, which is in turn aligned to show relationships between the cytoplasmic tyrosine kinases and transmembrane kinases typified by $gp74^{v\text{-}erbB}$, $gp140^{v\text{-}fms}$, and the β subunit of the insulin receptor.

The primary region of homology between FSV $P130^{gag\text{-}fps}$

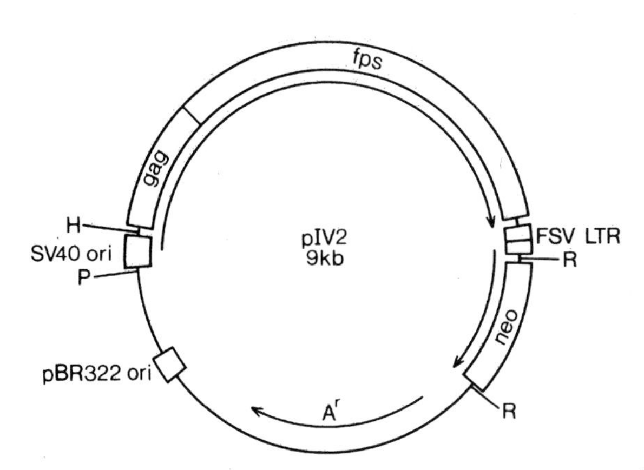

FIG. 3. pIV2 plasmid for the expression of *gag-fps* sequences in mammalian cells. Transcription of the *gag-fps* gene is initiated by the simian virus 40 (SV40) early promoter and terminated at the poly(A) site of the first tandemly repeated FSV long terminal repeat (LTR). The second long terminal repeat is used as a promoter for the bacterial neo^r gene. Termination of neo^r transcripts is mediated by the simian virus 40 early region poly(A) site. Abbreviations: H, *Hin*dIII; R, *Eco*RI; S, *Sst*I; P, *Pvu*II.

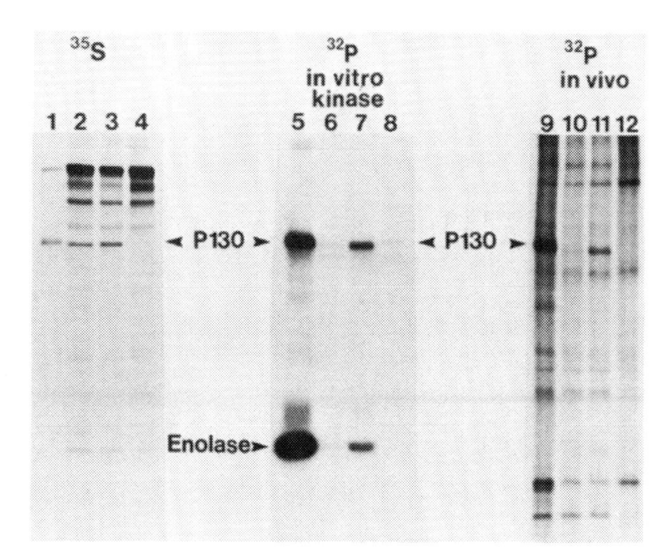

FIG. 4. Analysis of rat-2 cell lines transfected with wt or mutant FSV DNAs for synthesis, phosphorylation, and in vitro protein kinase activity of $P130^{gag\text{-}fps}$. G418-resistant rat-2 cells transfected with pIV2 containing wt FSV (lanes 1, 5, and 9), RX15m (lanes 2, 6, and 10), or AX9m (lanes 3, 7, and 11) coding sequences or normal rat-2 cells (lanes 4, 8, and 12) were analyzed for expression and activity of $P130^{gag\text{-}fps}$. To identify synthesis of $P130^{gag\text{-}fps}$, cells were labeled for 4 h with [^{35}S]methionine, lysed, and immunoprecipitated with anti-$p19^{gag}$ antibody (lanes 1 through 4). To assay $P130^{gag\text{-}fps}$ kinase activity, cells from duplicate unlabeled cultures were immunoprecipitated with anti-$p19^{gag}$ antibody, and the immune complexes were incubated with [γ-^{32}P]ATP and 5 μg of acid-denatured enolase (lanes 5 through 8). For analysis of in vivo $P130^{gag\text{-}fps}$ phosphorylation, cells were labeled for 18 h with $^{32}P_i$ and then immunoprecipitated with rabbit antibody to bacterial *fps* protein (lanes 9 through 12).

and other protein-tyrosine kinases, including transmembrane tyrosine kinases, is located between residues 912 and 1174 of $P130^{gag\text{-}fps}$ (residues 258 through 516 of $p60^{v\text{-}src}$) and is denoted in Fig. 2 as the catalytic domain. This region of sequence identity starts at position 133 in the consensus sequence for cytoplasmic tyrosine kinases and spans approximately 260 C-terminal residues of which only the first 37, including residues thought to participate in ATP binding, are shown in Fig. 2. Between residues 1 and 132 of the cytoplasmic tyrosine kinase consensus sequence there is no significant homology with the transmembrane tyrosine kinases. This is not surprising, since the hydrophobic segments of transmembrane kinases which link their internal catalytic sequences to extracellular domains are located only 34 residues N terminal to the proposed start of the kinase domain. In contrast the cytoplasmic tyrosine kinases evidently possess a related N-terminal sequence of about 100 residues (consensus 1-111; designated SH2) that is separated from the proposed catalytic domain by a variable stretch of 10 to 20 amino acids with a high proportion of proline and glycine residues. The first 50 residues of SH2 show particularly high sequence conservation; between consensus residues 15 and 47 ($P130^{gag\text{-}fps}$ residues 820 through 845) 9 of 33 amino acids are invariant. In contrast, within the cytoplasmic tyrosine kinase ATP-binding site (consensus residues 134 through 171, $P130^{gag\text{-}fps}$ residues 912 through 950) 8 of 37 amino acids are invariant. Therefore the primary structure of the N-terminal segment of SH2 has been conserved during evolution of cytoplasmic tyrosine kinases to an equivalent

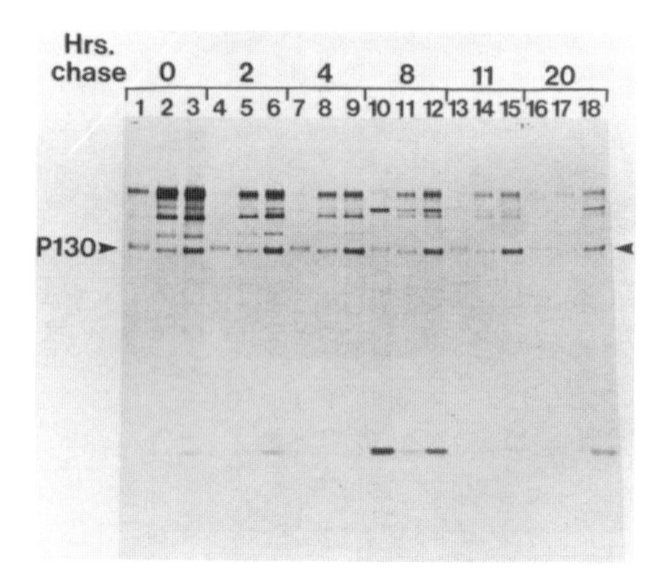

FIG. 5. Stability of wt, RX15m, and AX9m P130$^{gag\text{-}fps}$ proteins. Rat-2 cells expressing wt P130$^{gag\text{-}fps}$ (lanes 1, 4, 7, 10, 13, and 16), RX15m P130$^{gag\text{-}fps}$ (lanes 2, 5, 8, 11, 14, and 17) or AX9m P130$^{gag\text{-}fps}$ (lanes 3, 6, 9, 12, 15, and 18) were labeled with [^{35}S]methionine for 2 h. Labeled cells were then chased with fresh medium containing excess nonradioactive methionine for 0 h (lanes 1 through 3), 2 h (lanes 4 through 6), 4 h (lanes 7 through 9), 8 h (lanes 10 through 12), 11 h (lanes 13 through 15) or 20 h (lanes 16 through 18). To quantitate the radioactivity in P130$^{gag\text{-}fps}$ at each time point the appropriate band in each sample was excised from the gel and measured for ^{35}S counts per minute by liquid scintillation counting, from which values the half-lives of the wt and mutant proteins were estimated.

extent as the ATP-binding site. The segment of SH2 proximal to the kinase domain shows lower homology (consensus residues 48 through 111, P130$^{gag\text{-}fps}$ residues 846 through 900), with 4 of 63 invariant sites. The *fps/fes* and *abl* gene products show considerable divergence from the *src/yes/fgr/tck* family in this proximal region.

Thus the cytoplasmic tyrosine kinases share a 100-amino-acid region of sequence identity positioned N terminal to the proposed kinase domain which is completely absent from transmembrane kinases.

FSV mutants with insertions in the SH2 region of P130$^{gag\text{-}fps}$ are transformation defective. We have previously described the use of a plasmid (pJ2) containing the FSV genome to engineer and analyze in-frame linker insertion mutations in the FSV-coding sequence (53). The RX15m mutant, which was predicted to contain an in-frame dipeptide insertion near the amino terminus of SH2, was previously shown to be transformation defective (53). DNA sequencing of the mutation site confirmed that the insertion had resulted in the substitution of Tyr-821 with -Ser-Arg-Asp- (Fig. 2). Tyr-821 is within the highly conserved region of SH2 adjacent to an invariant tryptophan. We obtained an additional FSV mutant (AX9m) in this region of P130$^{gag\text{-}fps}$ by inserting an *Xho*I linker in pJ2 at an *Alu*I site at FSV nucleotide 2872. The mutation in AX9m was shown by sequence analysis to encode a dipeptide insertion of Leu-Glu between residues 832 and 833, of which Glu-832 is invariant (Fig. 2). The AX9m FSV mutant was tested for transforming activity by transfection into rat-2 cells by the calcium phosphate coprecipitation technique. No foci of transformed cells were induced by either AX9m or RX15m DNAs up to 8 weeks after transfection, whereas the parental pJ2 plasmid DNA containing the wt FSV genome induced numerous foci within 2 weeks. These mutants are therefore transformation defective in rat-2 cells.

To characterize the P130$^{gag\text{-}fps}$ encoded by the RX15m and AX9m mutants we isolated rat-2 cells expressing these proteins. For the purposes of these experiments we transferred the wt FSV, RX15m, and AX9m coding sequences into a vector (pIV2) in which FSV sequences are expressed from the simian virus 40 early promoter and a Neo^r gene is transcribed from an FSV long terminal repeat (Fig. 3). Rat-2 cells were transfected with these plasmids and selected for expression of the linked Neo^r gene by growth in G418. The resulting G418-resistant colonies were isolated, cloned, and tested for the synthesis of FSV P130$^{gag\text{-}fps}$ by metabolic labeling with [^{35}S]methionine and immunoprecipitation with a monoclonal antibody to p19^{gag} (Fig. 4). The NW-14, NR-7, and NA-9 cell lines synthesized wt, RX15m, and AX9m P130$^{gag\text{-}fps}$, respectively, in the ratio 1:1.2:1.3. It is interesting to note that the [^{35}S]methionine-labeled mutant proteins migrated more rapidly than wt P130$^{gag\text{-}fps}$ during electrophoresis (Fig. 4 and 5), a property which is explained by their poor phosphorylation (see below). Pulse-chase experiments showed that the stabilities of the wt and mutant proteins were equivalent (Fig. 5) with wt, RX15m, and AX9m P130$^{gag\text{-}fps}$ having calculated half-lives of 8.4, 7.5, and 10.1 h, respectively. These data confirmed that the RX15m and AX9m P130$^{gag\text{-}fps}$ were expressed to relatively abundant levels in the NR-7 and NA-9 cell lines. Morphologically NR-7 cells expressing RX15m P130$^{gag\text{-}fps}$ (Fig. 6C) were indistinguishable from untransfected rat-2 cells (Fig. 6A), whereas NA-9 cells expressing the AX9m protein closely resembled normal rat-2 cells but showed some evidence of morphological change (Fig. 6D). In contrast NW-14 cells containing wt P130$^{gag\text{-}fps}$ had a highly transformed morphology (Fig. 6B). In addition the NW-14 cells formed colonies in soft agar, whereas the NR-7 and NA-9 cells were unable to grow in agar (data not shown). Thus the high-level expression of these mutant proteins was unable to elicit transformation and anchorage-independent growth of rat-2 cells.

Mutant proteins from rat-2 cells are deficient in tyrosine kinase activity and are poorly phosphorylated. The protein-tyrosine kinase activities of wt, RX15m, and AX9m P130$^{gag\text{-}fps}$ were measured after their immunoprecipitation from the NW-14, NR-7, and NA-9 cell lines by incubation with acid-denatured enolase and [γ-^{32}P]ATP (Fig. 4, Table 1). Under these conditions wt P130$^{gag\text{-}fps}$ autophosphorylates (principally at Tyr-1073) and phosphorylates enolase at a single tyrosine (7, 64). Compared with wt P130$^{gag\text{-}fps}$ (Fig. 4, lane 5), the autophosphorylating activity of the AX9m protein was reduced approximately 10-fold, whereas its kinase activity against enolase showed a 20-fold decrease (Fig. 4, lane 7). The in vitro kinase activity of the RX15m protein was negligible (Fig. 4, lane 6).

The defective enzymatic activities of the RX15m and AX9m P130$^{gag\text{-}fps}$ in vitro suggested that they might be poorly phosphorylated at tyrosine in rat-2 cells. The cell lines expressing RX15m and AX9m P130$^{gag\text{-}fps}$ were metabolically labeled with $^{32}P_i$ and subjected to immunoprecipitation with anti-*fps* antiserum. The mutant proteins incorporated very little phosphate in comparison with wt P130$^{gag\text{-}fps}$ (Fig. 4, Table 1). Phosphoamino acid analysis of the RX15m mutant protein labeled in vivo yielded principally phosphoserine, but no phosphotyrosine, whereas the AX9m protein had some phosphotyrosine in addition to phosphoserine (data not shown). This contrasted with wt P130$^{gag\text{-}fps}$, which was highly phosphorylated in transformed cells

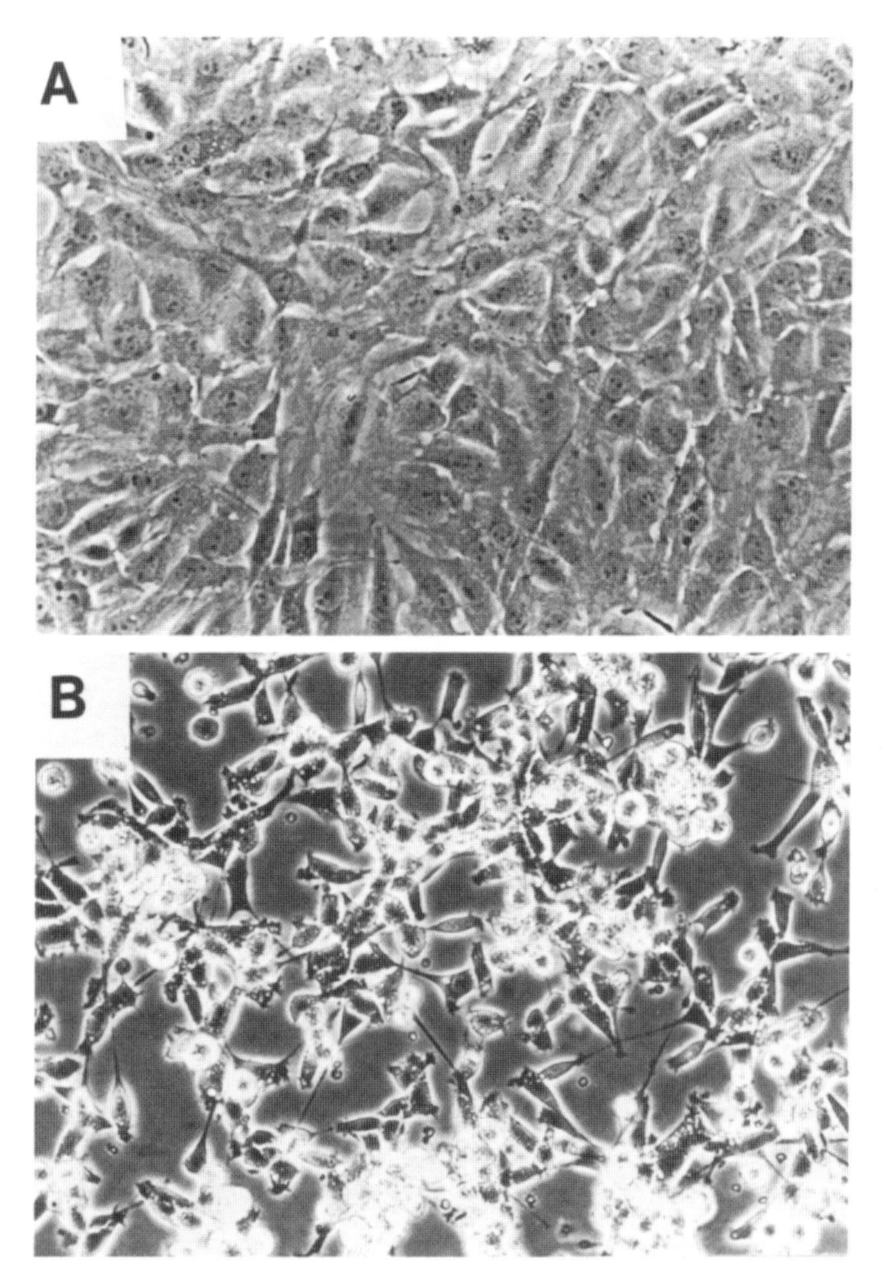

(Fig. 4). Tryptic phosphopeptide analysis of wt and RX15m P130$^{gag\text{-}fps}$ from ^{32}P-labeled cells revealed that only two minor peptides were labeled in the RX15m mutant protein (Fig. 7). None of the principal sites of tyrosine (spots 1, 3a-c, 4, and 7) or serine (spots 5, 6, and 8) phosphorylation characteristic of wt P130$^{gag\text{-}fps}$ were present. A similar phosphopeptide map is obtained when the product of a temperature-sensitive FSV is ^{32}P labeled at the nonpermissive temperature for transformation (35). These results suggested that the peptide insertions had abolished not only the ability of RX15m P130$^{gag\text{-}fps}$ to autophosphorylate, but also to a large extent its capacity to serve as a substrate for cellular serine-threonine-specific kinases in vivo, perhaps by affecting its location within the cell.

Expression of enzymatically active v-*fps* polypeptides in *E. coli*. Previous data have suggested that the SH2 region in FSV P130$^{gag\text{-}fps}$ lies outside the C-terminal catalytic domain (62). Paradoxically, dipeptide insertions within SH2 apparently interfered with P130$^{gag\text{-}fps}$ tyrosine kinase activity and transforming function in rat-2 cells. If the SH2 domain were not critical for enzymatic activity per se we reasoned that it might be possible to resurrect the enzyme activities of the proteins encoded by the RX15m and AX9m mutants by expressing them in a different environment. We have used bacterial expression of v-*fps* polypeptide fragments to investigate this possibility.

To achieve synthesis of FSV polypeptides in *E. coli* we fused v-*fps* coding sequences in frame to those for *trpE* in the pATH bacterial expression plasmid, as detailed in Materials and Methods. Foster and Hanafusa have shown that the *gag* region of P130$^{gag\text{-}fps}$ is dispensable for transformation, and that only FSV coding sequences 3′ to an *Nco*I site located just proximal to the *gag-fps* junction are required for transforming activity (13). By cutting FSV DNA at this *Nco*I site, we constructed *trpE*–v-*fps* fusions whose 100-kDa products contained 42 N-terminal *trpE* residues linked either to C-terminal wt v-*fps* amino acids (residues 359 through 1182 of P130$^{gag\text{-}fps}$ (pTd1359) or to the corresponding segments of the

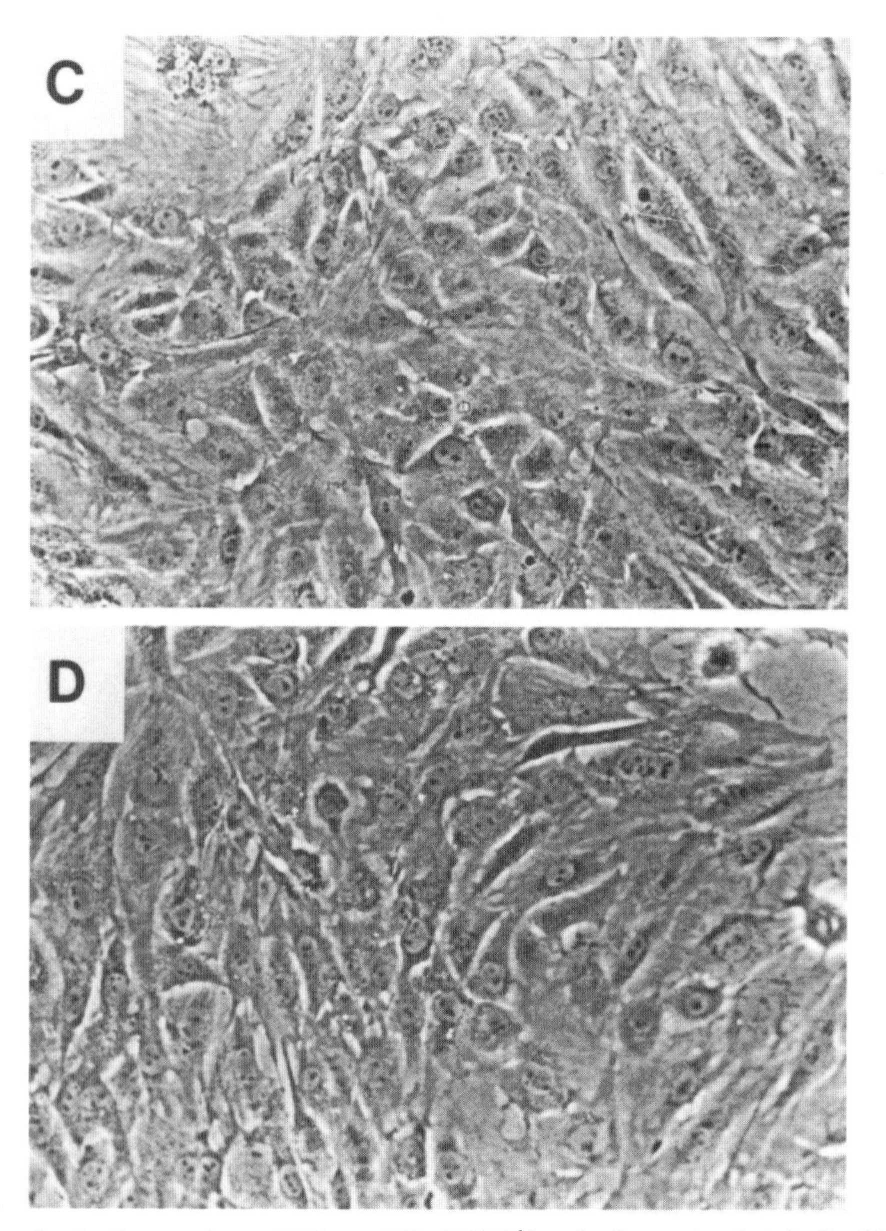

FIG. 6. Morphology of rat-2 cells expressing wt, RX15m, or AX9m P130$^{gag\text{-}fps}$. rat-2 cells were transfected with pIV2 plasmid containing wt FSV, RX15m or AX9m coding sequences, and G418-resistant colonies were isolated. Cell clones were screened for expression of P130$^{gag\text{-}fps}$ by labeling with [^{35}S]methionine and immunoprecipitation with anti-P19^{gag} antibody. Normal rat-2 cells (A), the NW-14 cell line expressing wt P130$^{gag\text{-}fps}$ (B), the NR-7 cell line expressing RX15m P130$^{gag\text{-}fps}$ (C), and the NA-9 cell line expressing AX9m P130$^{gag\text{-}fps}$ (D) are shown. These were the same cells used for the analysis of P130$^{gag\text{-}fps}$ shown in Fig. 4.

RX15m (pTd1359-RX15m) or AX9m (pTd1359-AX9m) proteins (Fig. 8). Additional expression vectors were constructed in which v-*fps* sequences were progressively deleted from the N-terminal end. For these a 37-kDa *trpE* leader sequence was linked to the C-terminal 454 (pTF729), 361 (pTF822), 350 (pTF833), or 289 (pTF893) amino acids of P130$^{gag\text{-}fps}$ (Fig. 8). The pTF729 *trpE*–v-*fps* protein contained both the SH2 and kinase domains of wt v-*fps*; this plasmid was also constructed from RX15m and AX9m v-*fps* sequences. The pTF822 and pTF833 plasmids were constructed from RX15m and AX9m, whereas pTF893 was obtained from wt v-*fps*.

After induction of the *trp* operon with indole acrylic acid, bacteria harboring these plasmids expressed novel polypeptides with the predicted molecular weights which were readily detected by Coomassie blue staining (Fig. 9). Incubation of induced bacteria with $^{32}P_i$ for 20 min resulted in radiolabeling of both wt and mutant *trpE*–v-*fps* proteins (Fig. 9). This suggested that the *trpE*–v-*fps* proteins were capable of autophosphorylation in bacteria. Phosphoamino acid analysis of the ^{32}P-labeled polypeptide encoded by the pTF822 *trpE*–v-*fps* vector yielded phosphotyrosine as the only phosphoamino acid (data not shown). The metabolically labeled pTF822 protein (P78$^{trpE\text{-}fps}$) was compared with P130$^{gag\text{-}fps}$ isolated from ^{32}P-labeled FSV-transformed rat cells by tryptic phosphopeptide analysis. In rat-2 cells wt P130$^{gag\text{-}fps}$ is autophosphorylated within its v-*fps* region at Tyr-1073 (contained within tryptic phosphopeptides 3a through c) (Fig. 7) and on at least one further tyrosine contained within peptides 4 and 7 (62, 64). The pTF822

TABLE 1. Relative phosphorylation and kinase activities of P130$^{gag\text{-}fps}$ and P100$^{trpE\text{-}fps}$ proteins from rat-2 cells and *E. coli*[a]

Protein source	Protein analyzed	Relative phosphorylation (%)			
		In vivo[b]	Autophosphorylation[c]	Enolase[d]	poly(Glu,Tyr)[e]
rat-2	FSV P130$^{gag\text{-}fps}$	100	100	100	ND
E. coli	pTd1359 P100$^{trpE\text{-}fps}$	100	100	100	100
rat-2	RX15m P130$^{gag\text{-}fps}$	4.6	1.3	2.5	ND
E. coli	pTd1359-RX15m P100$^{trpE\text{-}fps}$	106	84.1	24.5	91
rat-2	AX9m P130$^{gag\text{-}fps}$	17.7	12.6	5.7	ND
E. coli	pTd1359-AX9m P100$^{trpE\text{-}fps}$	97	113	44	126

[a] Values are expressed as a percentage of the activity obtained for wt P130$^{gag\text{-}fps}$ or P100$^{trpE\text{-}fps}$ in rat-2 cells or *E. coli*, respectively. All values were normalized to the amount of P130$^{gag\text{-}fps}$ or P100$^{trpE\text{-}fps}$ protein present in immunoprecipitates, determined by [^{35}S]methionine labeling of sibling rat-2 or bacterial cultures.
[b] ^{32}P metabolically incorporated into the indicated protein.
[c] In vitro autophosphorylation of rat-2 protein or *E. coli* protein immunoprecipitated with anti-p19^{gag} antibody or anti-*trpE* antiserum, respectively.
[d] In vitro phosphorylation of enolase by the indicated immunoprecipitated protein.
[e] In vitro phosphorylation of poly(Glu,Tyr) (80:20) polymer by cell extracts. ND, Not done.

protein from ^{32}P-labeled bacteria contained peptides 3a through c, 4, and 7, but lacked P130$^{gag\text{-}fps}$ peptides containing phosphoserine (i.e., 5, 6, and 8) or derived from the *gag* region (spot 1) (Fig. 7). The origin of several minor phosphopeptides specific to P78$^{trpE\text{-}fps}$ is unknown. It is apparent from these data that the bacterial *trpE*–v-*fps* proteins are enzymatically active, as manifested by their ability to autophosphorylate in vivo at physiological tyrosine sites.

Mutant v-*fps* bacterial polypeptides are enzymatically active. The 100-kDa *trpE*–v-*fps* (P100$^{trpE\text{-}fps}$) products of pTd1359, pTd1359-RX15m, and pTd1359-AX9m were phosphorylated to the same extent in *E. coli* (Fig. 9, Table 1). Since cells were only labeled for 20 min, this incorporation seemed likely to reflect the enzymatic activities of the bacterial proteins. By this measure there was no functional distinction between wt, RX15m, and AX9m v-*fps* polypeptides in bacteria, in contrast to the impaired autophosphorylation of RX15m and AX9m P130$^{gag\text{-}fps}$ in rat-2 cells. Similarly the phosphorylation of the pTF729 *trpE*-v-*fps* proteins was equivalent regardless of whether they contained wt, RX15m, or AX9m sequences (Fig. 9). We conclude that the insertions in SH2 do not materially affect v-*fps* kinase activity in the milieu of the bacterial cell.

To test the relative kinase activities of these bacterial proteins in vitro, cells expressing full-length wt, RX15m, or AX9m P100$^{trpE\text{-}fps}$ proteins were lysed, and insoluble material was removed by centrifugation. The soluble P100$^{trpE\text{-}fps}$ proteins were immunoprecipitated with antiserum to the N-terminal *trpE* region, and immunoprecipitates were incubated with [γ-^{32}P]ATP and acid-denatured enolase (Fig. 10, Table 1). Compared with wild-type P100$^{trpE\text{-}fps}$ the mutant proteins encoded by pTd1359-RX15m or pTd1359-AX9m were equivalent in autophosphorylation and showed a two- to fourfold decrease in enolase phosphorylation. Thus relative to wt the mutant proteins were at least 10-fold more active in both autophosphorylation and enolase phosphorylation in bacteria than in rat-2 cells. As a further test of the in vitro activities of the *trpE*–v-*fps* proteins, crude bacterial lysates were incubated with a poly(Glu,Tyr) polymer in the presence of [γ-^{32}P]ATP. The phosphorylated poly(Glu,Tyr) was isolated by electrophoresis on a non-SDS denaturing gel. No significant difference was found in the tyrosine kinase activities of bacterial extracts containing wt, RX15m, or AX9m pTd1359 *trpE*–v-*fps* proteins (Table 1).

The SH2 region is not required for kinase activity. The v-*fps* regions of the pTF822 and pTF833 expression plasmids (Fig.

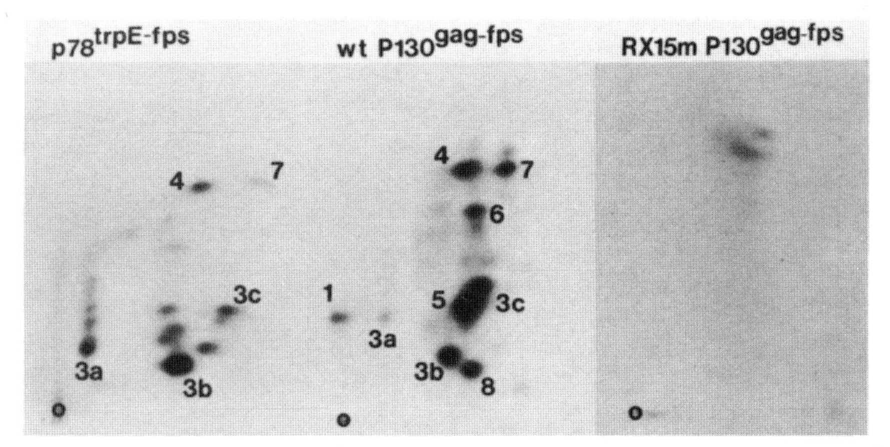

FIG. 7. Tryptic phosphopeptide analysis of P130$^{gag\text{-}fps}$ and bacterial v-*fps* proteins. wt or RX15m P130$^{gag\text{-}fps}$ were isolated by immunoprecipitation from ^{32}P-labeled rat-2 cells. The 78-kDa *trpE*-v-*fps* product of the pTF822 plasmid (P78$^{trpE\text{-}fps}$) was isolated from extracts of ^{32}P-labeled induced bacteria. v-*fps* proteins from mammalian and bacterial cells were eluted from SDS-polyacrylamide gels, oxidized, and digested with trypsin. Tryptic digests were separated in two dimensions by electrophoresis at pH 2.1 from left to right (anode to the left) and chromatography in *N*-butanol-acetic acid-water-pyridine (15:3:12:10) from bottom to top. Peptide maps were exposed to X-ray film in the presence of an intensifying screen for 14 days (wt and RX15m P130$^{gag\text{-}fps}$) or for 18 h (P78$^{trpE\text{-}fps}$). In the phosphopeptide map of wt FSV P130$^{gag\text{-}fps}$ spots 1, 3a through c, 4, and 7 contain phosphotyrosine, whereas spots 5, 6, and 8 contain phosphoserine (62–64). The identity of spots 3a through c, 4, and 7 in P78$^{trpE\text{-}fps}$ was confirmed by comigration with a tryptic digest of wt P130$^{gag\text{-}fps}$. The phosphopeptides of RX15m did not comigrate with any major spots of wt P130$^{gag\text{-}fps}$.

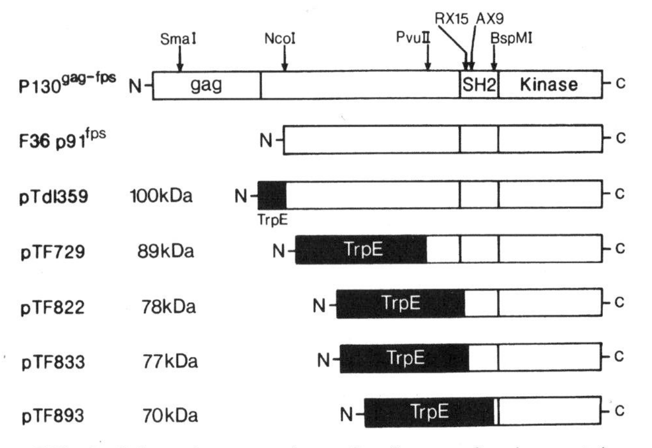

FIG. 8. Schematic comparison of v-*fps* transforming proteins and *trpE*-v-*fps* *E. coli* fusion proteins. The P130$^{gag\text{-}fps}$ protein of FSV, the p91$^{v\text{-}fps}$ protein of F36 virus (13), and *trpE*-v-*fps* *E. coli* proteins are shown. Bacterial expression plasmids were made as described in the text and encode proteins of the indicated molecular masses. The sites of the RX15m and AX9m *Xho*I linker insertions and the restriction sites used for construction of the expression vectors are indicated.

8) were derived from RX15m and AX9m, respectively, by cutting these DNAs at the insertion mutation *Xho*I sites. Their encoded products contain only v-*fps* sequences C terminal to the insertion sites and are missing the most N-terminal part of SH2. Nonetheless these *trpE*–v-*fps* proteins autophosphorylate efficiently in bacteria in comparison with the products of pTd1359 or pTF729 (Fig. 9). The plasmid pTF893 was derived from wt v-*fps* sequences and encodes a *trpE*–v-*fps* protein that is missing virtually all of SH2 (Fig. 8). This polypeptide was also phosphorylated in vivo (Fig. 9). Thus the SH2 region can be partially or almost completely removed without ablating or markedly affecting v-*fps* kinase activity in bacteria.

DISCUSSION

Here we describe a polypeptide sequence of approximately 100 residues which is shared between all known cytoplasmic protein-tyrosine kinases but is absent from tyrosine kinases which span the plasma membrane. This domain (here designated SH2) lies immediately N terminal to the catalytic sequences conserved between cytoplasmic protein-tyrosine kinases and other protein kinases and might be thought to interact with or constitute an ancillary part of the kinase domain. This notion is supported by the observation that mutations which introduce dipeptide insertions into the SH2 region of FSV P130$^{gag\text{-}fps}$ impair FSV transforming ability and P130$^{gag\text{-}fps}$ kinase activity in rat-2 cells. The Leu-Glu insertion in AX9m was expected to be less disruptive of protein structure than the substitution of Tyr-821 with Ser-Arg-Asp in RX15m (Fig. 2); indeed AX9m P130$^{gag\text{-}fps}$ proved to have residual kinase activity and to induce a minor morphological change in rat cells, whereas the RX15m protein was enzymatically and functionally inert. From this it might be assumed that these mutations are in a region required for catalytic activity and induce local conformational changes in the v-*fps* kinase domain which directly abolish enzymatic function. This simplistic view is challenged by the observation that mutant v-*fps* polypeptides containing the peptide insertions have full tyrosine kinase

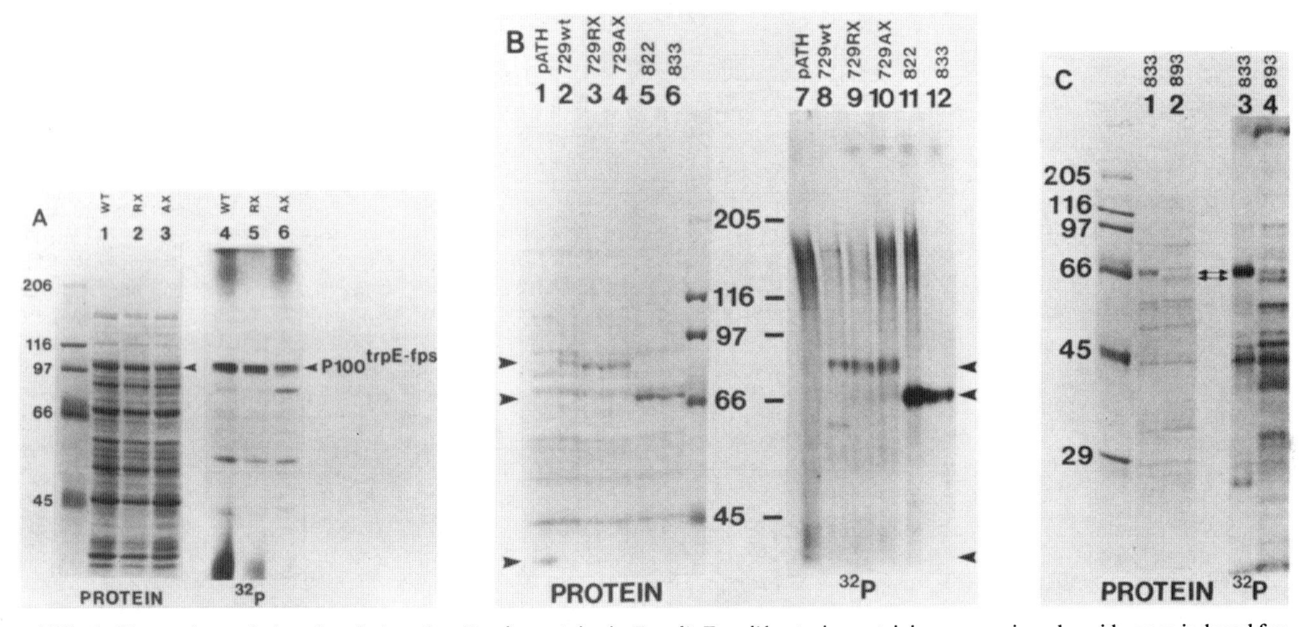

FIG. 9. Expression and phosphorylation of *trpE*-v-*fps* proteins in *E. coli*. *E. coli* bacteria containing expression plasmids were induced for 2 h with indole acrylic acid, metabolically labeled with $^{32}P_i$, lysed in SDS-urea buffer, and analyzed by electrophoresis through 7.5% (A and B) or 10% (C) SDS–polyacrylamide gels. Proteins were identified by staining with Coomassie blue and phosphoproteins on the same gel were located by autoradiography. The *trpE* and *trpE*-v-*fps* proteins are indicated by arrows, and the molecular weights of size markers ($\times 10^3$) are indicated. A, *E. coli* bacteria containing pTd1359 (lanes 1 and 4), pTd1359-RX15m (lanes 2 and 5), or pTd1359-AX9m (lanes 3 and 6) were analyzed for Coomassie blue-stained proteins (lanes 1 through 3) and by autoradiography for ^{32}P-labeled proteins (lanes 4 through 6). B, *E. coli* bacteria containing the pATH parental plasmid, which encodes a 37-kDa *trpE* polypeptide (lanes 1 and 7), pTF729 (lanes 2 and 8), pTF729-RX15m (lanes 3 and 9), pTF729-AX9m (lanes 4 and 10), pTF822 (lanes 5 and 11), or pTF833 (lanes 6 and 12) were analyzed as above for total protein (lanes 1 through 6) and phosphoproteins (lanes 7 through 12). C, *E. coli* bacteria containing pTF833 (lanes 1 and 3) or pTF893 (lanes 2 and 4) were analyzed for total protein (lanes 1 and 2) or phosphoproteins (lanes 3 and 4).

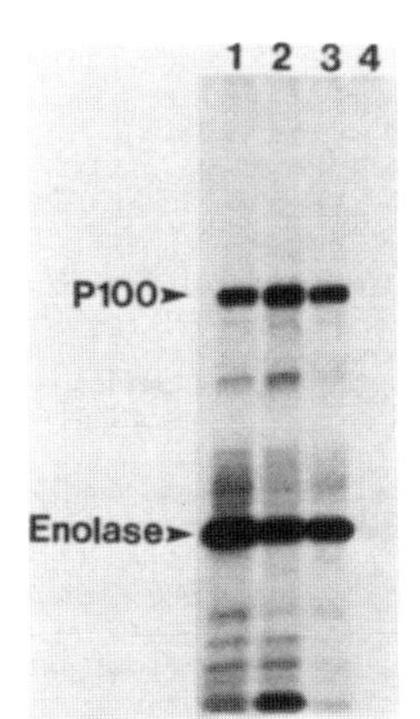

FIG. 10. In vitro kinase activities of wt, RX15m, and AX9m P100$^{trpE\text{-}fps}$ bacterial proteins. *E. coli* bacteria containing the pTd1359 (lane 1), pTd1359-RX15m (lane 2), or pTd1359-AX9m (lane 3) plasmids were grown, and expression of the P100$^{trpE\text{-}fps}$ proteins was induced for 2 h. Bacterial cells were lysed, and the clarified lysates were immunoprecipitated with rabbit anti-*trpE* antiserum. Immune complexes were incubated with [γ-^{32}P]ATP and 5 μg of acid-denatured enolase, and the in vitro reaction products were analyzed by SDS-polyacrylamide gel electrophoresis and autoradiography.

activity when expressed in bacteria, as do v-*fps* polypeptides lacking part or all of SH2. In contrast we have found that substitution of an essential lysine in the v-*fps* ATP-binding site with arginine destroys tyrosine kinase activity of *trpE*–v-*fps* proteins in *E. coli* (unpublished observation). Thus the integrity of the SH2 domain is apparently not of itself required for kinase activity, since the loss of kinase function is only observed when SH2 insertion mutants are expressed in rat-2 cells. It is likely that the SH2 domain of P130$^{gag\text{-}fps}$ can make contact with the catalytic region, and peptide insertions in SH2 might therefore directly disrupt the native conformation of the nearby kinase domain. However in this case we would have expected the bacterial *trpE*–v-*fps* proteins containing the insertions to also lack activity. Rather, these observations suggest that SH2 is a noncatalytic domain involved in an interaction with components specific to the vertebrate cell and by virtue of this interaction can modulate kinase activity.

A body of evidence supports the notion that SH2 is not essential for catalytic activity but may have some important role in directing the cellular actions of the kinase domain. Partial trypsinolysis of p60$^{v\text{-}src}$, P130$^{gag\text{-}fps}$, and P90$^{gag\text{-}yes}$ can yield enzymatically active fragments of 29 to 30 kDa which are apparently derived from the extreme C termini of these proteins and must therefore lack SH2 sequences (4, 25, 62). For RSV, a number of mutations in the coding region for the amino-terminal half of p60$^{v\text{-}src}$ impair transforming activity with little obvious effect on tyrosine kinase function. A v-*src* mutant protein with a deletion of residues 15 through 149 (to residue 16 in the consensus sequence in Fig. 1) induced a fusiform transformed morphology in chicken embryo fibroblasts, whereas deletions of p60$^{v\text{-}src}$ amino acids 149 through 169 (consensus residues 16 through 39) or 135 through 237 (consensus residues 2 through 107) had a more markedly deleterious effect on RSV transforming potential (9, 23). The proteins encoded by these mutants had essentially normal levels of tyrosine kinase activity in chicken cells. It is of interest that the v-*src* residues 149 through 169 lost in the RSV deletion mutant NY320 correspond to a region in P130$^{gag\text{-}fps}$ (amino acids 821 through 839) within which the RX15m and AX9m linker insertion mutations are located. A second class of RSV mutants with structural alterations in the amino-terminal half of p60$^{v\text{-}src}$ are temperature sensitive for transformation (6, 52). The p60$^{v\text{-}src}$ of one such mutant (LA32) showed no appreciable loss of kinase activity at the nonpermissive temperature (52), whereas another (tsCH119) with a deletion of p60$^{v\text{-}src}$ residues 173 through 227 (consensus residues 38 through 98) had only a 50% decline in kinase activity at 41°C (6). For Abelson murine leukemia virus, mutations which infringe on the SH2 domain largely abolished transforming activity (36). Deletion of the first 28 *abl*-encoded residues of P120$^{gag\text{-}abl}$ (consensus residues 4 through 32) resulted in diminished kinase activity and loss of transforming function, whereas peptide insertions between consensus residues 7 through 8 and 18 through 19 had little effect (36).

These observations for RSV and Abelson murine leukemia virus are generally consistent with the notion that the SH2 domain is not critical for kinase activity but plays an important role in transformation. The sequence comparison of cytoplasmic protein-tyrosine kinases provides two circumstantial arguments to validate the suggestion that the SH2 domain is important for the transforming activity of oncogenic proteins and the functions of their normal counterparts. First, the precise retention of the same SH2 sequences during the recombination of viral and cellular sequences in the formation of the Abelson murine leukemia virus and Gardner-Rasheed feline sarcoma virus genomes argues for their functional significance. Second, the SH2 noncatalytic domain has been retained in every known viral and cellular cytoplasmic tyrosine kinase and is highly conserved between such kinases from evolutionarily distant species. The concentration of invariant residues in the N-terminal half of SH2 suggests that functional constraints have operated to preserve the SH2 structural motif.

What are the actual functions of the SH2 region? It must fulfill a common requirement of cytoplasmic tyrosine kinases involving a functional interaction with the kinase domain. Indirect evidence suggests that SH2 does associate with the catalytic region. The principal protease-resistant fragment produced by mild chymotrypsin or trypsin digestion of P130$^{gag\text{-}fps}$ is an enzymatically active polypeptide of 45 kDa, which presumably represents a structural unit that includes both SH2 and kinase domains. This 45-kDa polypeptide can be converted to 29- to 33-kDa active fragments by further trypsinolysis which separates SH2 from the catalytic domain (62; unpublished results). As argued above, our data raise the possibility that the proposed interaction of SH2 with the catalytic domain is modified by cellular components that associate with SH2. These might be tyrosine kinase substrates or regulatory proteins. After their synthesis p60$^{v\text{-}src}$, P130$^{gag\text{-}fps}$, and P90$^{gag\text{-}yes}$ (but not transmembrane tyrosine kinases) all transiently associate with a 90-kDa heat shock protein and a 50-kDa tyrosine kinase substrate to form a complex which has been implicated in transport of the transforming proteins to the membrane (3, 5, 26, 34). In p60$^{v\text{-}src}$ a sequence encompassing residues 155 through 160 (consensus residues 22 through 27) within the SH2 region of p60$^{v\text{-}src}$ may be involved in binding of the 50-kDa cellular protein (56). Interestingly p60$^{v\text{-}src}$ is reported to have little kinase activity when associated with the complex (5). It is possible that in mammalian cells the interactions of the mutant P130$^{gag\text{-}fps}$ proteins with this complex or with other regulatory proteins are perturbed and they are therefore unable to assume a final active form. Since P130$^{gag\text{-}fps}$ kinase activity is stimulated by autophosphorylation at Tyr-1073 (64), a further possibility is that the SH2 domain influences the extent of P130$^{gag\text{-}fps}$ autophosphorylation and thereby indirectly regulates catalytic activity toward exogenous substrates. The insertions in the RX15m and AX9m proteins could decrease the stoichiometry of P130$^{gag\text{-}fps}$ autophosphorylation in rat cells, for example by increasing sensitivity to phosphotyrosyl phosphatases. The same mutant proteins in bacterial cells, which lack specific phosphotyrosyl phosphatases, might accumulate phosphorylated tyrosine and be converted to an enzymatically active form. Regarding substrate binding, it is of interest that the bacterial mutant proteins are equivalent to the wt proteins in autophosphorylation and poly(Glu,Tyr) phosphorylation but have somewhat decreased ability to phosphorylate the physiological substrate enolase in vitro. Experiments to explore the functions of the SH2 domain and to identify cellular proteins with which it interacts are in progress.

ACKNOWLEDGMENTS

We thank Ralph Zirngibl and Theresa Lee for technical assistance and Pauline Vine for help in the preparation of the manuscript.

I.S. holds a student fellowship from the Natural Science and Engineering Research Council. T.P. is a Research Associate of the National Cancer Institute of Canada. This work was supported by grants from the National Cancer Institute of Canada, the Medical Research Council of Canada, and the Leukemia Research Fund.

LITERATURE CITED

1. **Barker, W. C., and M. O. Dayhoff.** 1982. Viral *src* gene products are related to the catalytic chain of mammalian cAMP-dependent protein kinase. Proc. Natl. Acad. Sci. USA **79:** 2836–2839.
2. **Basu, M., R. Biswas, and M. Das.** 1984. 42,000-molecular weight EGF receptor has protein kinase activity. Nature (London) **311:**477–480.
3. **Brugge, J. S., and D. Darrow.** 1982. Rous sarcoma virus-induced phosphorylation of a 50,000 molecular weight cellular protein. Nature (London) **295:**250–253.
4. **Brugge, J. S., and D. Darrow.** 1984. Analysis of the catalytic domain of phosphotransferase activity of two avian sarcoma virus-transforming proteins. J. Biol. Chem. **259:**4550–4557.
5. **Brugge, J. S., E. Erikson, and R. L. Erikson.** 1981. The specific interaction of the Rous sarcoma virus transforming protein, pp60^{src}, with two cellular proteins. Cell **25:**363–372.
6. **Bryant, D., and T. Parsons.** 1982. Site-directed mutagenesis of the *src* gene of Rous sarcoma virus: construction and characterization of a deletion mutant temperature-sensitive for transformation. J. Virol. **44:**683–691.
7. **Cooper, J. A., F. S. Esch, S. S. Taylor, and T. Hunter.** 1984. Phosphorylation sites in enolase and lactate dehydrogenease utilized by tyrosine protein kinases in vivo and in vitro. J. Biol. Chem. **259:**7835–7841.
8. **Cotton, P. C., and J. S. Brugge.** 1983. Neural tissues express high levels of the cellular *src* gene product pp60^{src}. Mol. Cell. Biol. **3:**1157–1162.
9. **Cross, F. R., E. A Garber, and H. Hanafusa.** 1985. N-Terminal deletions in Rous sarcoma virus p60^{src}: effects on tyrosine kinase and biological activities and on recombination in tissue culture with the cellular *src* gene. Mol. Cell. Biol. **5:**2789–2795.
10. **Downward, J., Y. Yarden, E. Mayes, G. Scrace, N. Totty, P. Stockwell, A. Ullrich, J. Schlessinger, and M. D. Waterfield.** 1984. Close similarity of epidermal growth factor receptor and v-*erb*B oncogene protein sequences. Nature (London) **307:** 521–527.
11. **Ek, B., B. Westermark, A. Wasteson, and C. H. Heldin.** 1982. Stimulation of tyrosine-specific phosphorylation by platelet-derived growth factor. Nature (London) **295:**419–420.
12. **Feldman, R. A., J. L. Gabrilove, J. P. Tam, M. A. A. Moore, and H. Hanafusa.** 1985. Specific expression of the normal cellular *fps/fes*-encoded protein NCP92 in normal and leukemic myeloid cells. Proc. Natl. Acad. Sci. USA **82:**2379–2383.
13. **Foster, D., and H. Hanafusa.** 1984. A *fps* gene without *gag* gene sequences transforms cells in culture and induces tumors in chickens. J. Virol. **48:**744–751.
14. **Garber, A., F. Cross, and H. Hanafusa.** 1985. Processing of p60$^{v\text{-}src}$ to its myristylated membrane-bound form. Mol. Cell. Biol. **5:**2781–2788.
15. **Gentry, L. E., and L. R. Rohrschneider.** 1984. Common features of the *yes* and *src* gene products defined by peptide-specific antibodies. J. Virol. **51:**539–546.
16. **Goddard, J. M., J. J. Weiland, and M. R. Capecchi.** 1986. Isolation and characterization of *Caenorhabditis elegans* DNA sequences homologous to the v-*abl* oncogene. Proc. Natl. Acad. Sci. USA **83:**2172–2176.
17. **Golden, A., S. P. Nemeth, and J. S. Brugge.** 1986. Blood platelets express high levels of the pp60$^{c\text{-}src}$-specific tyrosine kinase activity. Proc. Natl. Acad. Sci. USA **83:**852–856.
18. **Hampe, A., M. Gobet, C. J. Sherr, and F. Galibert.** 1984. Nucleotide sequence of the feline retroviral oncogene v-*fms* shows unexpected homology with oncogenes encoding tyrosine-specific protein kinases. Proc. Natl. Acad. Sci. USA **81:**85–89.
19. **Hunter, T., and J. Cooper.** 1985. Protein-tyrosine kinases. Annu. Rev. Biochem. **54:**897–930.
20. **Hunter, T., N. Ling, and J. A. Cooper.** 1984. Protein kinase C phosphorylation of the EGF receptor at a threonine residue close to the cytoplasmic face of the plasma membrane. Nature (London) **311:**480–483.
21. **Ingman-Baker, J., E. Hinze, J. G. Levy, and T. Pawson.** 1984. Monoclonal antibodies to the transforming protein of Fujinami avian sarcoma virus discriminate between different *fps*-encoded proteins. J. Virol. **50:**572–578.
22. **Kitamura, N., A. Kitamura, K. Toyashima, Y. Hirayama, and M. Yoshida.** 1982. Avian sarcoma virus Y73 genome sequence and structural similarity of its transforming gene product to that of Rous sarcoma virus. Nature (London) **297:**205–207.
23. **Kitamura, N., and M. Yoshida.** 1983. Small deletion in *src* of Rous sarcoma virus modifying the transformation phenotypes: identification of 207-nucleotide deletion and its smaller product with protein kinase activity. J. Virol. **46:**985–992.
24. **Leppard, K., N. Totty, M. Waterfield, E. Harlow, J. Jenkins, and L. Crawford.** 1983. Purification and partial amino acid sequence analysis of the cellular tumor antigen, p53, from mouse SV40-transformed cells. EMBO J. **2:**1993–1999.
25. **Levinson, A. D., S. Courtneidge, and J. M. Bishop.** 1981. Structural and functional domains of the Rous sarcoma virus transforming protein (pp60^{src}). Proc. Natl. Acad. Sci. USA **78:**1624–1628.
26. **Lipsich, L. A., J. R. Cutt, and J. S. Brugge.** 1982. Association of the transforming proteins of Rous, Fujinami, and Y73 avian sarcoma viruses with the same two cellular proteins. Mol. Cell. Biol. **2:**875–880.
27. **MacDonald, I., J. Levy, and T. Pawson.** 1985. Expression of the mammalian c-*fes* protein in hematopoietic cells and identification of a distinct *fes*-related protein. Mol. Cell. Biol. **5:** 2543–2551.
28. **Manger, R., S. Rasheed, and L. Rohrschneider.** 1986. Localization of the feline sarcoma virus *fgr* gene product (P70$^{gag\text{-}actin\text{-}fgr}$): association with the plasma membrane and detergent-insoluble matrix. J. Virol. **59:**66–72.
29. **Marth, J. D., R. Peet, E. G. Krebs, and R. M. Perimutter.** 1984. A lymphocyte-specific protein-tyrosine kinase is rearranged and overexpressed in the murine T cell lymphoma LSTRA. Cell **43:**393–404.
30. **Martin-Zanca, D., S. H. Hughes, and M. Barbacid.** 1986. A human oncogene formed by the fusion of truncated tropomyosin and protein tyrosine kinase sequences. Nature (London) **319:**743–748.
31. **Moss, P., K. Radke, V. C. Carter, J. Young, T. Gilmore, and G. S. Martin.** 1984. Cellular localization of the transforming protein of wild-type and temperature-sensitive Fujinami sarcoma virus. J. Virol. **52:**557–565.

32. **Naharro, G., K. C. Robbins, and E. P. Reddy.** 1984. Gene product of v-*fgr* onc: hybrid protein containing a portion of actin and a tyrosine-specific protein kinase. Science **223**:63–66.
33. **Neckameyer, W. S., and L. H. Wang.** 1985. Nucleotide sequence of avian sarcoma virus UR2 and comparison of its transforming gene with other members of tyrosine protein kinase oncogene family. J. Virol. **53**:879–884.
34. **Opperman, H., W. Levinson, and J. M. Bishop.** 1981. A cellular protein that associates with a transforming protein of Rous sarcoma virus is also a heat-shock protein. Proc. Natl. Acad. Sci. USA **78**:1067–1071.
35. **Pawson, T., J. Guyden, T. H. Kung, K. Radke, T. Gilmore, and G. S. Martin.** 1980. A strain of Fujinami sarcoma virus which is temperature-sensitive in protein phosphorylation and cellular transformation. Cell **22**:767–775.
36. **Prywes, R., J. G. Foulkes, and D. Baltimore.** 1985. The minimum transforming region of v-*abl* is the segment encoding protein-tyrosine kinase. J. Virol. **54**:114–122.
37. **Schieven, G., J. Thorner, and G. S. Martin.** 1986. Protein-tyrosine kinase activity in *Saccharomyces cerevisiae*. Science **231**:390–393.
38. **Reddy, E. P., M. J. Smith, and A. Srinivasan.** 1983. Nucleotide sequence of Abelson murine leukemia virus genome: structural similarity of its transforming gene product to other onc gene products with tyrosine-specific kinase activity. Proc. Natl. Acad. Sci. USA **80**:3623–3627.
38a. **Roebrook, A. J. M., J. A. Schalken, J. S. Verbeek, A. M.W. Van der Ouweland, C. Onnekink, H. P. J. Bloemers, and W. J. M. Van de Ven.** 1985. The structure of the human c-*fes/fps* proto-oncogene. EMBO J. **4**:2897–2903.
39. **Rohrschneider, L. R.** 1979. Immunofluorescence on avian sarcoma virus-transformed cells: localization of the *src* gene product. Cell **16**:11–24.
40. **Rohrschneider, L. R.** 1980. Adhesion plaques of Rous sarcoma virus-transformed cells contain the *src* gene product. Proc. Natl. Acad. Sci. USA **77**:3514–3518.
41. **Rohrschneider, R. L., and L. E. Gentry.** 1984. Subcellular locations of retroviral transforming proteins define multiple mechanisms of transformation. Adv. Viral Oncol. **4**:269–306.
42. **Rohrschneider, R. L., and L. M. Najita.** 1984. Detection of the v-*abl* gene product at cell-substratum contact sites in Abelson murine leukemia virus-transformed fibroblasts. J. Virol. **51**: 547–552.
43. **Rosen, O. M., R. Herrera, Y. Olewe, L. M. Petruzzelli, and M. H. Cobb.** 1983. Phosphorylation activates the insulin receptor tyrosine protein kinase. Proc. Natl. Acad. Sci. USA **80**: 3237–3240.
44. **Sanger, F. G., S. Nicklen, and A. R. Coulson.** 1977. DNA sequencing with chain-terminating inhibitors. Proc. Natl. Acad. Sci. USA **74**:5463–5467.
45. **Schecter, A. L., D. F. Stern, L. Vaidyanathan, S. J. Decker, J. A. Drebin, M. I. Greene, and R. A Weinberg.** 1984. The *neu* oncogene: an *erb*B related gene encoding a 185,000-M_r tumour antigen. Nature (London) **312**:513–516.
46. **Sherr, C. J., C. W. Rettenmier, R. Sacca, M. F. Roussel, A. T. Look, and E. R. Stanley.** 1985. The c-*fms* proto-oncogene product is related to the receptor for the mononuclear phagocyte growth factor CSF-1. Cell **41**:665–676.
47. **Shibuya, M., and H. Hanafusa.** 1982. Nucleotide sequence of Fujinami sarcoma virus: evolutionary relationship of its transforming gene with transforming genes of other sarcoma viruses. Cell **30**:787–795.
48. **Simon, M. A., B. Drees, T. Kornberg, and J. M. Bishop.** 1985. The nucleotide sequence and the tissue-specific expression of Drosophila c-*src*. Cell **42**:831–840.
49. **Sorge, L. K., B. T. Levy, and P. F. Maness.** 1984. pp60$^{c\text{-}src}$ is developmentally regulated in the neural retina. Cell **32**:881–890.
50. **Southern, P. J., and P. Berg.** 1982. Transformation of mammalian cells to antibiotic resistance with a bacterial gene under control of the SV40 early region promoter. J. Mol. Appl. Genet. **1**:327–341.
51. **Spindler, K. R., D. S. E. Rosser, and A. J. Berk.** 1984. Analysis of adenovirus transforming proteins from early regions 1A and 1B with antisera to inducible fusion antigens produced in *Escherichia coli*. J. Virol. **49**:132–141.
52. **Stoker, A. W., P. J. Enrietto, and J. A. Wyke.** 1984. Functional domains of the pp60$^{v\text{-}src}$ protein as revealed by analysis of temperature-sensitive Rous sarcoma virus mutants. Mol. Cell. Biol. **4**:1508–1514.
53. **Stone, J. C., T. Atkinson, M. Smith, and T. Pawson.** 1984. Identification of functional regions in the transforming protein of Fujinami sarcoma virus by in-phase insertion mutagenesis. Cell **37**:549–558.
54. **Stone, J. C., and T. Pawson.** 1985. Correspondence between immunological and functional domains in the transforming protein of Fujinami sarcoma virus. J. Virol. **55**:721–727.
55. **Takeya, T., R. A. Feldman, and H. Hanafusa.** 1982. DNA sequence of the viral and cellular *src* gene of chickens: complete nucleotide sequence of an *Eco*RI fragment of recovered avian sarcoma virus which codes for gp37 and pp60^{src}. J. Virol. **44**:1–11.
56. **Tamura, T., H. Bauer, C. Birr, and R. Pipkorn.** 1983. Antibodies against synthetic peptides as a tool for functional analysis of the transforming protein pp60^{src}. Cell **34**:587–596.
57. **Tanese, N., M. Roth, and S. P. Goff.** 1985. Expression of enzymatically active reverse transcriptase in *Escherichia coli*. Proc. Natl. Acad. Sci. USA **82**:4944–4948.
58. **Topp, W. C.** 1981. Normal rat cell lines deficient in nuclear thymidine kinase. Virology **113**:408–411.
59. **Ullrich, A., J. R. Bell, E. Y. Chen, R. Herrera, L. M. Petruzzelli, T. J. Dull, A. Gray, L. Coussens, Y.-C. Liao, M. Tsubokawa, A. Mason, P. H. Seeburg, C. Grunfeld, O. M. Rosen, and J. Ramachandran.** 1985. Human insulin receptor and its relationship to the tyrosine kinase family of oncogenes. Nature (London) **313**:756–761.
60. **Ushiro, H., and S. Cohen.** 1980. Identification of phosphotyrosine as a product of epidermal growth factor-activated protein kinase in A-431 cell membranes. J. Biol. Chem. **255**:8363–8365.
61. **Voronova, A. F., and B. M. Sefton.** 1986. Expression of a new tyrosine protein kinase is stimulated by retrovirus promoter insertion. Nature (London) **319**:682–685.
62. **Weinmaster, G., E. Hinze, and T. Pawson.** 1983. Mapping of multiple phosphorylation sites within the structural and catalytic domains of the Fujinami avian sarcoma virus transforming protein. J. Virol. **45**:29–41.
63. **Weinmaster, G., and T. Pawson.** 1986. Protein kinase activity of FSV P130$^{gag\text{-}fps}$ shows a strict specificity for tyrosine residues. J. Biol. Chem. **261**:328–333.
63a. **Weinmaster, G., M. J. Zoller, and T. Pawson.** 1986. A lysine in the ATP-binding site of P130$^{gag\text{-}fps}$ is essential for protein-tyrosine kinase activity. EMBO J. **5**:69-76.
64. **Weinmaster, G., M. J. Zoller, M. Smith, E. Hinze, and T. Pawson.** 1984. Mutagenesis of Fujinami sarcoma virus: evidence that tyrosine phosphorylation of P130$^{gag\text{-}fps}$ modulates its biological activity. Cell **37**:559–568.

Phosphatidylinositol 3-kinase in signal transduction and intracellular membrane traffic

16

The turnover of phosphoinositide following cell activation was originally identified as involving primarily the hydrolysis of phosphatidylinositol 4,5-bisphosphate [PtdIns(4,5)P_2] to generate two second messengers, diacylglycerol (DAG) and inositol 1,4,5-trisphosphate [Ins(1,4,5)P_3] (Figure 16.1), which play well characterized roles in signal transduction. Stimulation of neutrophils, however, was found to increase the levels of another phosphoinositide, phosphatidylinositol 3,4,5-trisphosphate [1]. The mechanism by which this phosphoinositide is generated became clear following the discovery of an enzyme, phosphatidylinositol 3-kinase (PI 3-kinase), which can phosphorylate phosphoinositides (Figure 16.1) on the 3-position of the inositol ring (see selected paper [2]). PI 3-kinases are now known to act in a key position in cellular control pathways leading from activated protein tyrosine kinases that control cell proliferation, transformation and differentiation [3,4], and also in various steps in vesicular traffic within cells [5,6].

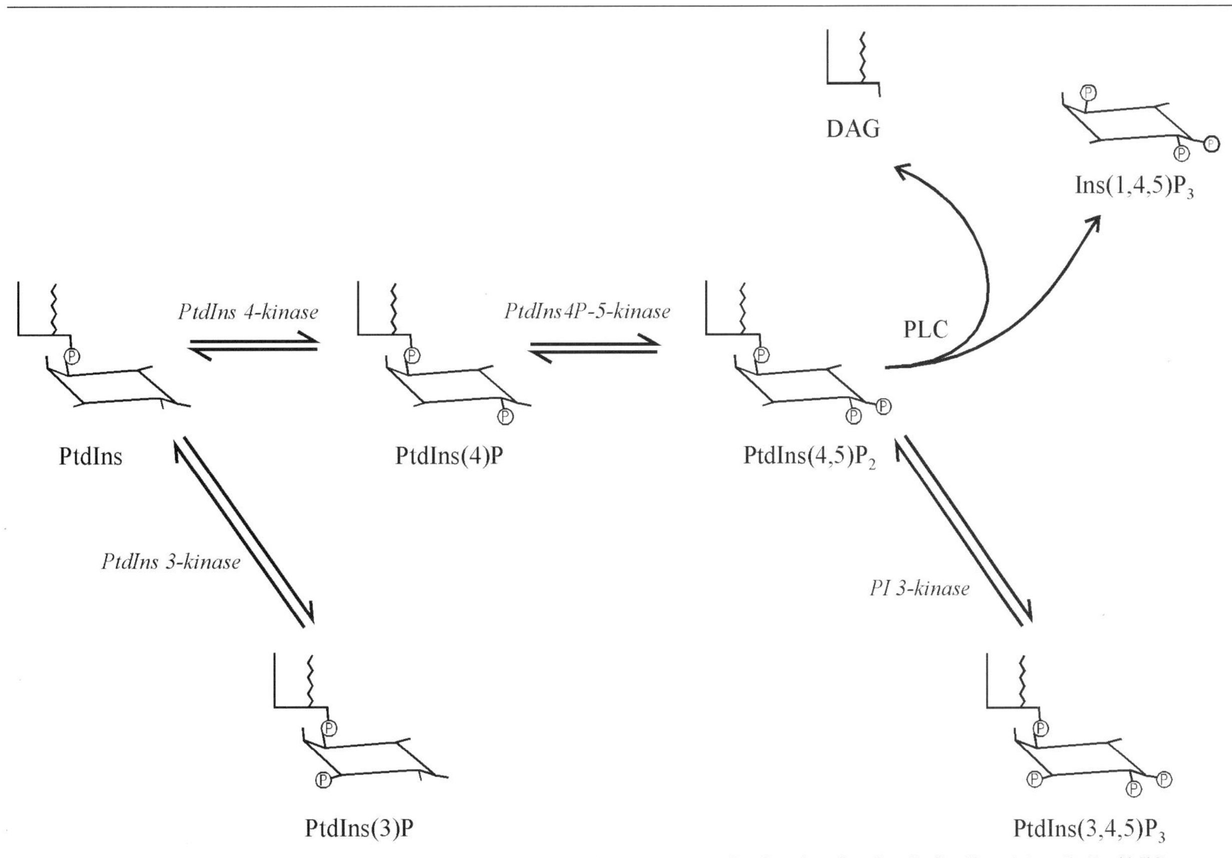

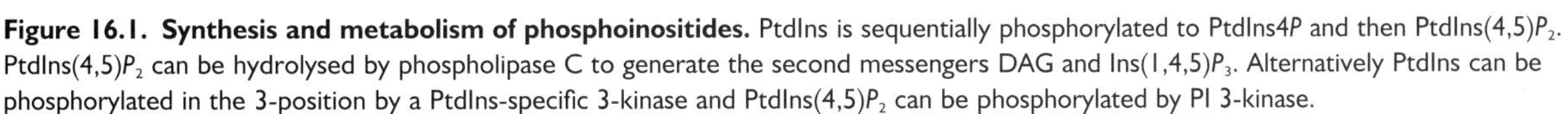

Figure 16.1. Synthesis and metabolism of phosphoinositides. PtdIns is sequentially phosphorylated to PtdIns4*P* and then PtdIns(4,5)P_2. PtdIns(4,5)P_2 can be hydrolysed by phospholipase C to generate the second messengers DAG and Ins(1,4,5)P_3. Alternatively PtdIns can be phosphorylated in the 3-position by a PtdIns-specific 3-kinase and PtdIns(4,5)P_2 can be phosphorylated by PI 3-kinase.

Two types of phosphoinositide kinase activities were identified in fibroblasts, and one of these, the type I, which is associated with activated tyrosine kinases, was characterized by Whitman et al. [2]. Partially purified preparations of the type I kinase were made by isolation of complexes containing polyoma virus middle T antigen, pp60$^{v\text{-}src}$ or growth factor receptors. In contrast to the partially purified type II kinase, which phosphorylated phosphatidylinositol (PtdIns) on the 4-position to generate PtdIns4*P*, the type I kinase catalysed phosphorylation on the 3-position. The presence of PtdIns3*P* in intact transformed fibroblasts was also confirmed in this paper [2]. Whitman et al. suggested that PtdIns3*P* itself, "rather than its metabolites, could be a critical mediator of mitogenic signals". At present the available evidence suggests that phosphoinositides phosphorylated on the 3-position are not hydrolysed by known phospholipases, and the suggestion made in this paper is likely to be correct, although the exact role of these lipids in cellular function is still unknown.

The evidence for a key role of PI 3-kinase in mitogenic responses has come from a variety of experimental approaches, including the demonstration that the enzyme is activated following treatment of cells with various growth factors [3,4] such as the platelet-derived growth factor (PDGF) [7]. In addition, certain mutations in the PDGF receptor, which prevent its ability to interact with PI 3-kinase, also prevent activation of cell division. Restoration of the defect allows PDGF to activate the mitogenic responses through its receptor [8].

Further molecular characterization of the function of PI 3-kinase came from the purification of the enzyme [9], the demonstration that it consists of two subunits, and the cloning of cDNAs encoding each of the subunits [10]. More recently, additional distinct 3-kinases with differing substrate specificities have been identified, one of which is activated by the $\beta\gamma$ subunits of heterotrimeric G-proteins [11]. The originally identified 3-kinase is likely to act on PtdIns(4,5)P_2 as its substrate within cells, but among those more recently identified is a type of 3-kinase (PtdIns 3-kinase) acting specifically on PtdIns to produce PtdIns3*P* (Figure 16.1). This lipid is present constitutively in cells [3,4].

The large 110 kDa subunit of PI 3-kinase contains the catalytic domain, whereas the 85 kDa subunit has regulatory functions. Important clues to the way in which PI 3-kinase interacts with protein tyrosine kinases, and other components of cellular signalling pathways, to become activated came from analysis of the amino acid sequence of the 85 kDa subunit. This regulatory subunit contains one src-homology 3 (SH3) domain, two SH2 domains and two proline-rich regions capable of making protein–protein interactions. The SH2 domains can mediate binding to phosphotyrosine motifs on auto-phosphorylated receptor protein tyrosine kinases and also on cytosolic protein tyrosine kinases or phosphorylated substrates [3]. SH3 domains interact with proline-rich sequences in their binding partners. The role of the SH3 domain in the 85 kDa subunit of PI 3-kinase is not clear, but the proline-rich regions are able to bind SH3 domains in other proteins, including various cytosolic protein tyrosine kinases [3], leading to activation of the PI 3-kinase activity. Thus, the function of the 85 kDa subunit appears to be to allow binding of the PI 3-kinase to specific targets in a range of signalling pathways leading from cell surface receptors, and this binding leads to activation of the PI 3-kinase activity of the 110 kDa subunit.

Following the determination of the amino acid sequence of the PI 3-kinase [12], the surprising finding was made that the catalytic subunit has sequence similarity to a yeast protein essential for one aspect of vesicular traffic. The yeast Vps34 protein was identified in studies on yeast mutants defective in the delivery of proteins to the vacuole and has two domains with 33% sequence similarity to the bovine 110 kDa PtdIns 3-kinase subunit [13]. Vps34 does not have a second regulatory subunit and has been shown to possess 3-kinase activity, with PtdIns being its specific substrate, and is thus related to the mammalian PtdIns 3-kinases [13]. Recent studies have also implicated PI 3-kinases in various vesicular transport steps in mammalian cells [5,6,14].

One question that remains to be resolved is the exact function of the 3-phosphorylated phosphoinositides generated by PI 3-kinases, and how their generation regulates cell growth, transformation and intracellular membrane traffic. There are currently two possibilities: first, that these lipids act as second messengers; and second, that they bind proteins with SH2 domains (see Section 15). In support of the first hypothesis, PtdIns(3,4)P_2 and to a lesser extent PtdIns(3,4,5)P_3 have been shown to activate one isoform of protein kinase C [15,16]. Recent evidence for the second theory is that SH2 domains bind PtdIns(3,4,5)P_3 [17], and this may be a mechanism for recruitment of SH2-containing proteins to membranes. Further work will be needed to resolve these issues, but it is clear that PI 3-kinases play central roles in the control of a number of aspects of cell function.

References

1. Traynor-Kaplan, A.E., Harris, A., Thompson, B., Taylor, P. and Sklar, I.A. (1988) *Nature (London)* **334**, 353–356
2. Whitman, M., Downes, C.P., Keeler, M., Keller, T. and Cantley, L. (1988) *Nature (London)* **332**, 644–646
3. Kapellar, R. and Cantley, L.C. (1994) *BioEssays* **16**, 565–576
4. Stephens, L. (1995) *Biochem. Soc. Trans.* **23**, 207–221
5. Shepherd, P.R., Reeves, B.J. and Davidson, H.W. (1996) *Trends Cell Biol.* **6**, 92–97
6. De Camilli, P., Emr, S.D., McPherson, P.S. and Novick, P. (1996) *Science* **271**, 1533–1539
7. Auger, K.R., Serunian, L.A., Soltoff, S.P., Libby, P. and Cantley, L.C. (1989) *Cell* **57**, 167–175
8. Valius, M. and Kazloukas, A. (1993) *Cell* **73**, 321–334
9. Carpenter, C.L., Duckworth, B.C., Auger, K.R., Cohen, B., Schaffhausen, B.S. and Cantley, L.C. (1990) *J. Biol. Chem.* **265**, 19704–19711
10. Zuelebil, M.J., MacDougall, L., Leevers, S., Volinia, S., Van Haesebroeck, B., Gout, I., Panyotou, G., Domin, J., Stein, R., Pages, F., Koya, H., Salim, K., Linacre, J., Das, P., Panaretou, C., Wetzker, R. and Waterfield, M. (1996) *Philos. Trans. R. Soc. London Ser. B.* **351**, 217–233
11. Stephens, L., Smrcka, A., Cooke, F.T., Jackson, T.R., Sternweis, P.C. and Hawkins, P.T. (1994) *Cell* **77**, 83–93
12. Hiles, L.D., Otsu, M., Volinia, S., Fry, M.J., Gout, I., Dhand, R., Panayotou, G., Ruiz, L.F., Thompson, A., Totty, N.F., Hsuan, J.J., Courtneidge, S.A., Parker, P.J. and Waterfield, M.D. (1992) *Cell* **70**, 419–429
13. Schu, P.V., Takegawa, K., Fry, M.J., Stack, J.H., Waterfield, M.D. and Emr, S.D. (1993) *Science* **260**, 88–91
14. Burgoyne, R.D. (1994) *Trends Biochem. Sci.* **19**, 55–57
15. Toker, A., Bachelot, C., Chen, C.S., Falck, J.R., Horting, J.H., Cantley, L.C. and Kovacsovics, T.J. (1995) *J. Biol. Chem.* **270**, 29525–29531
16. Moriya, S., Kazlauskas, N., Akimoto, K., Hirai, S., Mizuno, K., Takenawa, T., Fukui, Y., Watanabe, Y., Okazi, S. and Ohno, S. (1996) *Proc. Natl. Acad. Sci. U.S.A.* **93**, 151–155
17. Rameh, L.H., Chen, C.S. and Cantley, L.C. (1995) *Cell* **83**, 821–830

Whitman et al. (1988) Nature (London) **332**, 644–646

Type I phosphatidylinositol kinase makes a novel inositol phospholipid, phosphatidylinositol-3-phosphate

Malcolm Whitman*, C. Peter Downes†, Marilyn Keeler, Tracy Keller & Lewis Cantley

Department of Physiology, Tufts University School of Medicine, Boston, Massachusetts 02111, USA
† Smith Kline and French Research Ltd, The Frythe, Welwyn, Herts AL6 9AR, UK

The generation of second messengers from the hydrolysis of phosphatidylinositol-4,5-bisphosphate ($PtdInsP_2$) by phosphoinositidase C has been implicated in the mediation of cellular responses to a variety of growth factors and oncogene products[1–4]. The first step in the production of $PtdInsP_2$ from phosphatidylinositol (PtdIns) is catalysed by PtdIns kinase. A PtdIns kinase activity has been found to associate specifically with several oncogene products, as well as with the platelet-derived growth factor (PDGF) receptor[5–8]. We have previously identified two biochemically distinct PtdIns kinases in fibroblasts, and have found that only one of these, designated type I, specifically associates with activated tyrosine kinases[7]. We have now characterized the site on the inositol ring phosphorylated by type I PtdIns kinase, and find that this kinase specifically phosphorylates the D-3 ring position to generate a novel phospholipid, phosphatidylinositol-3-phosphate (PtdIns(3)P). In contrast, the main PtdIns kinase in fibroblasts, designated type II, specifically phosphorylates the D-4 position to produce phosphatidylinositol-4-phosphate (PtdIns(4)P), previously considered to be the only form of PtdInsP (ref. 9). We have also tentatively identified PtdIns(3)P as a minor component of total PtdInsP in intact fibroblasts. We propose that type I PtdIns kinase is responsible for the generation of PtdIns(3)P in intact cells, and that this novel phosphoinositide could be important in the transduction of mitogenic and oncogenic signals.

Type I PtdIns kinase was found to catalyse the phosphorylation of PtdIns to a product that at first appeared to comigrate with PtdIns(4)P in one and two-dimensional thin-layer chromatography (TLC) systems[5,7], but further comparison showed the type I product migrated slightly more slowly (Fig. 1). A mixture of the two reaction products moved as a broad spot coincident with the PtdIns(4)P standard. Reverse-phase high-pressure (liquid chromatography (HPLC) of the type I and type II PtdIns phosphates shows that they have a similar distribution of fatty acyl side chains (M.W., G. Patten and M.K., unpublished observations). We have therefore investigated whether there is a difference in the head group of the two products. The PtdIns phosphates were synthesized using both enzymes and [^{32}P]ATP as phosphate donor, and after deacylation the resulting glycerophosphoinositol phosphates (GroPInsPs) were separated in an anion-exchange HPLC system optimized for resolution of GroPInsP isomers (Fig. 2). The deacylated type I product eluted about one minute earlier than an internal standard of [^{3}H]GroPInsP prepared from turkey erythrocyte phospholipids, and presumed to be GroPIns(4)P (Fig. 2). Essentially all the PtdIns^{32}P produced by the type II kinase comigrated with the [^{3}H]PtdInsP standard on deacylation and HPLC whereas the single earlier peak of the deacylated PtdIns^{32}P produced by the type I kinase preparation separated completely from the standard. These results establish a structural difference between the two types of phosphorylated PtdIns, and indicate that it resides in the site of phosphorylation on the inositol head group. Each kinase was therefore quite specific for the characteristic head group structure generated.

To determine the different sites phosphorylated on the inositol ring by the type I and type II PtdIns kinases, PtdIns was labelled

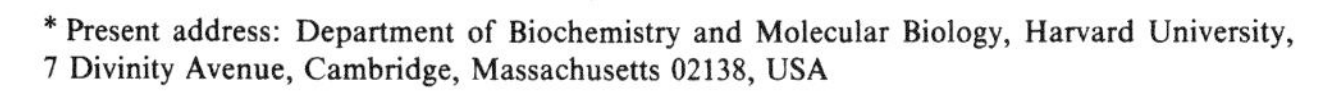

* Present address: Department of Biochemistry and Molecular Biology, Harvard University, 7 Divinity Avenue, Cambridge, Massachusetts 02138, USA

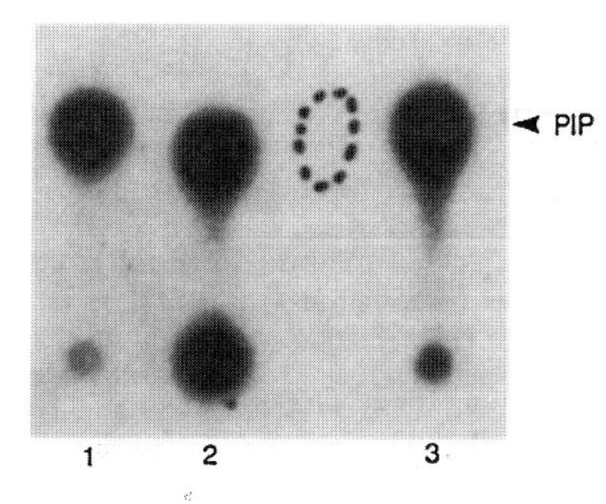

Fig. 1 Migration of PtdIns kinase products during TLC. Type I PtdIns kinase and type II PtdIns kinase were used to phosphorylate PtdIns with [^{32}P]ATP and the reaction products were extracted and separated by TLC. Lane 1: product of type II PtdIns kinase; lane 2: product of type I PtdIns kinase; lane 3: mixture of type I and Type II PtdIns kinase products. dotted line indicates migration of cold PtdIns(4)P standard.

Methods. Type I PtdIns kinase was prepared by immunoprecipitation of middle T/pp60$^{c\text{-}src}$ from middle T-transformed NIH 3T3 fibroblasts as described previously[3,5]. Immunoprecipitates were washed[5] and assayed in 0.2mg ml^{-1} sonicated detergent-free PtdIns (Avanti) with 10 μCi [^{32}P]ATP, 0.02 mM ATP, 10 mM $MgCl_2$, 0.1M NaCl, 20mM Tris, *p*H 7.6 in a final volume of 50μl for 5 min. Type II PtdIns kinase was prepared by precipitation with Sepharose Cl-4B from polyoma-transformed fibroblasts[5]. Precipitates were assayed as above except that 0.5% NP40 detergent was included and ATP was at 0.05 mM. After extraction, reactions were separated by TLC in $CHCl_3$:MeOH:2.5 M NH_4OH (9:7:2) and products were visualized by autoradiography; the PtdIns(4)P standard was visualized by staining with iodine vapour.

in the inositol ring with ^{3}H and phosphorylated by each enzyme using unlabelled ATP. The resulting [^{3}H]PtdIns phosphates were then TLC-purified and deacylated, and the glycerol moiety removed by mild periodate oxidation[10]. The resulting inositol bisphosphate ($InsP_2$) isomers were then identified according to the scheme depicted in Fig. 3. This involves periodate oxidation of the inositol ring, reduction with borohydride, and dephosphorylation with alkaline phosphatase to yield polyols characteristic of the different $InsP_2$ isomers[11,12]. These polyols can be separated (with the exception of enantiomeric pairs) by HPLC on a Brownlee polybore column, as shown in Fig. 3*a*. After this procedure, PtdIns(4)P yields two polyols with the chromatographic properties of altritol and iditol. These are the polyols expected from the partial cleavage of inositol-1,4-bisphosphate (Ins(1,4)P_2) by periodate. We obtained the same result using [^{3}H]$InsP_2$ derived from the type II PtdIns kinase product; in each case, the ratio of [^{3}H]altritol: [^{3}H]iditol was 0.29. This confirms that the type II PtdIns kinase converts PtdIns to PtdIns(4)P.

Preliminary experiments using the [^{3}H]polyol obtained from type I PtdInsP and ^{14}C-labelled polyol standards indicated that the tritum label ran as a single peak with a mobility comparable to the poorly separating erythritol and adonitol. We distinguished between these two possibilities in the experiment illustrated in Fig. 3*b*, in which the tritiated polyol was run separately with either ^{14}C-labelled erythritol or with ^{14}C-labelled adonitol as internal standards. The ^{3}H-labelled polyol was clearly separable from ^{14}C-labelled erythritol, but it co-chromatographed precisely with [^{14}C]adonitol. As shown in Fig. 3*c*, inositol-1,3-bisphosphate (Ins(1,3)P_2) is the only inositol bisphosphate isomer that yields adonitol after periodate oxidation, reduction, and dephosphorylation. As 60% of the HPLC-purified derived from the [^{3}H]$InsP_2$ type I kinase was recovered as a single peak of [^{3}H]adonitol and no other ^{3}H-labelled polyol was present, we conclude that the only significant product of the type I PtdIns kinase is the novel phosphoinositide PtdIns(3)P, whereas type II PtdIns kinase yields the conventional phosphoinositide, PtdIns(4)P.

Type I PtdIns kinase activity has been observed in a number of preparations from fibroblasts: immunoprecipitations of

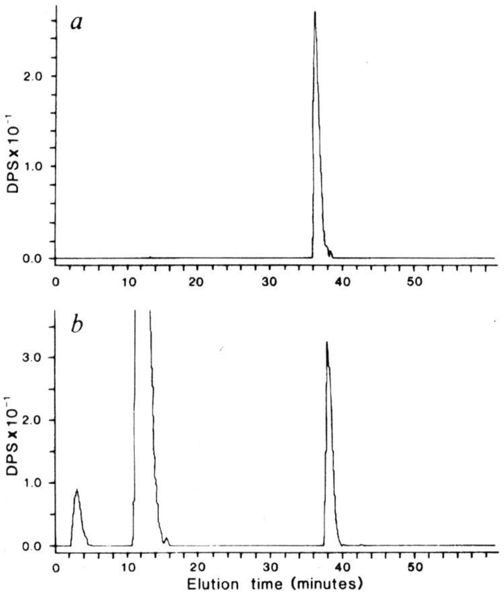

Fig. 2 Anion-exchange HPLC deacylated type I PtdIns kinase product. *a*, ^{32}P-labelled type I. *b*, Migration of a co-injected ^{3}H-labelled internal standard from turkey erythrocytes labelled with [2^3H]inositol[16]; [^{3}H]GroPIns4P elutes at 39 min, type I ^{32}P GroPInsP elutes slightly earlier than the standard, at ~38 min. DPS, disintegrations per second.

Methods. ^{32}P-labelled type I PtdIns kinase product was prepared and TLC-purified as described in Fig. 1. The region of the TLC plate containing the PtdInsP was cut out and incubated at 53 °C for 50 min in 25% methylamine/MeOH/butanol. The deacylated lipid mixture was then dried *in vacuo* overnight, redissolved in 2 ml water and extracted twice with butanol/petroleum ether/ethyl formate (20:4:1). The aqueous phase was dried *in vacuo* and redissolved in 0.5 ml water for HPLC. The sample was spiked with a standard mixture containing [^{3}H]GroPIns, [^{3}H]GroPInsP and [^{3}H]GroPInsP$_2$, prepared using a lipid extract from [^{3}H]inositol-labelled turkey erythrocytes[16] deacylated exactly as described above. Anion-exchange HPLC was carried out on a Whatman 5 Partisphere SAX column, the sample loaded in water, the column washed for 10 min with water and eluted with 0.42 M $(NH_4)_2HPO_4$, *p*H 3.8, at 1 ml min^{-1}. The linear gradient used was 0–0.252 M over 60 min, followed by 0.252–0.42M over 10 min.

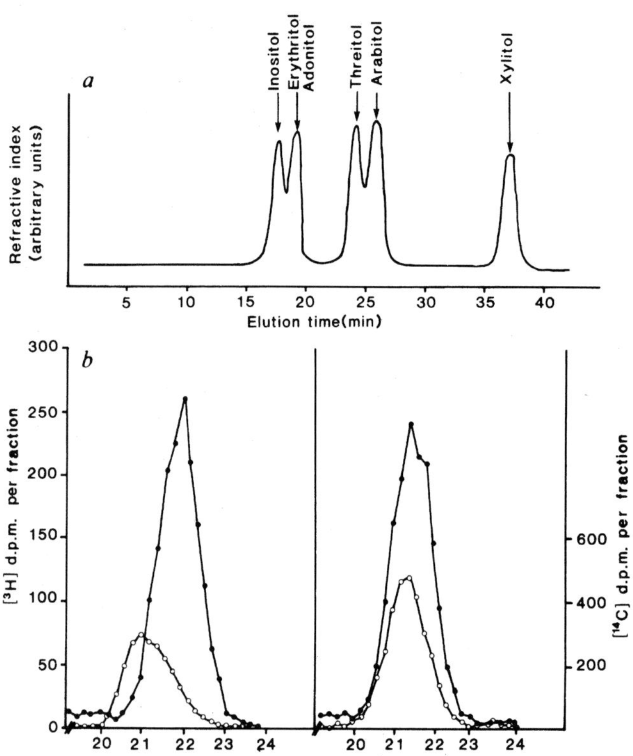

Fig. 3 Elucidation of the head group structure of type I PtdInsP. Panels *a–c* depict the strategy for the identification of InsP$_2$ isomers. Periodate oxidation of inositol bisphosphates, followed by reduction and dephosphorylation, yields a specific polyol that can be identified chromatographically. Of the 12 InsP$_2$ isomers not completely oxidized by periodate to yield CO_2 and P_i (so therefore not phosphorylated at opposite ends of the inositol ring), four should give threitol, three should yield xylitol, two should give erythritol and two arabitol. Only the optically inactive *meso*-compound, Ins(1,3)P$_2$, should give adonitol. *a*, HPLC separation of the five polyols and of myo-inositol, which is detectable when the periodate oxidation step is incomplete. *b*, Identification of [^{3}H]polyol derived from type I PtdInsP as adonitol. The elution of two similar samples of the [^{3}H]polyol is shown in the left and right panels (○)), with the internal standard ^{14}C-labelled erythritol in the left panel (●), and ^{14}C-labelled adonitol in the right (●). *c*, Structure and derivation of adonitol from Ins(1,3)P$_2$.

Methods. [^{3}H]inositol-labelled type I PtdInsP was prepared as described in Fig. 2, except that [^{3}H]PtdIns (NEN) (5μg ml^{-1}, 4 Ci mmol^{-1}) and unlabelled ATP (200 μM) were used as substrates. Fatty acids were removed from TLC-purified ^{3}H-labelled type I PtdInsP by transacylation with methylamine as described[18]. Glycerol was cleaved from the [^{3}H]GroPInsP sample by mild treatment with sodium periodate[10]. Remaining aldehyde was removed by 1,1-dimethylhydrazine (1.875 ml 1% *w/v*) and incubation at 25 °C for 4 h. The sample was desalted on a 3 ml column of Bio-Rad AG-50 (200–400 mesh), dried *in vacuo*, and the resulting [^{3}H]Type I InsP$_2$ purified by HPLC on a Whatman Partisphere FAX column eluted a linear gradient over 30 min, from 0–0.8M triethylammonium formate (*p*H 3.8). [^{3}H]InsP$_2$ eluted about 20 min after the start of the gradient. The sample was dried and triethylammonium formate removed *in vacuo*. Identification of inositol phosphate isomers was as described[11], and those from ^{3}H-labelled type I InsP$_2$ were identified as described previously for inositol tetrakisphosphates[19], except that the incubation with 100mM NaIO$_4$ was at *p*H 4.5. The resulting ^{3}H-labelled polyol was characterized by chromatography on a Brownlee polybore carbohydrate column[19], except that the column was equilibrated at 25 °C rather than at 90 °C. Standard polyols were from Sigma (myo-inositol); Aldrich (arabitol, erythritol and xylitol); and Lancaster Synthesis (D/L threitol); [^{14}C]erythritol was supplied by Amersham and [^{14}C]adonitol was prepared by reduction of [^{14}C]ribose (Amersham) using NaBH$_4$ (ref. 19). Unlabelled standards (25 μg) were injected using a 20-μl loop and detected using a differential refractometer (Waters, model 410). For separation and quantitation of ^{14}C-labelled standards and [^{3}H]polyols derived from [^{3}H]PtdIns samples, the refractometer was taken out of line and the column eluant redirected to a fraction collector, and peaks of [^{14}C] and [^{3}H] determined by liquid scintillation counting.

middle T/pp60$^{c\text{-}src}$ from middle T-transformed NIH-3T3 cells, anti-phosphotyrosine immunoprecipitations from PDGF-stimulated BALB/C-3T3 fibroblasts, wheat germ lectin-purified material from BALB/C-3T3 fibroblasts stimulated with PDGF, and whole-cell lysates of both normal and middle T-transformed NIH-3T3 fibroblasts[7]. The enzyme obtained by these various procedures was distinguished from type II PtdIns kinase by its preference for substrate presented as vesicles rather than in detergent micelles, by its K_m for ATP, and its resistance to inhibition by adenosine[7]. We routinely use middle T/pp60$^{c\text{-}src}$ immunoprecipitates as a convenient source of type I PtdIns kinase uncontaminated with significant type II activity. To verify that the same type of PtdIns kinase activity was present in these other preparations, their ability to phosphorylate PtdIns to PtdIns(3)P was tested. Each preparation was used to phosphorylate PtdIns with [^{32}P]ATP. The resulting PtdInsP was purified by TLC, deacylated and resolved by anion-exchange HPLC. The chromatographic properties of the deacylated products

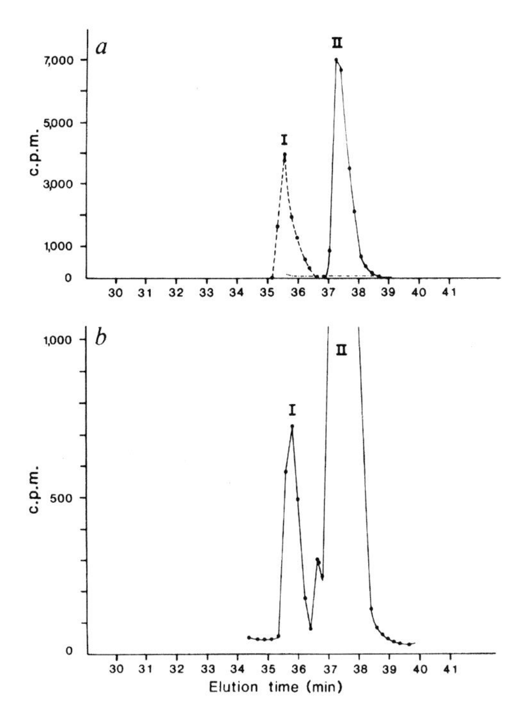

Fig. 4 PtdInsP from intact cells. Polyoma middle T-transformed NIH 3T3 fibroblasts were labelled for 72 h with 20 μCi ml^{-1}, [^{3}H]inositol. Labelled phosphoinositides were extracted[4], mixed with ^{32}P-labelled type I PtdInsP, deacylated and analysed by anion-exchange HPLC as described in Fig. 2 legend. *a*, Elution profile of [^{32}P]GroPIns(3)P co-injected with ^{3}H-labelled deacylation products (dotted line), (*I*); solid line is the elution profile of [^{32}P]GroPIns(4)P from a separate run with ^{3}H-labelled samples (II). *b*, elution profile of tritium counts: total ^{3}H in peak I, 2,063 c.p.m. and in peak II, 60,500 c.p.m.

indicate that middle T/pp60$^{c\text{-}src}$ immunoprecipitates produce 99% PtdIns(3)P, 1% PtdIns(4)P; anti-phosphotyrosine immunoprecipitates produce 92% PtdIns(3)P, 8% PtdIns(4)P; wheat germ lectin-precipitates produce 88% PtdIns(3)P, 12% PtdIns(4)P; cell lysates, when assayed under conditions which optimize type I PtdIns kinase activity[7], produce 30%–50% PtdIns(3)P. The production of PtdIns(4)P by these various type I PtdIns kinase preparations presumably results from contaminating type II PtdIns kinase, but the possibility that the specificity of type I PtdIns kinase differs between these various preparations has not been excluded.

To see whether PtdIns(3)P is present in intact cells, we labelled middle T-transformed fibroblasts for 48 hours with [^{3}H]inositol, and the phospholipids were extracted, deacylated and analysed by anion-exchange HPLC. About 95% of the deacylated PtdInsP co-eluted with ^{32}P-labelled GroPIns(4)P standard, whereas 3.3% of the deacylated PtdInsP co-eluted with the GroPIns(3)P standard (Fig. 4). We obtained identical results when cellular [^{3}H]PtdInsP was purified by TLC before deacylation (data not shown). Although this suggests that that PtdIns(3)P is present in intact fibroblasts, the identification of this compound requires a more rigorous analysis than is presented here. A PtdIns(3)P making up ~10% of the cellular PtdInsP pool has, however, been rigorously identified in an astrocytoma cell line (L. Stephens, P. T. Hawkins and C.P.D., manuscript in preparation). It is therefore likely that PtdIns(3)P is present in fibroblasts and is synthesized by the distinct subset of total cellular PtdIns kinase activity that we define as type I PtdIns kinase.

Our results indicate that type I PtdIns kinase can also catlyse in intact cells the reaction observed *in vitro*. Nothing is known, however, about how PtdIns(3)P is metabolized. Both fibroblast and human erythrocyte PtdInsP kinases can further phosphorylate PtdIns(3)P to PtdIns(3,*x*)P_2, the deacylation product of which is separable from GroPIns(4,5)P_2 by HPLC (data not shown). A possible structure for PtdIns(3, *x*)P_2 would be PtdIns(3,4)P_2 which would yield Ins(1,3,4)P_3 on hydrolysis by phosphoinositidase C. Ins(1,3,4)P_3 does accumulate in stimulated cells[12], but this can be accounted for by the sequential actions of Ins(1,4,5)P_3-3-kinase and Ins(1,3,4,5)P_4-5-phosphatase[17,18]. Ins(1,3,4)P_3 could also possibly be produced directly by action of phosphoinositidase C.

PtdIns(3)P itself, rather than its metabolites, could be a critical mediator of mitogenic signals. The ability of PtdIns(3) kinase to produce an isomer of PtdInsP that is not in the pathway for Ins(1,4,5)P_3 production may be analogous to the phosphofructokinase-catalysed production of fructose-2,6-bisphosphate, which is not an intermediate in glycolysis but is the most potent known regulator of glycolytic flux[17]. Like the two forms of phosphofructokinase, PtdIns kinase types I and II are not isozymes, but catalyse distinct phosphorylations of a common substrate. The possibility that PtdIns(3)P is a regulatory molecule in PtdIns turnover remains to be tested experimentally, as does the involvement of type I PtdIns kinase in the action of growth factor and oncogene products. The specific association between this kinase and immunoprecipitates of polyoma middle T/pp60$^{c\text{-}src}$ competent for transformation or of ligand-activated PDGF receptor indicates that it could be important in the transduction of signals by these tyrosine kinases.

We thank George Patten for extensive advice and assistance in the separation of intact PtdIns phosphates.

Received 23 November 1987; Accepted 18 February 1988.

1. Whitman, M., Fleischman, L., Chahwala, S. B., Cantley, L. & Rosoff, P. in *PI Turnover and Receptor Function* (ed. Putney, J.) 197–217 (Liss, New York, 1986).
2. Berridge, M. *Biochim. biophys. Acta* **907**, 33–45 (1987).
3. Downes, C. P. & Michell, R. H. in *Molecular Aspects of Cellular Regulation* Vol. 4 (eds Cohen, P. & Houslay, M. D.) 3–56 (Elsevier, Amsterdam, 1985).
4. Macara, I. G. *Am. J. Physiol.* **248**, C3–C11 (1985).
5. Whitman, M., Kaplan, D. R., Schaffhausen, B. S., Cantley, L. & Roberts, T. *Nature* **315**, 239–242 (1985).
6. Kaplan, D. R. *et al. Proc. natn. Acad. Sci. U.S.A.* **83**, 3,624–3,628 (1986).
7. Whitman, M., Kaplan, D. R., Roberts, T. M. & Cantley, L. C. *Biochem. J.* **247**, 165–174 (1987).
8. Kaplan, D. R. *et al. Cell* **50**, 1021–1029 (1987).
9. Chang, M. & Ballou, C. E. *Biochem. biophys. Res. Commun.* **26**, 199–205 (1967).
10. Brown, D. M. & Stewart, J. C. *Biochem. biophys. Acta* **125**, 413–421 (1966).
11. Grado, C. & Ballou, C. E. *J. biol. Chem.* **236**, 54–60 (1961).
12. Irvine, R. F., Letcher, A. J., Lander, D. J. & Downes, C. P. *Biochem. J.* **223**, 237–243 (1984).
13. Prottey, C. Salway, J. G. & Hawthorne, J. N. *Biochim. biophys. Acta* **164**, 238–251 (1968).
14. Endemann, G., Dunn, S. & Cantley, L. *Biochemistry* **26**, 6845–6851 (1987).
15. Hawkins, P. T., Stephens, L. & Downes, C. P. *Biochem. J.* **238**, 507–516 (1986).
16. Harden, T. K., Stephens, L., Hawkins, P. T. & Downes, C. P. *J. biol. Chem.* **262**, 9059–9061 (1987).
17. Hers, L. & Hue, S. *A. Rev. Biochem.* **52**, 617–654 (1983).
18. Hawkins, P. T., Stephens, L. & Downes, C. P. *Biochem. J.* **238**, 507–516 (1986).
19. Stephens, L. R. *et al. Biochem. J.* (in the press).

The Ras/MAP kinase pathway

17

Considerable efforts have been made to characterize the signal transduction pathways involved in the control of cell growth and proliferation by growth factors and oncogenes. An understanding of such pathways would give insights not only into the control of normal cell regulation but also into the events by which the expression of oncogenes triggers uncontrolled cell proliferation. Indeed, the study of cell transformation by oncogenes has led to many of the advances in the elucidation of normal cellular control pathways. One major pathway that has been described in detail is the Ras/MAP kinase pathway, which is stimulated by growth factor receptors and which involves the activation of a protein kinase cascade. A key to progress in this area was the discovery of the protein kinase known as MAP kinase (see selected paper [1]), which occupies a crucial position in the pathway.

It was known that treatment of cells with insulin resulted in autophosphorylation of the insulin receptor on tyrosine residues and subsequent phosphorylation of several proteins via a serine/threonine-specific protein kinase. The paper by Ray and Sturgill [1] showed that insulin treatment of 3T3-L1 adipocytes led to the activation of a novel protein kinase. The activity was detected on the basis of the ability of the kinase to phosphorylate a microtubule-associated protein (MAP-2) leading to the kinase being designated MAP-2 kinase. Insulin treatment was already known to result in phosphorylation of the ribosomal protein S6 [2], and the MAP kinase was shown to be distinct from the S6 kinase, based on the finding that it was activated before the S6 kinase in insulin-treated cells and that the two kinase activities could be separated by chromatography. An additional important observation was that the MAP-2 kinase activity declined following incubation of insulin-treated cell extracts and this decline was slowed by protein phosphatase inhibitors. These results suggested to the authors that the activity of the novel MAP-2 kinase or a regulatory component was controlled by phosphorylation. It was subsequently found that MAP-2 kinase must be phosphorylated to be in an active state.

The MAP-2 kinase has been given a number of different names, including ERK (extracellular-signal-regulated kinase), but is now generally known as MAP kinase (mitogen-activated protein kinase). A whole family of MAP kinases has now been characterized by molecular cloning [3]. The position of MAP kinase in a protein kinase cascade emerged from the important observation noted above that its activity was regulated by phosphorylation. In particular MAP kinase is unusual in that the active enzyme is phosphorylated on both tyrosine and threonine [4] residues that are separated by only a single amino acid [5]. Phosphorylation on both amino acids is required for activity of the MAP kinase [4], and it was initially thought that the enzyme could be a site of integration of signals from the already known tyrosine kinases and an unknown threonine kinase. In fact the earlier step in the kinase cascade is occupied by an unusual dual-specificity kinase (MAP kinase kinase or MAPKK; Figure 17.1) that phosphorylates both sites on MAP kinase [6]. Subsequent work identified a further kinase that phosphorylates MAPKK as a cellular homologue of the viral oncogene Raf [7].

Following the description of the early steps leading from growth factor signalling, particularly epidermal growth factor (EGF), an almost complete picture of the Ras/MAP kinase pathway has been built up. This pathway involves key interaction between the src-homology (SH)2- and SH3-domain-containing proteins [8], which lead to activation of Ras followed by Raf and thus the rest of the kinase cascade. The final steps of the pathway involve direct regulation of gene transcription through phosphorylation by MAP kinase of transcription factors [9].

The details of the mechanism that links receptor occupancy to Raf activation vary for different receptors. In the PDGF (platelet-derived growth factor) and insulin receptor pathways, different SH2-domain-containing proteins act in linking to Ras/Raf activation. In the case of insulin they interact with insulin receptor substrate-1, which is tyrosine-phosphorylated by the insulin receptor [10]. In addition, it is increasingly apparent that the pathway contains many more branch points and interconnections [8] than suggested by the linear pathway described in Figure 17.1. The exact mechanism by which guanine nucleotide exchange on Ras leads to activation of Raf is not yet understood. Ras appears

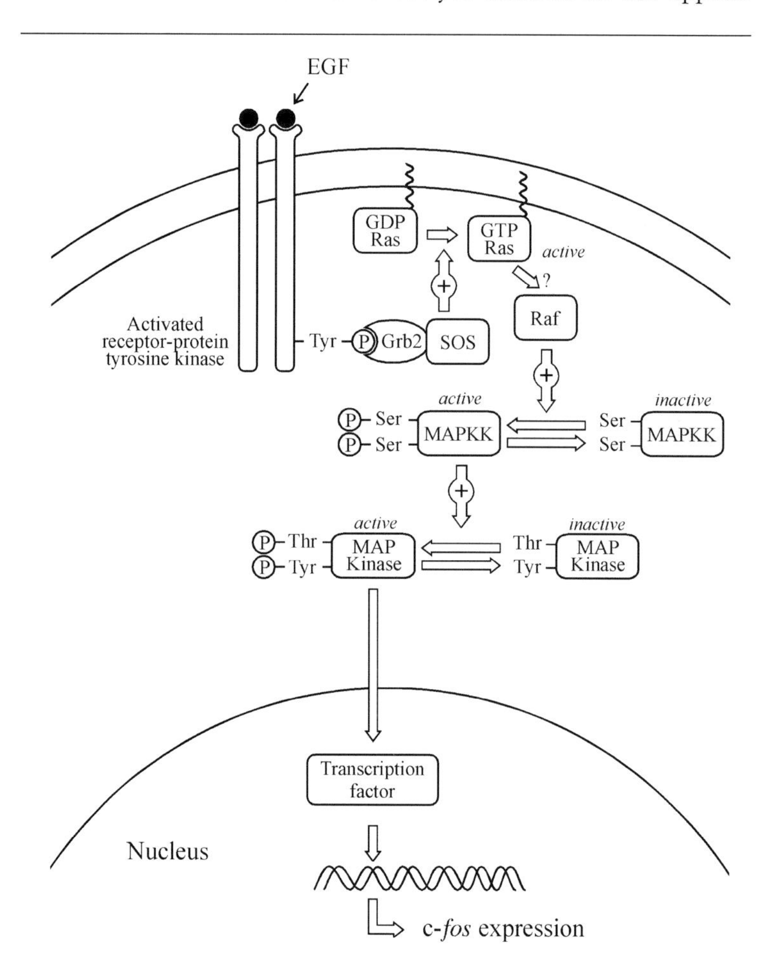

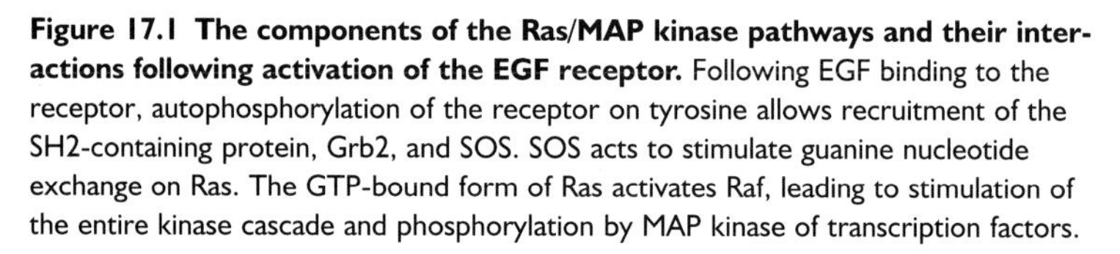
Figure 17.1 The components of the Ras/MAP kinase pathways and their interactions following activation of the EGF receptor. Following EGF binding to the receptor, autophosphorylation of the receptor on tyrosine allows recruitment of the SH2-containing protein, Grb2, and SOS. SOS acts to stimulate guanine nucleotide exchange on Ras. The GTP-bound form of Ras activates Raf, leading to stimulation of the entire kinase cascade and phosphorylation by MAP kinase of transcription factors.

to play a role in allowing recruitment of Raf kinase to the plasma membrane [11,12], but additional factors are needed for the activation of Raf kinase activity.

Studies on the Ras/MAP kinase pathway have aided our understanding of the control mechanisms for cell proliferation, and in addition can explain the role of the Ras and other oncogenes in cell transformation. Cell transformation can be brought about by viral Ras proteins, and, in many types of tumours, mutations in cellular Ras leading to its constitutive activation are extremely common. In addition, as these studies were progressing, it became clear that very similar pathways originate from cell surface receptors involved in diverse control mechanisms, including developmental regulation in the nematode *Caenorhabditis elegans* and in the signalling pathway for mating pheromones in yeast. This indicates that, despite its complexity, the Ras/MAP kinase pathway is, in evolutionary terms, an ancient and very basic pathway for cellular control.

References

1. Ray, L.B. and Sturgill, T.W (1987) *Proc. Natl. Acad. Sci. U.S.A.* **84**, 1502–1506
2. Cobb, M. and Rosen, O. (1993) *J. Biol. Chem.* **258**, 12472–12481
3. Leevers, S.J. and Marshall, C.J. (1992) *Trends Cell Biol.* **2**, 283–286
4. Anderson, N.G., Maller, J.L., Tonks, N.K. and Sturgill, T.W. (1990) *Nature (London)* **343**, 651–653
5. Nishida, E. and Gotoh, Y. (1993) *Trends Biochem. Sci.* **18**, 128–131
6. Crews, L.M., Alessandrini, A. and Erikson, R.L. (1992) *Science* **258**, 478–480
7. Kyriakis, J.M., Apps, H., Zhang, X.F., Banerjee, P., Brautigan, D.L., Rapp, U.R. and Aurich, J. (1992) *Nature (London)* **385**, 417–421
8. Pawson, T. (1995) *Nature (London)* **373**, 573–580
9. Karin, M. and Hunter, T. (1995) *Curr. Biol.* **5**, 747–757
10. Myers, M.G., Sun, X.J. and White, M.F. (1994) *Trends Biochem. Sci.* **19**, 289–293
11. Leevers, S.J., Paterson, H.F. and Marshall, L.J. (1994) *Nature (London)* **369**, 411–414
12. Stokoe, D., McDonald, S.G., Cadwallader, K., Symons, M. and Hancock, J.F. (1994) *Science* **264**, 1463–1467

Ray & Sturgill (1987) Proc. Natl. Acad. Sci. U.S.A. **84**, 1502–1506

Rapid stimulation by insulin of a serine/threonine kinase in 3T3-L1 adipocytes that phosphorylates microtubule-associated protein 2 *in vitro*

(ribosomal protein S6 kinase/phosphatase inhibitors)

L. BRYAN RAY AND THOMAS W. STURGILL

Departments of Internal Medicine and Pharmacology, University of Virginia School of Medicine, Charlottesville, VA 22908

Communicated by G. D. Aurbach, November 13, 1986 (received for review July 2, 1986)

ABSTRACT **Insulin treatment (K_{act}, 5 × 10^{-9} M) of serum-starved 3T3-L1 adipocytes stimulates a soluble serine/threonine kinase that catalyzes phosphorylation of microtubule-associated protein 2 (MAP-2) *in vitro*. Maximal activation of MAP-2 kinase activity by 80 nM insulin was observed after 10 min of hormonal stimulation, prior to maximal stimulation of S6 kinase activity (20 min). The insulin-stimulatable MAP-2 kinase activity is not adsorbed to phosphocellulose, whereas the principal S6 kinase activity is retained and elutes at ≈0.5 M NaCl. The insulin-stimulatable MAP-2 kinase is less stable during incubation at 30°C than S6 kinase activity. Inclusion of phosphatase inhibitors decreases the rate at which the stimulated MAP-2 kinase activity is lost from extract supernatants incubated at 30°C. *p*-Nitrophenyl phosphate is more effective than DL-phosphotyrosine, whereas DL-phosphoserine is without effect at the concentration used (40 mM). The difference in MAP-2 kinase activity in extract supernatants from control and insulin-treated cells is also preserved after rapid chromatography on Sephadex G-25. These results show that a soluble serine/threonine kinase is rapidly activated by insulin, possibly by phosphorylation of either the kinase itself or an interacting modulator.**

Phosphorylation of proteins is a fundamental regulatory mechanism involved in modulation of a variety of cellular processes by hormones (1). Treatment of intact cells, labeled with $^{32}PO_4$, with insulin causes specific increases or decreases in $^{32}PO_4$ incorporation into a number of cellular proteins (2). These changes in phosphorylation are presumably brought about by insulin-induced alterations in the activities of specific protein kinases and/or phosphatases.

The insulin receptor has intrinsic tyrosine kinase activity, which is augmented by insulin binding (3). In addition, insulin stimulates the activity of a serine/threonine kinase(s), which phosphorylates ribosomal protein S6 (2, 4–7). Regulation of that enzyme by a mechanism involving phosphorylation has been proposed based on evidence that the stimulated S6 kinase activity in cell extract supernatants is protected by the addition of phosphatase inhibitors (4). Furthermore, injection of *Xenopus* oocytes with active insulin receptors stimulates S6 kinase in extracts 2-fold (7). These observations have focused attention on the possible regulation of S6 kinase activity by sequential activation of protein kinases beginning with activation of receptor tyrosine kinase (8).

We have found (9) that extract supernatants (see *Materials and Methods*) from 3T3-L1 adipocytes treated with insulin catalyze increased incorporation of ^{32}P from [γ-^{32}P]ATP into an exogenous substrate, microtubule-associated protein 2 (MAP-2), purified from bovine brain. Here we present evidence that this MAP-2 kinase activity represents a previously unrecognized insulin-stimulatable enzyme distinct from the insulin receptor and activities phosphorylating S6.

MATERIALS AND METHODS

Materials. Bovine insulin was a gift from Eli Lilly; epidermal growth factor (receptor grade) was obtained from Collaborative Research (Waltham, MA). [γ-^{32}P]ATP was synthesized by the method of Johnson and Walseth (10). MAP-2 was purified from bovine brain as described by Kim *et al.* (11). Ribosomal 40S subunits were prepared from *Artemia salina* by the method of Zasloff and Ochoa (12). Membranes from human epidermal carcinoma cell line A431 were prepared as described by Thom *et al.* (13).

Cell Culture and Preparation of Extract Supernatants. 3T3-L1 cells were grown and differentiated as described by Rubin *et al.* (14) and used on days 5–8. Prior to hormonal treatment, two to four 100-mm plates per treatment were washed three times with 10 ml of Krebs–Ringer bicarbonate/Hepes buffer [120 mM NaCl/4.75 mM KCl/1.2 mM $MgSO_4$/1.2 mM $CaCl_2$/24 mM $NaHCO_3$/10 mM Hepes, pH 7.5 (37°C)] and incubated in 8 ml of this medium for 1 hr. Insulin (final concentration, 80 nM) or diluent was added to the medium and the incubation was continued for the appropriate time (5–40 min). After hormonal treatment, the plates were placed on ice and quickly washed once with 10 ml of ice-cold 0.15 M NaCl. This and all subsequent steps were carried out at 5°C. The cells were scraped into ice-cold homogenization buffer (400 μl per plate) [buffer A: 20 mM Hepes, pH 7.0 (at 22°C)/20 mM EGTA/15 mM magnesium acetate/1 mM dithiothreitol/40 mM *p*-nitrophenyl phosphate/0.1 mM phenylmethylsulfonyl fluoride; buffer B: 80 mM β-glycerol phosphate, pH 7.0 (at 22°C)/20 mM EGTA/15 mM MgCl/1 mM dithiothreitol/0.2 mM phenylmethylsulfonyl fluoride] as indicated in the text and legends. The cells were homogenized in a Potter–Elvehjem homogenizer with 25 strokes of a motor-driven Teflon pestle. The crude homogenate was centrifuged at 30,000 × *g* for 5 min at 4°C and the supernatant, beneath any lipid, was removed. This supernatant will hereafter be referred to as an "extract supernatant." Protein concentrations (0.4–0.6 mg/ml) of extract supernatants varied by <10% (15).

Phosphotransferase Assay. Assays were performed at 30°C in a final vol of 40 μl containing 50 mM β-glycerol phosphate (pH 7.0), 1 mM dithiothreitol, 10 mM magnesium acetate, 50 μM [γ-^{32}P]ATP (5 cpm/fmol), and either 0.33 mg of 40S ribosomal subunits per ml or 0.1 mg of MAP-2 per ml as substrates. The reaction was stopped (after 10 or 15 min for MAP-2 and S6, respectively) by the addition of 20 μl of 3× Laemmli $NaDodSO_4$ sample buffer (16). Phosphorylated products were resolved by $NaDodSO_4$/gel electrophoresis on 5% (for MAP-2) or 15% (for S6) polyacrylamide gels. Gels

The publication costs of this article were defrayed in part by page charge payment. This article must therefore be hereby marked "*advertisement*" in accordance with 18 U.S.C. §1734 solely to indicate this fact.

Abbreviation: MAP-2, microtubule-associated protein 2.

were stained with Coomassie brilliant blue R-250. The stained bands containing the substrates were excised, solubilized in 400 μl of 30% H_2O_2 (6–12 hr at 70°C), and ^{32}P incorporated was quantitated by liquid scintillation spectroscopy in 30% (vol/vol) Triton X-100/70% (vol/vol) toluene/Omnifluor (4 g/liter) (New England Nuclear).

Phosphocellulose Chromatography. Phosphocellulose (Whatman P11) was precycled according to the manufacturer's instructions and equilibrated with buffer C [25 mM Tris·HCl, pH 7.5 (at 5°C)/25 mM NaCl/2 mM EGTA/2 mM dithiothreitol/40 mM *p*-nitrophenyl phosphate/10% (vol/vol) glycerol/0.2 mM phenylmethylsulfonyl fluoride]. All procedures were carried out at 5°C. Extract supernatants (2 ml) from four plates of cells (per condition) were applied to phosphocellulose (1-ml bed volume) columns, and 10-drop (≈0.5 ml) fractions were collected. After a 4-ml wash with the starting buffer, a 0.025–1 M NaCl gradient in 8 ml, generated by a Pharmacia LCC500 Liquid Chromatography Controller and its pumps, was applied. The fractions were promptly assayed for S6 and MAP-2 kinase activities. Cells from each experimental treatment (insulin or diluent, 5 min or 40 min) were processed through the entire procedure, including homogenization, centrifugation, chromatography, and assay of fractions as rapidly as possible to avoid loss of activity upon storage. This sequence was then repeated with the cells from each of the remaining conditions in a single day.

Phosphoamino Acid Analysis. Proteins were recovered from dried gels as described by Beemon and Hunter (17), and phosphoamino acid analysis was performed by the method of Cooper *et al.* (18).

Data Analysis. Data points in Figs. 3, 4, and 6 are the average of values from two independent experiments, which varied by <10%.

RESULTS

Treatment of 3T3-L1 adipocytes with insulin (80 nM) for 10 min caused a 1.5- to 3-fold stimulation of a kinase activity that phosphorylates MAP-2 from bovine brain *in vitro* (9). This stimulation was dependent on the concentration of insulin applied to the cells (Fig. 1) and was half-maximal at 5 nM insulin. EGTA was included in all buffers in millimolar concentrations; thus, calcium is not essential for this activity. To verify the soluble nature of this activity, extract supernatants from insulin-treated or control cells homogenized in buffer A were subjected to centrifugation at 150,000 × *g* for 1 hr. Aliquots (5 μl) of initial extract supernatants (prepared by centrifugation of crude homogenates for 5 min at 30,000 × *g*) catalyzed the transfer of 11,600 cpm and 6900 cpm of ^{32}P from [γ-^{32}P]ATP to MAP-2, respectively, when assayed as described. Aliquots (5 μl) of 150,000 × *g* supernatants from the same insulin-treated and control cells catalyzed the transfer of 11,400 cpm and 5900 cpm of ^{32}P, respectively.

The stability of the enhanced activity to gel filtration was also studied. The activated MAP-2 kinase in extract supernatants from insulin-treated cells persisted after gel filtration on Sephadex G-25 by a centrifuge column procedure (20). In a representative experiment, recoveries of enzymatic activity applied to such columns were 93% and 112% for extracts of insulin-treated and control cells, respectively. Recoveries of total protein in the same samples were 68% and 69%. Tests of column performance showed that >90% of [γ-^{32}P]ATP added to extract supernatants was removed by this procedure.

Phosphoamino acid analysis of phosphorylated MAP-2 revealed that insulin induced phosphorylation primarily of serine and, to a lesser extent, threonine (Fig. 2). Phosphotyrosine was not detected in these samples. Phosphate esters of tyrosine are more susceptible than those of serine or threonine to partial acid hydrolysis used to release phosphoamino acids prior to chromatographic separation of the ^{32}P-labeled amino acids. To ensure that phosphotyrosine was not being artifactually lost during sample preparation, MAP-2 was incubated with [γ-^{32}P]ATP and a membrane fraction from A431 epidermoid cells rich in epidermal growth factor receptor kinase. An increase in phosphotyrosine content in MAP-2, induced by epidermal growth factor, was readily detected (Fig. 2).

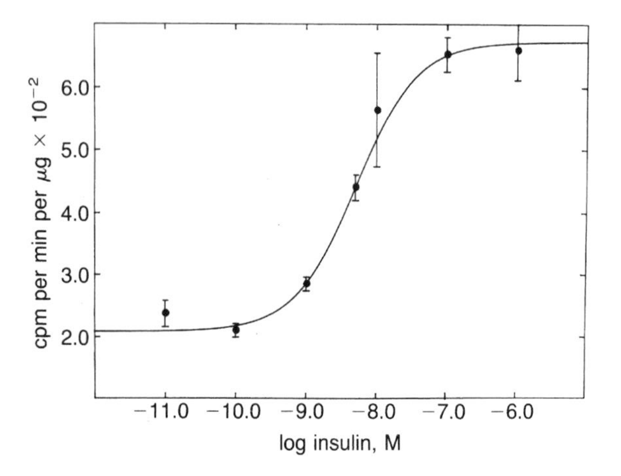

FIG. 1. Dose–response curve for MAP-2 kinase activity stimulated by insulin. 3T3-L1 adipocytes were incubated for 10 min with the indicated concentrations of insulin in the presence of 1.5 mM bacitracin (14). Extract supernatants from the cells were prepared as described in *Materials and Methods*. Protein concentrations in the supernatants were 0.75 ± 0.07 (SD) mg/ml. Results are expressed as cpm of ^{32}P incorporated into MAP-2 per min per μg of extract supernatant protein in a 10-min assay. The data were analyzed by a nonlinear least-squares procedure (19) assuming a single set of noninteracting binding sites. The estimated concentration of insulin required for half-maximal stimulation was 5 ± 1 (SD) × 10^{-9} M.

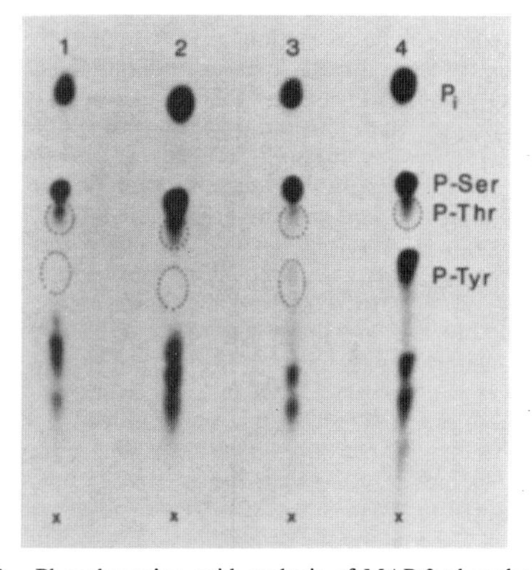

FIG. 2. Phosphoamino acid analysis of MAP-2 phosphorylated by incubation with extract supernatants from 3T3-L1 adipocytes (lanes 1 and 2) or A431 cell membranes (lanes 3 and 4). Extract supernatants from control (lane 1) or insulin-treated cells (lane 2) were prepared and incubated with MAP-2 and [γ-^{32}P]ATP as described in *Materials and Methods*. MAP-2 (1.5 μg) was phosphorylated by A431 cell membranes in the absence (lane 3) or presence (lane 4) of 200 nM epidermal growth factor essentially as described by Cohen (21). Phosphorylated MAP-2 was isolated by $NaDodSO_4$ gel electrophoresis and processed for phosphoamino acid analysis as described. The migration of phosphoamino acid standards visualized by reaction with ninhydrin is indicated. ^{32}P-labeled amino acids were detected by autoradiography using Kodak X-Omat AR-5 film and a DuPont Lightning Plus intensifying screen for 4 hr (lanes 1 and 2) or 7 days (lanes 3 and 4) at −70°C.

An insulin-stimulatable kinase that phosphorylates ribosomal protein S6 has been described (4–7). Therefore, we compared several properties of the kinase activities phosphorylating MAP-2 and S6. The time course of stimulation of these activities is shown in Fig. 3. Increased MAP-2 kinase activity was observed in extract supernatants of 3T3-L1 adipocytes within 5 min after exposure of cells to insulin. Incorporation of $^{32}PO_4$ into MAP-2 was stimulated 2.3-fold after 5 min and 3.4-fold after 10 min of exposure of 3T3-L1 adipocytes to insulin. This stimulation declined rapidly toward control values within 20–40 min after the application of insulin. In contrast, S6 kinase activity measured in these same extract supernatants was activated more slowly and was not maximal until 20 min after stimulation by insulin. S6 kinase activity remained high through 40 min of insulin treatment. The maximal observed stimulation of S6 kinase was 6.6-fold at 20 min.

The stability of the activation of the S6 and MAP-2 kinase activities was examined by assaying aliquots of extract supernatants from insulin-treated and control cells immediately after preparation or after incubation of the supernatant at 30°C for 10 or 30 min (Fig. 4). In this experiment, cells were homogenized in buffer containing 80 mM β-glycerol phosphate, a phosphatase inhibitor previously shown to protect the activation of an S6 kinase in Swiss 3T3 cells (4). Seventy-six percent of the initial increment in S6 kinase activity due to insulin remained after 10 min; 57% remained after 30 min. In contrast, 43% of the stimulation of MAP-2 kinase activity remained after 10 min; after 30 min at 30°C, the insulin-treated extract was indistinguishable from that of control cells.

Further distinction of the kinase activities phosphorylating MAP-2 and S6 was achieved by chromatography of extract supernatants from insulin-treated and control adipocytes on phosphocellulose columns. MAP-2 kinase activity was not adsorbed to phosphocellulose when applied in buffer C. Under these conditions, all but a minor fraction of the S6 kinase activity was retained and could be eluted with 0.5 M NaCl (data not shown; see Fig. 6).

Another experiment was devised to differentiate the activities both on the basis of their chromatographic behavior on phosphocellulose and on their inversely related patterns of activation in cells treated with insulin for 5 or 40 min (see Fig. 3). Extract supernatants were prepared from 3T3-L1 adipocytes treated with insulin or diluent for 5 or 40 min and

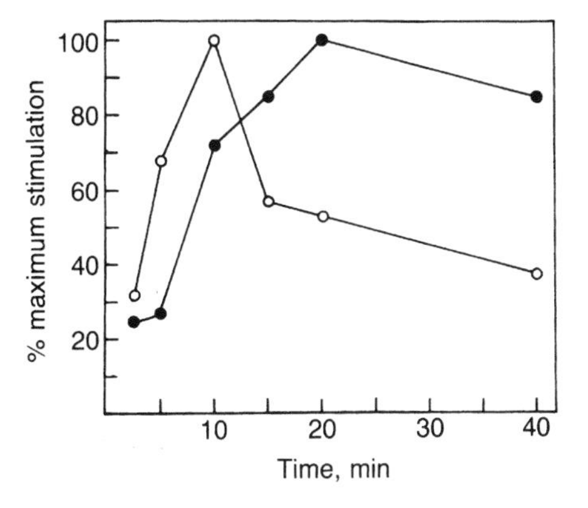

FIG. 3. Time course of activation of MAP-2 (○) and S6 (●) kinase activities. 3T3-L1 adipocytes were treated with insulin (80 nM) or diluent in the presence of 1.5 mM bacitracin for the times indicated. Extract supernatants were prepared as described in *Materials and Methods* and aliquots were assayed for phosphotransferase activities using MAP-2 and S6 as substrates. Data are presented as % maximum stimulation of kinase activity observed for each substrate; 100% corresponds to a 3.4-fold stimulation relative to control levels of activity for MAP-2 (12,500 cpm of ^{32}P incorporated) and a 6.6-fold stimulation in S6 kinase activity (2200 cpm of ^{32}P incorporated).

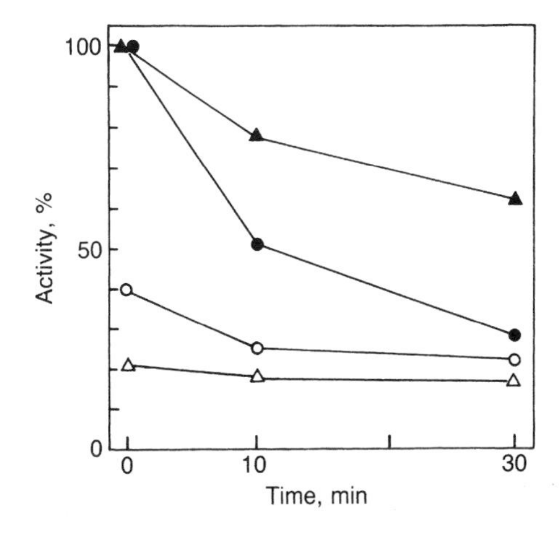

FIG. 4. Loss of MAP-2 and S6 kinase activities upon incubation of extract supernatants at 30°C. Extract supernatants from 3T3-L1 adipocytes treated with 80 nM insulin (●, ▲) or diluent (○, △) for 10 min were prepared as described in *Materials and Methods*. Aliquots were assayed for phosphotransferase activity using MAP-2 (●, ○) or S6 (▲, △) as substrates immediately after preparation (time 0) or after incubation at 30°C for the indicated times. Data for each substrate are presented as % phosphotransferase activity observed in extracts of insulin-treated cells at time 0; 100% corresponds to 15,500 cpm of ^{32}P incorporated into MAP-2 and to 1700 cpm of ^{32}P incorporated into S6.

subjected to phosphocellulose chromatography. Fractions were assayed for phosphotransferase activity using MAP-2 (Fig. 5*A*) and S6 (Fig. 5*B*) as substrates. Almost all of the MAP-2 kinase activity eluted from the column in fractions corresponding to the sample flow-through and the wash. The insulin-induced increase in incorporation of $^{32}PO_4$ into MAP-2 was found in this broad peak of activity. As predicted from the data presented in Fig. 3, the activation of MAP-2 kinase by insulin was greater after treatment of the adipocytes with insulin for 5 min than after 40 min of insulin treatment.

The major peak of S6 kinase activity (Fig. 5*B*) was retained by phosphocellulose and was eluted by 0.5 M NaCl. Unlike the MAP-2 kinase, S6 kinase activity was stimulated to a much greater extent in extracts of cells treated with insulin for 40 min than in those treated for 5 min. A minor fraction of the S6 kinase activity detected appeared in the unretained or wash fractions, which also contained MAP-2 kinase activity. However, the S6 kinase activity not adsorbed to the column was more intensely activated by treatment of cells with insulin for 40 min than for 5 min. The MAP-2 kinase activity eluting in this region was stimulated to a greater extent in extract supernatants from cells treated for 5 min. Thus, it appears that activated S6 kinase activity is not the same enzyme that catalyzes the increased incorporation of $^{32}PO_4$ into MAP-2.

Experiments were conducted to determine whether the rapid loss of the insulin-stimulated MAP-2 kinase activity during incubation of extract supernatants at 30°C (see Fig. 4) was affected by phosphatase inhibitors (Fig. 6). Less than 12% of the stimulated enzymatic activity (defined as the difference) remained after only 10 min of incubation without phosphatase inhibitors. With addition of 40 mM *p*-nitrophenyl phosphate, 81% and 50% of the stimulated activity persisted after 10 and 30 min, respectively. Phosphotyrosine (40 mM) preserved 44% and 17% of the stimulated activity at these times. β-Glycerol phosphate (80 nM, not shown) as used in the experiment shown in Fig. 4 was about as effective as phosphotyrosine. Phosphoserine (40 mM) was without effect.

DISCUSSION

We have found that insulin stimulates a kinase activity in extract supernatants from 3T3-L1 adipocytes that phospho-

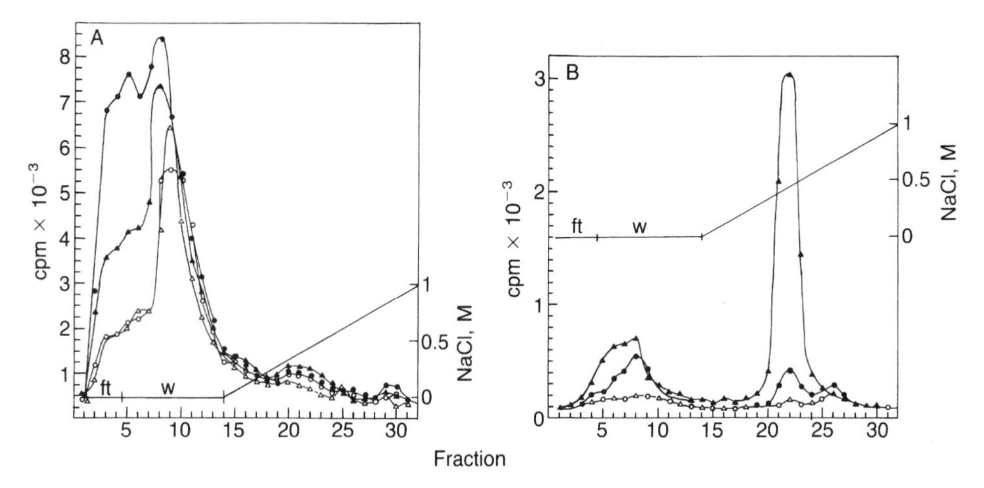

FIG. 5. Chromatography of MAP-2 and S6 kinase activities on phosphocellulose. Extract supernatants from insulin-treated (solid symbols) or control cells (open symbols) exposed to the hormone for 5 min (●, ○) or 40 min (▲, △) were applied to 1-ml phosphocellulose columns and eluted as described in *Materials and Methods*. Aliquots of each fraction were assayed for phosphotransferase activity using MAP-2 (*A*) or S6 (*B*) as substrates. Activity is expressed as cpm of ^{32}P incorporated into the substrate during the 10-min (MAP-2) or 15-min (S6) assay. The fractions eluted during sample flow-through (ft), wash with starting buffer (w), and the NaCl gradient are indicated. Alternate fractions are plotted for control samples in *B* for clarity.

rylates MAP-2 *in vitro*. This kinase activity phosphorylates MAP-2 on serine and threonine residues. It is soluble and is not removed from extract supernatants by centrifugation at 150,000 × *g* for 1 hr. These properties clearly distinguish the MAP-2 kinase from the membrane-associated tyrosine-specific insulin receptor kinase that has also been shown to phosphorylate MAP-2 *in vitro* (22).

Treatment of Swiss 3T3 fibroblasts with serum or epidermal growth factor stimulates a soluble protein kinase that phosphorylates ribosomal protein S6 (4). Similarly, insulin is known to activate a soluble S6 kinase in 3T3-L1 cells (6). Our results confirm these observations, but we find that the MAP-2 kinase activity can be distinguished from S6 kinase

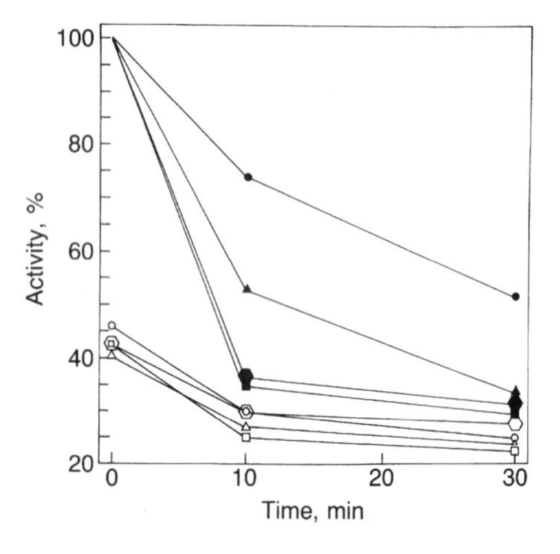

FIG. 6. Effect of phosphatase inhibitors on loss of MAP-2 kinase activity upon incubation at 30°C. Extract supernatants were prepared from cells treated with 80 nM insulin for 10 min (solid symbols) or control cells (open symbols) homogenized in a buffer A without *p*-nitrophenyl phosphate as described in *Materials and Methods*. These extract supernatants were divided into equal portions and supplemented with 40 mM (final concentrations) *p*-nitrophenyl phosphate (●, ○), 40 mM DL-phosphotyrosine (▲, △), 40 mM DL-phosphoserine (■, □), or homogenization buffer (control)(●, ○). Aliquots were assayed for phosphotransferase activity using MAP-2 as substrate immediately after preparation (time 0) or after incubation at 30°C for the indicated times. Results are means of two independent experiments and are expressed as % phosphotransferase activity observed in extracts from insulin-treated cells at time 0; 100% corresponds to 14,600 cpm of ^{32}P incorporated into MAP-2.

activity by three separate criteria. First, the time course of activation of the enzymes by insulin is different. MAP-2 kinase activity was activated more rapidly than S6 kinase attaining maximal activity during a narrow window of time 5–15 min after treatment of cells with insulin. S6 kinase became fully activated ≈10 min later and remained highly stimulated through 40 min of hormonal treatment. Second, MAP-2 kinase activity was much more labile during incubation of extract supernatants at 30°C than S6 kinase activity. Third, the MAP-2 and S6 kinase activities can be separated by chromatography of extract supernatants from 3T3-L1 adipocytes on phosphocellulose columns. We conclude from these results that insulin stimulates a MAP-2 kinase that is distinct from soluble activities phosphorylating S6.

The chromatographic properties on phosphocellulose of the major insulin-stimulated S6 kinase described here are similar to those reported by Cobb and Rosen (5) for an S6 kinase from 3T3-L1 cells. In that study, the S6 kinase characterized was obtained by extraction of a particulate fraction with 0.6 M KCl. The particulate fraction, containing plasma membranes, microsomal membranes, and ribosomes, was prepared from cells homogenized under conditions of low ionic strength, which may have favored adsorption of some cytoplasmic proteins to the particulate fraction. Therefore, it is possible that the S6 kinase activity characterized by Cobb and Rosen (5) is the same as that observed in the present study.

It is not clear from the present experiments whether the minor peak of S6 kinase activity, which is not retained by phosphocellulose, represents merely unadsorbed S6 kinase or a distinct enzyme. However, this minor peak of S6 kinase activity was stimulated to a greater extent in extracts of cells treated for 40 min with insulin than in those from cells exposed to the hormone for 5 min. Since the MAP-2 kinase showed an opposite degree of activation with respect to duration of insulin treatment, it appears that this minor peak of S6 kinase activity is not the result of phosphorylation of S6 by the MAP-2 kinase eluting in the same fractions from phosphocellulose. Thus, S6 seems to be a relatively poor substrate for the MAP-2 kinase. Similarly, there is little MAP-2 kinase activity associated with the major peak of S6 kinase activity eluted from phosphocellulose by 0.5 M NaCl. Only a slight stimulation in incorporation of ^{32}P into MAP-2 was catalyzed by fractions from this peak, suggesting that

MAP-2 is a relatively poor substrate for this S6 kinase activity.

Because the molecular basis of insulin action remains enigmatic, it is of particular interest to know the mechanism by which the MAP-2 kinase is stimulated in cells treated with insulin. Several implications of the results presented here are relevant in this regard. Increased synthesis of the enzyme is unlikely to account for the increased activity phosphorylating MAP-2 because of the rapid onset of this effect. The time course is similar to that of activation of glucose transport in 3T3-L1 adipocytes (23), one of the earliest cellular responses to insulin. Furthermore, the dose–response curve for activation of MAP-2 kinase (K_d, 5×10^{-9} M) approximates that reported by Kohanski *et al.* (23) for stimulation of glucose transport in 3T3-L1 adipocytes. These results suggest that activation of MAP-2 kinase may be closely linked to receptor activation.

The activation of MAP-2 kinase activity was retained after rapid gel filtration of extract supernatants. This result suggests that the stimulation of MAP-2 kinase more likely represents structural or covalent modification or subcellular relocalization rather than allosteric regulation involving a small molecule. Of course very tight binding of an activator cannot be excluded by experiments of this type.

Finally, the insulin-induced stimulation of MAP-2 kinase activity was preserved by addition of phosphatase inhibitors, in particular *p*-nitrophenyl phosphate, to the homogenization buffer. It is of interest to note that *p*-nitrophenyl phosphate is a structural analog of phosphotyrosine and a potent inhibitor of phosphotyrosine phosphatases (24). Also, phosphotyrosine itself was effective in maintaining MAP-2 kinase activity, whereas phosphoserine was without effect. These results are consistent with the hypothesis that MAP-2 kinase or a regulatory component of that activity might be modulated by phosphorylation, perhaps on tyrosine. However, other explanations for the effects of *p*-nitrophenyl phosphate, phosphotyrosine, and β-glycerol phosphate are not excluded.

Only purification of the kinase will allow direct analysis of the involvement of phosphorylation in its activation. Our efforts to this end have established the following properties of the enzyme, to be described in detail elsewhere: upon gel filtration chromatography, the activated enzyme migrates with an apparent molecular mass of ≈30 kDa; MAP-2 kinase activity in extract supernatants is also stimulated by treatment of 3T3-L1 cells with insulin-like growth factor I, epidermal growth factor, or phorbol esters; compared to MAP-2, casein and histone are poor substates for the insulin-stimulatable MAP-2 kinase.

A cascade of phosphorylation reactions initiated by the receptor tyrosine kinase has been proposed (e.g., see ref. 25) as a potential transducing mechanism for insulin and other growth factor receptors, but direct evidence for such an effect is lacking. Further characterization of the MAP-2 kinase and the manner in which it is stimulated by insulin may provide information useful in the evaluation of this hypothesis.

We thank Drs. J. Garrison and J. Larner for reading the manuscript, and Gail Garner for technical assistance. This research was supported by a Basil O'Connor Research Starter Grant (March of Dimes Birth Defects Foundation), Grant BC-546 from the American Cancer Society and the Jeffress Foundation. L.B.R. was supported by National Institutes of Health Training Grant AM-07320. Early studies were supported by a grant from the American Diabetes Association. Tissue culture media and [γ-^{32}P]ATP were provided by the University of Virginia Diabetes Research and Training Center (AM-22125).

Note Added in Proof. In experiments identical to those shown in Fig. 6, 2 mM sodium vanadate (26) was found to be effective in protecting the insulin-stimulated MAP-2 kinase activity.

1. Cohen, P. (1985) *Eur. J. Biochem.* **151,** 439–448.
2. Avruch, J., Nemenoff, R. A., Pierce, M., Kwok, Y. C. & Blackshear, P. J. (1985) in *Molecular Basis of Insulin Action*, ed. Czech, M. P. (Plenum, New York), pp. 263–296.
3. Pessin, J. E., Mottola, C., Kin-Tak, Y. & Czech, M. P. (1985) in *Molecular Basis of Insulin Action*, ed. Czech, M. P. (Plenum, New York), pp. 3–31.
4. Novak-Hofer, I. & Thomas, G. (1985) *J. Biol. Chem.* **260,** 10314–10319.
5. Cobb, M. & Rosen, O. (1983) *J. Biol. Chem.* **258,** 12472–12481.
6. Tabarini, D., Heinrich, J. & Rosen, O. (1985) *Proc. Natl. Acad. Sci. USA* **82,** 4369–4373.
7. Stefanovic, D., Erikson, E., Pike, L. J. & Maller, J. L. (1986) *EMBO J.* **5,** 157–160.
8. Petruzelli, L. M., Ganguly, S., Smith, C. J., Cobb, M. H., Rubin, C. S. & Rosen, O. (1982) *Proc. Natl. Acad. Sci. USA* **79,** 6792–6796.
9. Sturgill, T. W. & Ray, L. B. (1986) *Biochem. Biophys. Res. Commun.* **134,** 565–571.
10. Johnson, R. & Walseth, T. (1979) *Adv. Cyclic Nucleotide Res.* **10,** 135–168.
11. Kim, H., Binder, L. & Rosenbaum, J. (1979) *J. Cell Biol.* **80,** 266–276.
12. Zasloff, M. & Ochoa, S. (1974) *Methods Enzymol.* **30,** 197–206.
13. Thom, D., Powell, A. J., Lloyd, C. W. & Rees, D. A. (1977) *Biochem. J.* **168,** 187–194.
14. Rubin, C., Hirsch, A., Fung, C. & Rosen, O. (1978) *J. Biol. Chem.* **253,** 7570–7578.
15. Bensadoun, A. & Weinstein, D. (1976) *Anal. Biochem.* **70,** 241–249.
16. Laemmli, U. (1970) *Nature (London)* **227,** 680–685.
17. Beemon, K. & Hunter, T. (1978) *J. Virol.* **28,** 551–566.
18. Cooper, J., Sefton, B. & Hunter, T. (1983) *Methods Enzymol.* **99,** 387–402.
19. Johnson, M. L. & Frasier, S. G. (1985) *Methods Enzymol.* **117,** 301–342.
20. Christopherson, R. I., Jones, M. E. & Finch, L. R. (1979) *Anal. Biochem.* **100,** 184–187.
21. Cohen, S. (1983) *Methods Enzymol.* **99,** 379–387.
22. Kadowaki, T., Fujita-Yamaguchi, Y., Nishida, E., Takau, F., Akiyama, T., Kathuria, S., Akanuma, Y. & Kasuga, M. (1985) *J. Biol. Chem.* **260,** 4016–4020.
23. Kohanski, R. A., Frost, S. C. & Lane, M. D. (1986) *J. Biol. Chem.* **261,** 1272–1281.
24. Swarup, G., Cohen, S. & Garbers, D. L. (1981) *J. Biol. Chem.* **256,** 8197–8201.
25. Denton, R., Brownsey, R. W., Hopkirk, T. J. & Belsham, G. J. (1984) *Biochem. Soc. Trans.* **12,** 768–771.
26. Sparks, J. W. & Brautigan, D. L. (1986) *Int. J. Biochem.* **18,** 497–504.

Growth factor receptors and oncogenes

One of the most important and rapidly progressing areas in cell signalling has been the attempt to understand how oncogenes result in uncontrolled cell proliferation and the formation of tumours. Normal cellular proliferation is stimulated by growth factors that are generally peptides, and which act on the cells that secrete them in an autocrine fashion or on other nearby or distant cells. It is now apparent that oncogenes are able to subvert normal intracellular signalling pathways involved in the control of cell proliferation by acting in various ways as growth hormones, constitutively active growth hormone receptors, components of the pathways leading from growth hormone receptors, or as factors that regulate gene transcription [1]. Our chosen paper [2] showed for the first time that a viral oncogene could bring about cell transformation due to it being a 'hijacked' portion of the cellular receptor for the epidermal growth factor (EGF). This oncogene was suggested to function as a ligand-independent constitutively active receptor lacking an extracellular peptide-binding domain. The paper, therefore, clearly illustrated a new strategy whereby a viral oncogene could infiltrate normal cell proliferation control pathways.

The first suggestion of a link between growth factors and growth of cancer cells came from work suggesting that transformed cells might synthesize and secrete growth factors which would act in an autocrine or paracrine fashion to stimulate growth [3]. The presence of such factors in a growth medium of tumour cells was demonstrated, and two factors, transforming growth factor (TGF)-α and TGF-β identified [4,5]. This idea of autocrine growth factor production gained further ground with the demonstration from protein sequencing of a close similarity between the transforming oncogene ($p28^{sis}$) of simian sarcoma virus and the B subunit of a known growth factor, platelet-derived growth factor (PDGF) [6,7]. This important observation by two laboratories, including that of Mike Waterfield, generated considerable interest. This was equalled only by the subsequent finding by the Waterfield group that the partially determined protein sequence of the EGF receptor [2] was closely related to that of the protein encoded by the transforming oncogene v-Erb-B of avian erythroblastosis virus [8] (note that the product of the other transforming oncogene of this virus, v-Erb-A, turned out to be a thyroid hormone receptor; see Section 19). The relationships between PDGF and the Sis oncogene product and between the EGF receptor and v-Erb-B were confirmed later by the full sequencing of cDNA clones encoding the mammalian host cell proteins [9,10].

The initial characterization of the EGF receptor took advantage of the A431 cell line [11], which expresses very high levels of the receptor (10–50-fold higher than other cell types), and also made use of human placenta. The EGF receptor was purified by immunoaffinity chromatography using specific monoclonal antibodies which recognized the receptor [2]. The receptor was then subjected to trypsin digestion, the tryptic peptides were separated

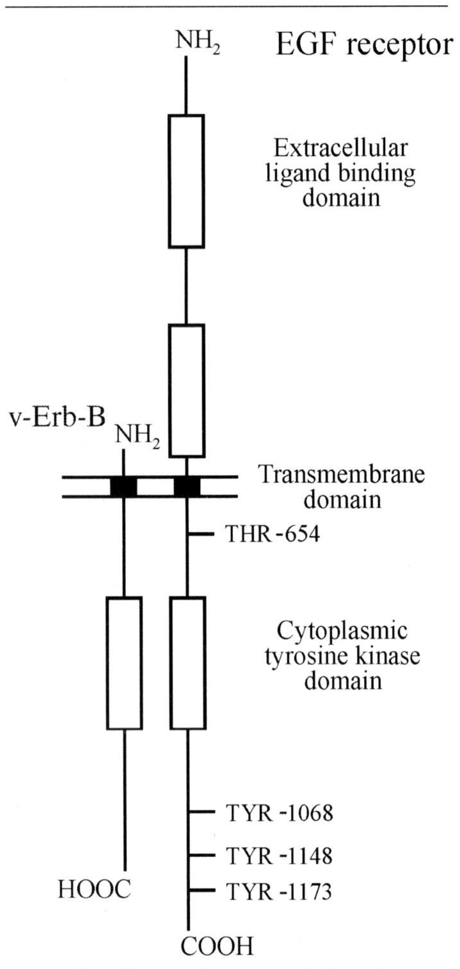

Figure 18.1 Schematic representation of the structure of the EGF receptor and the v-Erb-B protein. The EGF receptor possesses two cysteine-rich extracellular domains and a cytoplasmic tyrosine kinase domain. The v-Erb-B protein lacks most of the extracellular portion of the EGF receptor required for EGF binding but possesses all but the final 32 C-terminal cytoplasmic amino acids. The EGF receptor has sites for protein kinase C phosphorylation at Thr-654 [12]. The tyrosines that are autophosphorylated *in vitro* [13] are indicated.

by HPLC, and 14 peptides in total were sequenced. The derived amino acid sequences of six of the peptides were seen to be related to the sequence of the v-Erb-B oncogene product, and in particular 74 out of 83 amino acids were identical [2].

The sequence shared between the EGF receptor and v-Erb-B lay within the proposed cytoplasmic domain of the receptor, and contained the sites for autophosphorylation by the receptor's intrinsic tyrosine kinase activity (see Sections 15 and 17). Since the viral oncogene product was only around half the size of the EGF receptor polypeptide, it was proposed that, while the v-Erb-B protein was closely related to the EGF receptor, it was likely to be a truncated version lacking the extracellular EGF-binding domain (Figure 18.1). This suggestion was quickly confirmed when the full sequence of the EGF receptor from cDNA cloning was published only three months later [10].

The revelation of this relationship between the EGF receptor and the v-Erb-B protein was extremely important because it immediately suggested a novel mechanism for cell transformation [2]. In contrast to the earlier idea of autocrine effects of released growth factors, supported by the similarity of PDGF and the Sis oncogene product the new results suggested that the avian erythroblastosis virus had hijacked that part of the EGF receptor involved in signal transduction after growth factor binding. It was correctly suggested that this truncated receptor would be ligand-independent and constitutively active. Subsequent work on a variety of oncogenes has so far revealed only a few similar examples of constitutively active plasma membrane receptors. Instead, however, many oncogenes have been found to be activated homologues of other cellular proteins that function in the pathways leading from receptor activation to the control of cell proliferation (see Sections 15 and 17) [1].

References

1. Cantley, L.C., Auger, K.R., Carpenter, C., Duckworth, B., Graziani, A., Kapeller, R. and Soltoff, S. (1991) *Cell* **64**, 281–302
2. Downward, J. Yarden, Y., Mayes, E., Scrace, G., Totty, N. Stockwell, P., Ullrich, A., Schlessinger, J. and Waterfield, M.D. (1984) *Nature (London)* **307**, 521–527
3. Sporn, M.B., Todaro, G.J. (1980) *N. Engl. J. Med.* **303**, 878–880
4. Sporn, M.B. and Roberts, A.B. (1985) *Nature (London)* **313**, 745–747
5. Massague, J. (1987) *Cell* **49**, 437–438
6. Waterfield, M.D., Scrace, G.T., Whittle, N., Scroobant, P., Johnsson, A.,Westeson, A., Westermark, B., Heldin, C.H., Huang, J.S. and Deuel, T.F. (1993) *Nature (London)* **304**, 35–39
7. Doolittle, R.F., Hunkapiller, M.W., Hood, L.E., Devare, S.G., Robbins, K.C., Aaronson, S.A. and Antoniades, H.N. (1983) *Science* **221**, 273–277
8. Yamamoto, T. (1983) *Cell* **35**, 71–78
9. Johnsson, A., Heldin, C.H., Wasteson, A., Westermark, B., Deuel, T.F., Huang, J.S., Seeberg, P.H., Gray, A., Ullrich, A., Scrace, G., Stroobont, P. and Waterfield, M.D. (1984) *EMBO J.* **3**, 921–928
10. Ullrich, A., Coussens, L., Hayflick, J.S., Dull, T.J., Gray, A., Tam, A.V., Lee, J., Whittle, N., Waterfield, M.D. and Seeberg, P.H. (1984) *Nature (London)* **309**, 418–425
11. Fabriant, R.N., DeLarco, J.E. and Todaro, G.J. (1977) *Proc. Natl. Acad. Sci. U.S.A.* **74**, 565–569
12. Hunter, T., Ling, N. and Cooper, J.A. (1984) *Nature (London)* **311**, 480–483
13. Downward, J., Parker, P. and Waterfield, M.D. (1984) *Nature (London)* **311**, 483–485

Downward et al. (1984) Nature (London) **307**, 521–527

Close similarity of epidermal growth factor receptor and v-*erb-B* oncogene protein sequences

J. Downward*, Y. Yarden†, E. Mayes*, G. Scrace*, N. Totty*, P. Stockwell*§, A. Ullrich‡, J. Schlessinger† & M. D. Waterfield*

* Protein Chemistry Laboratory, Imperial Cancer Research Fund, Lincoln's Inn Fields, London WC2A 3PX, UK
† Department of Chemical Immunology, The Weizmann Institute of Science, Rehovot, Israel
‡Genentech Incorporated, 460 Point San Bruno Boulevard, San Francisco, California 94080, USA

*Each of six peptides derived from the human epidermal growth factor (EGF) receptor very closely matches a part of the deduced sequence of the v-*erb-B* transforming protein of avian erythroblastosis virus (AEV). In all, the peptides contain 83 amino acid residues, 74 of which are shared with v-*erb-B*. The AEV progenitor may have acquired the cellular gene sequences of a truncated EGF receptor (or closely related protein) lacking the external EGF-binding domain but retaining the transmembrane domain and a domain involved in stimulating cell proliferation. Transformation of cells by AEV may result, in part, from the inappropriate acquisition of a truncated EGF receptor from the c-*erb-B* gene.*

REGULATION of the proliferation of cells in culture can be influenced by a number of mitogens including a series of polypeptide growth factors which, acting alone or synergistically with other mitogens, can induce DNA synthesis and proliferation of specific target cells (for recent reviews see ref. 1). Epidermal growth factor (EGF) and platelet-derived growth factor (PDGF) are probably the best characterized growth factors: the precise function of these polypeptides *in vivo* is, however, unclear. EGF may have a role in cell proliferation and differentiation since it will induce early eyelid opening and incisor development in new born mice[2]; PDGF on the other hand, which is released from platelets during blood clot formation at wound sites, may have a role in repair processes[3]. These and other growth factors *in vitro* can trigger a variety of morphological and biochemical changes that resemble those characteristic of transformed cells, and have also been implicated in the abnormal regulation of proliferation shown by transformed and tumour-derived cell lines (reviewed in refs 4 and 5). Thus it has been suggested that transformed cells may both synthesise and respond to growth factors and consequently proliferate independently through 'autocrine' secretion[6]. Direct support for such an autocrine role for aberrantly expressed growth factors in the control of abnormal cell proliferation came recently from the discovery that the putative transforming protein (p28^{sis}) of simian sarcoma virus (SSV) is structurally related to the growth factor PDGF[7–9] and can also function like PDGF as a growth factor for cells in culture[10]. Other growth factors produced by transformed cells such as insulin-like growth factor (IGF)[11,12], fibroblast-derived growth factor[13,14] and the transforming growth factors (TGFs)[15–20], may also act as autocrine regulators of proliferation. Besides the specificity mediated by regulation of the production of growth factors, cellular specificity could also be controlled at several other levels—the most obvious being by binding of ligand to specific receptors present only on target cells. In addition the binding of one growth factor to its specific receptor can also alter the affinity of another growth factor for its receptor (for example, PDGF and the EGF receptor[21,22]). Conversely two growth factors may, as appears to be the case with αTGFs and EGF, bind to the same receptor[23,24].

It is clear that binding of different growth factors to their specific receptors can induce a cascade of biochemical events including rapid changes in ion movements and intracellular *p*H, stimulation of tyrosine specific protein kinases and several other changes which can culminate in DNA synthesis and proliferation of certain target cells[1,4–6]. It seems likely that at least in the case of the EGF receptor the primary function of EGF may be to induce cross-linking or conformational changes of receptors, and that following such an activation step, all the 'information' necessary for triggering a proliferative response may reside in the receptor itself (see ref. 1). One known function intrinsic to the EGF receptor is its ability to phosphorylate tyrosine residues[25–28], a property shared with five of the putative transforming proteins of the family of retroviruses whose oncogenes are structurally related to *src* but not by two others, the proteins encoded by *mos* and *erb-B*[29]. At present this tyrosine kinase activity provides the only functional activity associated with the oncogenes of this subset of retroviruses.

Here we report amino acid sequence analysis of six distinct peptides from human EGF receptors isolated by monoclonal immunoaffinity purification from A431 cells and placenta, and show that 74 out of 83 of the residues sequenced are identical to those of the transforming protein encoded by the v-*erb-B* oncogene of avian erythroblastosis virus (AEV)[30]. The amino acid sequences of several other peptides purified from the EGF receptor could not be aligned with sequences of the v-*erb-B* protein suggesting that AEV has acquired cellular sequences encoding only a portion of the avian EGF receptor. Several lines of evidence suggest that the v-*erb-B* oncogene encodes only the transmembrane region of the EGF receptor and the domain associated with the tyrosine kinase activity. These results suggest a hypothesis that the *src* related subset of oncogenes, which includes v-*erb-B*, may be derived from cellular sequences which encode growth factor receptors and produce transformation through expression of uncontrolled receptor functions.

Purification of EGF receptor

EGF receptors can be detected in a variety of cells either by measurement of EGF binding (reviewed in ref. 31), by cross-linking of labelled EGF to its receptor (reviewed in ref. 32) or by using monoclonal antibodies[33–38]. In this study the receptor has been purified from two sources: the human epidermoid carcinoma cell line A431, which expresses about 50 times more receptors than the majority of other cells[39,40], and human placenta[41] which is a readily available normal tissue. The recent isolation of monoclonal antibodies which recognize the human EGF receptor[34,38] has made it possible to use immunoaffinity chromatography for receptor purification. Here we compare by peptide mapping EGF receptor protein purified by either immunoaffinity or EGF affinity chromatography[26] and also compare the structures of the A431 and placental receptors.

A radioimmunoassay (RIA) which uses a monoclonal antibody (R1) has been used to quantitate various preparative

§ Present address: Department of Biochemistry, University of Otago, Dunedin, New Zealand.

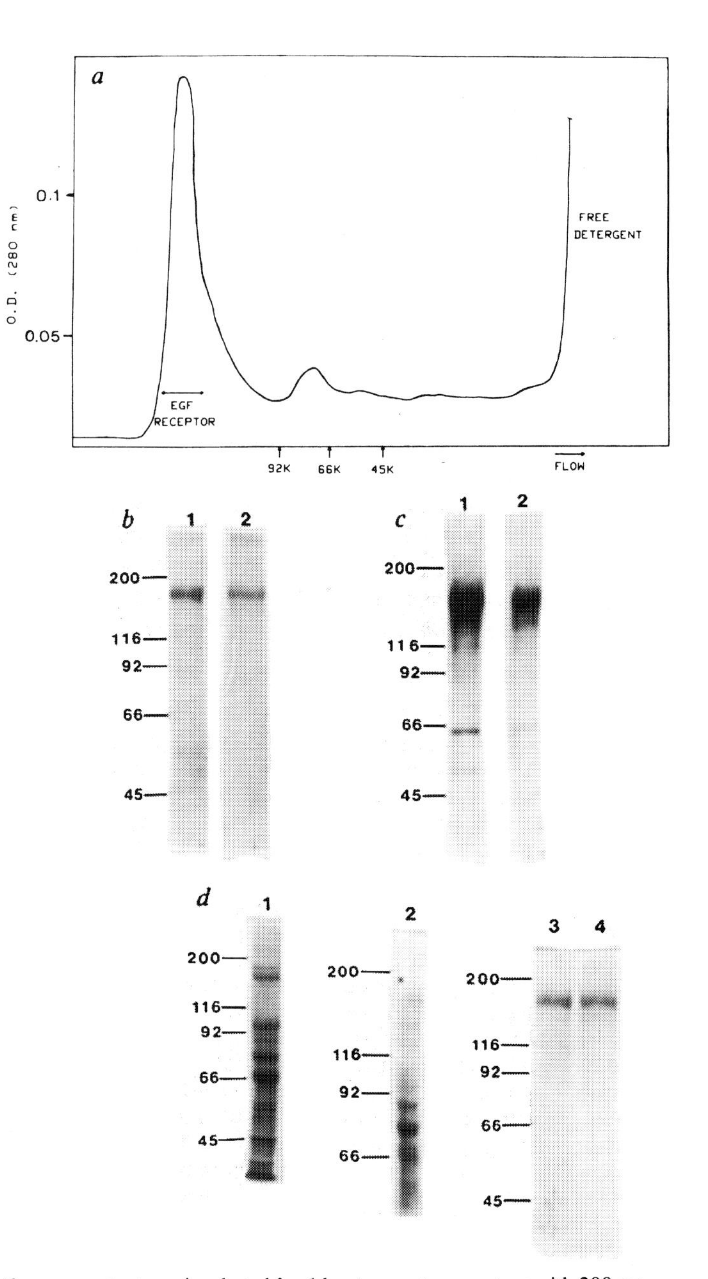

Fig. 1 Immunopurification of EGF receptor from A431 cells and human placenta. ***a***, Purification by gel permeation. Reduced and alkylated receptor was purified on a TSK4000 column (0.7 × 60 cm, LKB) using 0.1 M potassium dihydrogen phosphate buffer *p*H 4.5 containing 6 M guanidine HCl at a flow rate of 0.5 ml min^{-1} (ref. 43). The absorbance of the eluate was monitored at 280 nm (molecular weights of protein standards are indicated) and 0.25 ml fractions were collected. Fractions containing EGF receptor were dialysed against 10 mM ammonium bicarbonate. Panels ***b–d*** show 7% polyacrylamide SDS gels[72] used to monitor purification (MW × 10^{-3} of protein standards are indicated). *b*, R1 purified A431 EGF receptor: track 1, *p*H 3 eluate from R1 immunoaffinity matrix; track 2, eluate from the TSK4000 column. *c*, 29–1purified A431 EGF receptor: track 1, *p*H 3 eluate from 29–1 immunoaffinity matrix; track 2, eluate from SDS preparative gel electrophoresis. *d*, Lectin and R1 purified placental EGF receptor: track 1, placental vesicles; track 2, eluate from lectin affinity matrix; track 3, eluate from R1 immunoaffinity matrix; track 4, eluate from the TSK4000 column.

Methods: Approximately 2×10^9 A431 cells were washed in calcium and magnesium free phosphate-buffered saline (PBS) and solubilized in 400 ml lysis buffer (50 mM Tris-HCl *p*H 7.4, 0.15 M NaCl, 5 mM EGTA, 0.1% bovine serum albumin, 1% NP40, 25 mM benzamidine, 0.2 mM PMSF, 10 μg ml^{-1} leupeptin). After filtration through muslin the lysate was adjusted to *p*H 8.5 and centrifuged at $100{,}000g_{max}$ for 30 min. The supernatant was incubated for 2 h at 4 °C with immunoaffinity matrix, which consisted of 15 mg of monoclonal antibody R1[34] coupled to 15 ml of Affi-Gel 10 (BioRad). Unbound lysate was removed by suction through a 0.4 micron filter. The matrix was then washed by gentle agitation and filtration with 500 ml PBS, containing 0.65 M NaCl and 0.1% NP40, followed by 500 ml PBS, containing 0.1% NP40. The EGF receptor was eluted by gentle agitation and filtration of the matrix with 2 × 10 ml aliquots of 50 mM sodium citrate *p*H 3, containing 0.05% NP40 for 10 min each. Eluates were adjusted to *p*H 7. The yield of receptor was approximately 250 μg, measured by Bradford technique[70] or by amino acid analysis after gel permeation HPLC (see below). Alternatively EGF receptor was purified from A431 cells using monoclonal antibody 29-1[38] coupled to CNBr-activated Sepharose (Pharmacia) at 5–10 mg ml^{-1}. The purification procedure used was similar to that described for the R1 immunoaffinity matrix except that the EGF receptor was phosphorylated whilst bound to the matrix with 50 μCi [$\gamma-^{32}P$]ATP (3,000 Ci $mmol^{-1}$, Amersham International) in the presence of 3 mM $MnCl_2$. Eluted receptor was further purified by preparative SDS gel electrophoresis, followed by dialysis against 10% methanol at 4 °C. For purification of placental EGF receptor, vesicles were made from syncytiotrophoblast microvilli by a modification of the method of Smith *et al.*[71], using 2 mM EGTA in all buffers. Vesicles were solubilized by addition of an equal volume of 100 mM HEPES *p*H 7.4, 0.15 M NaCl and 5% Triton X-100. After centrifuging at $100{,}000g_{max}$ for 30 min the supernatant was incubated for 1 h at room temperature with 200 mg wheat germ agglutinin coupled to 10 g of Affi-Gel (BioRad). The lectin matrix was washed by filtration through a 0.4 μm filter with 100 ml of PBS, containing 0.1% Triton X-100. Bound protein was eluted by agitation and filtration with 2 × 15 ml aliquots of 0.25 M *N*-acetylglucosamine in 10 mM HEPES *p*H 7.4, 0.1% Triton X-100 for 15 min each. The eluate was incubated with R1 immuno-affinity matrix (15 mg antibody per placenta) for 2 h at 20 °C, followed by extensive washing and elution as described above for receptor purification from A431 cells. The yield of EGF receptor per placenta was 25 μg, measured by Bradford technique[70] and amino acid analysis after gel permeation HPLC (see below). Solutions containing EGF receptor were lyophilized and resuspended in 0.5 M Tris-HCl *p*H 8.5, 6 M guanidine hydrochloride (Schwarz-Mann) at 0.5–1 mg ml^{-1}. After incubation at 37 °C for 16 h with 10 mM dithiothreitol, cysteine residues were alkylated with ^{14}C-iodacetamide (40–60 mCi $mmol^{-1}$, Amersham International) as described previously[44].

techniques[42]. Receptors from A431 cells and placenta were both found to be unstable in detergent-solubilized whole cell or tissue lysates, perhaps as a result of the release of proteases from the cellular lysosomal compartment. This problem was overcome for placenta by the preparation of syncytiotrophoblast microvillus plasma membranes and as a result a 50-fold purification with a 30% yield of receptor was achieved. Unfortunately, with A431 cells the yield of receptor in plasma membrane preparations was impractically low. However, quantitative studies with the receptor RIA showed that rapid adjustment of the lysate *p*H to 8.5 followed by fast immunoaffinity chromatography of whole cell lysates minimised the effects of the proteases.

Placental membranes were solubilized and glycoproteins separated by wheat germ agglutinin (WGA) affinity chromatography to achieve a partial purification. EGF receptor was then purified from the placental glycoprotein fraction or from A431 cell lysates by immunoaffinity chromatography on either monoclonal antibodies R1[34] or 29-1[38] immobilised on agarose or Sepharose respectively. Nonspecifically bound protein was removed by washing the columns with a high salt buffer and the receptor was eluted at *p*H 3. The receptors were further purified either on preparative SDS-polyacrylamide gels or by gel permeation HPLC in guanidine solutions[43]. Details of the methods used and the yields of purified receptors are given in the legend to Fig. 1. Since the EGF binding and protein kinase activity were partially destroyed during purification, receptor was also purified by EGF affinity chromatography[26]. Comparative HPLC tryptic peptide maps were then carried out to establish the purity and structural similarity of receptor prepared by immunoaffinity chromatography from A431 cells and placental tissue. The peptide maps of the receptors (Fig. 2) showed that the elution profiles of the receptor tryptic peptides were very similar whether receptor was purified by EGF affinity or by immunoaffinity chromatography; from A431 cells or from placental tissue.

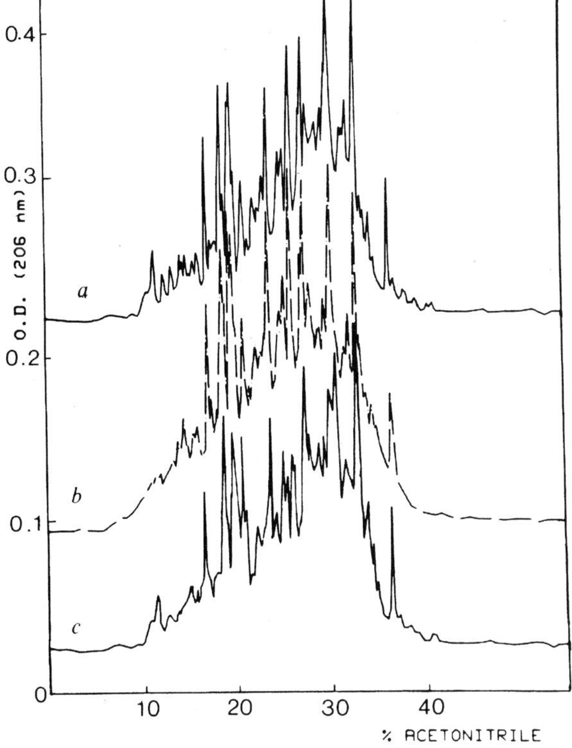

Fig. 2 Reverse phase HPLC analysis of tryptic peptides from EGF receptor purified by three different affinity methods. ***a***, Immunoaffinity purified A431 receptor. EGF receptor was purified from A431 cells using the monoclonal antibody R1 and gel permeation chromatography in guanidine solutions as described in Fig. 1. Pooled fractions containing the 175,000 MW EGF receptor were dialysed against 10 mM ammonium bicarbonate, lyophilized and resuspended in 500 μl of 100 mM ammonium bicarbonate, 10 mM $CaCl_2$. TPCK treated trypsin (Sigma) was then added (100:1, receptor:trypsin, w/w). This mixture was incubated at 37 °C for 12 h, a further identical aliquot of trypsin added and the incubation continued for another 12 h. Peptides were then loaded directly onto a Synchropak RPP C18 reverse phase HPLC column (Synchrom, Linden, Indiana, 4.6×75 mm) equilibrated in 0.1% trifluoroacetic acid (TFA, Rathburn, Scotland) and a gradient from zero to 45% acetonitrile was run over 45 min at 1 ml min^{-1} and 1 ml fractions collected[46]. A Waters HPLC system including two M6000 A pumps, a U6K manual injector and a 660 solvent programmer with 2 LKB 2138 Uvicord S absorbance detectors with filters at 206 nm and 280 nm was used[45]. The figure shows the optical density of the eluate at 206 nm plotted against acetonitrile concentration. ***b***, EGF affinity purified receptor. EGF receptor was purified from A431 cells using wheat germ agglutinin affinity chromatography followed by EGF affinity chromatography[26]. A431 cells were lysed as described in the legend to Fig. 1. Purification of this lysate on the WGA affinity column was identical to that described for the placental preparation (Fig. 1). The eluate from the WGA affinity column was mixed with 5 ml of Affi-Gel 10 (BioRad) having 1 mg of bound EGF. The mixture was tumbled for 4 h at room temperature before washing the immobilized EGF receptor with 100 ml of PBS, 0.1% Triton X-100. Receptor was then eluted with 10 mM ethanolamine, *p*H 9.7, 0.1% Triton X-100. This eluate was further purified on a TSK4000 gel permeation column as described in Fig. 1. Fractions containing EGF receptor were pooled, dialysed, lyophilized and trypsinized as described for *a* above. The resulting tryptic peptides were separated by reverse phase HPLC under identical conditions to those described above. ***c***, Immunoaffinity purified placental receptor. EGF receptor was purified from fresh term human placenta as described in the legend to Fig. 1. The receptor was digested with trypsin and peptides separated by reverse phase HPLC as described above. Peptides in the column breakthrough are not shown.

Amino acid sequence

Receptor was purified by immunoaffinity chromatography followed by either preparative SDS gel electrophoresis or by gel permeation HPLC in guanidine[43] after reduction and alkylation[44] to cleave disulphide bonds (see Fig. 1). Purified receptor was then digested with trypsin or cyanogen bromide (see Figs 2 and 3) and peptides were separated by preparative reverse phase HPLC[45,46] (Fig. 3). Amino acid sequences were determined with a gas phase sequencer constructed and operated as described by Hewick *et al.*[47], using the analytical techniques for the quantitation of phenyl thiohydantoin (PTH) amino acids described by M.D.W. *et al.* [48]. The quantitative data for analysis of six peptides are shown in Fig. 4.

The amino acid sequences of 14 different peptides from the human EGF receptor, three from placenta and 11 from A431 cells, were compared with sequences in an oncogene sequence data base (set up at ICRF using published sequences) by the rapid search techniques of Wilbur and Lipman[49]. A remarkable identity was found between the sequences of six of these peptides and regions of the predicted sequence of the putative transforming protein v-*erb-B* of the AEV-H isolate of avian erythroblastosis virus[30]. Of the 83 amino acid residues from these six sequenced peptides, 74 residues were identical and four showed conservative substitutions when they were aligned with the v-*erb-B* encoded protein sequence, as shown in Fig. 5. Peptide 1 was located near the amino terminus of the v-*erb-B* protein (residues 107–125) and peptide 6 at the carboxy terminus (residues 583–599), with the other four peptides in between. It was not necessary to introduce any deletions or insertions into the sequence to optimize the alignments.

Although the full extent of the similarity between the v-*erb-B* protein and EGF receptor sequences is not revealed by these limited sequence studies, it is likely that the region of the v-*erb-B* protein from residue 107 to the carboxy terminus has extensive homology to the EGF receptor. The degree of identity observed is very high and since the v-*erb-B* sequences of AEV were presumably of avian origin[30], while the EGF receptor sequences were from the human protein, it is likely that the v-*erb-B* sequences were mainly acquired by AEV from those cellular sequences which encode the avian EGF receptor. This suggests that the c-*erb-B* locus encodes the EGF receptor in humans and birds.

The amino acid sequence of 8 of the 14 peptides purified from the EGF receptor (data not shown) could not be aligned with the predicted sequences of the v-*erb-B* protein. Since the polypeptide backbone of the EGF receptor glycoprotein is thought to be about 1,250 amino acids[50] and the predicted v-*erb-B* protein is only 604 amino acids[30] the most likely explanation is that these eight peptides are encoded by a region of c-*erb-B* which has not been acquired by AEV. This could have arisen by a recombination event(s) which resulted in only a part of the EGF receptor coding sequences being acquired by AEV. Although it is possible that DNA rearrangements of receptor coding sequences occur similar to those found with immunoglobulins, it is more likely that differential mRNA splicing would be involved in any such recombination events. It has been shown that avian cells contain two c-*erb-B* related transcripts[51] and studies of the biosynthesis of the EGF receptor in A431 cells suggest that both a normal and a truncated receptor may be synthesized[50]. Alternatively two or more loci encoding polypeptides having very similar amino acid sequences to those of the EGF receptor exist on chromosome 7 (see below). An example of two closely related putative transforming proteins with tyrosine kinase activity has been reported in studies of the avian retroviruses Rous sarcoma virus (RSV) and Yamaguchi Y63[52]. The predicted amino acid sequences of the proteins encoded by the *src* and *yes* oncogenes were shown to have 82% match over a region covering 436 amino acid residues (while the DNA sequences showed only 31% overall match) and presumably the chicken genome contains both *src* and *yes* proto-oncogenes encoding separate proteins sharing extensive regions of sequence. It is not known whether human c-*src* and c-*yes*

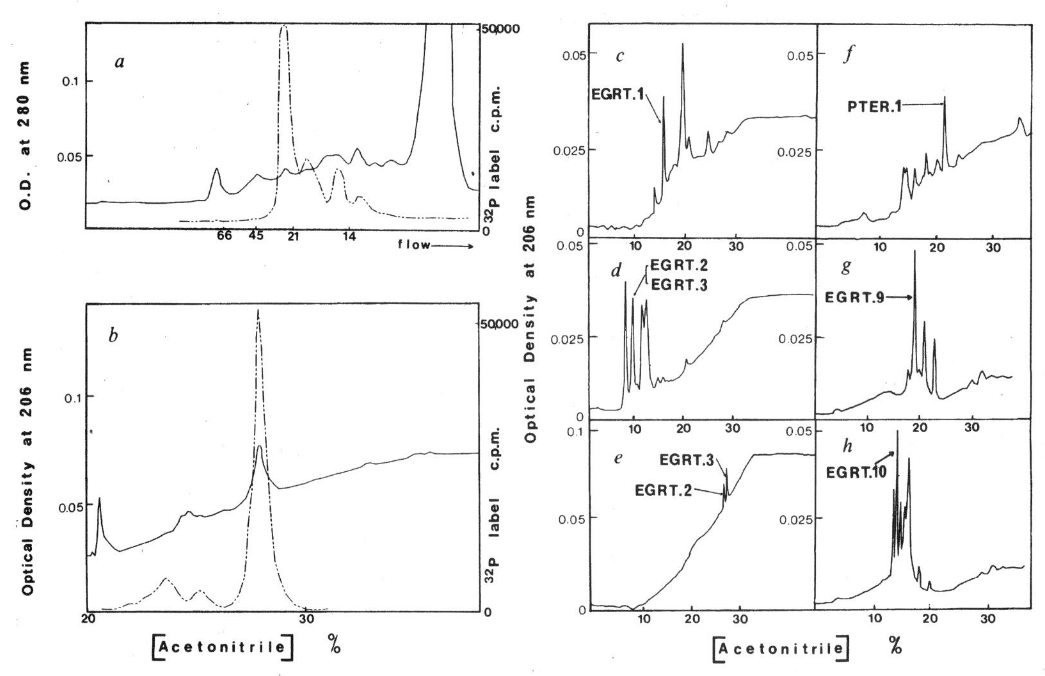

Fig. 3 Purification of peptides from EGF receptor for sequence analysis. ***a***, Cyanogen bromide cleavage and fractionation of peptides by size. ^{32}P-labelled EGF receptor in ammonium bicarbonate solution was lyophilized and resuspended in 70% formic acid. Cyanogen bromide was added under nitrogen, the tube sealed and incubated in the dark at room temperature for 24 h. Formic acid and excess cyanogen bromide were removed by repeated cycles of drying and resuspension in water using a Speed-vac concentrator (Savant). The dry sample was resuspended in 0.1 M potassium dihydrogen phosphate buffer, *p*H 4.5 containing 6 M guanidine HCl and the peptides separated by gel permeation HPLC on a TSK3000 column (0.7×60 cm, LKB) equilibrated in the same buffer at a flow rate of 0.3 ml min^{-1} [43]. The optical density of the eluate was monitored at 280 nm (---) and 0.3 ml fractions collected and counted for ^{32}P (-···-). Molecular weights ×10^{-3} of protein standards are indicated. ***b***, Subfractionation of cyanogen bromide fragments. The peak from the TSK3000 column containing most of the ^{32}P-label was pooled and dialysed against 10 mM ammonium bicarbonate. After lyophilization, the sample was redissolved in 0.1% TFA and peptides separated by reverse phase HPLC on a Synchropak RPP C18 column (see Fig. 2) equilibrated in 0.1% TFA, 10% acetonitrile[45,46]. A gradient of 10–40% acetonitrile run over 60 min was used to elute peptides, at a flow rate of 1 ml min^{-1}. The optical density of the eluate was monitored at 206 nm (--) and 1 ml fractions were collected and counted for ^{32}P-label (-···-). ***c***, The fractions corresponding to 23–24% acetonitrile from the HPLC analysis of A431 EGF receptor tryptic peptides were pooled. Peptides were further purified by reverse-phase HPLC on a Synchropak RPP C18 column equilibrated in 10 mM ammonium acetate buffer *p*H 6.5. A linear gradient of 0–45% acetonitrile was run over 45 min at a flow rate of 1 ml min^{-1}. The optical density of the eluate was monitored at 206 nm and 0.5 ml fractions collected. ***d***, The fractions corresponding to 19–20% acetonitrile from the reverse-phase HPLC purification of A431 EGF receptor tryptic peptides (Fig. 2*a*) were pooled. Peptides were separated as described in *c*. ***e***, The peak fractions arrowed in *d* were pooled and peptides subfractionated by reverse-phase HPLC on a μ Bondapak phenyl column (0.46×25 cm, Waters Assoc.) equilibrated in 0.1% TFA. A linear gradient of 0–45% acetonitrile over 45 min was used to elute peptides, at a flow rate of 1 ml min^{-1}. The optical density of the eluate was monitored at 206 nm and 0.2 ml fractions collected. ***f***, The fractions corresponding to 27–28% acetonitrile concentration from the reverse-phase HPLC analysis of placental EGF receptor tryptic peptides (Fig. 2*c*) were pooled. Peptides were subfractionated as described in *c*. ***g***, The fractions corresponding to 25–26% acetonitrile from the reverse-phase HPLC purification of A431 EGF receptor tryptic peptides (Fig. 2*a*) were pooled. Peptides were subfractionated as described in *c*. ***h***, The fractions corresponding to 21–22% acetonitrile from the reverse-phase HPLC analysis of A431 EGF receptor tryptic peptides (Fig. 2*a*) were pooled. Peptides were subfractionated as described in *c*.

are encoded by closely linked loci. However, analysis of human-mouse somatic cell hybrids has shown that the locus encoding the human EGF receptor is on chromosome 7 (7p13-7q22)[53–55] and that for c-*erb-B* is in the same region of this chromosome (7pter–7q22)[56]. This observation supports the concept that the c-*erb-B* locus encodes the human EGF receptor or is very closely linked to that encoding the EGF receptor.

Analysis of EGF receptor mRNA transcripts found in A431 and avian cells by cDNA cloning is currently in progress which, together with analysis of human and avian genomic clones, should define the relationship of c-*erb-B* and the locus encoding the EGF receptor (A.U. *et al.*, in preparation).

Shared regions of sequence

Several lines of evidence suggest that the EGF receptor protein can be divided into three major domains; an EGF binding domain which lies external to the plasma membrane, a transmembrane domain and a cytoplasmic kinase domain having both the kinase activity and the autophosphorylation sites.

Investigations of receptor biosynthesis show that the A431 receptor is a glycoprotein of apparent molecular weight (MW) 175,000, having ~37,000 MW of oligosaccharide side chains with a polypeptide backbone of ~138,000 MW (~1,250 amino acids). Limited proteolysis of the mature receptor suggests that the domain external to the plasma membrane which contains the oligosaccharide side chains and the antigenic sites for monoclonal R1 has a MW of ~115,000 (~710 amino acids)[50]. Several studies show that the EGF binding site is external to the plasma membrane[25,31,32].

The location of the tyrosine kinase enzymatic activity and the autophosphorylation sites on the cytoplasmic domain is supported by studies made using A431 and placental membrane vesicles (J.D. and M.D.W., in preparation). These show that EGF stimulated tyrosine kinase activity directed towards artificial substrates or towards autophosphorylation sites is significantly activated only after membrane permeabilization. Furthermore, the tyrosine kinase activity can phosphorylate pp36—a protein known to be located at the cytoplasmic side of the membrane[57]. In addition recent studies show that antibodies raised against synthetic peptides from pp60$^{v\text{-}src}$ recognize antigenic sites on the human EGF receptor that are from regions of sequence homologous to the sequence of v-*erb-B* (see below). These sites are only accessible in permeabilized cells (J.S. *et al.*, in preparation).

Autophosphorylation sites are located within peptide 5 (EGRC.1), a 20,000 MW cyanogen bromide fragment which contains 70% of the ^{32}P-label present in the autophosphorylated receptor (see Fig. 3*a*). Although the precise location of the residues phosphorylated has not been determined a consensus tyrosine phosphorylation sequence[58] was found near the amino

Downward et al. (1984) Nature (London) **307**, 521–527

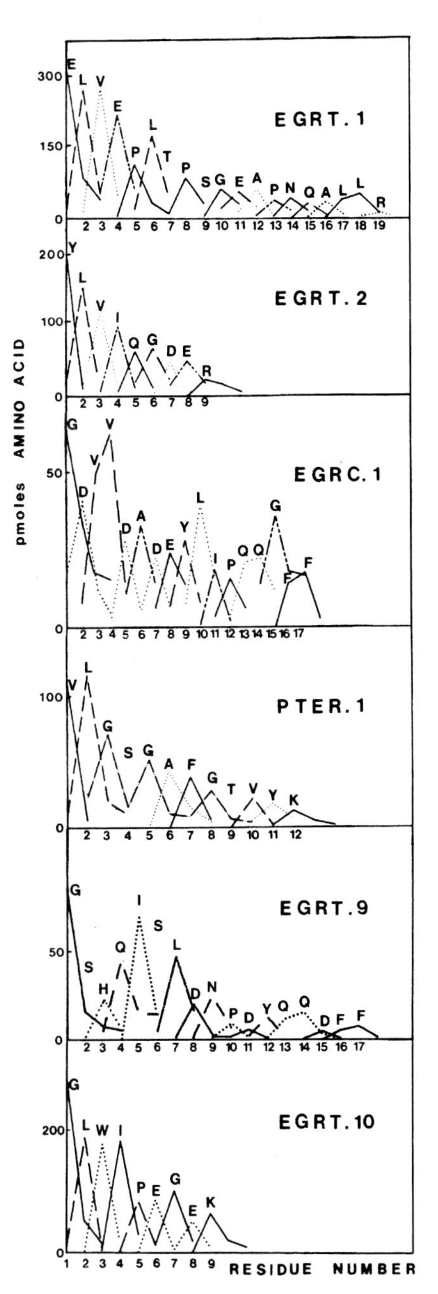

Fig. 4 Sequence analysis of peptides from the EGF receptor. Peptides were purified as described in Fig. 3. Sequence determination of each peptide was carried out using a gas phase sequencer assembled and operated as described by Hewick *et al.*[47]. PTH amino acids were analysed by HPLC using a Zorbax C8 column (4.6 × 150 mm, Dupont) at 43 °C with a linear gradient over 8 min of acetonitrile from 24% to 38% at a flow rate of 2 ml min^{-1} using 9 mM sodium acetate buffer *p*H 4.1[48]. A Waters HPLC system including two M6000 A pumps, a WISP autoinjector and system controller with a Beckman Model 160 detector was used. The recovery of PTH amino acids at each degradative cycle was measured using an integrative recorder (Waters Data module). The amounts of each peptide *analysed* (pmol) were measured by amino acid analysis (EGRT.1, 380; EGRT.2, 260; EGRC.1, 80; PTER.1, 145; EGRT.9, 100; EGRT.10, 325). The analysis for serine and threonine could not be accurately measured due to the presence of multiple peaks obtained during analysis of the PTH amino acids. The presence of these amino acids is thus indicated without quantitative data; these residues are assigned to the sequence using semi-quantitative recovery data based on peak heights rather than areas. Prior to loading peptides, fibre glass disks were treated with polybrene and glycylglycine and precycled for ten cycles. Each peptide was sequenced twice; on the second run of peptide EGRC.1 filters were treated with polybrene and cysteic acid and precycled ten times to clarify the assignment of an amino terminal glycine residue. However the background glycine at step 1 is still significant and this residue may be incorrect.

terminus of this peptide. Therefore we believe that the tyrosine phosphorylation sites lie within the cytoplasmic domain of the EGF receptor which is contained in the sequence shared with the v-*erb-B* protein.

Preliminary nucleotide sequence analysis (A.U. *et al.*, in preparation) of cDNA clones selected from a placental cDNA library using synthetic oligonucleotide probes synthesized on the basis of the receptor amino acid sequence shows that the predicted carboxyl terminus of the EGF receptor extends 32 amino acids from an amino acid equivalent to residue 601 (see Fig. 5) of the predicted v-*erb-B* protein sequence. This analysis when complete will show the precise size and sequence of the presumptive cytoplasmic domain of the EGF receptor. The approximate molecular weight of this domain would be 60,000 (or 545 amino acids) since that part which is external to the membrane is thought to have a molecular weight of 115,000 (see above)[50]. Thus the cytoplasmic domain would be predicted to be similar in size to that region of the v-*erb-B* protein which is carboxy terminal to a putative transmembrane sequence (see Fig. 5 and ref. 30) at residues 66–88. This carboxy terminal region of v-*erb-B* would have a molecular weight of 56,760 and would contain 516 amino acids.

The putative transmembrane sequence of the v-*erb-B* protein is not preceded by a signal sequence for membrane insertion. Nevertheless, immunofluorescence studies of AEV transformed cells show that the v-*erb-B* protein has antigenic sites external to the plasma membrane[59]. This external region probably corresponds to the 65 residue amino terminal section that precedes the putative transmembrane sequence and contains three asparagine residues which have the oligosaccharide attachment recognition sequence Asn-X-Ser or Thr. Some or all of these residues may be glycosylated since *in vitro* translation studies of mRNA from AEV infected cells show that post translational processing of nascent polypeptides occurs in the presence of membrane vesicles[59,60].

Together these studies suggest that the predicted v-*erb-B* transforming protein closely resembles the transmembrane region of the EGF receptor and the domain which is thought to be cytoplasmic. If the v-*erb-B* sequence was acquired from the gene encoding the EGF receptor then the v-*erb-B* protein represents a truncated receptor which lacks the EGF binding domain. It is particularly interesting that studies of EGF receptor biosynthesis in A431 cells have suggested that a polypeptide equivalent to the external domain of the receptor (of MW 115,000) is synthesized[50] in addition to the normal receptor. Further studies are necessary to understand the origin of this truncated receptor but the results show that defective receptors may be synthesized by this human tumour cell line.

Growth factors and transformation

Recently it has been shown that the transforming protein of simian sarcoma virus has a close structural and functional relationship to the growth factor PDGF[7–10] supporting the hypothesis that autocrine growth factor production may be involved in abnormal growth control and neoplasia. These observations together with those presented here illustrate two distinct but related mechanisms for subversion of normal growth regulation. In the case of SSV the oncogene encodes a growth factor which can act as a mitogen for target cells having PDGF receptors[10]. AEV on the other hand appears to utilize a different mechanism where a part of a growth factor receptor which is thought to be involved in transducing the EGF signal may be expressed in transformed cells. The absence of the EGF binding domain might remove the control generated by ligand binding and the result could be the continuous generation of a signal equivalent to that produced by EGF, causing cells to proliferate rapidly. How this could result in the block in differentiation observed in AEV infected haemopoietic cells[61] is unclear. However, EGF has been shown to promote proliferation while inhibiting terminal differentiation of human keratinocytes[31].

The ES4 strain of AEV has two oncogenes v-*erb-A* and v-*erb-B*, which are thought to encode proteins of MWs 75,000

```
Src      1  MGSSKSKPKDPSQRRHSLEPPDSTHHGGFPASQTPDETAAPDAHRNPSRS

Src     51  FGTVATEPKLFWGFNTSDTVTSPQRAGALAGGVTTFVALYDYESWTETDL

Src    101  SFKKGERLQIVNNTEGDWWLAHSLTTGQTGYIPSNYVAPSDSIQAEEWYF
                                                MKCAHFIDGPHCVKA

Src    151  GKITRRESERLLLNPENPRGTFLVRKSETAKGAYCLSVSDFDNAKGPNVK
Erb-B   16  CPAGVLGENDTLVRKYADANAVCQLCHPNCTRGCKGPGLEGCPNGSKTPS
                    ▲                    ▲               ▲

Src    201  HYKIYKLYSGGFYITSRTQFGSLQQLVAYYSKHADGLCHRLANVCPTSKP
Erb-B   66  IAAGVVGGLLCLVVVGLGIGLYLRRRHIVRKRTLRRLLQERELVEPLT-P
            ▬▬▬▬▬▬▬▬▬▬▬▬▬▬▬▬▬▬▬▬▬                    *ELVEPLT-P
                                                          1

Src    251  QTQGLAKDAWEIPRESLRLEAK-LGQGCFGEVWMGTWN--D---TTRVAI
Erb-B  115  SGEAPNQAHLRILKETEFKKVKVLGSGAFGTIYKGLWIPEGEKVKIPVAI
            SGEAPNQALLR          *VLGSGAFGTVYKGLWIPEGEK
                 1                      2        3

Src    295  KTLKPGTMSPEA---FLQEAQVMKKLRHEKLVQLYAVVSEEPIYIVIEYM
Erb-B  165  KELREAT-SPKANKEILDEAYVMASVDNPHVCRLLGICLTSTVQLITQLM

Src    342  SKGSLLDFLKGEMGKYLRLPQ-LVDMAAQIASGMAYVE-RMNYVHRDLRA
Erb-B  214  PYGCLLDY-IRE-HKDNIGSQYLLNWCVQIAKGMNYLEERR-LVHRDL-A

Src    390  A-NILVGENLVCKVADFGLARLIE-D-NEYTARQGAKFPIKWTAPEAALY
Erb-B  260  ARNVLVKTPQHVKITDFGLAKLLGADEKEYHAEGG-KVPIKWMALESILH
                                          †

Src    437  GRFTIKSDVWSFGILLTELTTKGRVPYPGMVNREVLDQVERGYRMPCPPE
Erb-B  309  RIYTHQSDVWSYGVTVWELMTFGSKPYDGIPASEISSVLEKGERLPQPPI

Src    487  CPESLHDLMCQCWRKDPEERPTFKYLQAQLLPACVLEVAE
Erb-B  359  CTIDVYMIMVKCWMIDADSRPKFRELIAEFSKMARDPPRYLVIQGDERMH
                                                  *YLVIQGDER
                                                       4

Erb-B  409  LPSPTDSKFYRTLMEEEDMEDIVDADEYLVPHQGFFNSPSTSRTPLLSSL
                                *GDVVDADEYLIPQQGFF
                                         5

Erb-B  459  SATSNNSATNCIDRNGQGHPVREDSFVQRYSSDPTGNFLEESIDDGFLPA

Erb-B  509  PEYVNQLMPKKPSTAMVQNQIYNFISLTAISKLPMDSRYQNSHSTAVDNP

Erb-B  559  EYLNTNQSPLAKTVFESSPYWIQSGNHQINLDNPDYQQDFLPTSCS
                                    GSHQISLDNPDYQQDFF
                                            6
```

Fig. 5 The relationship between the amino acid sequences of the EGF receptor peptides and the predicted amino acid sequences of the putative transforming proteins of v-*src* and v-*erb-B*. The predicted amino acid sequence of the v-*src* gene product (pp60$^{\text{v-src}}$) is translated from the presumptive initiation codon at nucleotide 7,129 of the Prague C strain of Rous sarcoma virus[73]. The predicted amino acid sequence of the v-*erb-B* gene product is translated from the presumptive initiation codon at nucleotide 155 of the v-*erb-B* gene in AEV-H[30]. The partial amino acid sequences of the six peptides purified from the EGF receptor are shown (underlined): 1, EGRT.1; 2, PTER.1; 3, EGRT.10; 4, EGRT.2; 5, EGRC.1; 6, EGRT.9. Letters in bold type represent residues shared by pp60$^{\text{v-src}}$ and the v-*erb-B* protein. Residues shared by the EGF receptor peptides and v-*erb-B* protein or pp60$^{\text{v-src}}$ are in bold type. ● indicates amino acid residues, which are common to the putative transforming proteins of v-*erb-B*, v-*src*, v-*fes*, v-*fps*, v-*yes* and v-*abl*. †, Phosphoacceptor tyrosine of pp60$^{\text{v-src}}$[74]. ▲, Possible *N*-linked glycosylation sites at the amino terminus of the v-*erb-B* protein. ▬ indicates the putative transmembrane sequence in the v-*erb-B* protein. *, Amino acid residues which would be expected to produce enzymatic or cyanogen bromide cleavages to generate the observed peptides. Numbers to the left of the sequences are residue numbers taking the presumptive initiation methionine as 1 in both cases. Sequences were aligned using a computer program[75] to optimize match.

and 65,000 respectively (for a recent review see ref. 62). Cells transformed by AEV *in vitro* and *in vivo* have the properties of erythroblasts which are late erythroid progenitors, although the target cells themselves may be earlier erythroid precursor cells. AEV can also transform fibroblasts and induce sarcomas. Evidence from deletion mutants[63] and from an isolate (AEV-H[30]) which lacks the v-*erb-A* gene suggests that the v-*erb-B* gene alone can induce transformation. This is supported by studies which show that RAV-1, a leucosis virus which has no oncogene, can activate the c-*erb-B* gene by a promoter insertion mechanism[64] perhaps similar to that presently being unravelled for c-*myc* activation[65–67]. It is possible that RAV-1 could induce expression of a normal receptor or a truncated receptor. EGF receptors have not generally been detected on haemopoietic cells by EGF binding studies but since these studies are limited in scope and sensitivity a more rigorous survey is needed before conclusions about normal receptor expression in different haemopoietic cell types can be made. Although many normal cells express 10–100,000 EGF receptors[31] only very low levels of c-*erb-B* transcripts have been found in normal chicken fibroblasts[51]. However a recent study of normal and neoplastic human lymphocytes suggests that both types of cells contain c-*erb-B* related transcripts. Clearly more detailed studies will be necessary to search for normal or altered c-*erb-B* or v-*erb-B* transcripts in normal, neoplastic or AEV transformed cells.

Previous reports have shown that the predicted amino acid sequences of the putative viral transforming proteins encoded by the oncogenes *erb-B*, *src*, *yes*, *fes*, *fps*, *mos* and *abl* show regions of similarity, implying homology[29,30,69]. In the case of *src*, *yes*, *fes*, *fps* and *abl* the putative transforming proteins have been shown to have tyrosine kinase activity (reviewed in ref. 29) but as yet those encoded by *erb-B*[30] and *mos* have not. Since the receptors for EGF and αTGF, PDGF, insulin and IGF-I also have associated tyrosine kinases, the structural relationship between the v-*erb-B* transforming protein and the EGF receptor observed here, suggests that other oncogenes from this subset of retroviruses could be derived in part from sequences encoding these or other growth factor receptors. Alternatively the other oncogenes in this group may be derived from sequences which encode proteins having a less direct functional relationship to the EGF and other receptors. Further study will clarify the precise function of growth factors, their receptors and *src* related oncogenes in normal and malignant growth.

M.D.W. thanks B. Ozanne and E. Rozengurt for introducing him to the field of study involving growth factors and their receptors. We acknowledge P. Parker, P. Stroobant, H. S. Earp and M. Fried for helpful discussions and advice, W. Gullick for help with receptor quantitation techniques and P. Bennett and R. Philp for technical assistance. We also thank W. J. Wilbur and D. J. Lipman for computer programs and N. Teich and the ICRF Computer Unit for their help. M.D.W. thanks S. J. for continued support. We are indebted to Queen Charlotte's Hospital and to E. J. Owen for help in providing placental tissue. This work was supported in part by grants from the USNIH (CA 25820) (J.S.) and from the US–Israel Binational Science Foundation (J.S.).

Received 18 January; accepted 23 January 1984.

1. Guroff, G. (ed.) *Growth and Maturation Factors* (Wiley, New York, 1983).
2. Cohen, S. *J. biol. Chem.* **237,** 1555–1562 (1962).
3. Ross, R. in *Tissue Growth Factors* (ed. Baserga, R.) 133–159 (Springer, Berlin, 1981).
4. Rozengurt, E. *Molec. Biol. Med.* **1,** 169–181 (1983).
5. James, R. & Bradshaw, R. A. *A. Rev. Biochem.* (in the press).
6. Sporn, M. B., & Todaro, G. J. *New Engl. J. Med.* **303,** 878–880, (1980).
7. Waterfield, M. D. *et al. Nature* **304,** 35–39 (1983).
8. Doolittle, R. F. *et al. Science* **221,** 275–277 (1983).
9. Robbins, K. C., Antoniades, H. N., Devare, S. G., Hunkapiller, M. W. & Aaronson, S. A. *Nature* **305,** 605–608 (1983).
10. Deuel, T. F., Huang, J. S., Huang, S. S., Stroobant, P. & Waterfield, M. D. *Science* **221,** 1348–1350 (1983).
11. Dulak, N. C. & Temin, H. M. *J. cell. Physiol.* **81,** 161–170 (1973).
12. De Larco, J. E. & Todaro, G. J. *Nature* **272,** 356–358 (1978).
13. Burk, R. R. *Expl Cell Res.* **101,** 193–298 (1976).
14. Bourne, H. & Rozengurt, E. *Proc. natn. Acad. Sci. U.S.A.* **73,** 4555–4559 (1976).
15. De Larco, J. E. & Todaro, G. J. *Proc. natn. Acad. Sci. U.S.A.* **75,** 4001–4005 (1978).
16. Roberts, A. B., Frolik, C. A., Anzano, M. A. & Sporn, M. B. *Fedn Proc.* **42,** 2621–2626 (1983).
17. Ozanne, B., Fulton, R. J. & Kaplan, P. L. *J. cell. Physiol.* **105,** 163–180 (1980).
18. Kaplan, P. L., Anderson, M. & Ozanne, B. *Proc. natn. Acad. Sci. U.S.A.* **79,** 485–489 (1982).
19. Kaplan, P. L. & Ozanne, B. *Cell* **33,** 931–938 (1983).
20. Marquardt, H. *et al. Proc. natn. Acad. Sci. U.S.A.* **80,** 4684–4688 (1983).
21. Bowen-Pope, D. F., Di Corletto, P. E. & Ross, R. *J. Cell Biol.* **96,** 679–683 (1983).
22. Collins, M. K. L., Sinnett-Smith, J. W. & Rozengurt, E. *J. biol. Chem.* **258,** 11689–11693 (1983).
23. Marquardt, H. & Todaro, G. J. *J. biol. Chem.* **257,** 5220–5225 (1982).
24. Carpenter, G., Stoscheck, C. M., Preston, Y. A. & De Larco, J. E. *Proc. natn. Acad. Sci. U.S.A.* **80,** 5627–5630 (1983).
25. Cohen, S., Ushiro, H., Stoscheck, C. & Chinkers, M. *J. biol. Chem.* **257,** 1523–1531 (1982).
26. Buhrow, S. A., Cohen, S. & Staros, J. V. *J. biol. Chem.* **257,** 4019–4022 (1982).
27. Cohen, S., Fava, R. A. & Sawyer, S. T. *Proc. natn. Acad. Sci. U.S.A.* **79,** 6237–6241 (1982).
28. Buhrow, S. A., Cohen, S., Garbers, D. L. & Staros, J. V. *J. biol. Chem.* **258,** 7824–7827 (1983).
29. Bishop, J. M., *Rev. Biochem.* **52,** 301–354 (1983).
30. Yamamoto, T., Nishida, T., Miyajima, N., Kawai, S., Ooi, T. & Toyoshima, K. *Cell* **35,** 71–78 (1983).
31. Adamson, E. D. & Rees, A. R. *Molec. cell. Biochem.* **34,** 129–152 (1981).
32. Linsley, P. S., Das, M. & Fox, C. F. in *Membrane Receptors* Vol. B11 (eds Jacobs, S. & Cuatrecasas, P.) 87–113 (Chapman & Hall, London, 1981).
33. Schreiber, A. B., Lax, I., Yarden, Y., Eshhar, Z. & Schlessinger, J. *Proc. natn. Acad. Sci. U.S.A.* **78,** 7535–7539 (1981).
34. Waterfield, M. D. *et al. J. Cell Biochem.* **20,** 149–161 (1982).

35. Kawamoto, T. *et al. Proc. natn. Acad. Sci. U.S.A.* **80,** 1337-1341 (1983).
36. Richert, N. D., Willingham, M. C. & Pastan, I. *J. biol. Chem.* **258,** 8902–8907 (1983).
37. Gregoriou, M. & Rees, A. R. *Cell Biol. Int. Rep.* **7,** 539-540 (1983).
38. Schlessinger, J., Lax, I., Yarden, Y., Kanety, H. & Libermann, T. A. in *Receptors and Recognition: Antibodies against Receptors* (ed. Greaves, M. F.) (Chapman and Hall, London, in the press).
39. Fabricant, R. N., De Larco, J. E. & Todaro, G. J. *Proc. natn. Acad. Sci. U.S.A.* **74,** 565–569 (1977).
40. Wrann, M. M. & Fox, C. F. *J. biol. Chem.* **254,** 8083–8086 (1979).
41. O'Keefe, E., Hollenberg, M. D. & Cuatrecasas, P. *Archs biochem. Biophys.* **164,** 518–526 (1974).
42. Gullick, W., Downward, D. J. H., Marsden, J. J. & Waterfield, M. D. *Analyt. Biochem.* (in the press).
43. Ui, N. *Analyt. Biochem.* **97,** 65–71 (1979).
44. Skehel, J. J. & Waterfield, M. D. *Proc. natn. Acad. Sci. U.S.A.* **72,** 93–97 (1975).
45. Waterfield, M. D. & Scrace, G. T. in *Biological/Biomedical Applications of Liquid Chromatography* Vol. 18 (ed. Hawk, G. L.) 135–157 (Dekker, New York, 1981).
46. Bennett, H. P. J., Browne, C. A. & Solomon, S. J. *Liquid Chromat.* **3,** 1353–1365 (1980).
47. Hewick, R. M., Hunkapiller, M. W., Hood, L. E. & Dreyer, W. J. *J. biol. Chem.* **256,** 7990–7997 (1981).
48. Waterfield, M. D., Scrace, G. & Totty, N. in *Practical Protein Biochemistry* (eds Darbre, A. & Waterfield, M. D.) (Wiley, New York, in the press).
49. Wilbur, W. J. & Lipman, D. J. *Proc. natn. Acad. Sci. U.S.A.* **80,** 726–730 (1983).
50. Mayes, E. L. V. & Waterfield, M. D. *EMBO J.* (in the press).
51. Vennström, B. & Bishop, J. M. *Cell* **28,** 135–143 (1983).
52. Kitamura, N., Kitamura, A., Toyoshima, K., Hirayama, Y. & Yoshida, M. *Nature* **297,** 205–208 (1982).
53. Goodfellow, P. N., Banting, G., Waterfield, M. D., Ozanne, B. *Cytogenet. cell. Genet.* **32,** 282 (1982).
54. Shimizu, N., Behzadian, M. A. & Shimizu, Y. *Proc. natn. Acad. Sci. U.S.A.* **77,** 3600–3604 (1980).
55. Kondo, I. & Shimizu, N. *Cytogenet, cell. Genet.* **35,** 9–14 (1983).
56. Spurr, N. *et al. EMBO J.* **3,** 159–163 (1984).
57. Greenberg, M. E. & Edelman, G. M. *Cell* **33,** 767–779 (1983).
58. Groffen, J., Heisterkamp, N., Reynolds, F. H. Jr. & Stephenson, J. R. *Nature* **304,** 167–169 (1983).
59. Hayman, M. J., Ramsay, G. M., Savin, K., Kitchener, G., Graf, T. & Beug, H. *Cell* **32,** 579–588 (1983).
60. Privalsky, M. L., Sealy, L., Bishop, J. M., McGrath, J. P. & Levinson, A. D. *Cell* **32,** 1257–1267 (1983).
61. Beug, H., Palmieri, S., Freudenstein, C., Zentgraf, H. & Graf, T. *Cell* **28,** 907–919 (1982).
62. Graf, T. & Beug, H. *Cell* **34,** 7–9 (1983).
63. Frykberg, L. *et al. Cell* **32,** 227–238 (1983).
64. Fung, Y-K. T., Lewis, W. G., Crittenden, L. B. & Kung, H-J. *Cell* **33,** 357–368 (1983).
65. Rabbitts, T. H., Forster, A., Baer, R. & Hamlyn, P. H. *Nature* **306,** 806–809 (1983).
66. Rabbitts, T. H., Hamlyn, P. H. & Baer, R. *Nature* **306,** 760–765 (1983).
67. Gelmann, E. P., Psallidopoulos, M. C., Papas, T. S. & Favera, R. D. *Nature* **306,** 799–803 (1983).
68. Roy-Burman, P., Devi, B. G. & Parker, J. W. *Int. J. Cancer* **32,** 185–191 (1983).
69. Reddy, E. P., Smith, M. J. & Srinivasan, A. *Proc. natn. Acad. Sci. U.S.A.* **80,** 3623–3627 (1983).
70. Bradford, M. M. *Analyt. Biochem.* **72,** 248–254 (1976).
71. Smith, C. M. *et al. Am. J. Obstet. Gynec.* **128,** 190–196 (1977).
72. Laemmli, U. K. *Nature* **227,** 680–685 (1970).
73. Schwartz, D., Tizzard, R. & Gilbert, W. *Cell* **32,** 853–869 (1983).
74. Smart, J. E. *et al. Proc. natn. Acad. Sci. U.S.A.* **78,** 6013–6017 (1981).
75. Orr, H. T., Lancet, D., Robb, R., Lopez de Castro, J. A. & Strominger, J. L. *Nature* **282,** 266–270 (1979).

Control of transcription by nuclear receptors

19

Many aspects of development and cellular homeostasis are regulated by steroid and thyroid hormones. These hormones are able to diffuse directly across the plasma membrane into the cell where they interact with specific intracellular receptors. The occupied receptors are translocated to the nucleus where they bind to high-affinity sites in chromatin and activate transcription of a specific subset of genes in a tissue-specific fashion leading to changes in cell function. The chosen paper [1] described the cloning of cDNA and determination of the first protein sequence of such a receptor, that for glucocorticoids. The analysis of the structure of the steroid and thyroid hormone receptors has provided important information on the molecular mechanisms by which they control gene transcription [2].

The physiological importance of steroids, retinoids and thyroid hormones has been known for a considerable time from the symptoms of their deficiency. During the 1970s biochemical characterization of the cytosolic receptors began using radiolabelled ligands, and this allowed purification of the receptors and generation of anti-receptor antibodies [3,4]. Through the 1970s and early 1980s the specific DNA sequences, or hormone-response elements (HREs), sensitive to each class of receptors were determined, and transfer of these HREs to DNA constructs showed that they are able to confer hormone inducibility [5,6]. The HREs act as position-independent enhancer elements [6] that activate transcription of genes under the control of tissue-specific promoters.

The availability of antibodies that recognized the purified glucocorticoid receptor was exploited for the selection of cDNA clones encoding the human receptor [7,8]. Hollenberg et al. [1] used such antibodies to select further cDNA clones, and then used the isolated clones for subsequent screening of cDNA libraries. This allowed determination of the full coding sequences of the α and β forms of the glucocorticoid receptor which differed at the C-terminus. The α form was found to be the major cellular form and the recombinant α-receptor protein was shown to be capable of binding to a radiolabelled synthetic glucocorticoid, ^{3}H-triamcinolone acetonide, thus confirming that the cDNA clone did indeed encode a functional glucocorticoid receptor.

The cloning of the glucocorticoid receptor was significant, since it provided the first complete information on the protein sequence of a steroid hormone receptor, which subsequently allowed its relatedness to other nuclear receptors to be determined, and the functional domains of these receptor proteins to be characterized. In addition, a significant sequence similarity between the human glucocorticoid receptor and the viral oncogene product Erb-A [9] was revealed. The Erb-A sequence was found to have 22% identity with the C-terminal 387 amino acids of the glucocorticoid receptor. From a proposed domain structure for the glucocorticoid receptor it was suggested that the Erb-A oncogene product contained DNA- and hormone-binding domains and

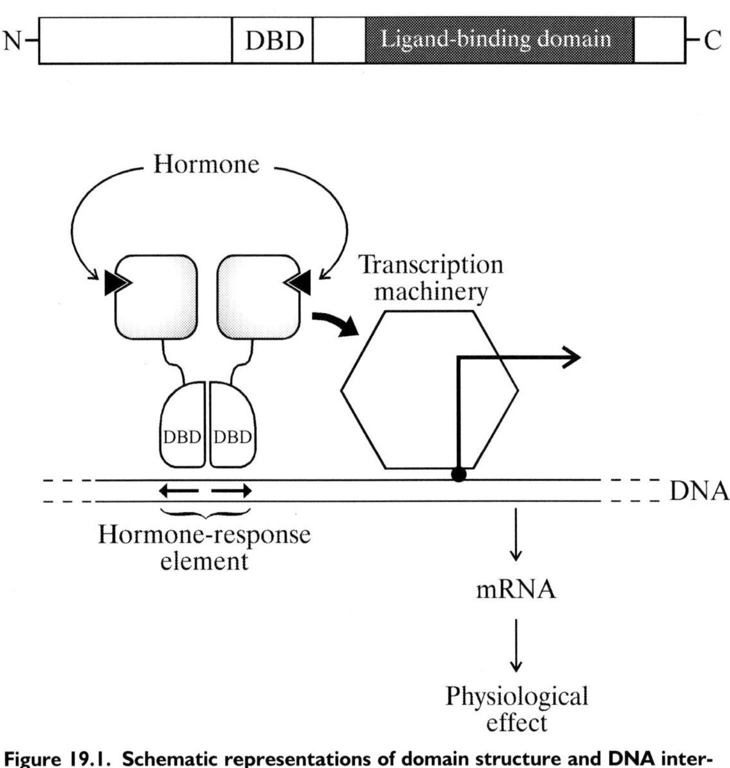

Figure 19.1. Schematic representations of domain structure and DNA interaction of members of the steroid hormone family. The top of the Figure shows the general organization of these proteins with a variable N-terminal domain, a central DNA-binding domain (DBD) and a C-terminal ligand-binding domain. The lower part of the Figure shows the interaction of a steroid receptor homodimer with the hormone-response element of DNA.

could represent a truncated steroid receptor. (It is intriguing that the two oncogenes of avian erythroblastosis, *erb-A* and *erb-B*, both encode truncated proteins related to mammalian proteins involved in hormone signalling. (For further discussion of *erb-B* see Section 18.) Subsequent work demonstrated that the cellular full-length counterpart of Erb-A is, in fact, the thyroid hormone receptor [10,11].

Following searches for related gene products, a number of additional receptors have been identified that form part of a superfamily of nuclear/steroid hormone receptors, including those for oestrogen, progesterone, aldosterone, vitamin D, the vitamin A metabolite retinoic acid, the insect moulting hormone ecdysone and other regulatory molecules. By 1995 over 150 members of this family of receptors in various species were known. All of these receptors have a central highly homologous DNA-binding domain, a C-terminal hormone-binding domain that is less similar, and a variable N-terminal domain (Figure 19.1). The DNA-binding domain is similar to that of other transcription factors, such as TFIIIA [12], in that it has two so-called zinc fingers containing cysteine and histidine residues that can co-ordinate zinc and are essential for DNA binding. The steroid receptors bind to DNA as homodimers, while the non-steroid receptors (e.g. for retinoic acid and thyroid hormones) bind as heterodimers. Analysis of the three-dimensional structure of receptors bound to DNA is now providing a detailed picture of the way in which these receptors interact with DNA [13,14] to regulate its transcription. In addition, structures of several ligand-binding domains, with or without ligand, have recently been solved [15].

Recent studies on the nuclear/steroid receptor family have begun to move away from analysis of their molecular action in the nucleus, which is understood (at least in principle) [15], and have begun to focus on the examination of their physiological and developmental roles. This new wave of functional analysis has been made possible by use of the powerful technology of targeted gene disruption and the generation of transgenic animals defective in receptor expression. These studies are giving surprising new insights into the function of the steroid hormone receptors and, particularly the novel non-steroid receptors [16,17]. This approach could only be undertaken following the logical developments from the pioneering work on the cloning of the first steroid hormone receptor in 1985.

References

1. Hollenberg, S.M., Weinberger, C., Ong, E.S., Cerelli, G., Oro, A., Lebo, R., Thompson, E.B., Rosenfeld, M.G. and Evans, R.M. (1985) *Nature (London)* **318**, 635–641
2. Evans, R.M. (1988) *Science* **240**, 889–898
3. Yamamoto, K. (1985) *Annu. Rev. Genet.* **19**, 209–215
4. Ringold, G.M. (1985) *Annu. Rev. Pharmacol. Toxicol.* **25**, 529–566
5. Beato, M. (1989) *Cell* **56**, 335–344
6. Chandler, V.L., Maler, B.A. and Yamamoto, K.R. (1983) *Cell* **33**, 489–499
7. Weinberger, C., Hollenberg, S.M., Ong, E.S., Hamon, J.M., Brower, S.T., Cidlowski, J., Thompson, E.B., Rosenfeld, M.G. and Evans, R.M. (1985) *Science* **228**, 740–742
8. Miesfeld, R. Okret, S., Wikstrom, A.C., Wrange, O., Gustafsson, J.A. and Yamamoto, K.R. (1984) *Nature (London)* **312**, 779–781
9. Weinberger, C., Hollenberg, S.M., Rosenfeld, M.G. and Evans, R.M. (1985) *Nature (London)* **318**, 670–672
10. Sap, J., Murioz, A., Damm, K., Golberg, Y., Ghysdael, J., Leutz, A., Beug, H. and Vennstrom, B. (1986) *Nature (London)* **324**, 635–640
11. Weinberger, C., Thompson, C.C., Ong, E.S., Lebo, R., Gruol, D.G. and Evans, R.M. (1986) *Nature (London)* **324**, 641–646
12. Miller, J., McLachlan, A. and Klug, A. (1985) *EMBO J.* **4**, 1609–1614
13. Luisa, B.F., Xu, W.X., Otwinowski, Z., Freedman, L.P., Yamamoto, K.R. and Sigler, P.B. (1991) *Nature (London)* **352**, 497–505
14. Rastinejad, F., Perlmann, T., Evans, R.M. and Sigler, P.B. (1995) *Nature (London)* **375**, 203–211
15. Mangelsdorf, D.J.,Thummel, C., Beato, M., Herrlich, P., Schutz, G., Umesone, K., Blumberg, B., Kaster, P., Mark, M., Chambon, P. and Evans, R.M. (1995) *Cell* **83**, 835–839
16. Cole, T.J. Blendy, J.A., Monaghan, A.P., Krieglstein, K., Schmid, W., Agguzzi, A., Fantuzzi, G., Hummler, E., Unsicker, K. and Schultz, G. (1995) *Genes Dev.* **9**, 1608–1621
17. Kastner, P., Mark, M. and Chambon, P. (1995) *Cell* **83**, 859–869

Hollenberg et al. (1985) Nature (London) 318, 635–641

Primary structure and expression of a functional human glucocorticoid receptor cDNA

Stanley M. Hollenberg*†, Cary Weinberger*, Estelita S. Ong*, Gail Cerelli*, Anthony Oro*, Roger Lebo‡, E. Brad Thompson§, Michael G. Rosenfeld‖ & Ronald M. Evans*

* Howard Hughes Medical Institute, Gene Expression Laboratory, The Salk Institute, 10010 North Torrey Pines Road, La Jolla, California 92037, USA
† Department of Biology, ‖ Howard Hughes Medical Institute, Eukaryotic Regulatory Biology Program, School of Medicine, University of California, San Diego, California 92093, USA
‡ Howard Hughes Medical Institute, Department of Medicine, University of California, San Francisco, California 94143, USA
§ University of Texas Medical School, Galveston, Texas 77550, USA

Identification of complementary DNAs encoding the human glucocorticoid receptor predicts two protein forms, of 777 (α) and 742 (β) amino acids, which differ at their carboxy termini. The proteins contain a cysteine/lysine/arginine-rich region which may define the DNA-binding domain. Pure radiolabelled glucocorticoid receptor, synthesized in vitro, *is immunoreactive and possesses intrinsic steroid-binding activity characteristic of the native glucocorticoid receptor.*

TRANSCRIPTIONAL regulation of development and homeostasis is controlled in complex eukaryotes by a wide variety of regulatory substances, including steroid hormones. The latter exert potent effects on development and differentiation in phylogenetically diverse organisms and their actions are mediated as a consequence of their interactions with specific, high-affinity binding proteins referred to as receptors[1-6]. Many of the primary effects of steroid hormones involve increased transcription of a subset of genes in specific cell types[7,8]. The structural characterization of the steroid receptor protein and the definition of the molecular actions of the steroid hormone/receptor complex would help to define the biochemical events which modulate gene transcription. A series of receptor proteins, each specific for one of several known steroid hormones, are distributed in a tissue-specific fashion, although many cell types simultaneously express receptors for several steroid hormones[9,10]. It is postulated that steroids enter cells by facilitated diffusion and bind to the specific receptor protein, initiating an allosteric alteration of the complex. As a result of this transformation, the steroid hormone/receptor complex appears capable of binding high-affinity sites on chromatin and modulating transcription of specific genes by a mechanistically unknown process[11,12].

The glucocorticoid receptor is widely distributed and expressed in many cultured cell lines, and the control of gene expression by glucocorticoids, therefore, has been widely studied as a model for transcriptional regulation. A number of glucocorticoid-responsive transcription units, including mouse mammary tumour virus (MMTV)[13,14], mouse and human metal-

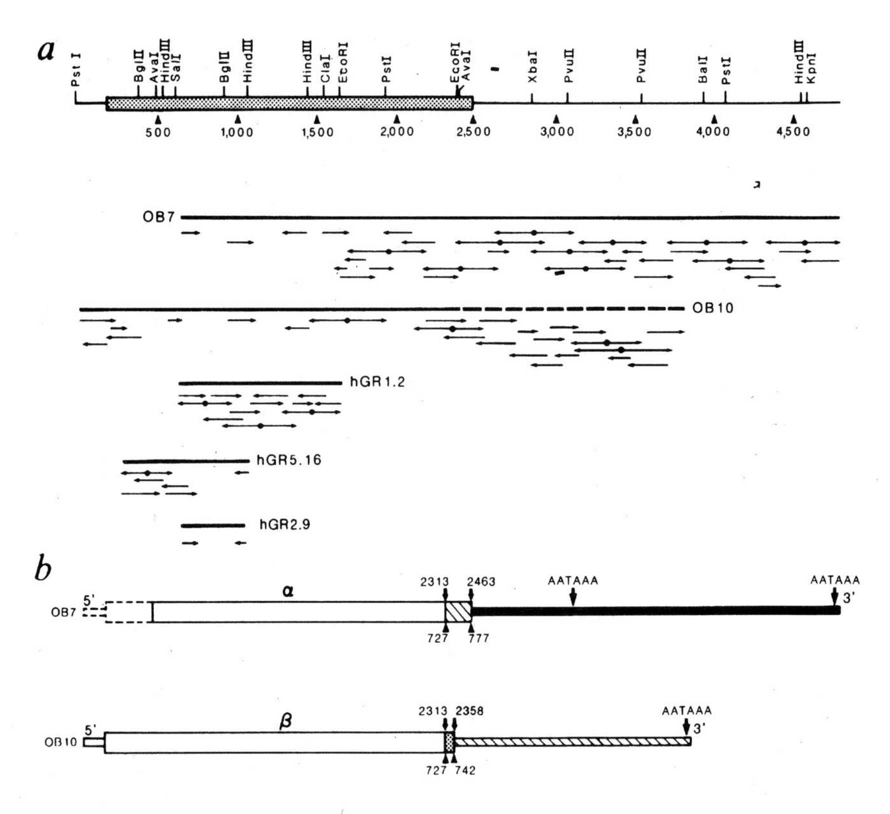

Fig. 1 Human glucocorticoid receptor cDNA sequencing strategy and schematic representation of cDNA clones. *a*, The composite cDNA for the α glucocorticoid receptor is represented at the top, with noncoding (lines) and coding (stippled portion) sequences indicated. Common 6-nucleotide restriction enzyme sites are shown. Overlapping cDNA inserts used to determine the sequence are shown: arrows beneath the regions sequenced show the direction and extent of sequencing. The dashed line at the 3′ end of OB10 indicates divergent sequence. Numbers refer to nucleotide positions in OB10 relative to the 5′-most transcribed sequence. *b*, cDNAs encoding the α and β forms of the receptor (OB7 and OB10, respectively). The 5′ end of OB7 (broken lines) is contributed by the OB10 clone. Protein-coding information is represented by wide bars; untranslated sequences are indicated by thin bars. Nucleotides and amino acids are numbered above and below the coding sequence, respectively. Common DNA sequences extend to nucleotide 2,313 (amino-acid residue 727), at which point the α- and β-receptor forms diverge, with the α cDNAs (OB12, OB7) continuing in an open reading frame for 150 nucleotides (50 amino acids) and the β cDNA (OB10) continuing for 45 nucleotides (15 amino acids; see Fig. 3). Hexanucleotide signals (AATAAA) just upstream of the poly(A) in the clones are indicated, with the first hexanucleotide in OB7 serving as poly(A) in OB12.

Methods. The inserts hGR1.2, hGR2.9 and hGR5.16 were isolated from a λgt11 IM-9 lymphoid cell cDNA library as described previously[42]. Two clones were isolated from cDNA libraries constructed by H. Okayama in pcD (ref. 44) using poly(A)$^+$ mRNA from GM637 human fibroblasts (OB7) and primary human fibroblasts (OB10). Screening was performed with the hGR1.2 cDNA, radiolabelled by nick-translation with ^{32}P-dCTP. Sequences were determined by the chemical cleavage method of Maxam and Gilbert[45].

lothionein[15,16], rat α_{2u}-globulin[17] and rat and human growth hormone[18–20] genes have been identified. DNA sequences mediating transcriptional stimulation of several of these genes have been localized. For MMTV, these sequences are discrete genomic regions upstream of the transcriptional start site which appear to exert their actions independently of orientation and position[21,22]. The steroid/receptor complex appears to bind to these regulatory sequences and purified receptor has been used to define the specific binding sites[23–26]. Based on the footprinting analyses of several responsive genes, a consensus DNA binding sequence sharing the core sequence 5′ TGT/CTCT 3′ has been proposed[27].

The ability of the glucocorticoid-responsive element (GRE) to alter its position and orientation yet still maintain promoter inducibility suggests that it resembles the class of *cis*-acting regulatory sequences termed enhancers[21]. First discovered in viruses and subsequently in cellular genes, these sequences are necessary for efficient transcription *in vivo*[28–31]. It has been suggested that enhancers are recognized by *trans*-acting factors that mediate regulatory effects by tissue-specific transcriptional control. Although the enhancer factors have not been well characterized, the glucocorticoid receptor may serve as a paradigm for these putative gene activator proteins.

The availability of radiolabelled high-affinity glucocorticoid analogues such as dexamethasone and triamcinolone acetonide has led to the development of purification strategies resulting in the isolation of nearly pure rat and human receptors[32,33]. Although the receptor migrates as a dimer in sucrose gradients, analysis on denaturing SDS-polyacrylamide gels detects a single polypeptide of relative molecular mass (M_r) ~94,000 (94K)[34,35]. The native polypeptide contains intrinsic specificity for steroid binding and DNA sequence recognition. By using as probes monoclonal and polyclonal antibodies raised against the purified rat and human receptors[36–38], it has been possible to identify a major immunogenic region in the receptor residing on a portion of the molecule that is distinct from the steroid- and DNA-binding regions[39–41]. To gain further information about the structure of this molecule and to begin an analysis of the molecular mechanisms by which it regulates gene transcription, we set out to clone receptor cDNA sequences. By using receptor-specific antibodies as probes, we and others have isolated clones containing human or rat glucocorticoid receptor cDNA inserts[42,43].

Here we report the complete amino-acid sequence of the human glucocorticoid receptor (hGR), deduced from human lymphoid and fibroblast cDNA clones. The sequence reveals various structural features of the receptor, including the major immunogenic domain and a cysteine/arginine/lysine-rich region which may constitute a portion of the DNA-binding domain. We describe the use of the SP6 transcription vector system to generate analytical amounts of full-length protein, and demonstrate that the cell-free translated protein is both immunoreactive and possesses steroid-binding properties characteristic of the native glucocorticoid receptor. An accompanying paper describes the homology of the hGR sequence to that of the oncogene *v-erb-A*[64].

Glucocorticoid receptor cDNA

A library of cDNA clones was constructed in the phage expression vector λgt11 using poly(A)$^+$ RNA from the human lymphoid cell line IM-9 as template, as described previously[42]. This library was initially screened with a rabbit polyclonal antiserum to the purified glucocorticoid receptor, resulting in the isolation of several immunopositive candidate clones from $\sim 2.5 \times 10^5$ plaques. The β-galactosidase fusion proteins generated from these clones were used to affinity-purify receptor epitope-specific antibody, which was subsequently recovered and identified by binding to protein blots of cellular extracts. Three clones containing inserts expressing antigenic determinants of the human glucocorticoid receptor were isolated. The inserts of these clones, although of different sizes, cross-hybridized, indicating that they contained a common sequence which presumably delimits the major immunogenic domain of the receptor. Together, these clones spanned 1.4 kilobase pairs (kbp) but were clearly not long enough to code for the entire

```
TTTTTAGAAAAAAAAAATATATTTCCCTCCTGCTCCTTCTGCGTTCACAAGCTAAGTTGTTTATCTCGGCTGCGGCGGGAACTGCGGACGGTGGCGGGCGAGCGGCTCCTCTGCCAGAGT  120

                      10                                      20                                      30
   MetAspSerLysGluSerLeuThrProGlyArgGluGluAsnProSerSerValLeuAlaGlnGluArgGlyAspValMetAspPheTyrLysThrLeuArgGlyGly
TGATATTCACTGATGGACTCCAAAGAATCATTAACTCCTGGTAGAGAAGAAAACCCCAGCAGTGTGCTTGCTCAGGAGAGGGGAGATGTGATGGACTTCTATAAAACCCTAAGAGGAGGA  240

   40                                      50                                      60                                      70
AlaThrValLysValSerAlaSerSerProSerLeuAlaValAlaSerGlnSerAspSerLysGlnArgArgLeuLeuValAspPheProLysGlySerValSerAsnAlaGlnGlnPro
GCTACTGTGAAGGTTTCTGCGTCTTCACCCTCACTGGCTGTCGCTTCTCAATCAGACTCCAAGCAGCGAAGACTTTTGGTTGATTTTCCAAAAGGCTCAGTAAGCAATGCGCAGCAGCCA  360

   80                                      90                                      100                                     110
AspLeuSerLysAlaValSerLeuSerMetGlyLeuTyrMetGlyGluThrGluThrLysValMetGlyAsnAspLeuGlyPheProGlnGlnGlyGlnIleSerLeuSerSerGlyGlu
GATCTGTCCAAAGCAGTTTCACTCTCAATGGGACTGTATATGGGAGAGACAGAAACAAAAGTGATGGGAAATGACCTGGGATTCCCACAGCAGGGCCAAATCAGCCTTTCCTCGGGGGAA  480

   120                                     130                                     140                                     150
ThrAspLeuLysLeuLeuGluGluSerIleAlaAsnLeuAsnArgSerThrSerValProGluAsnProLysSerSerAlaSerThrAlaValSerAlaAlaProThrGluLysGluPhe
ACAGACTTAAAGCTTTTGGAAGAAAGCATTGCAAACCTCAATAGGTCGACCAGTGTTCCAGAGAACCCCAAGAGTTCAGCATCCACTGCTGTGTCTGCTGCCCCCACAGAGAAGGAGTTT  600

   160                                     170                                     180                                     190
ProLysThrHisSerAspValSerSerGluGlnGlnHisLeuLysGlyGlnThrGlyThrAsnGlyGlyAsnValLysLeuTyrThrThrAspGlnSerThrPheAspIleLeuGlnAsp
CCAAAAACTCACTCTGATGTATCTTCAGAACAGCAACATTTGAAGGGCCAGACTGGCACCAACGGTGGCAATGTGAAATTGTATACCACAGACCAAAGCACCTTTGACATTTTGCAGGAT  720

   200                                     210                                     220                                     230
LeuGluPheSerSerGlySerProGlyLysGluThrAsnGluSerProTrpArgSerAspLeuLeuIleAspGluAsnCysLeuLeuSerProLeuAlaGlyGluAspAspSerPheLeu
TTGGAGTTTTCTTCTGGGTCCCCAGGTAAAGAGACGAATGAGAGTCCTTGGAGATCAGACCTGTTGATAGATGAAAACTGTTTGCTTTCTCCTCTGGCGGGAGAAGACGATTCATTCCTT  840

   240                                     250                                     260                                     270
LeuGluGlyAsnSerAsnGluAspCysLysProLeuIleLeuProAspThrLysProLysIleLysAspAsnGlyAspLeuValLeuSerSerProSerAsnValThrLeuProGlnVal
TTGGAAGGAAACTCGAATGAGGACTGCAAGCCTCTCATTTTACCGGACACTAAACCCAAAATTAAGGATAATGGAGATCTGGTTTTGTCAAGCCCCAGTAATGTAACACTGCCCCAAGTG  960

   280                                     290                                     300                                     310
LysThrGluLysGluAspPheIleGluLeuCysThrProGlyValIleLysGlnGluLysLeuGlyThrValTyrCysGlnAlaSerPheProGlyAlaAsnIleIleGlyAsnLysMet
AAAACAGAAAAAGAAGATTTCATCGAACTCTGCACCCCTGGGGTAATTAAGCAAGAGAAACTGGGCACAGTTTACTGTCAGGCAAGCTTTCCTGGAGCAAATATAATTGGTAATAAAATG  1,080

   320                                     330                                     340                                     350
SerAlaIleSerValHisGlyValSerThrSerGlyGlyGlnMetTyrHisTyrAspMetAsnThrAlaSerLeuSerGlnGlnGlnAspGlnLysProIlePheAsnValIleProPro
TCTGCCATTTCTGTTCATGGTGTGAGTACCTCTGGAGGACAGATGTACCACTATGACATGAATACAGCATCCCTTTCTCAACAGCAGGATCAGAAGCCTATTTTTAATGTCATTCCACCA  1,200

   360                                     370                                     380                                     390
IleProValGlySerGluAsnTrpAsnArgCysGlnGlySerGlyAspAspAsnLeuThrSerLeuGlyThrLeuAsnPheProGlyArgThrValPheSerAsnGlyTyrSerSerPro
ATTCCCGTTGGTTCCGAAAATTGGAATAGGTGCCAAGGATCTGGAGATGACAACTTGACTTCTCTGGGGACTCTGAACTTCCCTGGTCGAACAGTTTTTTCTAATGGCTATTCAAGCCCC  1,320

   400                                     410                                     420                                     430
SerMetArgProAspValSerSerProProSerSerSerSerThrAlaThrThrGlyProProProLysLeuCysLeuValCysSerAspGluAlaSerGlyCysHisTyrGlyValLeu
AGCATGAGACCAGATGTAAGCTCTCCTCCATCCAGCTCCTCAACAGCAACAACAGGACCACCTCCCAAACTCTGCCTGGTGTGCTCTGATGAAGCTTCAGGATGTCATTATGGAGTCTTA  1,440

   440                                     450                                     460                                     470
ThrCysGlySerCysLysValPhePheLysArgAlaValGluGlyGlnHisAsnTyrLeuCysAlaGlyArgAsnAspCysIleIleAspLysIleArgArgLysAsnCysProAlaCys
ACTTGTGGAAGCTGTAAAGTTTTCTTCAAAAGAGCAGTGGAAGGACAGCACAATTACCTATGTGCTGGAAGGAATGATTGCATCATCGATAAAATTCGAAGAAAAAACTGCCCAGCATGC  1,560

   480                                     490                                     500                                     510
ArgTyrArgLysCysLeuGlnAlaGlyMetAsnLeuGluAlaArgLysThrLysLysLysIleLysGlyIleGlnGlnAlaThrThrGlyValSerGlnGluThrSerGluAsnProGly
CGCTATCGAAAATGTCTTCAGGCTGGAATGAACCTGGAAGCTCGAAAAACAAAGAAAAAAATAAAAGGAATTCAGCAGGCCACTACAGGAGTCTCACAAGAAACCTCTGAAAATCCTGGT  1,680

   520                                     530                                     540                                     550
AsnLysThrIleValProAlaThrLeuProGlnLeuThrProThrLeuValSerLeuLeuGluValIleGluProGluValLeuTyrAlaGlyTyrAspSerSerValProAspSerThr
AACAAAACAATAGTTCCTGCAACGTTACCACAACTCACCCCTACCCTGGTGTCACTGTTGGAGGTTATTGAACCTGAAGTGTTATATGCAGGATATGATAGCTCTGTTCCAGACTCAACT  1,800

   560                                     570                                     580                                     590
TrpArgIleMetThrThrLeuAsnMetLeuGlyGlyArgGlnValIleAlaAlaValLysTrpAlaLysAlaIleProGlyPheArgAsnLeuHisLeuAspAspGlnMetThrLeuLeu
TGGAGGATCATGACTACGCTCAACATGTTAGGAGGGCGGCAAGTGATTGCAGCAGTGAAATGGGCAAAGGCAATACCAGGTTTCAGGAACTTACACCTGGATGACCAAATGACCCTACTG  1,920

   600                                     610                                     620                                     630
GlnTyrSerTrpMetPheLeuMetAlaPheAlaLeuGlyTrpArgSerTyrArgGlnSerSerAlaAsnLeuLeuCysPheAlaProAspLeuIleIleAsnGluGlnArgMetThrLeu
CAGTACTCCTGGATGTTTCTTATGGCATTTGCTCTGGGGTGGAGATCATATAGACAATCAAGTGCAAACCTGCTGTGTTTTGCTCCTGATCTGATTATTAATGAGCAGAGAATGACTCTA  2,040

   640                                     650                                     660                                     670
ProCysMetTyrAspGlnCysLysHisMetLeuTyrValSerSerGluLeuHisArgLeuGlnValSerTyrGluGluTyrLeuCysMetLysThrLeuLeuLeuLeuSerSerValPro
CCCTGCATGTACGACCAATGTAAACACATGCTGTATGTTTCCTCTGAGTTACACAGGCTTCAGGTATCTTATGAAGAGTATCTCTGTATGAAAACCTTACTGCTTCTCTCTTCAGTTCCT  2,160

   680                                     690                                     700                                     710
LysAspGlyLeuLysSerGlnGluLeuPheAspGluIleArgMetThrTyrIleLysGluLeuGlyLysAlaIleValLysArgGluGlyAsnSerSerGlnAsnTrpGlnArgPheTyr
AAGGACGGTCTGAAGAGCCAAGAGCTATTTGATGAAATTAGAATGACCTACATCAAAGAGCTAGGAAAAGCCATTGTCAAGAGGGAAGGAAACTCCAGCCAGAACTGGCAGCGGTTTTAT  2,280

   720                                     730                                     740                                     750
GlnLeuThrLysLeuLeuAspSerMetHisGluValValGluAsnLeuLeuAsnTyrCysPheGlnThrPheLeuAspLysThrMetSerIleGluPheProGluMetLeuAlaGluIle
CAACTGACAAAACTCTTGGATTCTATGCATGAAGTGGTTGAAAATCTCCTTAACTATTGCTTCCAAACATTTTTGGATAAGACCATGAGTATTGAATTCCCCGAGATGTTAGCTGAAATC  2,400

   760                                     770
IleThrAsnGlnIleProLysTyrSerAsnGlyAsnIleLysLysLeuLeuPheHisGlnLysSTOP
ATCACCAATCAGATACCAAAATATTCAAATGGAAATATCAAAAAACTTCTGTTTCATCAAAAGTGACTGCCTTAATAAGAATGGTTGCCTTAAAGAAAGTCGAATTAATAGCTTTTATTG  2,520

TATAAACTATCAGTTTGTCCTGTAGAGGTTTTGTTGTTTTATTTTTTATTGTTTTCATCTGTTGTTTTGTTTTAAATACGCACTACATGTGGTTTATAGAGGGCCAAGACTTGGCAACAG  2,640
AAGCAGTTGAGTCGTCATCACTTTTCAGTGATGGGAGAGTAGATGGTGAAATTTATTAGTTAATATATCCCAGAAATTAGAAACCTTAATATGTGGACGTAATCTCCACAGTCAAAGAAG
GATGGCACCTAAACCACCAGTGCCCAAAGTCTGTGTGATGAACTTTCTCTTCATACTTTTTTTCACAGTTGGCTGGATGAAATTTTCTAGACTTTCTGTTGGTGTATCCCCCCCCTGTAT  2,880
AGTTAGGATAGCATTTTTGATTTATGCATGGAAACCTGAAAAAAAAGTTTACAAGTGTATATCAGAAAAGGGAAGTTGTGCCTTTTATAGCTATTACTGTCTGGTTTTAACAATTTCCTTT
ATATTTAGTAGAACTACGCTTGCTCATTTTTTCTTACATAATATTTTTATTCAAGTTATTGTACAGCTGTTTAAGATGGGCAGCTAGTTCGTAGCTTTCCCAAATAAACTCTAAACATTAAT  3,120
CAATCATCTGTGTGAAAATGGGTTGGTGCTTCTAACCTGATGGCACTTAGCTATCAGAAGACCACAAAAATTGACTCAAATCTCCAGTATTCTTGTCAAAAAAAAAAAAAAAAAAAGCTCA
TATTTTGTATATATCTGCTTCAGTGGAGAATTATATAGGTTGTGCAAATTAACAGTCCTAACTGGTATAGAGCACCTAGTCCAGTGACCTGCTGGGTAAACTGTGGATGATGGTTGCAAA  3,360
AGACTAATTTAAAAAATAACTACCAAGAGGCCCTGTCTGTACCTAACGCCCTATTTTTGCAATGGCTATATGGCAAGAAAGCTGGTAAACTATTTGTCTTTCAGGACCTTTTGAAGTAGT
TTGTATAACTTCTTAAAAGTTGTGATTCCAGATAACCAGCTGTAACACAGCTGAGAGACTTTTAATCAGACAAAGTAATTCCTCTCACTAAACTTTACCCAAAAACTAAATCTCTAATAT  3,600
GGCAAAAATGGCTAGACACCCATTTTCACATTCCCATCTGTCACCAATTGGTTAATCTTTCCTGATGGTACAGGAAAGCTCAGCTACTGATTTTTGTGATTTAGAACTGTATGTCAGACA
TCCATGTTTGTAAAACTACACATCCCTAATGTGTGCCATAGAGTTTAACACAAGTCCTGTGAATTTCTTCACTGTTGAAAATTATTTTAAACAAAATAGAAGCTGTAGTAGCCCTTTCTG  3,840
TGTGCACCTTACCAACTTTCTGTAAACTCAAAACTTAACATATTTACTAAGCCACAAGAAATTTGATTTCTATTCAAGGTGGCCAAATTATTTGTGTAATAGAAAACTGAAAATCTAATA
TTAAAAATATGGAACTTCTAATATATTTTTATATTTAGTTATAGTTTCAGATATATATCATATTGGTATTCACTAATCTGGGAAGGGAAGGGCTACTGCAGCTTTACATGCAATTTATTA  4,080
AAATGATTGTAAAATAGCTTGTATAGTGTAAAATAAGAATGATTTTTAGATGAGATTGTTTTATCATGACATGTTATATATTTTTTGTAGGGGTCAAAGAAATGCTGATGGATAACCTAT
ATGATTTATAGTTTGTACATGCATTCATACAGGCAGCGATGGTCTCAGAAACCAAACAGTTTGCTCTAGGGGAAGAGGGAGATGGAGACTGGTCCTGTGTGCAGTGAAGGTTGCTGAGGC  4,320
TCTGACCCAGTGAGATTACAGAGGAAGTTATCCTCTGCCTCCCATTCTGACCACCCTTCTCATTCCAACAGTGAGTCTGTCAGCGCAGGTTTAGTTTACTCAATCTCCCCTTGCACTAAA
GTATGTAAAGTATGTAAACAGGAGACAGGAAGGTGGTGCTTACATCCTTAAAGGCACCATCTAATAGCGGGTTACTTTCACATACAGCCCTCCCCCAGCAGTTGAATGACAACAGAAGCT  4,560
TCAGAAGTTTGGCAATAGTTTGCATAGAGGTACCAGCAATATGTAAATAGTGCAGAATCTCATAGGTTGCCAATAATACACTAATTCCTTTCTATCCTACAACAAGAGTTTATTTCCAAA
TAAAATGAGGACATGTTTTTGTTTTCTTTGAATGCTTTTTGAATGTTATTTGTTATTTTCAGTATTTTGGAGAAATTATTTAATAAAAAAACAATCATTTGCTTTTTGAAAAAAAAAAAA  4,800
```

Fig. 2 cDNA and predicted protein sequence of human glucocorticoid receptor. The complete α coding sequence and OB7 3′-untranslated region are shown, with the deduced amino acids given above the long open reading frame. An upstream in-frame stop codon at nucleotides 121–123 and putative additional polyadenylation signals in OB7 are underlined.

receptor, which was estimated to require ~2,500 nucleotides to encode a polypeptide of M_r 94K.

To isolate additional cDNA clones we again screened the original library and also examined a second library (given by H. Okayama) prepared with poly(A)$^+$ RNA from human fibroblasts in the vector described by Okayama and Berg[44]. Using one of the immunopositive cDNA inserts (hGR1.2) as probe, 12 clones were isolated that, together, covered more than 4.0 kbp. The nucleotide sequences of these clones were determined by the procedure of Maxam and Gilbert[45] according to the strategy

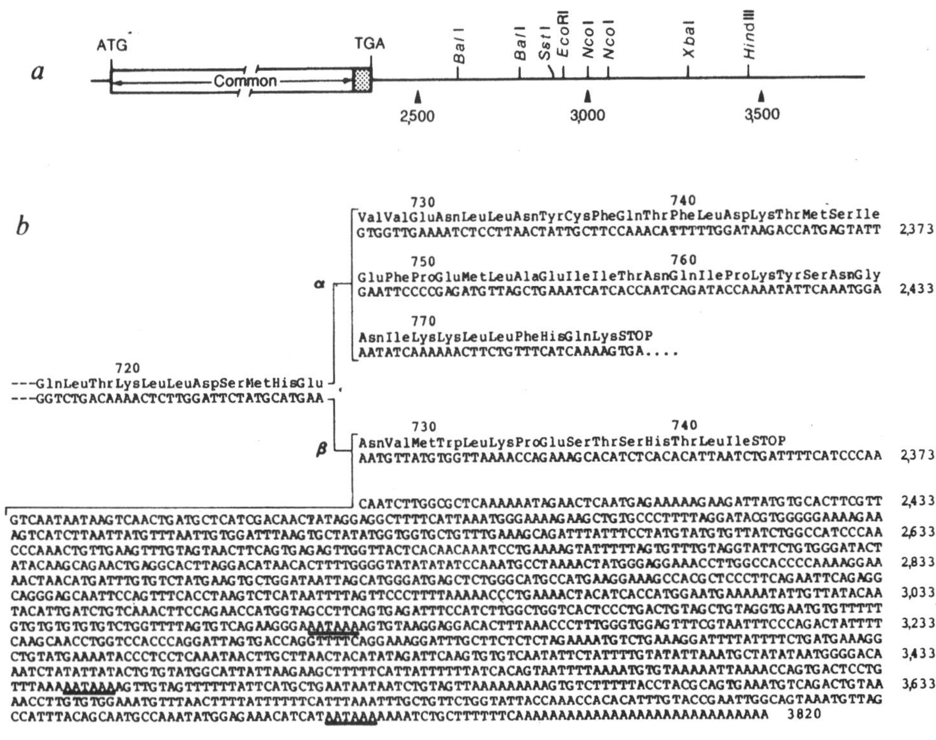

Fig. 3 Restriction map and nucleotide sequence of the 3′ end of the human glucocorticoid receptor β cDNA. *a*, The common 6-nucleotide restriction enzyme sites are shown for the 3′-untranslated region of OB10. *b*, The cDNA sequence of the β form (OB10) from nucleotide 2,281 to 3,820 compared with the protein-coding information found in the 3′-terminal coding portion of the α form (OB7). Amino acids encoded by each of the cDNAs are presented above the nucleotide sequences. Putative polyadenylation signals (AATAAA) in the 3′-untranslated sequence of OB10 are underlined.

indicated in Fig. 1*a*. RNA blot analysis indicated that a cDNA insert of 5-7 kilobases (kb) would be necessary to obtain a full-length clone and sequence analysis indicated that the overlapping clones OB7 and hGR5.16 spanned an open reading frame of 720 amino acids, not large enough to encode the complete receptor. Therefore, a second human fibroblast cDNA library of $\sim 2 \times 10^6$ transformants was screened, yielding a clone (OB10) containing a large insert that extended 150 base pairs (bp) upstream of the putative translation initiation site (see Fig. 1). Sequence analysis predicts two protein forms, termed α and β, which diverge at amino acid 727 and contain additional distinct open reading frames of 50 and 15 amino acids, respectively, at their carboxy termini (see Fig. 1*b*). The α form, represented by clone OB7, is the predominant form of glucocorticoid receptor because eight cDNA clones isolated from various libraries contain this sequence.

cDNA and protein sequences

Figure 2 shows the 4,800-nucleotide sequence encoding the human α glucocorticoid receptor, determined using clones hGR1.2, hGR5.16, OB7 and OB10. The translation initiation site was assigned to the methionine codon corresponding to nucleotides 133-135 because this is the first ATG triplet that appears downstream from the in-frame terminator TGA (nucleotides 121-123). However, in the absence of amino-terminal peptide sequence information, unequivocal determination of the initiation site is not yet possible. The codon specifying the lysine at position 777 is followed by the translation termination codon TGA. The remainder of the coding sequence is covered by multiple overlapping clones, with OB7 containing a 4.3-kb insert that continues to the poly(A) addition site and OB10 containing the putative initiator methionine. The 3′ regions of clones OB7 and OB10 diverge at nucleotide 2,314, as shown by both restriction endonuclease and DNA sequence analysis. At this junction, the α-receptor continues with a unique sequence encoding an additional 50 amino acids whereas the β-receptor continues for only 15 additional amino acids (Fig. 3). The 3′-untranslated region of OB7 is 2,325 nucleotides long, while that of OB10 is 1,433 nucleotides. There is no significant homology between these two regions, as indicated by direct sequence comparison (Figs 2, 3) or by hybridization analysis under stringent conditions (data not shown).

In addition, we have isolated from a human primary fibroblast library another cDNA clone, OB12 (data not shown), which contains sequences identical to OB7 but uses the polyadenylation signal at nucleotide 3,101 (Figs 1*b*, 2), giving rise to a shorter 3′-untranslated region. Use of probes specific for the 3′-untranslated region of OB7 to screen a human placenta cDNA library reveals that most clones terminate at the first poly(A) site in OB7. Thus, messenger RNA variation is the apparent consequence of both alternative polyadenylation and alternative RNA splicing (see below). The fact that the human fibroblast library contained both cDNAs suggests that both receptor forms may be present in the same cell.

Analysis of α- and β-receptor protein

Sequence analysis reveals that the α and β forms of the human glucocorticoid receptor are 777 and 742 residues long, respectively; the two forms are identical up to residue 727, after which they diverge. To examine the receptor levels *in vivo*, cytoplasmic extracts from several human and mouse cell lines were probed by immunoblot analysis with a polyclonal antibody directed against the human glucocorticoid receptor[37]. α- And β-receptor cDNAs were inserted into the SP6 transcription vector to create synthetic mRNA for *in vitro* translation (Fig. 4*a*). The RNAs were separately added to a rabbit reticulocyte lysate system and the unlabelled products analysed by SDS-polyacrylamide gel electrophoresis (SDS-PAGE). The two RNAs programme the synthesis of distinct translation products whose migration differences are consistent with the predicted polypeptide lengths of the two forms (Fig. 4*b*, lanes 2, 3). Cytoplasmic extracts from untreated IM-9 cells and IM-9 cells treated with 1 μM triamcinolone acetonide serve as markers (Fig. 4*b*, lanes 4, 5) for the 94K receptor (the 79K form represents a putative receptor degradation product)[40]. Note that after steroid treatment, the intensity of the 94K band is reduced, corresponding to tighter receptor/chromatin binding and, therefore, receptor translocation to the nucleus. The α form co-migrates with the 94K band of the native receptor while the β form migrates more rapidly (Fig. 4*b*, compare lanes 2,3 with lanes 4,5). A comparison of cytoplasmic extracts from various human and mouse cell lines reveals the presence of only the α-receptor (Fig. 4*b*, lanes 6-9). Interestingly, the mouse ADR6 lymphoma cell line[46], selected for resistance to steroid-induced lysis, contains no steroid-binding activity and shows no immunoreactive receptor (Fig. 4*b*, lane 7). Therefore, based on characterization of multiple recep-

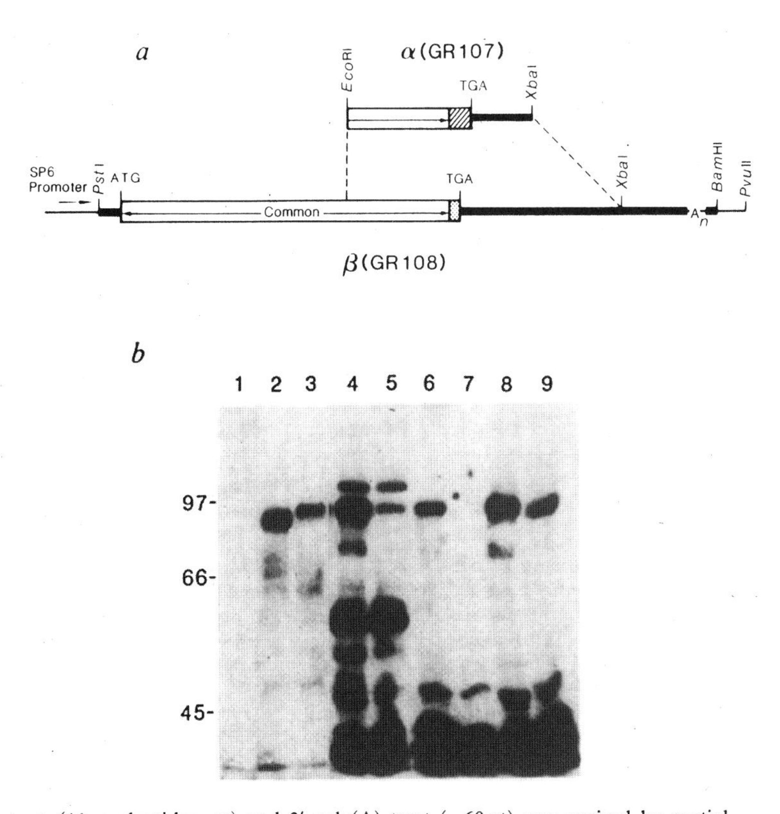

Fig. 4 Immunoblot comparison of hGR translated *in vitro* with *in vivo* hGR from cell extracts. *a*, The vectors constructed for *in vitro* transcription of the hGR cDNA sequence. The complete α (pGR107) and β (pGR108) coding sequences were placed under the transcriptional control of the SP6 promoter in pGEM1. Vector sequences, noncoding cDNA sequences and coding sequences are indicated by thin lines, thick bars and boxed regions, respectively. The poly(A) tract of ~60 nucleotides is indicated by A_n. Divergent coding sequences are indicated by striped and stippled regions. *b*, Western blot analysis of *in vitro* translation products and cell extracts. Unlabelled translation products synthesized in a rabbit reticulocyte lysate system with no added RNA (lane 1) or with RNA synthesized from pGR108 (β, lane 2) or pGR107 (α, lane 3) were fractionated on a 7.5% SDS-polyacrylamide gel. Additional lanes are: cytoplasmic extracts from IM-9 (lane 4), IM-9 treated with 1 μM triamcinolone acetonide (lane 5), HeLa (lane 6), ADR6.M1890.AD1 mouse lymphoma (lane 7), S49 mouse lymphoma (lane 8) and EL4 lymphoma (lane 9). Proteins were transferred to nitrocellulose and probed with anti-hGR antibody, followed by ^{125}I-labelled *Staphylococcus aureus* protein A as described previously[42].

Methods. To construct an expression vector containing the entire α coding sequence shown in Fig. 2, the 3′ coding sequence of OB7 was fused to OB10 5′ coding information. OB7 was partially digested with *Eco*RI, completely digested with *Xba*I, and the 1.2-kbp fragment was gel-purified and ligated with *Eco*RI/*Xba*I-digested OB10 to produce the intermediate pOB107. The entire pOB107 cDNA sequence including the 5′ poly(G) tract (11 nucleotides, nt) and 3′ poly(A) tract (~60 nt) was excised by partial *Pst*I/complete *Bam*HI digestion. The resultant 3.5-kb fragment was gel-purified and inserted between the *Pst*I and *Bam*HI sites of pGEM1 (Promega Biotec) to yield pGR107. Plasmid GR108 was directly constructed from pOB10 by partial *Pst*I/complete *Bam*HI digestion and insertion of the resulting cDNA insert into the corresponding sites of pGEM1. Capped SP6 transcripts were synthesized from *Pvu*II-linearized pGR107 and pGR108, as described by Krieg and Melton[62], with simultaneous capping effected by reduction of the GTP concentration from 400 to 100 μM and addition of m^{7}GppG (Pharmacia) to 500 μM. Transcripts were purified by P60 chromatography and translated with micrococcal nuclease-treated rabbit reticulocyte lysate (Promega Biotec) in conditions suggested by the manufacturer. Preparation of IM-9 cytosol from steroid-treated cells was as described previously[42]. Size markers are phosphorylase B (97K), bovine serum albumin (66K) and ovalbumin (45K).

tor cDNA clones and receptor protein by immunoblot analysis, we conclude that the predominant physiological form of the glucocorticoid receptor is the α (94K) species.

Expression of hGR *in vitro*

To provide additional evidence that the cloned receptor is functional, we investigated the possibility that the *in vitro*-translated products might be able to selectively bind corticosteroids. Accordingly, the rabbit reticulocyte lysate was incubated with the radiolabelled synthetic glucocorticoid analogue ^{3}H-triamcinolone acetonide (^{3}H-TA) before or after addition of *in vitro*-synthesized α or β hGR RNA. As shown in Fig. 5, those lysates programmed with α-hGR RNA acquired selective steroid-binding capacity; unexpectedly, the β-receptor synthesized *in vitro* failed to bind competable ^{3}H-TA. The *in vitro-synthesized* α-hGR bound radiolabelled steroid which could be competed with by addition of excess unlabelled cortisol or dexamethasone; however, binding of ^{3}H-TA was not effectively competed with by addition of excess unlabelled oestrogen or testosterone. In contrast, excess progesterone constituted an effective competitor, consistent with the previously reported anti-glucocorticoid activities of progesterone[47]. To confirm these results, the competition experiments were repeated with native glucocorticoid receptor prepared from extracts of human lymphoid cells. Both the *in vitro*-translated receptor and the natural *in vivo* receptor have nearly identical properties with regard to steroid binding and competition with excess unlabelled steroid analogue (Fig. 5).

hGR sequences map to at least two genes

The human glucocorticoid receptor gene has been functionally mapped to chromosome 5. Analysis of somatic cell hybrids constructed by fusing receptor-deficient mouse T cells (EL4) with human receptor-containing T cells (CEM-C7) indicated that segregants expressing the wild-type CEM-C7 receptor maintained human chromosome 5 while dexamethasone-resistant segregants had lost this chromosome[48].

To confirm the authenticity of our cDNA clones, we mapped receptor cDNA sequences using Chinese hamster/human somatic cell hybrids containing only human chromosome 5 (HHW454). DNAs extracted from human placenta, HHW454 hybrid cells and Chinese hamster ovary (CHO) cells were digested with *Eco*RI or *Hin*dIII restriction endonucleases and separated on a 0.8% agarose gel. DNA fragments transferred to nitrocellulose were probed with a portion of the receptor-coding region derived from nucleotides 570–1,640 (hGR1.2A; Fig. 1). In addition to CHO-specific *Eco*RI bands of 6.8 and 17 kbp (Fig. 6*a*, lanes 2,3), DNA from the hybrid cell line also contains human-specific bands of 3.0 and 5.0 kbp. Unexpectedly, a DNA fragment of 9.5 kbp is found in total human DNA but not in the hybrid line (Fig. 6*a*, lane 1). Similarly, *Hin*dIII digestion revealed a 7.5-kbp band that is not present in the chromosome 5 hybrid cell DNA (Fig. 6*a*, lane 4). These results indicate that the receptor cDNA maps to human chromosome 5, but that there are additional receptor-related sequences elsewhere in the genome. To map these sequences, we used a dual-laser fluorescence-activated cell sorter (FACS) to sort mitotic chromosome suspensions stained with DIPI/chromomycin in conjunction with Hoechst 33258 chromomycin; this technique allows separation of the 24 human chromosome types into 22 fractions[49]. After the chromosomes were sorted directly onto nitrocellulose, the chromosomal DNA was denatured and hybridized to the hGR cDNA probe. In addition to confirming the chromosome 5 localization, additional sequences were found on chromosome 16 (Fig. 6*b*). To confirm this localization, DNAs from mouse erythroleukaemia cells and a mouse erythroleukaemia cell line containing human chromosome 16 (ref. 50) were digested with

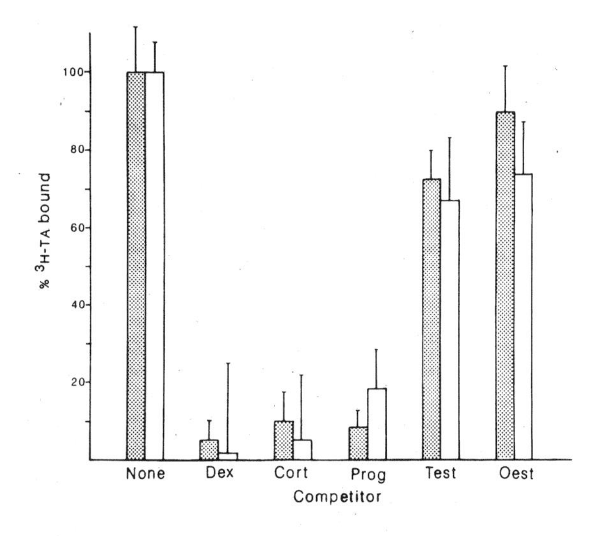

Fig. 5 Steroid-binding of α-hGR (GR107) translated *in vitro*. Binding to IM-9 cytosol extract (stippled bars) and to reticulocyte lysate containing SP6-generated α-hGR RNA (GR107; open bars) are shown. Bars represent bound ^{3}H-triamcinolone acetonide (TA) determined with a 100-fold excess of various steroid competitors; 100% competition was determined using unlabelled TA as competitor. The values represent the mean of triplicate determinations, with error bars showing $P<0.05$. Steroid competitors are dexamethasone (Dex), cortisol (Cort), progesterone (Prog), testosterone (Test), and oestradiol (Oest).
Methods. Binding assays were performed in 100 μl containing 10 mM Tris-HCl *p*H 7.4, 100 mM NaCl, 1 mM EDTA, 10 mM sodium molybdate, 10 dithiothreitol, 150 mM ^{3}H-TA (20 Ci mmol^{-1}; Amersham) and 10 μl translation mixture or 100 μg fresh IM-9 cytosol. Unlabelled steroid competitor (15 μM) was added as indicated. After 2 h at 0 °C, samples were extracted twice for 5 min each with 5 μl of 50% dextran-coated charcoal to remove unbound steroid, and counted. Uncompeted and fully competed values for the α glucocorticoid receptor (GR107) were 490 and 290 c.p.m., respectively. Reticulocyte lysate translation mixtures without added transcript or programmed with β-receptor SP6 RNA (GR108) contained no competable ^{3}H-TA binding.

*Hin*dIII and probed with hGR cDNA (Fig. 6*c*); as predicted, the only DNA fragment found in the hybrid and not in the control was the 7.5-kbp DNA fragment, thus establishing the chromosome 16 assignment (Fig. 6*c*, lanes 1, 2).

Additional Southern blot analyses using the *Eco*RI-*Xba*I fragments from OB7 and OB10 3′-untranslated regions revealed hybridization only to chromosome 5 (data not shown). We conclude that both the α- and β-receptor cDNAs are probably encoded by a single gene on chromosome 5 and suggest that the two cDNA forms are generated by alternative splicing. In addition, we conclude that another gene residing on human chromosome 16 contains homology to the glucocorticoid receptor gene, at least between nucleotides 570 and 1,640. It is not clear whether these sequences on chromosome 16 represent a related steroid receptor gene, a processed gene or pseudogene, or a gene that shares a common domain with the gene for the glucocorticoid receptor. Genomic cloning and DNA sequencing may provide the answer.

To determine the size of the mRNA encoding the glucocorticoid receptor, Northern blot hybridization[51] experiments were performed using cytoplasmic mRNA isolated from a human fibroblast cell line, HT1080. Using the hGR1.2 coding sequence as probe, multiple mRNAs of 5.6, 6.1 and 7.1 kb were detected. Treatment of these cells with glucocorticoids for 24 h leads to a 2–3-fold reduction in receptor mRNAs, suggesting a potential negative feedback regulation.

Discussion

Structural analysis of the glucocorticoid receptor is a prerequisite for gaining insight into the mechanisms by which this regulatory molecule exerts its effects on gene transcription. Here, we have presented the primary sequence of the human glucocorticoid receptor deduced from nucleotide sequence analysis of cDNA clones.

Isolation of hGR cDNAs has revealed the existence of multiple mRNAs encoding at least two forms of the polypeptide. The predicted proteins differ at their carboxy termini by the substitution of 50 amino acids in the case of α-hGR and 15 amino acids in the case of β-hGR. The α glucocorticoid receptor is the major

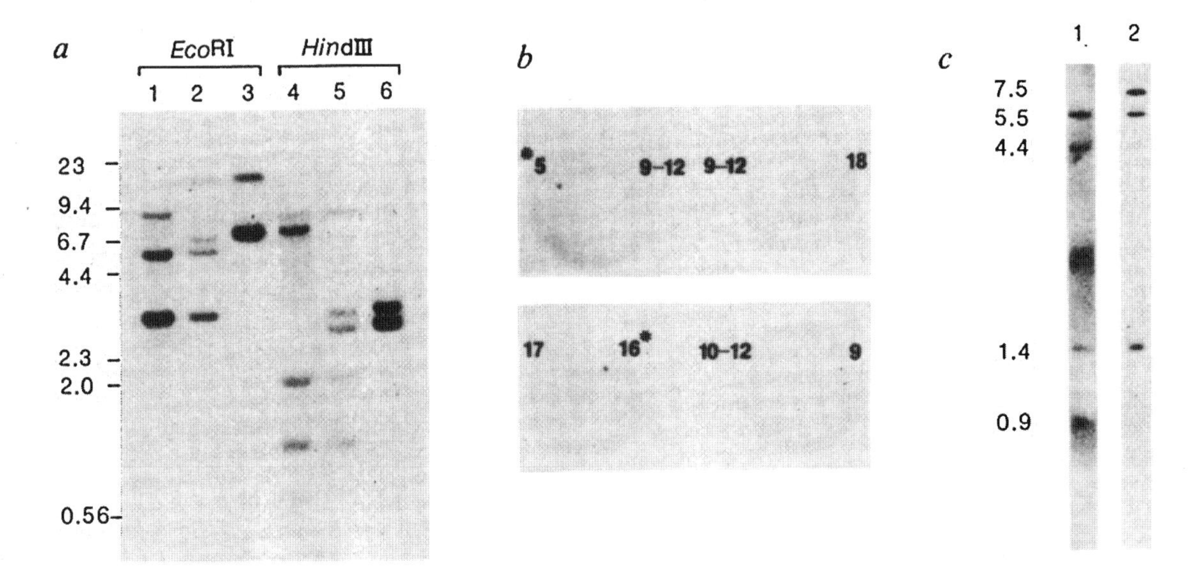

Fig. 6 Chromosome mapping analysis of hGR cDNA. *A*, 10 μg of DNA from human placenta (lanes 1, 4), CHO/human somatic cell hybrid (HHW454, lanes 2, 5) containing chromosome 5 as its only human complement, or CHO (lanes 3, 6), was digested with *Eco*RI (lanes 1–3) or *Hin*dIII (lanes 4–6) to completion, fractionated on a 0.8% agarose gel and transferred to nitrocellulose paper. *b*, Chromosomes (3×10^{4}) prepared from a human lymphocyte cell line, stained with 4,6-bis(2″-imadazolinyl-4H,5H)-2-phenylindole (DIPI)/chromomycin A3 and sorted using a dual-laser custom FACS IV chromosome sorter[63], were denatured and neutralized on nitrocellulose paper. Note that Hoechst/chromomycin-stained chromosome 9 was sorted with chromosomes 10–12. *c*, 10 μg of DNA from the parental mouse cell line MEL (lane 1) or the parentally derived somatic cell hybrid carrying human chromosome 16 (ref. 50; lane 2) was digested with *Hin*dIII and separated on a 0.8% agarose gel, then transferred to nitrocellulose paper. All filters were probed with the 1,100-bp insert from hGR1.2, nick-translated to a specific activity of 3×10^{8} c.p.m. μg^{-1} and hybridized in 5 × SSPE, 1 × Denhardt's, 0.1% SDS, 50% formamide, 100 μg ml^{-1} denatured salmon sperm DNA, 50% dextran sulphate at 42 °C for 18 h. Filters were washed twice (for 30 min each) in 2 × SSC at 68 °C and exposed to X-ray film at −70 °C with an intensifying screen.

Fig. 7 Northern blot analysis of hGR mRNA. 10 μg of poly(A)$^{+}$ mRNA from human HT1080 fibroblast cells, collected after 24 h without (−) or with (+) treatment with 10 μM dexamethasone, was electrophoresed through a 0.8% agarose/1% formaldehyde gel, stained with acridine orange (120 μg ml^{-1}) and transferred to nitrocellulose. The filter was hybridized overnight with nick-translated hGR1.2 (10^6 c.p.m. ml^{-1}, specific activity 10^8 c.p.m. μg^{-1}) and washed with 2×SSC at 68 °C. Sizes were estimated from human fibronectin mRNA (7.8 kb), and 28 S (5.0 kb) and 18 S (2.1 kb) ribosomal RNAs.

form identified in several human cell lines and cDNA libraries. However, a recent report by Northrop *et al.* characterizes two forms of the receptor in mouse lymphoid cells[52]. The relationship of α- and β-hGR to the mouse doublet species remains to be established. Also, the cellular distribution and potential function of β-hGR are unclear, although it is possible that variant receptors are used for tissue-specific functions. We are now generating antisera to synthetic peptides specific for each human receptor form to elucidate their tissue-specific expression.

Among the cDNAs selected using the immunopositive phage DNA insert hGR1.2A as probe were those containing 3′ ends similar to OB7, except that polyadenylation was signalled earlier by the use of an AATAAA at nucleotide 3,101. These clones have been isolated from both human fibroblast and placental libraries (data not shown). Alternative poly(A) site selection is a feature of many eukaryotic transcription units[53]. In some instances, selection of poly(A) sites specifies particular polypeptide products[54–57] while in other cases, alternative poly(A) site selection produces no change in the primary structure of the polypeptide[58]. The selection of poly(A) sites during receptor transcription may (1) alter the stability of the mRNA in a particular tissue, (2) lead to splicing changes, or (3) be random, with no physiological consequence.

The *in vitro* translation studies described here provide direct evidence that the cloned molecule encodes the complete glucocorticoid receptor. First, the *in vitro*-translated product is identical in size to the native glucocorticoid receptor and is immunologically reactive with receptor-specific antiserum. Second, the *in vitro*-translated protein acts functionally as a glucocorticoid receptor in that it is capable of selectively binding the synthetic glucocorticoid triamcinolone acetonide. This binding is specifically competed with by glucocorticoids, glucocorticoid analogues and progesterone but is not competed with by the sex steroids testosterone and oestrogen. In this respect, the *in vitro*-translated receptor behaves identically to the *in vivo* receptor from human lymphoid cells, providing the first evidence for a function of the cloned molecule. The acquisition of steroid-binding properties does not appear to require any specific modifications or, if it does, these modifications can occur in the *in vitro* translation mix.

The results presented here provide the information necessary for studying the molecular interactions of a eukaryotic transcriptional regulatory protein with its target genes. These structural studies provide a basis from which the glucocorticoid receptor, its gene, and its RNA products can be analysed. Furthermore, the ability to express receptor *in vitro* provides a novel means by which the consequence of specific *in vitro* mutagenesis can be rapidly tested.

In addition to the *in vitro* studies, the analysis of several existing rodent cell lines[49,59–61] with well-characterized receptor defects in both the DNA- and steroid-binding domains should facilitate future analysis. Furthermore, the isolation of genes responsive to glucocorticoids and specific regulatory elements by both mutagenic and protein-binding studies suggests that this protein will serve as a very useful model for analysis of inducible eukaryotic gene regulation.

We thank Drs Kelly Mayo, Geoffrey Wahl, Michael Wilson, Noellyn Oliver, Tony Hunter and Donald Gruol for advice and discussion; our collaborator Dr John Wasmuth for providing human/hamster somatic cell hybrids; Hiroto Okayama for providing human fibroblast cDNA libraries; Noellyn Oliver for HT1080 cells; and Kevin Struhl for suggestions on *in vitro* translation studies. We also thank Marijke ter Horst for preparation of artwork and secretarial assistance. S.M.H. is a predoctoral trainee, supported by a training grant to the Department of Biology, University of California, San Diego. C.W. is a fellow of the Damon Runyon-Walter Winchell Cancer Fund (DRG-755). R.L. is an associate and M.G.R. and R.M.E. are investigators of the Howard Hughes Medical Institute. This work was supported by grants from the NIH and the Howard Hughes Medical Institute.

Received 27 August; accepted 7 November 1985.

1. Jensen, E. V. & De Sombre, E. R. *A. Rev. Biochem.* **41,** 203–230 (1972).
2. Gorski, J. & Gannon, F. *A. Rev. Physiol.* **38,** 425–450 (1976).
3. Yamamoto, K. R. & Alberts, B. M. *A. Rev. Biochem.* **45,** 721–746 (1976).
4. O'Malley, B. W., McGuire, W. L., Kohler, P. O. & Kornman, S. G. *Recent Prog. Horm. Res.* **25,** 105–160 (1969).
5. Hayward, M. A., Brock, M. L. & Shapiro, D. J. *Nucleic Acids Res.* **10,** 8273–8284 (1982).
6. Ashburner, M. & Berendes, H. D. in *The Genetics and Biology of* Drosophila, Vol. 2 (eds Ashburner, M. & Wright, T. R. F.) 315–395 (Academic, London, 1978).
7. Peterkofsky, B. & Tomkins, G. *Proc. natn. Acad. Sci. U.S.A.* **60,** 222–228 (1968).
8. McKnight, G. S. & Palmiter, R. D. *J. biol. Chem.* **254,** 9050–9058 (1979).
9. Horwitz, K. B. & McGuire, W. L. *J. biol. Chem.* **253,** 2223–2228 (1978).
10. Palmiter, R. D., Moore, P. B., Mulvihill, E. R. & Emtage, S. *Cell* **8,** 557–572 (1976).
11. Yamamoto, K. R. & Alberts, B. M. *Proc. natn. Acad. Sci. U.S.A.* **69,** 2105–2109 (1972).
12. Jensen, E. V. *et al. Proc. natn. Acad. Sci. U.S.A.* **59,** 632–638 (1968).
13. Ringold, G. M., Yamamoto, K. R., Tomkins, G. M., Bishop, J. M. & Varmus, H. E. *Cell* **6,** 299–305 (1975).
14. Parks, W. P., Scolnick, E. M. & Kozikowski, E. H. *Science* **184,** 158–160 (1974).
15. Hager, L. J. & Palmiter, R. D. *Nature* **291,** 340–342 (1981).
16. Karin, M., Anderson, R. D., Slater, E., Smith, K. & Herschman, H. R. *Nature* **286,** 295–297 (1980).
17. Kurtz, D. T. & Feigelson, P. *Proc. natn. Acad. Sci. U.S.A.* **74,** 4791–4795 (1977).
18. Spindler, S. R., Mellon, S. H. & Baxter, J. D. *J. biol. Chem.* **257,** 11627–11632 (1982).
19. Evans, R. M., Birnberg, N. C. & Rosenfeld, M. G. *Proc. natn. Acad. Sci. U.S.A.* **79,** 7659–7663 (1982).
20. Robins, D. M., Paek, I., Seeburg, P. H. & Axel, R. *Cell* **29,** 623–631 (1982).
21. Chandler, V. L., Maler, B. A. & Yamamoto, K. R. *Cell* **33,** 489–499 (1983).
22. Ostrowski, M. C., Huang, A. L., Kessel, M., Woolford, R. G. & Hager, G. L. *EMBO J.* **3,** 1891–1899 (1984).
23. Govindan, M. V., Spiess, E. & Majors, J. *Proc. natn. Acad. Sci. U.S.A.* **79,** 5157–5161 (1982).
24. Scheidereit, C., Geisse, S., Westphal, H. M. & Beato, M. *Nature* **304,** 749–752 (1983).
25. Pfahl, M. *Cell* **31,** 475–482 (1982).
26. Payvar, F. *et al. Cell* **35,** 381–392 (1983).
27. Karin, M. *et al. Nature* **308,** 513–519 (1984).
28. Laimonis, L. A., Khoury, G., Gorman, C., Howard, B. & Gruss, P. *Proc. natn. Acad. Sci. U.S.A.* **79,** 6453–6457 (1982).
29. Benoist, C. & Chambon, P. *Nature* **290,** 304–310 (1981).
30. Banerji, J., Olson, L. & Schaffner, W. *Cell* **33,** 729–740 (1983).
31. Grosschedl, R. & Birnstiel, M. L. *Proc. natn. Acad. Sci. U.S.A.* **77,** 7102–7106 (1980).
32. Simons, S. S. & Thompson, E. B. *Proc. natn. Acad. Sci. U.S.A.* **78,** 3541–3545 (1981).
33. Gehring, U. & Hotz, A. *Biochemistry* **22,** 4013–4018 (1983).
34. Westphal, H. M., Moldenhauer, G. & Beato, M. *EMBO J.* **1,** 1467–1471 (1982).
35. Wrange, O., Carlstedt-Duke, J. & Gustafsson, J.-A. *J. biol. Chem.* **254,** 9284–9290 (1979).
36. Okret, S., Carlstedt-Duke, J., Wrange, O., Carlstrom, K. & Gustafsson, J.-A. *Biochim. biophys. Acta* **677,** 205–219 (1981).
37. Harmon, J. M. *et al. Cancer Res.* **44,** 4540–4547 (1984).
38. Gametchu, B. & Harrison, R. W. *Endocrinology* **114,** 274–279 (1984).
39. Carlstedt-Duke, J., Okret, S., Wrange, O. & Gustafsson, J.-A. *Proc. natn. Acad. Sci. U.S.A.* **79,** 4260–4264 (1982).
40. Wrange, O., Okret, S., Radojcic, M., Carlstedt-Duke, J. & Gustafsson, J.-A. *J. biol. Chem.* **259,** 4534–4541 (1984).
41. Dellweg, H. G., Hotz, A., Mugele, K. & Gehring, U. *EMBO J.* **1,** 285–289 (1982).
42. Weinberger, C. *et al. Science* **228,** 740–742 (1985).
43. Miesfeld, R. *et al. Nature* **312,** 779–781 (1984).
44. Okayama, H. & Berg, P. *Molec. cell. Biol.* **3,** 280–289 (1983).
45. Maxam, A. & Gilbert, W. *Proc. natn. Acad. Sci. U.S.A.* **74,** 560–564 (1977).
46. Danielsen, M. & Stallcup, M. R. *Molec. cell. Biol.* **4,** 449–453 (1984).
47. Rousseau, G. G., Baxter, J. D. & Tomkins, G. M. *J. molec. Biol.* **67,** 99–115 (1972).
48. Gehring, U., Segnitz, B., Foellmer, B. & Francke, U. *Proc. natn. Acad. Sci. U.S.A.* **82,** 3751–3755 (1985).
49. Lebo, R. V. *et al. Science* **225,** 57–59 (1984).
50. Bode, U., Deisseroth, A. & Hendrik, D. *Proc. natn. Acad. Sci. U.S.A.* **78,** 2815–2819 (1981).
51. Thomas, P. S. *Proc. natn. Acad. Sci. U.S.A.* **77,** 5201–5205 (1980).
52. Northrop, J. P., Gametchu, B., Harrison, R. W. & Ringold, G. M. *J. biol. Chem.* **260,** 6398–6403 (1985).
53. Darnell, J. E. *Nature* **297,** 365–371 (1982).
54. Amara, S. G., Jonas, V., Rosenfeld, M. G., Ong, E. S. & Evans, R. M. *Nature* **298,** 240–244 (1982).
55. Rosenfeld, M. G. *et al. Nature* **304,** 129–135 (1983).
56. Alt, F. W. *et al. Cell* **20,** 293–301 (1980).
57. Schwarzbauer, J. E., Tamkun, J. W., Lemischka, I. R. & Hynes, R. O. *Cell* **35,** 421–431 (1983).
58. Setzer, D. R., McGrogan, M. & Schimke, R. T. *J. biol. Chem.* **257,** 5143–5147 (1982).
59. Yamamoto, K. R., Stampfer, M. R. & Tomkins, G. M. *Proc. natn. Acad. Sci. U.S.A.* **71,** 3901–3905 (1974).
60. Bourgeois, S. & Newby, R. F. *Cell* **11,** 423–430 (1977).
61. Grove, J. R., Dieckmann, B. S., Schroer, T. A. & Ringold, G. M. *Cell* **21,** 47–56 (1980).
62. Krieg, P. A. & Melton, D. A. *Nucleic Acids Res.* **12,** 7057–7070 (1984).
63. Lebo, R. V. & Bastian, A. M. *Cytometry* **3,** 213–219 (1982).
64. Weinberger, C., Hollenberg, S. M., Rosenfeld, M. G. & Evans, R. M. *Nature* **318,** 670–672 (1985).

Cyclic AMP regulation of gene transcription

The second messenger cyclic AMP exerts its effects on cell function by activation of cyclic AMP-dependent protein kinase (PKA) as described in Section 1. PKA phosphorylates many substrate proteins, leading to rapid cellular effects, but it can also affect specific gene expression. Binding of cyclic AMP to the regulatory subunit of PKA results in dissociation of the catalytic subunit which can translocate to the nucleus [1]. The target for the PKA catalytic subunit in the nucleus, which exerts its effect on transcription, was found to be a nuclear protein, cyclic AMP-response element-binding protein (CREB) first described by Montminy and Bilezikjian (see selected paper [2]).

Among those genes whose transcription is regulated by cyclic AMP is that for the hypothalamic peptide, somatostatin. Characterization of the sequence requirements in the promoter region of the gene that would allow cyclic AMP responsiveness identified a 30-nucleotide cyclic AMP-response element (CRE) [3]. Montminy and Bilezikjian [2] looked for a nuclear protein that bound to this element and were able to identify and purify a 43 kDa protein based on its interaction with a specific DNA affinity-column. This protein, termed CREB, was phosphorylated in intact cells following elevation of cyclic AMP levels and was a substrate for PKA *in vitro*, suggesting that it is the link between cyclic AMP elevation and regulation of gene transcription.

Further analysis of CREB revealed that its phosphorylation enhanced formation of a dimer of the protein and increased its efficiency in stimulating gene transcription [4]. Cloning of cDNAs [5,6] revealed that CREB possesses a cluster of sites for phosphorylation within the N-terminal domain of the protein [5]. The specific site of phosphorylation of CREB required for its activation was found to be Ser-133, and mutagenesis of this amino acid abolished the ability of CREB to stimulate transcription [7]. These biochemical studies suggested that phosphorylation of CREB by the catalytic subunit of PKA is a key step in the activation of transcription of genes controlled by CREs. *In vivo* information on the crucial nature of CREB came from the study of transgenic mice expressing a mutated form of CREB containing a Ser-133-to-Ala substitution which prevented it from becoming phosphorylated [8]. The mutant gene was expressed under the control of the rat growth hormone promoter so that its expression was directed to pituitary somatotrophs which synthesize this hormone. The idea behind this experiment was based on the knowledge that cyclic AMP stimulates proliferation of somatotrophs. If CREB phosphorylation was required for this effect of cyclic AMP, then the targeted expression of the mutated CREB should inhibit somatotroph proliferation. The results demonstrated that this was indeed the case, and the transgenic mice showed reduced numbers of somatotrophs, and, as a consequence of reduced production of growth hormone, exhibited a dwarf phenotype [8].

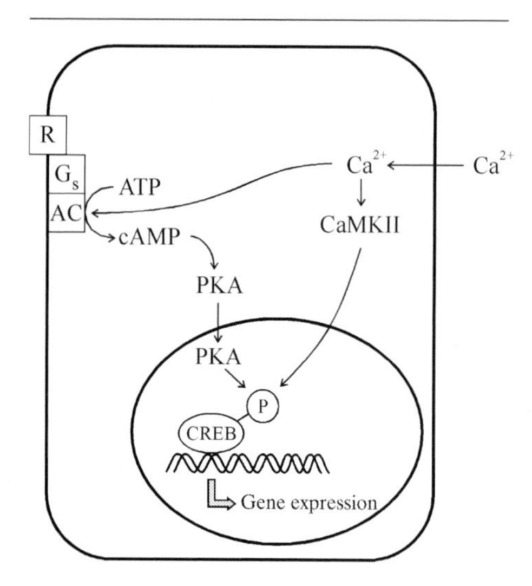

Figure 20.1 Pathways from cyclic AMP and Ca^{2+} to gene transcription via protein kinases acting on CREB. Activation of plasma membrane receptors (R) coupled via G_s to adenylate cyclase (AC) leads to increased levels of cellular cyclic AMP (cAMP). Cyclic AMP binds to PKA and the catalytic subunit translocates into the nucleus where it phosphorylates CREB on Ser-133. A rise in cytosolic Ca^{2+} concentration activates the calmodulin-dependent protein kinase CaMKII which also phosphorylates CREB to activate its transcriptional activity.

The molecular mechanisms by which CREB acts as a transcriptional regulator are beginning to be understood. Following the phosphorylation of Ser-133, CREB shows increased affinity for a nuclear protein termed CREB-binding protein (CBP) [9,10]. It appears that binding to CBP is required for the effects of CREB [11,12] on transcription, perhaps by linking CREB to other general cellular transcription factors.

An additional important aspect of CREB function emerged from studies on Ca^{2+}-responsive elements required for activation of expression of the intermediate-early gene c-*fos*, following Ca^{2+} entry in neurons and neuronal cell lines [13]. This work led to the discovery of a pathway in which CREB is also activated in a Ca^{2+}-dependent manner (Figure 20.1), most likely due to phosphorylation of Ser-133 by calmodulin-dependent protein kinases, including CaMKII and CaMKIV, but also by other novel kinases [14]. CREB, therefore, acts at a junction point in cellular signalling pathways where it can translate cyclic AMP and Ca^{2+} signals into effects on gene transcription.

An additional emphasis on the physiological importance of CREB has come from studies on various organisms that have suggested a key role for CREB in the establishment of long-term memory. The most striking findings have arisen from the generation by targeted mutation of mice that did not express two of the three isoforms of CREB [15], and by the expression of an inhibitory mutant of CREB in *Drosophila* [16]. In both cases short-term memory in the organism was intact but long-term learning was impaired. It appears, therefore, that CREB plays a crucial role in activating the transcription of genes required for the formation of long-term memory.

References

1. Nigg, E.A., Hilz, H., Eppenberger, H.M. and Dutley, F. (1985) *EMBO J.* **4**, 2801–2806
2. Montminy, M.R. and Bilezikjian, L.M. (1987) *Nature (London)* **328,** 175–178
3. Montminy, M.R., Sevarino, K.A., Wagner, J.A., Mandel, G. and Goodman, R.H. (1986) *Proc. Natl. Acad. Sci. U.S.A.* **83**, 6682–6686
4. Yamamoto, K.A., Gonzalez, G.A., Biggs, W.H. and Montminy, M.R. (1988) *Nature (London)* **334**, 494–498
5. Gonzalez, G.A., Yamamoto, K.K., Fischer, W.H., Karr, D., Menzel, P., Biggs, W., Vale, W.W. and Montminy, M.R. (1989) *Nature (London)* **337**, 749–752
6. Hoeffler, J.P., Meyer, T.E., Yun, Y., Jameson, J.L. and Habener, J.L. (1988) *Science* **242,** 1430–1432
7. Gonzalez, G.A. and Montminy, M.R. (1989) *Cell* **59**, 675–680
8. Struthers, R.S., Vale, W.W., Arias, C., Sawchenko, P.E. and Montminy, M.R. (1991) *Nature (London)* **350**, 622–624
9. Chrivia, J.C., Kwok, R.P.S., Lamb, N., Hagiwara, M., Montminy, M.R. and Goodman, R.H. (1993) *Nature (London)* **365**, 855–859
10. Parker, D., Ferreri, K., Nakajima, T., Morte, V.J.L., Evans, R., Koerber, S.C., Hoeger, C. and Montminy, M.R. (1996) *Mol. Cell. Biol.* **16**, 694–705
11. Arias, J., Alberts, A.S., Brindle, P., Claret, F.X., Smeal, T., Kavin, M., Feramisco, J. and Montminy, M. (1994) *Nature (London)* **370**, 226–229
12. Kwok, R.P.S., Lundlblad, J.R., Chrivia, J.C., Richards, J.P., Bachinger, H.P., Brennan, R.G., Roberts, S.G.E., Green, A.R. and Gooman, R.H. (1994) *Nature (London)* **370**, 223–226
13. Sheng, M., McFadden, G. and Greenberg, M.E. (1990) *Neuron* **4**, 571–582
14. Ghosh, A. and Greenberg, M.E. (1995) *Science* **268**, 239–247
15. Bourtchuladze, R., Frenguelli, B., Blendy, J., Cioffi, D., Schutz, G., and Silva, A.J. (1994) *Cell* **79**, 59–68
16. Yin, J.C.P., Wallach, J.S., Delvecchio, M., Wilder, E.L., Zhou, A., Quinn, W.G. and Tully, T. (1994) *Cell* **79**, 49–58

Montminy & Bilezikjian (1987) Nature (London) **328**, 175–178

Binding of a nuclear protein to the cyclic-AMP response element of the somatostatin gene

Marc R. Montminy & Louise M. Bilezikjian

The Clayton Foundation Laboratories for Peptide Biology, The Salk Institute, 10010 N. Torrey Pines Road, La Jolla, California 92037, USA

Many hormones act on neuroendocrine cells by activating second messenger pathways. Two of these, the phosphoinositol and cAMP-dependent pathways, cause changes in cellular activity through specific protein kinases. By phosphorylating cytoplasmic and nuclear proteins, these kinases apparently coordinate cellular processes, including the biosynthesis and release of neuropeptides. Somatostatin biosynthesis and release, for example, are both positively regulated by the second messenger cAMP in hypothalamic cells[1], and cAMP also induces somatostatin gene transcription 8–10-fold in transfected PC12 pheochromocytoma cells[2]. Transcriptional induction requires a 30-nucleotide cAMP response element (CRE) which is conserved in other cAMP-responsive genes[2–4]. This element also confers cAMP responsiveness when placed upstream of the heterologous simian virus 40 (SV40) promoter. The somatostatin gene does not, however, respond to cAMP in mutant PC12 cells which lack cAMP-dependent protein kinase type II activity[2]. Activation of somatostatin gene transcription may consequently require the phosphorylation of a nuclear protein which binds to the CRE. Using a DNase I protection assay, we have characterized a nuclear protein in PC12 cells which binds selectively to the CRE in the somatostatin gene. We have purified this protein which is of relative molecular mass 43,000 (M_r 43K) by sequence-specific DNA affinity chromatography. This 43K CRE binding protein (CREB) is phosphorylated *in vitro* when it is incubated with the catalytic subunit of cAMP-dependent protein kinase. Stimulating PC12 cells with forskolin, an activator of adenyl cyclase, causes a 3–4-fold increase in the phosphorylation of this protein. We conclude that the cAMP-dependent pathway may regulate gene transcription in response to hormonal stimulation by phosphorylating this CREB protein.

Somatostatin is a hypothalamic peptide that inhibits the release of several pituitary hormones, including growth hormone, prolactin, and thyrotropin[5]. Somatostatin release from hypothalamic neurons is, in turn, regulated by a number of classical neurotransmitters and neuromodulators which activate adenyl cyclase[6–8]. Somatostatin gene transcription is also stimulated by agents which increase intracellular cAMP (ref. 2). The observation that transcriptional regulation by cAMP depends on a conserved promoter element has prompted us to ask whether a specific nuclear protein binds to this sequence.

We used a DNase I footprinting assay on nuclear extracts of PC12 cells to detect DNA-binding proteins which interact with the CRE. Increasing amounts of protein from the extract were incubated with a 5′ end-labelled somatostatin promoter fragment. Samples were partially digested with DNase I and the products of digestion were resolved on 8% denaturing polyacrylamide gels. We observed a single footprint which covered 23 nucleotides on each strand of the DNA (Fig. 1). When run alongside DNA sequence markers (not shown), this footprint, extending from −55 to −32 relative to the transcription initiation site, mapped to the same region as the somatostatin CRE sequence. Approximately 10 μg protein from unfractionated extracts was required to see complete protection. Nuclear extracts prepared from either forskolin-treated or untreated PC12 cells contained the same amount of CRE-binding activity relative to total nuclear protein. To determine whether binding of this nuclear factor *in vitro* correlated with transcriptional activation by cAMP *in vivo*, we examined the somatostatin promoter deletion mutant Δ(−48)CAT (ref. 2). We have shown elsewhere[2] that this fusion gene, extending to position −48 of the somatostatin promoter, is not transcriptionally responsive to cAMP. Likewise, we were unable to observe any protein binding to the mutant CRE sequence (data not shown).

To purify CREB, we constructed a sequence-specific DNA

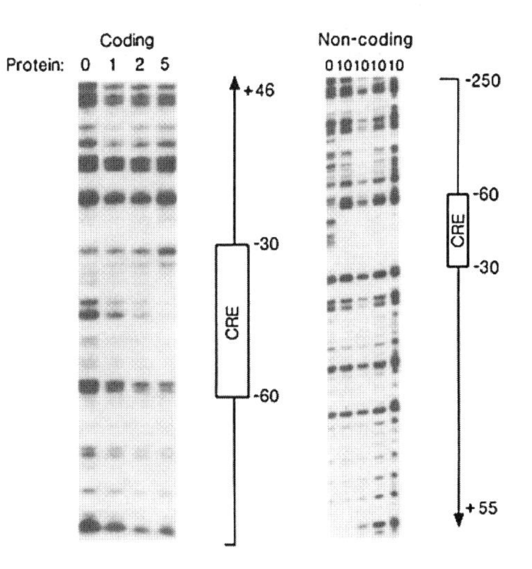

Fig. 1 DNase I footprinting of somatostatin promoter fragments using unfractionated PC12 nuclear extract with (left) coding and (right) non-coding strand probes. Position of cAMP response element (CRE) compared with footprinted region is shown diagrammatically. Nucleotide positions relative to somatostatin transcription initiation point are listed alongside. Coding strand probe was prepared from plasmid Δ(−71) (ref. 2). Following digestion with *Hin*dIII, Δ(−71) plasmid was 5′ end-labelled with T4 polynucleotide kinase and redigested with *Ava*II. The 140-bp *Hin*dIII-*Ava*II fragment was purified by polyacrylamide gel electrophoresis. Non-coding strand probe was isolated from plasmid Δ(−250) (ref. 2). DNA from this plasmid was digested with *Xba*I, 5′ end-labelled, and redigested with *Sal*I. The 305-bp *Xba*I-*Sal*I fragment was also gel purified. Numbers above each lane, amount of unfractionated PC12 nuclear extract (in μg) used for each footprint reaction. Protein concentration was determined according to Bradford[15]. After incubating nuclear extract with labelled DNA fragments, samples were treated with DNase I, and the products of digestion were resolved on 8% urea-polyacrylamide gels.

Methods. Nuclear extracts were prepared using a protocol adapted from Dignam *et al.*[16]. Cells (5×10^9) were harvested and spun at 2,000*g* for 10 min. The cell pellet was then resuspended in four packed cell volumes of buffer H (10 mM Tris-HCl, *p*H 7.9, 10 mM KCl, 1.5 mM $MgCl_2$, 1 mM dithiothreitol (DTT), 1 mM phenylmethylsulphonylfluoride (PMSF), 10 mM NaF, 1,000 U ml^{-1} Trasylol (aprotinin). PC12 cells were allowed to swell on ice for 15 min. Cells were lysed with 20 strokes of a Dounce homogenizer (B pestle) and nuclei were subsequently recovered by centrifugation at 3,000*g* for 8 min. The cloudy supernatant was decanted and the nuclear pellet was washed with an additional 25 ml buffer H. The nuclei were resuspended in four packed nuclear volumes of buffer D (50 mM Tris-HCl, *p*H 7.5, 10% sucrose, 0.42 M KCl, 5 mM $MgCl_2$, 0.1 mM EDTA, 20% glycerol, 2 mM DTT, 0.1 mM PMSF, 20 mM NaF, 1,000 U ml^{-1} Trasylol) and were incubated on ice for 30 min with stirring. The viscous nuclear suspension was then centrifuged at 25,000*g* for one hour. The supernatant was collected, and the nuclear protein from this fraction was precipitated in 53% saturated NH_4SO_4. After centrifugation at 16,000*g* for 20 min (4 °C), the ammonium sulphate pellet was resuspended in 2-3 ml TM buffer (50 mM Tris-HCl, *p*H 7.9, 12.5 mM $MgCl_2$, 1 mM EDTA, 1 mM DTT, 20% glycerol, 0.1% Nonidet-P40, 1 mM PMSF, 0.1 M KCl) and dialysed for 3 h (4 °C) against 500 vol TM buffer. A stringy precipitate which formed during dialysis was removed by centrifuging the extract at 10,000*g* for 10 min. Typically, 20–30 mg nuclear protein were recovered per 5×10^9 PC12 cells. For footprint reactions, nuclear extract protein was incubated with 10 fmol ^{32}P-labelled DNA probe for 15 min on ice, 1 μg poly (dI-dC) competitor DNA (no competitor DNA was used with purified fractions), in 2% polyvinyl alcohol, 25 mM Tris-HCl (*p*H 7.9), 6 mM $MgCl_2$, 0.5 mM EDTA, 0.5 mM DTT, 10% glycerol, 0.05% Nonidet-P40, O.1 M KCl. Samples were then warmed to 25 °C and diluted with an equal volume of 10 mM $MgCl_2$, 10 mM $CaCl_2$. Freshly diluted DNase I was added and reactions were allowed to proceed for 1 min at 25 °C. DNase digestion was terminated by the addition of 90 μl 20 mM EDTA, 1% SDS, 0.2 M NaCl, 25 μg transfer RNA. Samples were deproteinized by phenol-chloroform extraction and then ethanol precipitated.

affinity column[9]. Affinity resin was prepared by self-ligating a double-stranded synthetic oligonucleotide containing a high-affinity CRE site to produce concatemers with an average of eight copies. The ligated DNA was attached covalently to cyanogen-bromide-activated CL-2B Sepharose beads. Unfractionated PC12 nuclear extracts were then passed over 1-ml affinity columns. After washing the columns repeatedly with low salt buffer (0.1 M KCl), the specific CRE binding protein was eluted with salt buffer containing 1.0 M KCl. Recovery of CREB protein, followed by footprint assay, averaged 50% for each pass over the affinity column (Fig. 2*a*).

We monitored the purification of CREB protein by analysing the high salt eluates from successive column passes on SDS-polyacrylamide gels (Fig. 2*b*). Aliquots containing equivalent footprinting activity (100 footprint units; one unit is defined as the amount of CREB protein required to bind to the CRE sequence and protect it from digestion by DNase I) from each affinity pass were precipitated with trichloroacetic acid, acetone washed, and electrophoresed under denaturing conditions. We observed a band of relative molecular mass 43,000 (M_r 43K) which became the predominant protein in the extract after two passes over the affinity column (Fig. 2*b*). After four column passes, all protein bands were visibly reduced except for the 43K band.

To verify that the 43K protein which we had purified was CREB, we performed UV-cross-linking experiments (Fig. 3). Using this technique, individual DNA-binding proteins can be identified in crude nuclear extracts[10]. The DNA probe for cross-linking experiments was prepared by uniformly labelling a 160 base pair (bp) somatostatin promoter fragment with [α-^{32}P]dCTP and BUdR triphosphate. After incubating the DNA probe with unfractionated nuclear extract, the mixture was exposed to ultraviolet light of wavelength 300 nm for one hour. Samples were digested with DNase I and micrococcal nuclease and then analysed on an SDS-polyacrylamide gel. We observed a single protein–DNA adduct of 43K which was resistant to further increases in DNase digestion (Fig. 3*a*). No DNase-resistant bands were observed when either extract protein or ultraviolet irradiation were omitted (data not shown).

We performed competition experiments to determine whether the 43K complex represented the specific binding of a nuclear protein to its target DNA sequence. When a 50-fold molar excess of unlabelled competitor DNA containing the CRE was added to the reactions, no cross-linked proteins were observed (Fig. 3*a*). By contrast, similar addition of competitor DNA containing the Sp1 binding site had no effect on the 43K complex (Fig. 3*b*). Because the migration of proteins does not appear to be significantly altered when cross-linked to small oligonucleotide fragments[11] we estimate that the actual M_r of the CRE-binding protein is near 43K.

The dependence of cAMP-regulated transcription on cAMP-dependent protein kinase activity[2] led us to examine the phosphorylation of CRE-binding protein by the catalytic subunit of cAMP-dependent protein kinase. Affinity purified extract was incubated with catalytic subunit of cAMP-dependent protein kinase and [γ-^{32}P]ATP. After electrophoresis, two major phosphoproteins of 30K and 43K were observed in addition to the autophosphorylated 41K catalytic subunit (Fig. 4). Affinity chromatography of this labelled extract showed that 43K phosphoprotein was eluted from the column only with high salt buffer. In contrast, the 41K catalytic subunit, although in large excess compared with CREB, was completely eluted from the column with low salt. Thus 43K protein retained high CRE binding affinity after phosphorylation with catalytic subunit.

To determine whether phosphorylation of CREB protein is stimulated by cAMP *in vivo*, we labelled PC12 cells overnight with ^{32}P-orthophosphate and then incubated these in the presence or absence of forskolin for one hour. Nuclear extracts were prepared, and equal numbers of TCA-precipitable counts from control and forskolin-treated samples were incubated with affinity resin. After washing the resin repeatedly with low salt

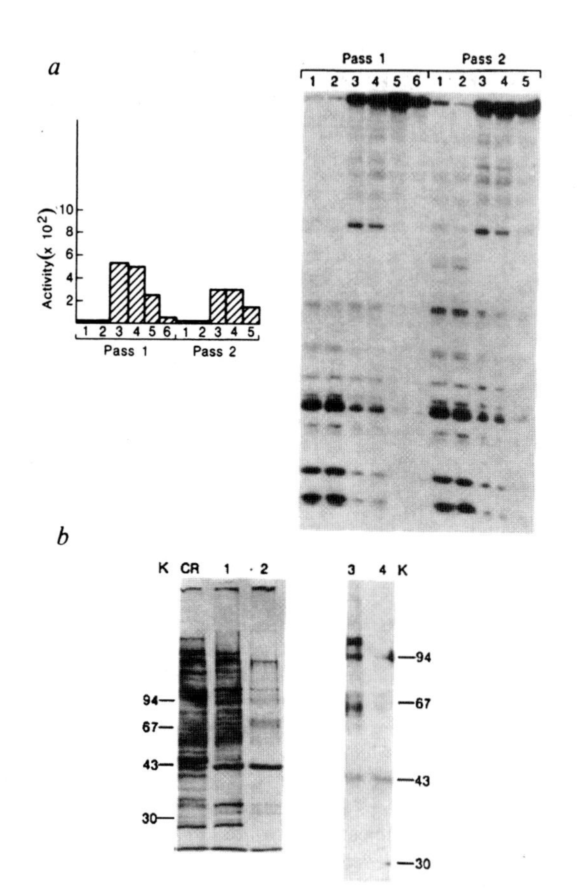

Fig. 2 Purification of CREB protein from PC12 nuclear extract by DNA affinity chromatography. Crude nuclear extract (25 mg protein) was preincubated with 125 μg calf thymus DNA, 125 μg poly(dI-dC) for 10 min on ice. The extract was passed over a 1-ml DNA affinity column equilibrated in buffer Z with 0.1 M KCl. Buffer Z contained 25 mM HEPES *p*H 7.8, 12.5 mM $MgCl_2$, 1 mM DTT, 20% glycerol, 0.1% Nonidet-P40. The column was washed with 6 ml buffer Z containing 0.1 M KCl. After collecting the wash fractions, the column was capped and 1 ml buffer Z containing 1.0 M KCl was added to the affinity resin. The resin was briefly stirred and then incubated for 20 min to allow dissociation of bound protein. The eluate was collected, an additional 1 ml of buffer Z with 1.0 M KCl was added to the resin and residual CRE-binding protein activity was collected. Additional passes over the affinity column using purified extracts were as described in Fig. 1 legend. *a*, CRE binding activity recovered from the column was monitored by DNase I footprinting assay using coding strand probe (Fig. 1). Right, elution profiles for two successive affinity column passes are shown. Lanes marked 1 denote flow-through fractions. Lanes marked 2 represent 0.1 M KCl wash fractions. Lanes marked 3, 4, 5, 6 represent successive 1-ml high salt (1.0 M KCl) eluate fractions. Bar graph on left, CRE-binding activity recovered from each corresponding fraction expressed in footprint units. *b*, SDS-polyacrylamide gel analysis of eluate fractions from DNA affinity column. Ticks alongside each gel, protein markers. Sizes are M_r in thousands. CR, crude (unfractionated) extract; lanes marked 1, 2, 3, 4 represent first, second, third and fourth pass high salt eluates, respectively. Third and fourth pass high salt fractions shown on right panel originated from a different nuclear extract preparation from those on left panel. Samples containing 100 footprints each (except crude extract sample which contained 20 footprint units) were precipitated with 20% trichloroacetic acid plus 1 mM deoxycholate, washed with acetone and resuspended in sample buffer. Following electrophoresis over 8% SDS-polyacrylamide gels under reducing conditions, proteins were silver-stained according to Morissey[17]. Arrow, putative 43K CRE-binding protein.
Methods. The sequence-specific DNA affinity column was constructed according to Kadonaga and Tjian[9]. A double-stranded oligonucleotide extending from −62 to −37 and containing *Bam*HI cohesive ends was synthesized. The oligomers were annealed, 5′ phosphorylated, and ligated to produce concatemers with an average length of 300 bases. The DNA (400 μg) was covalently attached to 10 ml of CnBr-activated CL-2B Sepharose beads[9].

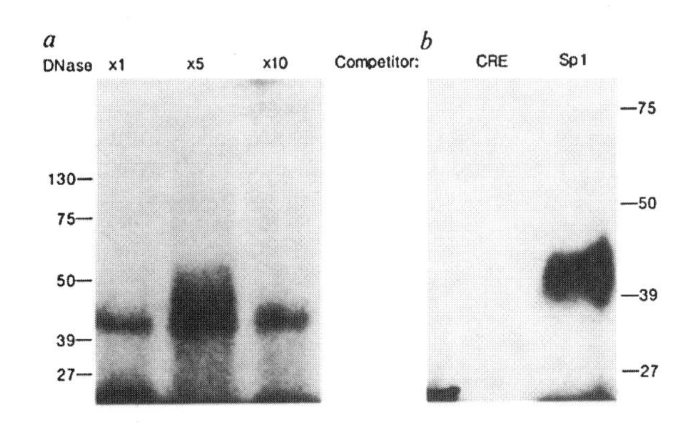

Fig. 3 Cross-linking of CRE-binding protein to somatostatin CRE by ultraviolet light. Cross-linking experiments were performed using a protocol similar to that of ref. 10. A uniformly labelled probe for cross-linking reactions was obtained by replacement synthesis of a DNA fragment containing the somatostatin CRE. After labelling this fragment with [α-^{32}P]deoxycytosine and bromodeoxyuridine (BUdR), 1 ng (5×10^5 c.p.m.) DNA probe was incubated with 60 μg unfractionated nuclear extract and 5 μg competitor poly(dI-dC) DNA. The samples were incubated on ice for 15 min to allow binding of proteins to the DNA. The samples were then irradiated under a Fotodyne ultraviolet lamp (emission wavelength 300 nm, maximum intensity 7,500 μW cm^{-2}) at 25 °C for one hour. $CaCl_2$ was added to 10 mM and the samples were digested at 37 °C for 30 min with 5-50 U micrococcal nuclease and 2.5-25 μg DNase I. The samples were then precipitated with 20% trichloroacetic acid and 1 mM deoxycholate, and analysed by electrophoresis through 8% SDS-polyacrylamide gels. The gels were dried and exposed to X-ray film. *a*, Cross-linking reactions were treated with increasing amounts of DNase (1 × DNase equals 5 units micrococcal nuclease plus 2.5 μg DNase I). *b*, For competition experiments 25 ng unlabelled DNA competitor was added to samples during the initial binding step. CRE and Sp1 competitors are synthetic oligonucleotides containing binding sites for CREB and Sp1 proteins, respectively. Size markers show M_r in thousands.
Methods. To prepare DNA probe for ultraviolet cross-linking, $\Delta(-71)$ plasmid was digested with *Hin*dIII to release a 160-bp fragment containing somatostatin promoter sequences from −71 to +55. After phenol-chloroform extraction and ethanol precipitation, 0.1 pmol *Hin*dIII-digested $\Delta(-71)$ DNA was digested with 1 U T4 DNA polymerase for 2.5 min. The fill-in reaction was started by adding 250 μCi [α-^{32}P]dCTP (3,000 Ci $mmol^{-1}$), 50 μM each dATP, dGTP and BUdR triphosphate. After 15 min, the reaction was terminated by adding EDTA to 3 mM and heating to 65 °C for 10 min. The 160-bp labelled fragment was resolved on a 5% polyacrylamide gel, excised and electroeluted. The incorporation of [^{32}P]dCTP by the 160-bp fragment was used to estimate the efficiency of replacement synthesis. Approximately 75 nucleotides were removed and replaced on each strand by T4 polymerase. The noncoding strand, therefore, was uniformly labelled over the entire CRE sequence.

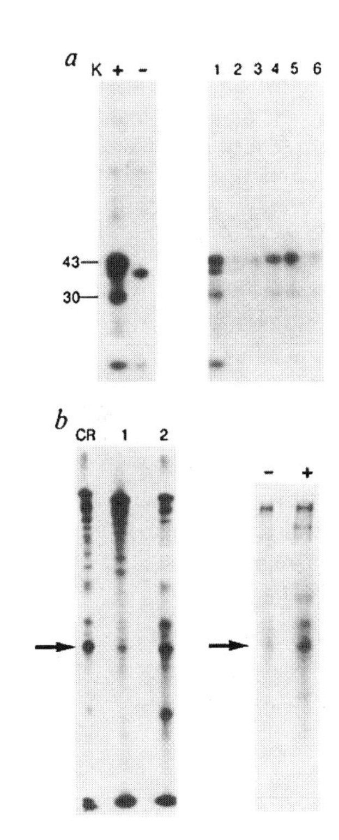

Fig. 4 Phosphorylation of CREB protein *in vitro* and *in vivo*. *a*, Affinity-purified extract (third pass) containing 50 footprint units was incubated with purified catalytic subunit of cAMP-dependent protein kinase in 50 mM potassium orthophosphate (*p*H 7.2), 10 mM $MgCl_2$, 5 mM NaF. The reaction was initiated by the addition of 2 μM (20 μCi) [γ-^{32}P]ATP and incubation was at 30 °C for 5 min. Samples were precipitated with 20% trichloroacetic acid and phosphorylation was assessed by autoradiography following SDS-polyacrylamide gel electrophoresis. Left panel: +, affinity purified extract plus catalytic subunit; −, catalytic subunit only. Size markers show M_r in thousands. Right panel, DNA affinity chromatography of phosphorylated extract. Following incubation with catalytic subunit, purified CREB protein was passed over a 1-ml DNA affinity column (Fig. 2) and 1-ml fractions were collected. Proteins were eluted with increasing concentrations of KCl in buffer Z and precipitated with 20% trichloroacetic acid. Lane 1, flow-through fraction; lane 2, 0.1 M KCl; lane 2, 0.2 M KCl; lanes 4–6, 1.0 M KCl. *b*, Phosphorylation of CREB protein *in vivo*. PC12 cells (5×10^6 per 100-mm dish) were labelled with inorganic ^{32}P (0.5 mCi ml^{-1}) in phosphate-free medium for 15 h. Cells were then treated with 10 μM forskolin or ethanol vehicle for one hour. Nuclear extracts (Fig. 1) from the labelled cells were incubated with DNA affinity resin (Fig. 2) and proteins were eluted as above. Left panel: CR, crude nuclear extract from forskolin-treated cells; lane 1, 0.1 M KCl fraction; lane 2, 1.0 M KCl fraction. Labelled fractions were precipitated with 20% TCA, resuspended in sample buffer and analysed on a 10% SDS-polyacrylamide gel. More than 95% of TCA precipitable counts incubated with affinity resin were in the 0.1 M KCl fraction (lane 1). Autoradiographic exposures for each lane were adjusted to allow comparison between fractions. Right panel: 1.0 M KCl fractions from control (−) and forskolin-treated (+) cells. Arrows, 43K phosphoprotein.

buffer, affinity-bound proteins were eluted as before in high salt. We observed a 43K phosphoprotein which was specifically retained by the affinity resin. When compared with labelled extracts from untreated cells, the 43K protein appears to be phosphorylated about 3-4 times more in cells treated with forskolin. Several other bands retained by the affinity resin also seem to be induced by forskolin. These may represent other DNA-binding proteins which are also phosphorylated in response to cAMP.

Constantinou *et al.*[12,13] have proposed that the regulatory subunit of cAMP-protein kinase binds to and stimulates the expression of cAMP responsive genes. The type-2 regulatory subunit appears to possess intrinsic topoisomerase activity[12]. By altering the chromatin structure in responsive genes, the regulatory subunit may consequently regulate gene expression. The M_r of the CRE-binding protein characterized in our report, 43K, differs considerably from that of the regulatory subunit (54K). Genes responsive to cAMP like somatostatin, therefore, appear to be regulated by a nuclear factor distinct from the regulatory subunit.

Lee and coworkers[14] have recently identified a transcription factor, AP-1, which recognizes the sequence -TGACTCA- in the SV40 and human metallothionine II promoters. The AP-1 binding site is homologous to the 'core' motif in the CRE, namely -TGACGTCA-, suggesting that CREB and AP-1 proteins may be closely related transcription factors. The characterization of cDNAs encoding these proteins will reveal the extent to which CREB and AP-1 are homologous.

Our results indicate that cAMP regulates expression of the somatostatin gene through a 43K nuclear protein which binds selectively to the CRE. Cyclic AMP appears to stimulate the phosphorylation of this protein *in vivo*. Furthermore, the purified protein is a substrate for phosphorylation by catalytic subunit of cAMP-dependent protein kinase *in vitro*. Consequently, phosphorylation of CREB *in vivo* may occur directly after activation of catalytic subunit and not by a cascade of protein kinases. CRE-binding activity is not affected by treatment of PC12 cells with forskolin. Cyclic AMP may therefore regulate the transcriptional activity of CREB rather than its binding activity. Alternatively, CREB may interact with another protein which ultimately regulates transcription in response to cAMP. *In vitro* transcription experiments using purified CREB protein will clarify the mechanism by which cAMP regulates gene expression.

We thank Katherine Jones for discussions and Sp1 oligonucleotide DNA, Susan Taylor for purified catalytic subunit, and Nancy Karpinski and Amy Blount for technical assistance. We also thank Jean Rivier, Ron Kaiser and Richard McClintock for synthesis of oligonucleotides. This work was supported by NIH grants, and conducted in part by the Clayton Foundation for Research, California Division.

Received 6 May; accepted 12 May 1987.

1. Montminy, M. R. *et al. J. Neurosci.* **6**, 1171–1176 (1986).
2. Montminy, M. R., Sevarino, K. A., Wagner, J. A., Mandel, G. & Goodman, R. H. *Proc. natn. Acad. Sci. U.S.A.* **83**, 6682–6686 (1986).
3. Comb, M., Birnberg, N. C., Seasholtz, A., Herbert, E. & Goodman, H. M. *Nature* **323**, 353–356 (1986).
4. Short, J. M., Wynshaw-Boris, A., Short, H. P. & Hanson, R. W. *J. biol. Chem.* **261**, 9721–9726 (1986).
5. Reichlin, S. *New Engl. J. Med.* **309**, 1495–1501, 1556–1563 (1983).
6. Tapia-Arancibia, L. & Reichlin, S. *Brain Res.* **336**, 67–72 (1985).
7. Shimatsu, A. *et al. Endocrinology* **110**, 2113–2117 (1982).
8. Chihara, K., Arimura, A. & Schally, A. *Endocrinology* **104**, 1656–1682 (1979).
9. Kadonaga, J. T. & Tjian, R. *Proc. natn. Acad. Sci. U.S.A.* **83**, 5889–5893 (1986).
10. Chodosh, L. A., Carthew, R. W. & Sharp, P. A. *Molec. cell. Biol.* **6**, 4723–4733 (1986).
11. Hillel, Z. & Wu, C. W. *Biochemistry* **17**, 2954–2961 (1978).
12. Constantinou, A., Squinto, S. & Jungmann, R. *Cell* **42**, 429–437 (1985).
13. Squinto, S. R., Kelley-Gerraghty, D. C., Kuettel, M. R. & Jungmann, R. A. *J. Cyclic Nucleotide Protein Phosphoryl. Res.* **10**, 65–73 (1985).
14. Lee, W., Haslinger, A., Karin, M. & Tjian, R. *Nature* **325**, 368–372 (1987).
15. Bradford, M. M. *Analyt. Biochem.* **72**, 248–254 (1976).
16. Dignam, J. D., Lebovitz, R. M. & Roeder, R. G. *Nucleic Acids Res.* **11**, 1475–1489 (1983).
17. Morissey, J. H. *Analyt. Biochem.* **117**, 307–313 (1981).

Subject index

5′-Adenosine monophosphate, 159
α-Adrenergic agonist, 131
α-Adrenergic receptor, 2, 48
β-Adrenergic receptor, 47, 48
β-Adrenergic receptor kinase, 2, 48, 49
βARK (*see* β-Adrenergic receptor kinase)
A431 cell line, 237
Acetylcholine, 77, 113, 188
Acetylcholine esterase, 77
Acetylcholine receptor, 77–79
Action potential, 57, 77, 93, 169
Active zone, 94
Acute pancreatitis, 132
Adenyl cyclase, 31, 32, 47, 48
Adenylate cyclase, 1, 2
ADP, 188
ADP-ribosyl transferase, 32
Adrenaline, 1, 2
Aldosterone, 248
Antibiotic, 57
L-Arginine, 187, 188
ATP, 188
ATP-dependent Ca^{2+} re-uptake, 132

Bimolecular lipid membrane, 57
Bradykinin, 188

C2 motif, 140
c-*fos*, 260
Calcium,
  ATP-dependent re-uptake of, 132
  channel, 93, 94
  concentration, 159, 160
  -dependent ionic current, 132
  influx factor, 115
  ionophore, 170
  -dependent interaction, 160
  extrusion, 100, 131
  free intracellular concentration, 59, 93, 188
  inflow, 86, 93, 113, 188
  oscillation, 131
  puff, 114
  pump, 99, 131
  pumping into store, 188
  release of intracellular store, 113–115, 131
  -responsive element, 260
  signal, 100, 132
  site of action, 169
  spark, 114
  spike, 131, 132
  toxicity, 132
  uptake, 99
  wave, 189
Ca^{2+}-ATPase, 160
Ca^{2+}-binding protein, 132, 159, 160, 171
Ca^{2+}-calmodulin-NADPH-dependent synthase, 188
Ca^{2+}-dependent kinase, 114
Ca^{2+}-dependent phosphatase, 114
Ca^{2+}-release channel, 100, 188
Ca^{2+}-sensitive fluorescent probe, 59
Ca^{2+}-sensitive photo-protein, 131
Ca^{2+}-sensitive protease, 132
Ca^{2+},Mg^{2+}-ATPase, 99
CaBP (*see* Ca^{2+}-binding protein)
Calmodulin, 2, 159, 160
Calmodulin-dependent protein kinase, 260
Capacitance measurement, 131
Capacitative Ca^{2+} entry, 115
Carrier, 58
Catecholamine, 31, 170
CBP (*see* CREB-binding protein)
Cell surface receptor, 139
Cerebellar granule cell, 189
Chimaeric subunit, 78
Cholera toxin, 32
Chromaffin, 169–171
Citrulline, 188
Coagulation product, 187
Cone photoreceptor, 85
Contraction, 99, 100
CREB (*see* Cyclic-AMP-response element-binding protein)
CREB-binding protein, 260
Cross-talk, 2
Cyclic adenosine 3′,5′-monophosphate (*see* Cyclic AMP)
Cyclic AMP, 1, 2, 31, 47, 85, 94, 159, 260
Cyclic AMP-dependent phosphorylation, 1, 2, 85, 94
Cyclic AMP-dependent protein kinase, 49, 259
Cyclic AMP-response element, 259
Cyclic-AMP-response element-binding protein, 259, 260
Cyclic GMP, 85, 86, 187, 188
Cyclic GMP-activated cation channel, 85, 86
Cyclic GMP-dependent cation channel, 86
Cyclic GMP-dependent protein kinase, 86
Cyclic nucleotide binding site, 86
Cyclic nucleotide-gated conductance, 85
Cyclic 3′,5′-nucleotide phosphodiesterase, 2, 159
Cyclic nucleotide-regulated ion channel, 86
Cysteine-rich sequence, 140
Cytokine, 187
Cytosolic Ca^{2+} signal, 115
Cytosolic Ca^{2+} spike, 100, 114

DAG (*see* Diacylglycerol)
Dehydrogenase, 131
Dephosphophosphorylase, 1
Diacylglycerol, 113, 139, 140, 219
Diffusible messenger, 189
Dihydropyridine, 94
Diolein, 139
Dopamine, 2
Drosophila trp protein, 115

Ecdysone, 248
EDRF (*see* Endothelium-derived relaxing factor)
E-F hand, 160
EGF (*see* Epidermal growth factor)
EGF receptor, 195, 196, 237, 238
Electrical capacitance, 58
Electrical noise, 57, 58
Electrical photoresponse, 86
End-plate membrane, 77
Endonuclease, 132
Endoplasmic reticulum, 100, 114
Endothelial cell, 131, 187–189
Endothelium-derived relaxing factor, 187, 188
Epidermal growth factor, 195, 197, 228, 237
Erb-A, 248
Erb-A oncogene, 247
ERK (*see* Extracellular-signal regulated kinase)
Exocrine gland, 131, 132
Exocytosis, 2, 94, 113, 131, 132, 169–171
Extracellular-signal regulated kinase, 227

Fibroblast, 131
Fluid secretion, 2, 132

Gene transcription, 228, 237, 247, 259, 260
Glucagon, 1, 2, 31
Glucocorticoid, 247
Glucocorticoid receptor, 247
Glucose-evoked oscillation, 131
Glycogen breakdown, 160
G-protein, 31–33, 47–49, 85, 113, 220
Growth factor receptor, 227, 237
GTP, 31
GTP-binding protein (*see* G-protein)
GTP-binding transducer protein, 2
GTPase, 31–33
Guanosine 3′,5′-cyclic monophosphate (see Cyclic GMP)
Guanyl nucleotide, 31
Guanyl cyclase, 86
Guanyl nucleotide-binding protein, 31

Heart contraction, 94
HeLa cell, 131
Hepatocyte, 131, 132

High voltage-activated Ca^{2+} current, 93
Histamine, 2, 188
Hormone-response element, 247

Inner nuclear membrane, 114
Inositol polyphosphate production, 114
Inositol ring, 219
Inositol 1,4,5-trisphosphate, 2, 113–115, 132, 139, 219
Ins(1,4,5)P_3 (see Inositol 1,4,5-trisphosphate)
Ins(1,4,5)P_3 receptor, 114, 115
Insulin, 227
Insulin receptor, 195, 197, 228
Intracellular Ca^{2+} concentration, 77
Intracellular Ca^{2+} store, 99, 100
Intracellular messenger, 85
Intra-nucleoplasmic Ca^{2+} signal, 114
Ion channel, 57, 77, 85
Ion selectivity, 78

K^+ channel, 85, 86

L-type Ca^{2+} channel, 93, 94
Light adaptation, 86
Light-dependent conductance, 85
Liver cell, 2
Local Ca^{2+} spike, 114
Long-term memory, 260
Low voltage-activated Ca^{2+} current, 93

M2 segment, 77, 78, 79
Macrophage, 131
MAP (see Microtubule-associated protein)
MAP kinase, 227
MAP-2 kinase, 227
MAP kinase kinase, 227
Measurement of free intracellular Ca^{2+} concentration, 59
Membrane capacitance, 58, 170
Membrane conductance, 57
Membrane permeability, 57
Membrane receptor occupation, 113
Messenger pathway, 2
Metabolism, 100
N-Methyl-D-aspartate glutamate receptor, 189
Microtubule-associated protein, 227
Mitochondria, 131
Muscle contraction, 159
Muscle fibre, 93
Muscle relaxation, 99, 100
Mutagenesis, 48, 49, 78, 259
Myosin light chain kinase, 160

N-type Ca^{2+} channel, 93
Na^+,K^+-pump, 58
Na^+–amino acid carrier, 58
Na^+–amino acid co-transporter, 58
Na^+–Ca^+ exchanger, 58
Na^+–K^+ ATPase, 58
Negative feedback, 114
Nerve terminal, 169
Neuroendocrine cell, 171
Neuromuscular junction, 77, 169
Neuron, 189
Neurotransmission, 169–171
Neurotransmitter, 94, 169, 170
Nicotinic acetylcholine receptor, 77
Nifedipine, 94
Nitric oxide, 187–189
Nitric oxide synthase, 187, 189
Nitrovasodilator, 188
NMDA (see *N*-methyl-D-aspartate)
Non-selective cation channel, 85, 114
Noradrenaline, 2, 94
Nuclear envelope, 114
Nucleus, 249, 259
5′-Nucleotidase, 159

Oestrogen, 248
Olfactory receptor cilia, 85
Oncogene, 195, 227–229, 237, 238, 247, 248
Oocyte, 131
Opsin, 47
Outer nuclear membrane, 114

Pancreatic β-cell, 131
Pancreatic acinar cell, 113, 114
Parathyroid hormone, 2
Patch-clamp technique, 57–59, 77, 85, 93, 171
Patch pipette, 59
PDGF (see Platelet-derived growth factor)
PDGF receptor, 220
Pertussis toxin, 32
Phenylephrine, 131
Phorbol 12-myristate 13-acetate (see 12-*O*-Tetradecanoylphorbol-13-acetate)
Phorbol ester, 139, 140
Phosphatidic acid, 113
Phosphatidylinositol 4,5-bisphosphate, 113, 114, 139, 219, 220
Phosphatidylinositol 3-kinase, 219–221
Phosphatidylinositol 3,4,5-trisphosphate, 219
Phosphatidylinositol breakdown, 113
Phosphatidylserine, 139
Phosphodiesterase, 85, 86, 159
Phosphoinositide, 49, 113, 139, 219, 220
Phospholipase C, 2, 113, 132, 139, 220
Phospholipid turnover, 113
Phosphoprotein, 195
Phosphorylase, 1
Phosphorylase kinase, 160
Phosphorylation, 49, 196, 197, 219, 227, 259, 260
Phosphotyrosine, 195, 220
Photoreceptor, 85
Phototransduction, 47
PI 3-kinase (see Phosphatidylinositol 3-kinase)
Pituitary somatotroph, 259
PKC (see Protein kinase C)
Plasma membrane receptor, 47
Plasma membrane Ca^{2+}-ATPase, 99, 100
Platelet, 187
Platelet-derived growth factor, 220, 228, 237, 238
PMA (see 12-*O*-Tetradecanoylphorbol-13-acetate)
PMCA pump (see Plasma membrane Ca^{2+} ATPase)
Point mutation analysis, 78
Positive feedback, 114
Potassium channel, 58, 85
Pp60$^{v\text{-}src}$, 195, 220
Presynaptic nerve terminal, 77
Progesterone, 248
Prostacyclin, 188
Protein kinase, 1, 49
Protein kinase C, 2
Protein phosphorylation, 49
Protein secretion, 2
Protein kinase, 160, 195, 227
Protein kinase C, 139–141, 221
Protein phosphatase inhibitor, 227
Protein tyrosine kinase, 196, 220
Protein tyrosine phosphorylation, 195, 196
Pseudo-substrate inhibitory region, 140
PtdIns(4,5)P_3 (see Phosphatidylinositol 4,5-bisphosphate)

Raf, 228, 229
Ras/MAP kinase pathway, 197, 227–229
Relaxation, 187, 189
Relaxing factor, 99
Retinoic acid, 248
Retinoid, 247
Rhodopsin, 47, 48
Rod photoreceptor, 85
Rous sarcoma virus, 195
Ryanodine receptor, 114

Salivary gland cell, 2, 131
Sarcoplasmic/endoplasmic reticulum Ca^{2+}-ATPase, 99, 100
Sarcoplasmic reticulum, 99
Secretin, 2
Secretion, 100
Secretory pathway, 170
Selectivity filter, 78
Sensory neuron, 85
SERCA pump (see Sarcoplasmic/endoplasmic reticulum Ca^{2+}-ATPase)
Serotonin, 2
Seven-transmembrane-segment receptor, 47, 48
SH2 domain, 196, 197, 220, 221, 228
SH3 domain, 196, 197, 220, 228
Shear stress, 188
Signal amplification, 86
Single calcium channel, 93
Single-channel current, 57, 78, 93
Sis oncogene, 237
Site-directed mutagenesis, 77
Skeletal muscle, 77
Smooth muscle cell, 187, 189

Somatostatin, 259
Spatio-temporal organization, 132
Squid giant synapse, 170
Steroid, 247
Steroid hormone receptor, 249
Stimulus–permeability coupling, 113
Structure–function analysis, 77, 79
Synapse, 77, 170
Synaptic event, 58
Synaptic terminal, 94, 171
Synaptic vesicle, 77, 169
Synaptotagmin I, 171

T-type Ca^{2+} channel, 93
12-*O*-Tetradecanoylphorbol-13-acetate, 139
Tetrodotoxin, 169
TGF (*see* Transforming growth factor)
Thapsigargin, 99
Thrombin, 188
Thyroid hormone, 247
Thyroid hormone receptor, 248
Thyroid-stimulating hormone, 2
Tissue-slice preparation, 58
TPA (*see* 12-*O*-Tetradecanoylphorbol-13-acetate)
Transcription, 247, 249, 259, 260
Transcriptional regulator, 260
Transcription factor, 248
Transducin, 33, 85
Transforming growth factor, 237
Transgenic animal, 249
Transgenic mouse, 171, 259
Transmembrane protein, 47
Troponin C, 160
Trypsin, 188
Tyrosine kinase, 195–197, 219, 238
Tyrosine phosphorylation, 197, 228
Tyrosine-specific protein kinase, 195

v-Erb-B, 237, 238
Vascular muscle strip, 188
Vasoactive intestinal polypeptide, 2
Vasopressin, 2
Vasorelaxation response, 188
Vesicular transport, 219–221
Vitamin A, 248
Vitamin D, 248
Voltage-gated Ca^{2+} channel, 93
Voltage-sensitive Ca^{2+} channel, 94, 114
Voltage sensitive Na^{2+} channel, 169

Whole-cell current, 93
Whole-cell patch-clamp recording, 170
Whole-cell recording technique, 59

X-ray crystallography, 33, 160
Xenopus oocyte, 78, 115

Zinc-binding domain, 140
Zinc finger, 248
Zymogen granule, 115